Lecture Notes in Mathematics

Edited by A. Dold, F. Takens and B. Teissier

Editorial Policy
for the publication of monographs

1. Lecture Notes aim to report new developments in all areas of mathematics – quickly, informally and at a high level. Monograph manuscripts should be reasonably self-contained and rounded off. Thus they may, and often will, present not only results of the author but also related work by other people. They may be based on specialized lecture courses. Furthermore, the manuscripts should provide sufficient motivation, examples and applications. This clearly distinguishes Lecture Notes from journal articles or technical reports which normally are very concise. Articles intended for a journal but too long to be accepted by most journals, usually do not have this "lecture notes" character. For similar reasons it is unusual for doctoral theses to be accepted for the Lecture Notes series.

2. Manuscripts should be submitted (preferably in duplicate) either to one of the series editors or to Springer-Verlag, Heidelberg. In general, manuscripts will be sent out to 2 external referees for evaluation. If a decision cannot yet be reached on the basis of the first 2 reports, further referees may be contacted: the author will be informed of this. A final decision to publish can be made only on the basis of the complete manuscript, however a refereeing process leading to a preliminary decision can be based on a pre-final or incomplete manuscript. The strict minimum amount of material that will be considered should include a detailed outline describing the planned contents of each chapter, a bibliography and several sample chapters.
Authors should be aware that incomplete or insufficiently close to final manuscripts almost always result in longer refereeing times and nevertheless unclear referees' recommendations, making further refereeing of a final draft necessary.
Authors should also be aware that parallel submission of their manuscript to another publisher while under consideration for LNM will in general lead to immediate rejection.

3. Manuscripts should in general be submitted in English.
Final manuscripts should contain at least 100 pages of mathematical text and should include
– a table of contents;
– an informative introduction, with adequate motivation and perhaps some historical remarks: it should be accessible to a reader not intimately familiar with the topic treated;
– a subject index: as a rule this is genuinely helpful for the reader.

Lecture Notes in Mathematics
1709

Editors:
A. Dold, Heidelberg
F. Takens, Groningen
B. Teissier, Paris

Subseries:
Institut de Mathématiques, Université de Strasbourg
Adviser: J.-L. Loday

Springer
Berlin
Heidelberg
New York
Barcelona
Hong Kong
London
Milan
Paris
Singapore
Tokyo

J. Azéma   M. Émery
M. Ledoux   M. Yor (Eds.)

Séminaire
de Probabilités XXXIII

 Springer

Editors

Jacques Azéma
Laboratoire de Probabilités
Université Pierre et Marie Curie
Tour 56, 3ème étage
4, Place Jussieu
F-75252 Paris cedex 05, France
E-mail: jaze@ccr.jussieu.fr

Michel Émery
Institut de Recherche Mathématique
  Avancée
Université Louis Pasteur
7, rue René Descartes
F-67084 Strasbourg, France
E-mail: emery@math.u-strasbg.fr

Michel Ledoux
Laboratoire de Statistiques et Probabilités
Université Paul Sabatier
118, route de Narbonne
F-31601 Toulouse cedex, France
E-mail: ledoux@proba.jussieu.fr

Marc Yor
Laboratoire de Probabilités
Université Pierre et Marie Curie
Tour 56, 3ème étage
4, Place Jussieu
F-75252 Paris cedex 05, France

Cataloging-in-Publication Data applied for

Die Deutsche Bibliothek - CIP-Einheitsaufnahme

Séminaire de probabilités .... - Berlin ; Heidelberg ; New York ; ;
Barcelona ; Hong Kong ; London ; Milan ; Paris ; Singapore ; Tokyo
: Springer
   ISSN 0720-8766
   33 (1999)
   (Lecture notes in mathematics ; Vol. 1709)
   ISBN 3-540-66342-8

Mathematics Subject Classification (1991): 60GXX, 60HXX, 60JXX

ISSN 0075-8434
ISBN 3-540-66342-8 Springer-Verlag Berlin Heidelberg New York

Typesetting: Camera-ready TeX output by the author
SPIN: 10650247   41/3143-543210 - Printed on acid-free paper

Beside topics belonging to the tradition of the Séminaire de Probabilités, the present volume XXXIII also proposes three texts originating from advanced courses on the following subjects:

Dynamics of stochastic algorithms;

Simulated annealing algorithms and Markov chains with rare transitions;

Concentration of measure and logarithmic Sobolev inequalities.

They are meant to be accessible to the probabilist community at large, and hopefully will become reference texts.

J. Azéma, M. Émery, M. Ledoux, M. Yor

SÉMINAIRE DE PROBABILITÉS XXXIII

TABLE DES MATIÈRES

COURS SPÉCIALISÉS

M. Benaïm : Dynamics of stochastic algorithms. 1

O. Catoni : Simulated annealing algorithms and Markov chains with rare transitions. 69

M. Ledoux : Concentration of measure and logarithmic Sobolev inequalities. 120

QUESTIONS DE FILTRATIONS

B. De Meyer : Une simplification de l'argument de Tsirelson sur le caractère non-brownien des processus de Walsh. 217

W. Schachermayer : On certain probabilities equivalent to Wiener measure, d'après Dubins, Feldman, Smorodinsky and Tsirelson. 221

S. Beghdadi-Sakrani, M. Émery : On certain probabilities equivalent to coin-tossing, d'après Schachermayer. 240

J. Warren : On the joining of sticky Brownian motion. 257

M. Émery, W. Schachermayer : Brownian filtrations are not stable under equivalent time-changes. 267

S. Watanabe : The existence of a multiple spider martingale in the natural filtration of a certain diffusion in the plane. 277

M. Émery, W. Schachermayer : A remark on Tsirelson's stochastic differential equation. 291

M. Arnaudon : Appendice à l'exposé précédent : La filtration naturelle du mouvement brownien indexé par **R** dans une variété compacte. 304

J. Kallsen : A stochastic differential equation with a unique (up to indistinguishability) but not strong solution. 315

Théorie des martingales

K. Takaoka : Some remarks on the uniform integrability of continuous martingales.   327

M. Pratelli : An alternative proof of a theorem of Aldous concerning convergence in distribution for martingales.   334

M. Morayne, K. Tabisz : A short proof of decomposition of strongly reduced martingales.   339

P. Grandits : Some remarks on L^∞, H^∞, and BMO.   342

Autres exposés

W. Brannath, W. Schachermayer : A bipolar theorem for $L^0_+(\Omega, \mathcal{F}, \mathbb{P})$.   349

A. Es-Sahib, H. Heinich : Barycentres canoniques pour un espace métrique à courbure négative.   355

N. Belili : Dualité du problème des marges et ses applications.   371

J. Pitman : The distribution of local times of a Brownian bridge.   388

L. Dubins : Paths of finitely additive Brownian Motion need not be bizarre.   395

H. Tsukahara : A limit theorem for the prediction process under absolute continuity.   397

K. Chrétien, D. Kurtz, B. Maisonneuve : Processus gouvernés par des noyaux.   405

A. Bentaleb : Sur l'hypercontractivité des semi-groupes ultrasphériques.   410

Corrections et addenda à des volumes antérieurs

F. Delbaen : An addendum to a remark on Slutsky's theorem.   415

N. Eisenbaum : Quelques précisions sur "Théorèmes limites pour les temps locaux d'un processus stable symétrique" (Volume XXXI, LNM. 1655, 1997).   417

Dynamics of Stochastic Approximation Algorithms

Michel Benaïm

Abstract

These notes were written for a D.E.A course given at *Ecole Normale Supérieure* de Cachan during the 1996-97 and 1997-98 academic years and at *University Toulouse III* during the 1997-98 academic year. Their aim is to introduce the reader to the dynamical system aspects of the theory of stochastic approximations.

Contents

1 **Introduction**   3
   1.1 Outline of Contents 4

2 **Some Examples**   6
   2.1 Stochastic Gradients and Learning Processes 6
   2.2 Polya's Urns and Reinforced Random Walks 6
   2.3 Stochastic Fictitious Play in Game Theory 8

3 **Asymptotic Pseudotrajectories**   9
   3.1 Characterization of Asymptotic Pseudotrajectories 10

4 **Asymptotic Pseudotrajectories and Stochastic Approximation Processes**   11
   4.1 Notation and Preliminary Result 11
   4.2 Robbins-Monro Algorithms 14
   4.3 Continuous Time Processes 18

5 **Limit Sets of Asymptotic Pseudotrajectories**   20
   5.1 Chain Recurrence and Attractors 20
   5.2 The Limit Set Theorem 24

6 **Dynamics of Asymptotic Pseudotrajectories**   25
   6.1 Simple Flows, Cyclic Orbit Chains 25
   6.2 Lyapounov Functions and Stochastic Gradients 27
   6.3 Attractors 28
   6.4 Planar Systems 29

7 **Convergence with positive probability toward an attractor**   30
   7.1 Attainable Sets 31
   7.2 Examples 32
   7.3 Stabilization 34

8 **Shadowing Properties**   35
   8.1 λ-Pseudotrajectories 36
   8.2 Expansion Rate and Shadowing 40
   8.3 Properties of the Expansion Rate 43

9 **Nonconvergence to Unstable points, Periodic orbits and Normally Hyperbolic Sets**   47
   9.1 Proof of Theorem 9.1 50

10 **Weak Asymptotic Pseudotrajectories**   60
   10.1 Stochastic Approximation Processes with Slow Decreasing Step-Size   64

1   Introduction

Stochastic approximation algorithms are discrete time stochastic processes whose general form can be written as

$$x_{n+1} - x_n = \gamma_{n+1} V_{n+1} \qquad (1)$$

where x_n takes its values in some euclidean space, V_{n+1} is a random variable and $\gamma_n > 0$ is a "small" step-size.

Typically x_n represents the parameter of a system which is adapted over time and $V_{n+1} = f(x_n, \xi_{n+1})$. At each time step the system receives a new information ξ_{n+1} that causes x_n to be updated according to a *rule* or *algorithm* characterized by the function f. Depending on the context f can be a function designed by a user so that some goal (estimation, identification, ...) is achieved, or a model of adaptive behavior.

The theory of *stochastic approximations* was born in the early 50s through the works of Robbins and Monro (1951) and Kiefer and Wolfowitz (1952) and has been extensively used in problems of signal processing, adaptive control (Ljung, 1986; Ljung and Söderström, 1983; Kushner and Yin, 1997) and recursive estimation (Nevelson and Khaminski, 1974). With the renewed and increased interest in the *learning paradigm* for artificial and natural systems, the theory has found new challenging applications in a variety of domains such as neural networks (White, 1992; Fort and Pages, 1994) or game theory (Fudenberg and Levine, 1998).

To analyse the long term behavior of (1), it is often convenient to rewrite the noise term as

$$V_{n+1} = F(x_n) + U_{n+1} \qquad (2)$$

where $F : \mathbb{R}^m \to \mathbb{R}^m$ is a deterministic vector field obtained by suitable averaging. The examples given in Section 2 will illustrate this procedure. A natural approach to the asymptotic behavior of the sequences $\{x_n\}$ is then, to consider them as approximations to solutions of the ordinary differential equation (ODE)

$$\frac{dx}{dt} = F(x). \qquad (3)$$

One can think of (1) as a kind of Cauchy-Euler approximation scheme for numerically solving (3) with step size γ_n. It is natural to expect that, owing to the fact that γ_n is small, the noise washes out and that the asymptotic behavior of $\{x_n\}$ is closely related to the asymptotic behavior of the ODE. This method called the *ODE method* was introduced by Ljung (1977) and extensively studied thereafter. It has inspired a number of important works, such as the book by Kushner and Clark (1978), numerous articles by Kushner and coworkers, and more recently the books by Benveniste, Métivier and Priouret (1990), Duflo (1996) and Kushner and Yin (1997).

However, until recently, most works in this direction have assumed the simplest dynamics for F (for example that F is the negative of the gradient of a cost function), and little attention has been payed to dynamical system issues.

The aim of this set of notes is to show how dynamical system ideas can be fully integrated with probabilistic techniques to provide a rigorous foundation to the ODE method beyond gradients or other dynamically simple systems. However it is not intended to be a comprehensive presentation of the theory of stochastic approximations. It is principally focused on the *almost sure dynamics of stochastic approximation processes with decreasing step sizes*. Questions of weak convergence, large deviation, or rate of convergence, are not considered here. The assumptions on the "noise" process are chosen for simplicity and clarity of the presentation.

These notes are partially based on a DEA course given at Ecole Normale Supérieure de Cachan during the 1996-1997 and 1997-1998 academic years and at University Paul Sabatier during the 1997-1998 academic year. I would like to especially thank Robert Azencott for asking me to teach this course and Michel Ledoux for inviting me to write these notes for *Le Séminaire de Probabilités*.

An important part of the material presented here results from a collaboration with Morris W. Hirsch and it is a pleasure to acknowledge the fundamental influence of Moe on this work. I have also greatly benefited from numerous discussions with Marie Duflo over the last months which have notably influenced the presentation of these notes. Finally I would like to thank Odile Brandière, Philippe Carmona, Laurent Miclo, Gilles Pages and Sebastian Schreiber for valuable insights and informations.

Although most of the material presented here has already been published, some results appear here for the first time and several points have been improved.

1.1 Outline of Contents

These notes are organized as follows.

Section 2 presents simple motivating examples of stochastic approximation processes.

Section 3 introduces the notion of asymptotic pseudotrajectories for a semiflow. This a purely deterministic notion due to Benaïm and Hirsch (1996) which arises in many dynamical settings but turns out to be very well suited to stochastic approximations.

In section 4 classical results on stochastic approximations are (re)formulated in the language of asymptotic pseudotrajectories. Attention is restricted to the classical situation where U_{n+1} (see (2)) is a sequence of martingale differences. It is shown that, under suitable conditions, the continuous time process obtained by a convenient interpolation of $\{x_n\}$ is almost surely an asymptotic pseudotrajectory of the semiflow induced by the associated ODE. This section owes much to the ideas and techniques developped by Kushner and his co-workers (Kushner and Clark, 1978; Kushner and Yin, 1997), Métivier and Priouret (1987) (see also Benveniste, Métivier and Priouret (1990)) and Duflo (1990, 1996, 1997). The case of certain diffusions and jump processes is also considered.

Section 5 characterizes the limit sets of asymptotic pseudotrajectories. It begins with a comprehensive introduction to chain-recurrence and chain-transitivity. Several properties of chain transitive sets are formulated. The main result of the section establishes that limit sets of precompact asymptotic pseudotrajectories are internally chain-transitive. This theorem was originally proved in (Benaïm, 1996) but I have chosen to present here the proof of Benaïm and Hirsch (1996). I find this proof conceptually attractive and it is somehow more directly related to the original ideas of Kushner and Clark (1978).

Section 6 applies the abstract results of section 5 in various situations. It is shown how assumptions on the deterministic dynamics can help to identify the possible limit sets of stochastic approximation processes with a great deal of generality. This section generalizes and unifies many of the results which appear in the literature on stochastic approximation.

Section 7 establishes simple sufficient conditions ensuring that a given attractor of the ODE has a positive probability to host the limit set of the stochastic approximation process. It also provides lower bound estimates of this probability. This section is based on unpublished works by Duflo (1997) and myself.

Section 8 considers the question of *shadowing*. The main result of the section asserts that when the step size of the algorithm goes to zero at a suitable rate (depending on the expansion rate of the ODE) trajectories of (1) are almost surely asymptotic to forward trajectories of (3). This section represents a synthesis of the works of Hirsch (1994), Benaïm (1996), Benaïm and Hirsch (1996) and Duflo (1996) on the question of shadowing. Several properties and estimates of the expansion rate due to Hirsch (1994) and Schreiber (1997) are presented. In particular, Schreiber's ergodic characterization of the expansion rate is proved.

Section 9 pursues the qualitative analysis of section 7. The focus is on the behavior of stochastic approximation processes near "unstable" sets. The centerpiece of this section is a theorem which shows that stochastic approximation processes have zero probability to converge toward certain repelling sets including linearly unstable equilibria and periodic orbits as well as normally hyperbolic manifolds. For unstable equilibria this problem has often been considered in the literature but, to my knowledge, only the works by Pemantle (1990) and Brandière and Duflo (1996) are fully satisfactory. I have chosen here to follow Pemantle's arguments. The geometric part contains new ideas which allow to cover the general case of normally hyperbolic manifolds but the probability part owes much to Pemantle.

Section 10 introduces the notion of a stochastic process being a *weak asymptotic pseudotrajectory* for a semiflow and analyzes properties of its empirical occupation measures. This is motivated by the fact that any stochastic approximation process with decreasing step size is a weak asymptotic pseudotrajectory of the associated ODE regardless of the rate at which $\gamma_n \to 0$.

2   Some Examples

2.1   Stochastic Gradients and Learning Processes

Let $\{\xi_i\}_{i \geq 1}$, $\xi_i \in E$ be a sequence of independent identically distributed random inputs to a system and let $x_n \in \mathbb{R}^m$ denote a parameter to be updated, $n \geq 0$. We suppose the updating to be defined by a given map $f : \mathbb{R}^m \times E \to \mathbb{R}^m$, and the following stochastic algorithm:

$$x_{n+1} - x_n = \gamma_{n+1} f(x_n, \xi_{n+1}). \tag{4}$$

Let μ be the common probability law of the ξ_n. Introduce the average vector field

$$F(x) = \int f(x, \xi) d\mu(\xi)$$

and set

$$U_{n+1} = f(x_n, \xi_{n+1}) - F(x_n).$$

It is clear that this algorithm has the form given by [(1) (2)]. Such processes are classical models of adaptive algorithms.

A situation often encountered in "machine learning" or "neural networks" is the following: Let I and O be euclidean spaces and $M : \mathbb{R}^m \times I \to O$ a smooth function representing a system (e.g a neural network). Given an *input* $y \in I$ and a parameter x the system produces the *output* $M(x, y)$.

Let $\{\xi_n\} = \{(y_n, o_n)\}$ be a sequence of i.i.d random variables representing the *training set* of M. Usually the law μ of ξ_n is unknown but many samples of ξ_n are available. The goal of learning is to adapt the parameter x so that the output $M(x, y_n)$ gives a good approximation of the *desired output* o_n. Let $e : O \times O \to \mathbb{R}_+$ be a smooth error function. For example $e(o, o') = ||o - o'||^2$. Then a basic training procedure for M is given by (4) where

$$f(x, \xi) = -\frac{\partial}{\partial x} e(M(x, y), o), \text{ and } \xi = (y, o).$$

Assuming that derivation and expectation commute, the associated ODE is the gradient ODE given by

$$F(x) = -\nabla C(x) \text{ with } C(x) = \int e(M(x, y), o) d\mu(y, o).$$

2.2   Polya's Urns and Reinforced Random Walks

The *unit m-simplex* $\Delta^m \subset \mathbb{R}^{m+1}$ is the set

$$\Delta^m = \{v \in \mathbb{R}^{m+1} : v_i \geq 0, \sum v_i = 1\}.$$

We consider Δ^m as a differentiable manifold, identifying its tangent space at any point with the linear subspace

$$E^m = \{z \in \mathbb{R}^{m+1} : \sum z_j = 0\}.$$

An urn initially (i. e. ,at time $n = 0$) contains $n_0 > 0$ balls of colors $1, \ldots, m+1$. At each time step a new ball is added to the urn and its color is randomly chosen as follows:

Let $x_{n,i}$ be the proportion of balls having color i at time n and denote by $x_n \in \Delta^m$ the vector of proportions $x_n = (x_{n,1}, \ldots, x_{n,m+1})$. The color of the ball added at time $n+1$ is chosen to be i with probability $f_i(x_n)$, where the f_i are the coordinates of a function $f : \Delta^m \to \Delta^m$.

Such processes, known as *generalized Polya urns*, have been considered by Hill, Lane and Sudderth (1980) for $m = 1$; Arthur, Ermol'ev and Kaniovskii (1983); Pemantle (1990). Arthur (1988) used this kind of model to describe competing technologies in economics.

An urn model is determined by the initial *urn composition* (x_0, n_0) and the *urn function* $f : \Delta^m \to \Delta^m$. We assume that the initial composition (x_0, n_0) is fixed one for all. The σ-field $\mathcal{F}_n$ is the field generated by the random variables $x_0, \ldots, x_n$. One easily verifies that the equation

$$x_{n+1} - x_n = \frac{1}{n_0 + n + 1}(-x_n + f(x_n) + U_{n+1}) \tag{5}$$

defines random variables $\{U_n\}$ that satisfy $\mathsf{E}(U_{n+1}|\mathcal{F}_n) = 0$. We can identify the affine space $\{v \in \mathbb{R}^{m+1} : \sum_{j=1}^{m+1} v_j = 1\}$ with E^m by parallel translation, and also with $\mathbb{R}^m$ by any convenient affine isometry. Under the latter identification, we see that process (5) has exactly the form of [(1) (2)], taking F to be any map which equals $-Id + f$ on Δ^m, and setting $\gamma_n = \frac{1}{n_0 + n}$.

Observe that f being arbitrary, the dynamics of $F = -Id + f$ can be arbitrarily complicated.

The next example is a generalization of urn processes which I call *Generalized Vertex Reinforced Random Walks* after Diaconis and Pemantle. These are non-Markovian discrete time stochastic processes living on a finite state space for which the transition probabilities at each step are influenced by the proportion of time each state has been visited.

Let $\mathcal{M}_{m+1}(\mathbb{R})$ denote the space of real $(m+1) \times (m+1)$ matrices and let $M : E^m \to \mathcal{M}_{m+1}(\mathbb{R})$ be a smooth map such that for all $v \in \Delta^m$, $M(v) = \{M_{i,j}(v)\}$ is Markov transition matrix. Given a point $x_0 \in Int(\Delta^m)$, a vertex $y \in \{1, \ldots, m+1\}$ and a positive integer $n_0 \in \mathbb{N}$, consider a stochastic process $\{(Y_n, (S_1(n), \ldots, S_{m+1}(n))\}_{n \geq 0}$ defined on $\{1, \ldots, m+1\} \times \mathbb{R}_+^{m+1}$ by

- $S_i(0) = n_0 x_{0,i}$, $Y_0 = y$.

- $S_i(n) = S_i(0) + \sum_{k=1}^{n} \delta_{Y_k,i}$, $n \geq 0$.

- $P(Y_{n+1} = j|\mathcal{F}_n) = M_{Y_n,j}(x_n)$

where $\mathcal{F}_n$ denotes the σ-field generated by $\{Y_j : 0 \le j \le n\}$ and $x_n = \dfrac{S(n)}{n + n_0}$ is the empirical occupation measure of $\{Y_n\}$.

Suppose that for each $v \in \Delta^m$ the Markov chain $M(v)$ is indecomposable (i.e has a unique recurrence class), then by a standard result of Markov chains theory, $M(v)$ has a unique invariant probability measure $f(v) \in \Delta^m$. As for Polya's urns, equation (5) defines a sequence of random variables $\{U_n\}$. Here the $\{U_n\}$ are no longer martingale differences but the ODE governing the long term behavior of $\{x_n\}$ is still given by the vector field $F(x) = -x + f(x)$ (see Benaïm, 1997).

The original idea of these processes is due to Diaconis who introduced the process defined by

$$M_{i,j}(v) = \frac{R_{i,j}v_j}{\sum_k R_{i,k}v_k}$$

with $R_{i,j} > 0$. For this process called a *Vertex Reinforced Random Walk* the probability of transition to site j increases each time j is visited. The long term behavior of $\{x_n\}$ has been analyzed by Pemantle (1992) for $R_{ij} = R_{ji}$ and by Benaïm (1997) in the non-symmetric case. With a non-symmetric R the ODE may have nonconvergent dynamics and the behavior of the process becomes highly complicated (Benaïm, 1997).

2.3 Stochastic Fictitious Play in Game Theory

Our last example is an adaptive learning process introduced by Fudenberg and Kreps (1993) for repeated games of incomplete information called *stochastic fictitious play*. It belongs to a flourishing literature which develops the explanation that equilibria in games may arise as the result of learning rather than from rationalistic analysis. For more details and economics motivation we refer the reader to the recent book by Fudenberg and Levine (1998).

For notational convenience we restrict attention to a two-players and two-strategies game. The players are labeled $i = 1, 2$ and the set of strategies is denoted $\{0, 1\}$.

Let $\{\xi_n\}_{n \ge 1}$ be a sequence of identically distributed random variables describing the *states of nature*. The *payoff* to player i at time n is a function $U^i(.,\xi_n) : \{0,1\}^2 \to \mathbb{R}$. We extend $U^i(.,\xi_n)$ to a function $U^i(.,\xi_n) : [0,1]^2 \to \mathbb{R}$ defined by $U^i(x^1, x^2, \xi_n) =$

$$x^1[x^2 U^i(1,1,\xi_n) + (1-x^2)U^i(1,0,\xi_n)] + (1-x^1)[x^2 U^i(0,1,\xi_n) + (1-x^2)U^i(0,0,\xi_n)].$$

Consider now the repeated play of the game. At round n player i chooses an action $s_n^i \in \{0,1\}$ independently of the other player. As a result of these choices player i receives the payoff $U^i(s_n^1, s_n^2, \xi_n)$. The basic assumption is that $U^i(.,\xi_n)$ is known to player i at time n but the strategy chosen by her opponent is not. At the end of the round, both players observe the strategies played.

Fictitious play produces the following adaptive process: At time $n+1$ player 1 (respectively 2) knowing her own payoff function $U^1(.,\xi_{n+1})$ and the strategies played by her opponent up to time n computes and plays the action which maximizes her expected payoff under the assumption that her opponent will play an action whose probability distribution is given by historical frequency of past plays. That is

$$s_{n+1}^1 = Argmax_{s \in \{0,1\}} U^1(s, x_n^2, \xi_{n+1})$$

where

$$x_n^i = \frac{1}{n} \sum_{k=1}^{n} s_k^i.$$

A simple computation shows that the vector of empirical frequencies $x_n = (x_n^1, x_n^2)$ satisfies a recursion of type $[(1),(2)]$ with $\gamma_n = \frac{1}{n}$, $E(U_{n+1}|\mathcal{F}_n) = 0$ and F is the vector field given by

$$F(x^1, x^2) = (-x^1 + h^1(x^2), -x^2 + h^2(x^1)) \qquad (6)$$

where

$$h^1(x^2) = P(U^1(1, x^2, \xi) > U^1(0, x^2, \xi)),$$
$$h^2(x^1) = P(U^2(x^1, 1, \xi) > U^2(x^1, 0, \xi)).$$

The mathematical analysis of stochastic fictitious play has been recently conducted by (Benaïm and Hirsch, 1994) and (Kaniovski and Young, 1995). We will give in section 6.4 (see Example 6.16) a simple argument ensuring the convergence of the process.

3 Asymptotic Pseudotrajectories

A *semiflow* Φ on a metric space (M, d) is a continuous map

$$\Phi : \mathbb{R}_+ \times M \to M,$$

$$(t, x) \mapsto \Phi(t, x) = \Phi_t(x)$$

such that

$$\Phi_0 = \text{Identity}, \quad \Phi_{t+s} = \Phi_t \circ \Phi_s$$

for all $(t, s) \in \mathbb{R}_+ \times \mathbb{R}_+$. Replacing $\mathbb{R}_+$ by $\mathbb{R}$ defines a *flow*.

A continuous function $X : \mathbb{R}_+ \to M$ is an *asymptotic pseudotrajectory* for Φ if

$$\lim_{t \to \infty} \sup_{0 \le h \le T} d(X(t + h), \Phi_h(X(t))) = 0$$

for any $T > 0$. Thus for each fixed $T > 0$, the curve

$$[0, T] \to M : h \mapsto X(t + h)$$

shadows the Φ-trajectory of the point $X(t)$ over the interval $[0, T]$ with arbitrary accuracy for sufficiently large t. By abuse of language we call X *precompact* if its image has compact closure in M.

The notion of asymptotic pseudotrajectories has been introduced in Benaïm and Hirsch (1996) and is particularly useful for analyzing the long term behavior of stochastic approximation processes.

3.1 Characterization of Asymptotic Pseudotrajectories

Let $C^0(\mathbb{R}, M)$ denote the space of continuous M-valued functions $\mathbb{R} \to M$ endowed with the topology of uniform convergence on compact intervals. If $X : \mathbb{R}_+ \to M$ is a continuous function, we consider X as an element of $C^0(\mathbb{R}, M)$ by setting $X(t) = X(0)$ for $t < 0$. The space $C^0(\mathbb{R}, M)$ is metrizable. Indeed, a distance is given by: for all $f, g \in C^0(\mathbb{R}, M)$,

$$d(f,g) = \sum_{k \in \mathbb{N}} \frac{1}{2^k} \min(1, d_k(f,g))$$

where $d_k(f,g) = \sup_{t \in [-k, k]} d(f(t), g(t))$.

The *translation flow* $\Theta : C^0(\mathbb{R}, M) \times \mathbb{R} \to C^0(\mathbb{R}, M)$ is the flow defined by:

$$\Theta^t(X)(s) = X(t + s).$$

Let Φ be a flow or a semiflow on M. For each $p \in M$, the trajectory $\Phi^p : t \to \Phi_t(p)$ is an element of $C^0(\mathbb{R}, M)$ (with the convention that $\Phi^p(t) = p$ if $t < 0$ and Φ is for a semiflow). The set of all such Φ^p defines a subspace $\mathbf{S}_\Phi \subset C^0(\mathbb{R}, M)$.

It is easy to see that the map $H : M \to \mathbf{S}_\Phi$ defined by $H(p) = \Phi^p$ is an homeomorphism which conjugates $\Theta | S_\Phi$ (Θ restricted to S_Φ) and Φ. That is

$$\Theta^t \circ H = H \circ \Phi_t$$

where $t \in \mathbb{R}$ if Φ is a flow and $t \geq 0$ if Φ is a semiflow. This makes $\mathbf{S}_\Phi$ a closed set invariant under Θ. Define the retraction $\hat{\Phi} : C^0(\mathbb{R}, M) \to \mathbf{S}_\Phi$ as

$$\hat{\Phi}(X) = H(X(0)) = \Phi^{X(0)}$$

Lemma 3.1 *A continuous function $X : \mathbb{R}_+ \to M$ is an asymptotic pseudotrajectory of Φ if and only if:*

$$\lim_{t \to \infty} d(\Theta^t(X), \hat{\Phi} \circ \Theta^t(X)) = 0.$$

Proof Follows from definitions. **QED**

Roughly speaking, this means that an asymptotic pseudotrajectory of Φ is a point of $C^0(\mathbb{R}_+, M)$ whose forward trajectory under Θ is attracted by $\mathbf{S}_\Phi$. We also have the following result:

Theorem 3.2 *Let $X : \mathbb{R}_+ \to M$ be continuous function whose image has compact closure in M. Consider the following assertions*

(i) *X is an asymptotic pseudotrajectory of Φ*

(ii) *X is uniformly continuous and every limit point[1] of $\{\Theta^t(X)\}$ is in $\mathbf{S}_\Phi$ (i.e a fixed point of $\hat{\Phi}$).*

---

[1] By a limit point of $\{\Theta^t(X)\}$ we mean the limit in $C^0(\mathbb{R}, M)$ of a convergent sequence $\Theta^{t_k}(X), t_k \to \infty$.

(iii) *The sequence* $\{\Theta^t(X)\}_{t\geq 0}$ *is relatively compact in* $C^0(\mathbb{R}, M)$.

Then (*i*) *and* (*ii*) *are equivalent and imply* (*iii*).

Proof Suppose that assertion (*i*) holds. Let K denote the closure of $\{X(t) : t \geq 0\}$. Let $\epsilon > 0$. By continuity of the flow and compactness of K there exists $a > 0$ such that $d(\Phi_s(x), x) < \epsilon/2$ for all $|s| \leq a$ uniformly in $x \in K$. Therefore $d(\Phi_s(X(t)), X(t)) < \epsilon/2$ for all $t \geq 0$, $|s| \leq a$.

Since X is an asymptotic pseudotrajectory of Φ, there exists $t_0 > 0$ such that $d(\Phi_s(X(t)), X(t+s)) < \epsilon/2$ for all $t > t_0$, $|s| \leq a$. It follows that $d(X(t+s), X(t)) < \epsilon$ for all $t > t_0$, $|s| \leq a$. This proves uniform continuity of X. On the other hand, Lemma 3.1 shows that any limit point of $\{\Theta^t(X)\}$ is a fixed point of $\hat{\Phi}$. This proves that (*i*) implies (*ii*).

Suppose now that (*ii*) holds. Since $\{X(t) : t \geq 0\}$ is relatively compact and X is uniformly continuous, $\{\Theta^t(X)\}_{t\geq 0}$ is equicontinuous and for each $s \geq 0$ $\{\Theta^t(X)(s)\}_{t\geq 0}$ is relatively compact in M. Hence by the Ascoli Theorem (see e.g Munkres 1975, Theorem 6.1), $\{\Theta^t(X)\}$ is relatively compact in $C^0(\mathbb{R}, M)$.

Therefore $\lim_{t\to\infty} d(\Theta^t(X), \hat{\Phi}(\Theta^t(X))) = 0$ which by Lemma 3.1 implies (*i*). The above discussion also shows that (*ii*) implies (*iii*). **QED**

Remark 3.3 Let $D(\mathbb{R}, M)$ be the space of functions which are right continuous and have left-hand limits (*càd làg* functions). The definition of asymptotic pseudotrajectories can be extended to elements of $D(\mathbb{R}, M)$. Since the convergence of a sequence $\{f_n\} \in D$ toward a continuous function f is equivalent to the uniform convergence of $\{f_n\}$ toward f on compact intervals, Lemma 3.1 continues to hold and Theorem 3.2 remains valid provided that we replace the statement that X is uniformly continuous by the weaker statement:

$\forall \epsilon > 0$ there exists $a > 0$ such that

$$\limsup_{t\to\infty} \sup_{|s|\leq a} d(X(t+s), X(t)) \leq \epsilon.$$

4 Asymptotic Pseudotrajectories and Stochastic Approximation Processes

4.1 Notation and Preliminary Result

Let $F : \mathbb{R}^m \to \mathbb{R}^m$ be a continuous map. Consider here a discrete time process $\{x_n\}_{n\in\mathbb{N}}$ living in $\mathbb{R}^m$ (an algorithm) whose general form can be written as

$$x_{n+1} - x_n = \gamma_{n+1}(F(x_n) + U_{n+1}) \tag{7}$$

where

- $\{\gamma_n\}_{n\geq 1}$ is a given sequence of nonnegative numbers such that

$$\sum_k \gamma_k = \infty, \quad \lim_{n\to\infty} \gamma_n = 0.$$

- $U_n \in \mathbb{R}^m$ are (deterministic or random) perturbations.

Formula (7) can be considered to be a perturbed version of a variable step-size Cauchy-Euler approximation scheme for numerically solving $dx/dt = F(x)$:

$$y_{k+1} - y_k = \gamma_{k+1} F(y_k).$$

It is thus natural to compare the behavior of a sample path $\{x_k\}$ with trajectories of the flow induced by the vector field F. To this end we set

$$\tau_0 = 0 \text{ and } \tau_n = \sum_{i=1}^{n} \gamma_i \text{ for } n \geq 1,$$

and define the continuous time *affine* and *piecewise constant interpolated processes* $X, \overline{X} : \mathbb{R}_+ \to \mathbb{R}^m$ by

$$X(\tau_n + s) = x_n + s \frac{x_{n+1} - x_n}{\tau_{n+1} - \tau_n}, \text{ and } \overline{X}(\tau_n + s) = x_n$$

for all $n \in \mathbb{N}$ and $0 \leq s < \gamma_{n+1}$. The "inverse" of $n \to \tau_n$ is the map $m : \mathbb{R}_+ \to \mathbb{N}$ defined by

$$m(t) = \sup\{k \geq 0 : t \geq \tau_k\} \tag{8}$$

let $\overline{U}, \overline{\gamma} : \mathbb{R}_+ \to \mathbb{R}^m$ denote the continuous time processes defined by

$$\overline{U}(\tau_n + s) = U_{n+1}, \overline{\gamma}(\tau_n + s) = \gamma_{n+1}$$

for all $n \in \mathbb{N}$, $0 \leq s < \gamma_{n+1}$. Using this notation (7) can be rewritten as

$$X(t) - X(0) = \int_0^t [F(\overline{X}(s)) + \overline{U}(s)] ds \tag{9}$$

The vector field F is said to be *globally integrable* if it has unique integral curves. For instance a bounded locally Lipschitz vector field is always globally integrable. We then have

Proposition 4.1 *Let F be a continuous globally integrable vector field. Assume that*

A1 *For all $T > 0$*

$$\lim_{n \to \infty} \sup\{\|\sum_{i=n}^{k-1} \gamma_{i+1} U_{i+1}\| : k = n+1, \ldots, m(\tau_n + T)\} = 0.$$

or equivalently

$$\lim_{t \to \infty} \Delta(t, T) = 0$$

with

$$\Delta(t, T) = \sup_{0 \leq h \leq T} \|\int_t^{t+h} \overline{U}(s) ds\|. \tag{10}$$

A2 $\sup_n \|x_n\| < \infty$, or

A2' F is Lipschitz and bounded on a neighborhood of $\{x_n : n \geq 0\}$.

Then the interpolated process X is an asymptotic pseudotrajectory of the flow ϕ induced by F. Furthermore, under assumption **A2'**, for $t \geq 0$ large enough we have the estimate

$$\sup_{0 \leq h \leq T} \|X(t+h) - \Phi_h(X(t))\| \leq C(T)[\Delta(t-1, T+1) + \sup_{t \leq s \leq t+T} (\overline{\gamma}(s))] \quad (11)$$

where $C(T)$ is a constant depending only on T and F.

Proof By continuity of F and assumption **A2** there exists $K > 0$ such that $\|F(X(t))\| \leq K$ for all $t \geq 0$. Thus (9) and **A1** imply

$$\limsup_{t \to \infty} \sup_{0 \leq h \leq T} \|X(t+h) - X(t)\| \leq KT.$$

Hence X is uniformly continuous.

On the other hand, a simple computation shows that

$$\Theta^t(X) = L_F(\Theta^t(X)) + A_t + B_t \quad (12)$$

where $L_F : C^0(\mathbb{R}, \mathbb{R}^m) \to C^0(\mathbb{R}, \mathbb{R}^m)$ is the continuous function defined as

$$L_F(X)(s) = X(0) + \int_0^s F(X(u)) du.$$

and

$$A_t(s) = \int_t^{t+s} [F(\overline{X}(u)) - F(X(u))] du,$$

$$B_t(s) = \int_t^{t+s} \overline{U}(u) du.$$

By assumption **A1**, $\lim_{t \to \infty} B_t = 0$ in $C^0(\mathbb{R}, \mathbb{R}^m)$.

For any $T > 0$ and $t \leq u \leq t + T$, (9) implies

$$\|X(u) - \overline{X}(u)\| = \| \int_{\tau_{m(u)}}^u F(\overline{X}(s)) + \overline{U}(s) ds \|$$

$$\leq K\overline{\gamma}(u) + \| \int_{\tau_{m(u)}}^u \overline{U}(s) ds \|$$

For t large enough $\overline{\gamma}(u) < 1$, therefore

$$\| \int_{\tau_{m(u)}}^u \overline{U}(s) ds \| \leq \| \int_{t-1}^{\tau_{m(u)}} \overline{U}(s) ds \| + \| \int_{t-1}^u \overline{U}(s) ds \| \leq 2\Delta(t-1, T+1).$$

Thus

$$\sup_{t \leq u \leq t+T} \|X(u) - \overline{X}(u)\| \leq 2\Delta(t-1, T+1) + \sup_{t \leq u \leq t+T} K\overline{\gamma}(u).$$

Under assumption **A2** F is uniformly continuous on a neighborhood of $\{x_n\}$, therefore $\lim_{t\to\infty} A_t = 0$ in $C^0(\mathbb{R}, \mathbb{R}^m)$.

Let X^* denote a limit point of $\{\Theta^t(X)\}$. Then

$$X^* = L_F(X^*).$$

By uniqueness of integral curves, this implies

$$X^* = \hat{\Phi}(X^*)$$

Therefore Theorem 3.2 shows that X is an asymptotic pseudotrajectory of Φ.

To prove the estimate in case F is Lipschitz with Lipschitz constant L observe that for $0 \le s \le T$

$$\|A_t(s)\| \le LT(2\Delta(t-1, T+1) + \sup_{t \le s \le t+T} \bar{\gamma}(s)K),$$

$$\|B_t(s)\| \le \Delta(t-1, T+1),$$

$$\Delta(t, T) \le 2\Delta(t-1, T+1)$$

and by equation (12)

$$\|X(t+s) - \Phi_s(X(t))\| \le L \int_0^s \|X(t+u) - \Phi_u(X(t))\|du + \|A_t(s)\| + \|B_t(s)\|.$$

Then use Gronwall's inequality. **QED**

4.2 Robbins-Monro Algorithms

In application of Proposition 4.1 to stochastic approximation algorithms one usually tries to verify assumption **A1** by use of maximal inequalities and martingale techniques.

To illustrate this idea let us consider here the simplest case of stochastic approximation algorithms.

Let $(\Omega, \mathcal{F}, P)$ be a probability space and $\{\mathcal{F}_n\}$ a nondecreasing sequence of sub-σ-algebras of $\mathcal{F}$. We say that a stochastic process $\{x_n\}$ given by (7) satisfies the *Robbins-Monro* or *Martingale difference Noise* (Kushner and Yin, 1997) condition if

(i) $\{\gamma_n\}$ is a deterministic sequence.

(ii) $\{U_n\}$ is *adapted:* U_n is measurable with respect to $\mathcal{F}_n$ for each $n \ge 0$.

(iii) $E(U_{n+1}|\mathcal{F}_n) = 0$.

The next proposition is a particular case of a general theorem due to Métivier and Priouret (1987). The proof contains several inequalities that will be used later.

Proposition 4.2 *Let $\{x_n\}$ given by (7) be a Robbins-Monro algorithm. Suppose that for some $q \geq 2$*

$$\sup_n E(||U_{n+1}||^q) < \infty$$

and

$$\sum_n \gamma_n^{1+q/2} < \infty.$$

Then assumption A1 of proposition 4.1 holds with probability 1.

Proof For any $t \geq 0$ Burkholder's inequality (see e.g Stroock, 1993) implies

$$E\{ \sup_{n < k \leq m(\tau_n+T)} || \sum_{i=n}^{k-1} \gamma_{i+1}U_{i+1}||^q \} \leq C_q E\{ [\sum_{i=n}^{m(\tau_n+T)-1} \gamma_{i+1}^2 ||U_{i+1}||^2]^{q/2} \} \quad (13)$$

for some universal constant $C_q > 0$.

To go further we need the following inequality:
For any $\alpha_i \geq 0$, $\beta_i \in \mathbb{R}$, $u > 1$ and $0 < \delta < 1$

$$(\sum_i |\alpha_i \beta_i|)^u \leq (\sum_i \alpha_i^{\delta u/(u-1)})^{u-1} \sum_i \alpha_i^{(1-\delta)u} |\beta_i|^u \quad (14)$$

The proof of (14) is a consequence of the familiar Hölder inequality

$$\sum_i x_i y_i \leq (\sum_i x_i^u)^{1/u} (\sum_i y_i^{u/(u-1)})^{(u-1)/u}$$

obtained with $x_i = \alpha_i^{1-\delta}|\beta_i|$ and $y_i = \alpha_i^\delta$.

Suppose $q > 2$. We now apply (14) with $u = q/2$, $\delta = (q-2)/2q$, $\alpha_i = \gamma_{i+1}^2$ and $\beta_i = ||U_{i+1}||^2$. Hence (13) yields

$$E(\sup_{n < k \leq m(\tau_n+T)} || \sum_{i=n}^{k-1} \gamma_{i+1}U_{i+1}||^q) \leq C_q E((\sum_{i=n}^{m(\tau_n+T)-1} \gamma_{i+1})^{q/2-1} \sum_{i=n}^{m(\tau_n+T)-1} \gamma_{i+1}^{1+q/2}||U_{i+1}||^q$$

$$\leq C_q T^{q/2-1} E(\sum_{i=n}^{m(\tau_n+T)-1} \gamma_{i+1}^{1+q/2}||U_{i+1}||^q) \quad (15)$$

$$\leq C(q,T) \sum_{i=n}^{m(\tau_n+T)-1} \gamma_{i+1}^{1+q/2} \leq C(q,T) \int_{\tau_n}^{\tau_n+T} \overline{\gamma}^{q/2}(s)ds$$

for some constant $C(q,T) > 0$.

From the preceding inequality we get that

$$E(\Delta(t,T)^q) \leq C(q,T) \int_t^{t+T} \overline{\gamma}^{q/2}(s)ds \quad (16)$$

where $\Delta(t, T)$ is as in (10). If $q = 2$ inequality (16) follows directly from (13). Hence for $q \geq 2$

$$\sum_{k \geq 0} E(\Delta(kT, T)^q) \leq C(q, T) \int_0^\infty \overline{\gamma}^{q/2}(s) ds = \sum \gamma_{i+1}^{1+q/2} < \infty. \qquad (17)$$

By the Borel-Cantelli Lemma this proves that

$$\lim_{k \to \infty} \Delta(kT, T) = 0$$

with probability one. On the other hand for $kT \leq t < (k+1)T$

$$\Delta(t, T) \leq 2\Delta(kT, T) + \Delta((k+1)T, T).$$

Hence assumption **A1** is satisfied **QED**

Remark 4.3 Suppose that $\{\gamma_n\}$ is a sequence of random variables such that γ_{n+1} is $\mathcal{F}_n$ measurable. Then the conclusion or corollary 4.2 remains valid provided that we strengthen the assumption on $\{U_n\}$ to

$$\sup_n E(\|U_{n+1}\|^q | \mathcal{F}_n) \leq C$$

for some deterministic constant $C < \infty$, and replace the assumption on $\{\gamma_n\}$ by

$$E(\sum_n \gamma_n^{1+q/2}) < \infty.$$

The sequence $\{U_n\}$ is said to be *subgaussian* if there exists a positive number Γ such that for all $\theta \in \mathbb{R}^m$

$$E(\exp(\langle \theta, U_{n+1} \rangle | \mathcal{F}_n)) \leq \exp(\frac{\Gamma}{2} \|\theta\|^2).$$

This is for instance the case if $\|U_n\|$ is bounded by $\sqrt{\Gamma}$.

The following result follows from Duflo (1997) (see also Kushner and Yin (1997) or Benaïm and Hirsch (1996))

Proposition 4.4 *Let $\{x_n\}$ given by (7) be a Robbins-Monro algorithm. Suppose $\{U_n\}$ is subgaussian and $\{\gamma_n\}$ is a deterministic sequence such that*

$$\sum_n e^{-c/\gamma_n} < \infty$$

*for each $c > 0$. Then assumption **A1** of proposition 4.1 is satisfied with probability 1. Therefore if **A2** and **A2**′ hold almost surely and F has unique integral curves the interpolated process X is almost surely an asymptotic pseudotrajectory of the flow*

Proof Let

$$Z_n(\theta) = \exp[\sum_{i=1}^{n} \langle \theta, \gamma_i U_i \rangle - \frac{\Gamma}{2} \sum_{i=1}^{n} \gamma_i^2 ||\theta||^2], \ n \geq 1.$$

By the assumption on $\{U_n\}$, $\{Z_n(\theta)\}$ is a supermartingale. Thus for any $\beta > 0$

$$P(\sup_{n < k \leq m(\tau_n + T)} \langle \theta, \sum_{i=n}^{k-1} \gamma_{i+1} U_{i+1} \rangle \geq \beta)$$

$$\leq P(\sup_{n < k \leq m(\tau_n + T)} Z_k(\theta) \geq Z_n(\theta) \exp(\beta - \frac{\Gamma}{2}||\theta||^2 \sum_{i=n}^{m(\tau_n+T)-1} \gamma_{i+1}^2)$$

$$\leq \exp(\frac{\Gamma}{2}||\theta||^2 \sum_{i=n}^{m(\tau_n+T)-1} \gamma_{i+1}^2 - \beta).$$

Let $e_1, \ldots, e_m$ be the canonical basis of $\mathbb{R}^m$, $\alpha > 0$ and $e \in \{e_1, \ldots, e_m\} \cup \{-e_1, \ldots, -e_m\}$.
Set

$$R = \alpha / (\Gamma \sum_{i=n}^{m(\tau_n+T)-1} \gamma_{i+1}^2),$$

$\beta = R\alpha$ and $\theta = Re$. Then

$$P(\sup_{n < k \leq m(\tau_n+T)} \langle e, \sum_{i=n}^{k-1} \gamma_{i+1} U_{i+1} \rangle \geq \alpha) = P(\sup_{n < k \leq m(\tau_n+T)} \langle \theta, \sum_{i=n}^{k-1} \gamma_{i+1} U_{i+1} \rangle \geq \beta)$$

$$\leq exp(\frac{-\alpha^2}{2\Gamma \sum_{i=n}^{m(\tau_n+T)-1} \gamma_{i+1}^2}).$$

It follows that

$$P(\Delta(t,T) \geq \alpha) \leq C \exp(\frac{-\alpha^2}{C' \int_t^{t+T} \overline{\gamma}(s)ds}) \leq C \exp(\frac{-\alpha^2}{C'T\overline{\gamma}(u)}) \qquad (18)$$

for some $t \leq u \leq t + T$ and some positive constants C, C' depending on m (the dimension of $\mathbb{R}^m$) and Γ. The end of the proof is now exactly as in proposition 4.2. **QED**

Remark 4.5 Propositions 4.2 and 4.4 assume a Robbins Monro type algorithm. However it is not hard to verify that the conclusions of these propositions continue to hold if $\{x_n\}$ satisfies the more general recursion

$$x_{n+1} - x_n = \gamma_{n+1}(F(x_n) + U_{n+1} + b_{n+1})$$

where U_n is a martingale difference noise and $\lim_{n \to \infty} b_n = 0$ almost surely.

4.3 Continuous Time Processes

The technique used in the proof of corollary 4.4 can be easily adapted to analyse a class of continuous time stochastic processes which include certain diffusion and jump processes.

Let $\epsilon : \mathbb{R}_+ \to \mathbb{R}_+^*$ be a continuous non-increasing function. Consider the families of operators $\{L_t^d\}_{t \geq 0}$, $\{L_t^j\}_{t \geq 0}$ and $\{L_t\}$ acting on C^2 functions $f : \mathbb{R}^m \to \mathbb{R}$ according to the formulas

$$L_t^d f(x) = \sum_{i=1}^m G_i(x) \frac{\partial f}{\partial x_i}(x) + \frac{\epsilon(t)}{2} \sum_{i,j} a_{i,j}(x) \frac{\partial^2 f}{\partial x_i \partial x_j}(x) \tag{19}$$

$$L_t^j f(x) = \frac{1}{\epsilon(t)} \int_{\mathbb{R}^m} (f(x + \epsilon(t)v) - f(x)) \mu_x(dv) \tag{20}$$

and

$$L_t = L_t^d + L_t^j \tag{21}$$

where

(i) G is a bounded continuous vector field on $\mathbb{R}^m$,

(ii) $a = (a_{ij})$ is a $m \times m$ matrix-valued continuous bounded function such that $a(x)$ is symmetric and nonnegative definite for each $x \in \mathbb{R}^m$.

(iii) $\{\mu_x\}_{x \in \mathbb{R}^m}$ is a family of positive measures on $\mathbb{R}^m$ such that

- $x \to \mu_x(A)$ is measurable for each Borel set $A \subset \mathbb{R}^m$.
- The support of μ_x is contained in a compact set independent of x.

Under these assumptions there exists a nonhomogeneous Markov process $X = \{X(t) : t \geq 0\}$ with sample paths in $D(\mathbb{R}, \mathbb{R}^m)$ (the space of càd lag functions) and initial condition $X(0) = x_0 \in \mathbb{R}^m$ which solves the martingale problem for $\{L_t\}$ (Ethier and Kurtz, 1975; Stroock and Varadhan 1997). That is for each C^∞ function $f : \mathbb{R}^m \to \mathbb{R}$ with compact support,

$$f(X_t) - \int_0^t L_s f(X_s) ds$$

is a martingale with respect to $\mathcal{F}_t = \sigma\{X(s) : s \leq t\}$.

Define the vector field

$$F(x) = G(x) + \int_{\mathbb{R}^m} v \mu(x)(dv) \tag{22}$$

Proposition 4.6 *Suppose that*

(i) *F is a continuous globally integrable vector field*

(ii)

$$\int_0^\infty exp(-\frac{c}{\epsilon(t)})dt < \infty$$

for all c > 0

(iii) *$P(\sup_t ||X(t)|| < \infty) = 1$ or F is Lipschitz.*

Then X is almost surely an asymptotic pseudotrajectory of the flow induced by F. Furthermore when F is Lipschitz we have the estimate: There exist constant $C, C(T) > 0$ such that for all $\alpha > 0$

$$P(\sup_{0 \le h \le T} ||X(t+h) - \Phi_t(X(t))|| \ge \alpha) \le C \exp(-\alpha^2 C(T)/\epsilon(t)).$$

Proof Set $\Delta(t, T) = \sup_{0 \le h \le T} ||X(t+h) - X(t) - \int_t^{t+h} F(X(s))ds||$

Let $f(x) = exp\langle \theta, x - x_0 \rangle$ and $\tau_n = \inf\{t \ge 0 \; \overline{X[0,t]} \cap B(0,n)^c \ne \emptyset\}$ where $B(0,n) = \{x \in \mathbb{R}^m \; ||x|| < n\}$. Since the measure μ_x has uniformly bounded support there exists $r > 0$ such that $f(X(t \wedge \tau_n)) = f_n(X(t \wedge \tau_n))$ where f_n is a C^∞ function with compact support which equals f on $B(0, n + r)$. Since $f_n(X(t)) - \int_0^t L_s f_n(X(s))ds$ is a martingale and τ_n a stopping time $f(X(t \wedge \tau_n)) - \int_0^{t \wedge \tau_n} L_s f(X(s))ds$ is a martingale. Hence

$$f(X(t \wedge \tau_n))exp[-\int_0^{t \wedge \tau_n} \frac{L_s f(X(s))}{f(X(s))}ds]$$

is a martingale (Ethier and Kurtz, 1975).

Let $g(u) = e^u - u - 1$. Then

$$\frac{L_s f(x)}{f(x)} = \langle F(x), \theta \rangle + \frac{\epsilon(s)}{2}\langle \theta, a(x)\theta \rangle + \frac{1}{\epsilon(s)}\int_{\mathbb{R}^m} g(\epsilon(s)\langle \theta, v \rangle)\mu_x(dv).$$

Now using the facts that $g(|u|) \le g(u)$, g is non-decreasing on $\mathbb{R}^+$, $g(u) = u^2/2 + o(u)$ and the boundedness assumptions on x and the support of μ_x it is not hard to verify that there exist a constant $\Gamma > 0$ and $t_0 > 0$ such that for $s \ge t_0$

$$\frac{L_s f(x)}{f(x)} - \langle F(x), \theta \rangle \le \Gamma ||\theta||^2 \epsilon(s).$$

Therefore

$$exp[\langle \theta, X(t \wedge \tau_n) - X(t_0 \wedge \tau_n) - \int_{t_0 \wedge \tau_n}^{t \wedge \tau_n} F(X(s)) \rangle ds - \int_{t_0 \wedge \tau_n}^{t \wedge \tau_n} \Gamma ||\theta||^2 \epsilon(s)ds]$$

is an $\mathcal{F}_t$ supermartingale for $t \ge t_0$. As is Proposition 4.4 we obtain

$$P(\sup_{0 \le h \le T} ||X((t+h) \wedge \tau_n) - X(t \wedge \tau_n) - \int_{t \wedge \tau_n}^{(t+h) \wedge \tau_n} F(X(s))|| \ge \alpha) \le C \exp(-\frac{\alpha^2}{C'T\epsilon(t)})$$

for $t \geq t_0$ and by Fatou's lemma we conclude that

$$P(\Delta(t, T) \geq \alpha) \leq C \exp(-\frac{\alpha^2}{C' T \epsilon(t)}).$$

The rest of the proof is now exactly as in Proposition 4.4. Details are left to the reader. **QED**

5 Limit Sets of Asymptotic Pseudotrajectories

5.1 Chain Recurrence and Attractors

In this section we introduce some basic terminology and a few results from (topological) dynamics that will be useful to understand the behavior of asymptotic pseudotrajectories and stochastic approximation processes. In particular we introduce the notion of *chain recurrence* and emphasize its relation with the notion of *attractors* (thanks to Moe Hirsch who taught me the importance of this relation). The material of this section is fairly standard to dynamicists and can be found in numerous places. However since the students for which these notes have been written (as well as the typical reader of the *Séminaire*) may not be familiar with these notions we have tried to give a self contained and comprehensive presentation.

The main and original reference for this section is Conley (1978). The books by Shub (1987) and Robinson (1995) also contain most of the material here.

Basic Notions of Recurrence

Let Φ be a flow or semiflow on the metric space (M, d). We let $\mathbb{T} = \mathbb{R}_+$ if Φ is a semiflow and $\mathbb{T} = \mathbb{R}$ if Φ is a flow.

A subset $A \subset M$ is said *positively invariant* if $\Phi_t(A) \subset A$ for all $t \geq 0$. It is said *invariant* if $\Phi_t(A) = A$ for all $t \in \mathbb{T}$.

A point $p \in M$ is an *equilibrium* if $\Phi_t(p) = p$ for all t. When M is a manifold and Φ is the flow induced by a vector field F, equilibria coincide with zeros of F. A point $p \in M$ is a periodic point of period $T > 0$ if $\Phi_T(p) = p$ for some $T > 0$ and $\Phi_t(p) \neq p$ for $0 < t < T$.

The *forward orbit* of $x \in M$ is the set $\gamma^+(x) = \{\Phi_t(x) : t \geq 0\}$ and the orbit of x is $\gamma(x) = \{\Phi_t(x) : t \in \mathbb{T}\}$. A point $p \in M$ is an *omega limit point* of x if $p = \lim_{t_k \to \infty} \Phi_{t_k}(x)$ for some sequence $t_k \to \infty$. The *omega limit set* of x denoted $\omega(x)$ is the set of omega limit points of x.

If $\gamma^+(x)$ has compact closure, $\omega(x)$ is a compact connected invariant set (It is a good warm up exercise for the reader unfamiliar with these notions) and $\overline{\gamma^+(x)} = \gamma^+(x) \cup \omega(x)$.

If Φ is a flow the *alpha limit set* of x is defined as the omega limit set of x for the reversed flow $\Psi = \{\Psi_t\}$ with $\Psi_t = \Phi_{-t}$.

Further we set $Eq(\Phi)$ the set of equilibria, $Per(\Phi)$ the closure of the set of periodic orbits $\mathcal{L}_+(\Phi) = \bigcup_{x \in M} \omega(x), \mathcal{L}_-(\Phi) = \bigcup_{x \in M} \alpha(x)$ and

$$\mathcal{L}(\Phi) = \mathcal{L}_+(\Phi) \cup \mathcal{L}_-(\Phi).$$

Chain Recurrence and Attractors

Equilibria, periodic and omega limit points are clearly "recurrent" points. In general, we may say that a point is recurrent if it somehow returns near where it was under time evolution.

A notion of recurrence related to slightly perturbed orbits and well suited to analyse stochastic approximation processes is the notion of *chain recurrence* introduced by Bowen (1975) and Conley (1978).

Let $\delta > 0$, $T > 0$. A (δ, T)-*pseudo-orbit* from $a \in M$ to $b \in M$ is a finite sequence of partial trajectories

$$\{\Phi_t(y_i) : 0 \le t \le t_i\}; \ i = 0, \ldots, k - 1; \ t_i \ge T$$

such that

$$
\begin{aligned}
d(y_0, a) &< \delta, \\
d(\Phi_{t_j}(y_j), y_{j+1}) &< \delta, \ j = 0, \ldots, k - 1; \\
y_k &= b.
\end{aligned}
$$

We write $(\Phi : a \hookrightarrow_{\delta,T} b)$ (or simply $a \hookrightarrow_{\delta,T} b$ when there is no confusion on Φ) if there exists a (δ, T)-*pseudo-orbit* from a to b. We write $a \hookrightarrow b$ if $a \hookrightarrow_{\delta,T} b$ for every $\delta > 0$, $T > 0$. If $a \hookrightarrow a$ then a is a *chain recurrent* point. If every point of M is chain recurrent then Φ is a chain recurrent semiflow (or flow).

If $a \hookrightarrow b$ for all $a, b \in M$ we say the flow Φ is *chain transitive*.

We denote by $R(\Phi)$ the set of chain recurrent points for Φ. It is easy to verify (again a good warm up exercise) that $R(\Phi)$ is a closed, positively invariant set and that

$$Eq(\Phi) \subset Per(\Phi) \subset \mathcal{L}(\Phi) \subset R(\Phi.)$$

We will see below that $R(\Phi)$ is always invariant when it is compact (Theorem 5.5).

Let $\Lambda \subset M$ be a nonempty invariant set. Φ is called *chain recurrent on* Λ if every point $p \in \Lambda$ is a chain recurrent point for $\Phi|\Lambda$, the restriction of Φ to Λ. In other words, $\Lambda = R(\Phi|\Lambda)$.

A compact invariant set on which Φ is chain recurrent (or chain transitive) is called an *internally chain recurrent* (or *internally chain transitive*) set.

Example 5.1 Consider the flow on the unit circle $S^1 = \mathbb{R}/2\pi\mathbb{Z}$ induced by the differential equation

$$\frac{d\theta}{dt} = f(\theta)$$

Figure 1: $\dot{\theta} = f(\theta)$

where f is a 2π-periodic smooth nonnegative function such that

$$f^{-1}(0) = \{k\pi : k \in \mathbf{Z}\}.$$

We have

$$Eq(\Phi) = \{0, \pi\} = L_+(\Phi) = L(\Phi)$$

and

$$R(\Phi) = S^1.$$

Internally chain recurrent sets are $\{0\}, \{\pi\}$ and S^1. Remark that the set $X = [0, \pi]$ is a compact invariant set consisting of chain recurrent points. However, X is not internally chain recurrent.

A subset $A \subset M$ is an *attractor* for Φ provided:

(i) A is nonempty, compact and invariant ($\Phi_t A = A$); and

(ii) A has a neighborhood $W \subset M$ such that dist($\Phi_t x, A$) $\to 0$ as $t \to \infty$ uniformly in $x \in W$.

The neighborhood W is usually called a *fundamental neighborhood* of A. The *basin* of A is the positively invariant open set comprising all points x such that dist($\Phi_t x, A$) $\to 0$ as $t \to \infty$. If $A \neq M$ then A is called a *proper* attractor. A *global attractor* is an attractor whose basin is all the space M. An equilibrium (= stationary point) which is an attractor is called *asymptotically stable*.

The following Lemma due to Conley (1978) is quite useful.

Lemma 5.2 *Let $U \subset M$ be an open set with compact closure. Suppose that $\Phi_T(\overline{U}) \subset U$ for some $T > 0$. Then there exists an attractor $A \subset U$ whose basin contains $\overline{U}$.*

Proof By compactness of $\Phi_T(\overline{U})$ there exists an open set V such that $\Phi_T(\overline{U}) \subset V \subset \overline{V} \subset U$. By continuity of the flow there exists $\epsilon > 0$ such that $\Phi_t(\overline{U}) \subset V$ for $T - \epsilon \leq t \leq T + \epsilon$. Let $t_0 = T(T+1)/\epsilon$. For $t > t_0$ write $t = k(T + r/k)$ with $k \in \mathbb{N}$ and $0 \leq r/k < \epsilon$. Therefore for all $x \in \overline{U}$ $\Phi_t(x) = \Phi_{T+r/k} \circ \ldots \circ \Phi_{T+r/k}(x) \in V$.

Then $A_t = \overline{\cup_{s \geq t} \Phi_s(U)} \subset \overline{V} \subset U$. and $A = \cap_{t \geq 0} A_t \subset U$. It is now easy to verify that A is an attractor. Details are left to the reader. **QED**

The following proposition originally due to Bowen (1975) makes precise the relation between the different notions we have introduced.

Proposition 5.3 *Let $\Lambda \subset M$. The following assertions are equivalent*

(i) Λ *is internally chain-transitive*

(ii) Λ *is connected and internally chain-recurrent*

(iii) Λ *is a compact invariant set and $\Phi|\Lambda$ admits no proper attractor.*

Proof $(i) \Rightarrow (ii)$ is easy and left to the reader. $(ii) \Rightarrow (iii)$. Let $A \subset \Lambda$ be a nonempty attractor. To prove that $\Lambda = A$ it suffices to show that A is open and closed in Λ. Let W be an open (in Λ) fundamental neighborhood of A. We claim that $W = A$. Suppose to the contrary that there exists $p \in W \setminus A$. Let $U_\delta = \{x \in \Lambda : d(x, A) \leq \delta\}$. Choose δ small enough so that $U_\delta \subset U_{2\delta} \subset W$ and $B(p, \delta) \subset W \setminus U_{2\delta}$. For T large enough and $t \geq T$ $\Phi_t(W) \subset U_\delta$. Therefore it is impossible to have $p \hookrightarrow_{\delta,T} p$. A contradiction.
$(iii) \Rightarrow (i)$. Let $x \in \Lambda, \delta > 0, T > 0$ and $V = \{y \in \Lambda : (\Phi|\Lambda : x \hookrightarrow_{\delta,T} y)\}$. The set V is open (by definition) and satisfies $\Phi_T(\overline{V}) \subset V$. It then follows from Lemma 5.2 that V contains an attractor but since there are no proper attractors $V = \Lambda$. Since this is true for all $x \in \Lambda, \delta > 0$ and $T > 0$ it follows that Λ is internally chain transitive. **QED**

Corollary 5.4 *If an internally chain transitive set K meets the basin of an attractor A, it is contained in A.*

Proof By compactness, $K \cap A$ is nonempty, hence an attractor for the $\Phi|K$ Since $\Phi|K$ has no proper attractors, being chain transitive, it follows that $K \subset A$. **QED**

The following theorem was proved by Conley (1978) for flows but the proof given here is adapted from a proof given by Robinson (1977) for diffeomorphims.

Theorem 5.5 *If M is compact then $R(\Phi)$ is internally chain recurrent.*

Proof First observe that $R(\Phi)$ is obviously a compact subset of M. Let $p \in R(\Phi)$. For $n \in \mathbb{N}$ and $T > 0$ there exist points $p = p_0^n, \ldots, p_{k_n}^n$ in M and times $t_1, \ldots, t_{k_n}$ with $t_i > T$ such that $p_0^n = p$, $d(\Phi_{t_i}(p_i^n), p_{i+1}^n) < 1/n$ for $i = 0, \ldots k_n - 1$ and $d(p_{k_n}^n, p) < 1/n$. Further we can always assume (by adding points to the sequence) that $t_i \leq 2T$ and since $C_n = \{p_0^n, \ldots, p_{k_n}^n\}$ is a compact subset of M we can also assume (by replacing C_n by a subsequence C_{n_j}) that $\{C_n\}$ converges toward some compact set C for the Hausdorff topology[2]. By

---

[2] If A and B are closed subsets of M the Hausdorff distance $D(A, B)$ is defined as $D(A, B) = \inf\{\epsilon > 0 \; A \subset U_\epsilon(B) \text{ and } B \subset U_\epsilon(A)\}$. This distance makes the space of closed subsets of M a compact space (se e.g Munkres exercise 7 page 279).

construction $C \subset R(\Phi)$ and $p \in C$. Fix $\epsilon > 0$. By uniform continuity of Φ : $[0, 2T] \times M \to M$ there exists $0 < \delta < \epsilon/3$ so that $d(a, b) \leq \delta \Rightarrow d(\Phi_t(a), \Phi_t(b)) \leq \epsilon$ for all $0 \leq t \leq 2T$. Now for n large enough $C \subset U_\delta(C_n)$ and $C_n \subset U_\delta(C)$. Therefore there exist points $q_i^n \in C$ such that $d(q_i^n, p_i^n) \leq \delta$. It follows that

$$d(\Phi_{t_i}(q_i^n), q_{i+1}^n) \leq d(\Phi_{t_i}(p_i^n), \Phi_{t_i}(q_i^n)) + d(\Phi_{t_i}(p_i^n), p_{i+1}^n) + d(p_{i+1}^n, q_{i+1}^n) \leq \epsilon.$$

Thus we have constructed an (ϵ, T) pseudo orbit from p to p which lies entirely in $C \subset R(\Phi)$. To conclude the proof it remains to show that $R(\Phi)$ is invariant. It is clearly positively invariant. Let p and C_n as above. By extracting convergent subsequences from $\{t_{k_n - 1}\}$ and $\{p_{k_n - 1}\}$ we obtain points $\tau \in [T, 2T]$ and $p* \in R(\Phi)$ such that $\Phi_\tau(p*) = p$. Hence $p \in \Phi_t(R(\Phi))$ for all $0 \leq t \leq T$ and since p is arbitrary $R(\Phi) \subset \Phi_t(R(\Phi))$ for all $0 \leq t \leq T$. By the semiflow property this implies $R(\Phi) \subset \Phi_t(R(\Phi))$ for all $t \geq 0$.   **QED**

Corollary 5.6 *Let $x \in M$ (non-necessarily compact). If $\overline{\gamma(x)^+}$ is compact then $\omega(x)$ is internally chain transitive.*

Proof   Let $T = [0, 1] \times \overline{\gamma^+(x)}$ and Ψ the semiflow on T defined by $\Psi_t(u, y) = (e^{-t}u, \Phi_t(y))$. Clearly $\{0\} \times \omega(x)$ is a global attractor for Ψ and points of $\{0\} \times \omega(x)$ are chain recurrent for Ψ. Therefore $R(\Psi) = \{0\} \times \omega(x)$. By Theorem 5.5 $R(\Psi|R(\Psi)) = R(\Psi)$. This implies $R(\Phi|\omega(x)) = \omega(x)$ and $\omega(x)$ being connected it is internally chain transitive by Proposition 5.3.   **QED**

5.2   The Limit Set Theorem

Let $X : \mathbb{R}_+ \to M$ be an asymptotic pseudotrajectory of a semiflow Φ. The *limit set* $L(X)$ of X, defined in analogy to the omega limit set of a trajectory, is the set of limits of convergent sequences $X(t_k), t_k \to \infty$. That is

$$L(X) = \bigcap_{t \geq 0} \overline{X([t, \infty))}.$$

Theorem 5.7

(i) *Let X be a precompact asymptotic pseudotrajectory of Φ. Then $L(X)$ internally chain transitive.*

(ii) *Let $L \subset M$ be an internally chain transitive set, and assume M is locally path connected. Then there exists an asymptotic pseudotrajectory X such that $L(X) = L$.*

Proof   We only give the proof of (i). We refer the reader to Benaïm and Hirsch (1996) for a proof of (ii) and further results. Since $\{X(t) : t \geq 0\}$ is relatively compact, Theorem 3.2 shows that $\{\Theta^t(X) : t \in \mathbb{R}\}$ is relatively compact in $C^0(\mathbb{R}, M)$ and $\lim_{t \to \infty} d(\Theta^t(X), S_\Phi) = 0$. Therefore by Corollary 5.6 the omega limit set of X for Θ, denoted by $\omega_\Theta(X)$, is internally chain transitive for the semiflow $\theta|S_\Phi$.

The homeomorphism $H : M \to \mathbf{S}_\Phi$, defined by $H(x)(t) = \Phi_t(x)$ conjugates $\Theta | \mathbf{S}_\Phi$ and Φ:

$$(\Theta^t | \mathbf{S}_\Phi) \circ H = H \circ \Phi_t$$

where $t \geq 0$ for a semiflow Φ, and $t \in \mathbb{R}$ for a flow. Since the property of being chain transitive is (obviously) preserved by conjugacy it suffices to verify that

$$H(L(X)) = \omega_\Theta(X)$$

to prove assertion (i). Let $p \in L(X)$. Then $p = \lim_{t_k \to \infty} X(t_k)$. By relative compactness of $\{\Theta^t(X)\}_{t \geq 0}$ we can always suppose that $\{\Theta^{t_k}(X)\}$ converges toward some point $Y \in C^0(\mathbb{R}, M)$. By lemma 3.1 $Y = \hat{\Phi}(Y) = H(Y(0)) = H(p)$. This shows that $H(L(X)) \subset \omega_\Theta(X)$. The proof of the converse inclusion is similar. **QED**

Remark 5.8 Our proof of Theorem 5.7 follows from Benaïm and Hirsch (1996). It has the nice interpretation that the limit set $L(X)$ can be seen as an omega limit set for an extension of the flow to some larger space. A more direct proof in the spirit of Theorem 5.5 can be found in Benaïm (1996) (see also Duflo 1996).

6   Dynamics of Asymptotic Pseudotrajectories

Theorem 5.7 and its applications in later sections show the importance of understanding the dynamics and topology of internally chain recurrent sets (which in most dynamical settings are the same as limit sets of asymptotic pseudotrajectories). Many of the results which appear in the literature on stochastic approximation can be easily deduced (and generalized) from properties of chain recurrent sets. While there is no general structure theory for internally chain recurrent sets, much can be said about many common situations. Several useful results are presented in this section. The main source of this section are the papers (Benaïm, 1996) and Benaïm and Hirsch (1996) but some results have been improved. In particular we give an elementary proof of the convergence of stochastic gradient algorithms with possibly infinitely many equilibria. Several results by Fort and Pages (1996) are similar to those of this section.

We continue to assume that $X : \mathbb{R}_+ \to M$ is an asymptotic pseudotrajectory for a flow or semiflow Φ in a metric space M. Remark that we do not *a priori* assume that X is precompact.

6.1   Simple Flows, Cyclic Orbit Chains

A flow on M is called *simple* if it has only a finite set of alpha and omega limit points (necessarily consisting of equilibria). This property is inherited by the restriction of Φ to invariant sets.

A subset $\Gamma \subset M$ is a *orbit chain* for Φ provided that for some natural number $k \geq 2$, Γ can be expressed as the union

$$\Gamma = \{e_1, \ldots, e_k\} \bigcup \gamma_1 \bigcup \ldots \bigcup \gamma_{k-1}$$

of equilibria $\{e_1, \ldots, e_k\}$ and nonsingular orbits $\gamma_1, \ldots, \gamma_{k-1}$ connecting them: this means that γ_i has alpha limit set $\{e_i\}$ and omega limit set $\{e_{i+1}\}$. Neither the equilibria nor the orbits of the orbit chain are required to be distinct. If $e_1 = e_k$, Γ is called a *cyclic orbit chain*. A homoclinic loop is an example of a cyclic orbit chain.

Concerning cyclic orbit chains, Benaïm and Hirsch (1995a, Theorem 3.1) noted the following useful consequence of the important Akin-Nitecki-Shub Lemma (Akin 1993).

Proposition 6.1 *Let $L \subset M$ be an internally chain recurrent set. If $\Phi|L$ is a simple flow, then every non-stationary point of L belongs to a cyclic orbit chain in L.*

From Theorem 5.7 we thus get:

Corollary 6.2 *Assume that X is precompact and $\Phi|L(X)$ is a simple flow. Then every point of $L(X)$ is an equilibrium or belongs to a cyclic orbit chain in $L(X)$.*

Corollary 6.3 *Assume $L \subset M$ is an internally chain recurrent set such that*

$$\mathcal{L}(\Phi|L) \subset \Lambda = \bigcup_{j=1}^{n} \Lambda_j$$

where $\Lambda_1, \ldots, \Lambda_n$ are compact invariant subsets of L. Then for every point $p \in L$ either $p \in \Lambda$ or there exists a finite sequence $x_1, \ldots, x_k \in L \setminus \Lambda$ and indices $i_1, \ldots, i_k$ such that

(i) $\{i_1, \ldots, i_{k-1}\} \subset \{1, \ldots, n\}$ *and* $i_k = i_1$

(ii) $\alpha(x_{i_l}) \subset \Lambda_{i_l}, \; \omega(x_{i_l}) \subset \Lambda_{i_{l+1}}$ *for* $l = 1, \ldots, k-1$.

In particular if there is no cycle among the Λ_j then $L \subset \Lambda$.

Proof Let $\hat{L}$ be the topological quotient space obtained by collapsing each Λ_i to a point. Let π denote the quotient map $\pi : L \to \hat{L}$. We claim that $\hat{L}$ is metrizable. By the Urysohn metrization Theorem, it suffices to verify that $\hat{L}$ is a regular space with a countable basis.

We first construct a countable basis $\{\hat{U}_n(\hat{x})\}_{n \geq 1}$ at each point $\hat{x} = \pi(x) \in \hat{L}$ as follows. If $x \notin \Lambda$ choose $0 < d_x < d(x, \Lambda)$ and set $\hat{U}_n(\hat{x}) = \pi(B(x, \frac{d_x}{n}))$.

Let $0 < \epsilon < \inf_{i \neq j} d(\Lambda_i, \Lambda_j)$. For $x \in \Lambda_i$ set $\hat{U}_n(\hat{x}) = \{\pi(U_{\epsilon/n}(\Lambda_i))\}$.

Using this basis it is immediate to verify that $\hat{L}$ is Hausdorff, and since it is compact (by continuity of π) it is a regular space. Now let $\{x_i\}_{i \geq 1}$ be a countable dense set in L. The family $\{\hat{U}_n(\hat{x}_i))\}_{n, i \geq 1}$ is a countable basis of $\hat{L}$.

The flow Φ induces a flow $\hat{\Phi}$ on $\hat{L}$ defined by $\hat{\Phi} \circ \pi = \pi \circ \Phi$ which has simple dynamics, and the Λ_j as equilibria.

Let $x \in L$. It is clear, by definition of chain recurrence and uniform continuity of π, that $\pi(x)$ is chain recurrent for $\hat{\Phi}$. Hence $\hat{L}$ is internally chain recurrent and the result follows from Proposition 6.1. **QED**

6.2 Lyapounov Functions and Stochastic Gradients

Let $\Lambda \subset M$ be a compact invariant set of the semiflow Φ. A continuous function $V : M \to \mathbb{R}$ is called a *Lyapounov function* for Λ if the function $t \in \mathbb{R}_+ \to V(\Phi_t(x))$ is constant for $x \in \Lambda$ and strictly decreasing for $x \in M \setminus \Lambda$. If Λ equals the equilibria set $Eq(\Phi)$, V is called a *strict Lyapounov function* and Φ a *gradientlike system*.

Proposition 6.4 *Let $\Lambda \subset M$ be a compact invariant set and $V : M \to \mathbb{R}$ a Lyapounov function for Λ. Assume that $V(\Lambda) \subset \mathbb{R}$ has empty interior. Then every internally chain transitive set L is contained in Λ and $V|L$ is constant.*

Proof Let $L \subset M$ be an internally chain transitive set. Let $v* = \inf\{V(x) : x \in L\}$. We claim that $L \cap \Lambda \neq \emptyset$ and

$$v* = \inf\{V(x) : x \in L \cap \Lambda\}.$$

Let $x \in L$. The function $t \to V(\Phi_t(x))$ being non-increasing and bounded the limit $V_\infty(x) = \lim_{t \to \infty} V(\Phi_t(x))$ exists. Therefore $V(p) = V_\infty(x) \leq V(x)$ for all $p \in \omega(x)$. By invariance of $\omega(p)$, V is constant along trajectories in $\omega(x)$. Hence $\omega(x) \subset \Lambda$. This proves the claim.

By continuity of V and compactness of $L \cap \Lambda$, $v* \in V(L \cap \Lambda)$. Since $V(\Lambda)$ has empty interior there exists a sequence $\{v_n\}_{n \geq 1}$, $v_n \in \mathbb{R} \setminus V(\Lambda)$ decreasing to $v *$. For $n \geq 1$ let $L_n = \{x \in L : V(x) < v_n\}$. Because V is a Lyapounov function for Λ $\Phi_t(\overline{L_n}) \subset L_n$ for any $t > 0$. Hence by Lemma 5.2 and Proposition 5.3 $L = L_n$. Then $L = \cap_{n \geq 1} L_n = \{x \in L : V(x) = v*\}$. This implies $L = \Lambda$ and $V(L) = \{v*\}$. **QED**

Remark 6.5 The following example shows that the assumption that $V(\Lambda)$ has empty interior is essential in Proposition 6.4.

Consider the flow on the unit circle $S^1 = \mathbb{R}/2\pi\mathbb{R}$ induced by the differential equation $\dfrac{d\theta}{dt} = f(\theta)$ where f is a 2π periodic smooth nonnegative function such that $f^{-1}(0) = \{[k\pi, k(\pi + 1)] : k \in 2\mathbb{Z}\}$. Then S^1 is clearly internally chain transitive. However any 2π periodic smooth nonnegative function $V : S^1 \to \mathbb{R}$ strictly increasing on $]0, \pi[$ and strictly decreasing on $]\pi, 2\pi[$ is a strict Lyapounov function.

Corollary 6.6 *Assume that X is precompact, Φ admits a strict Lyapounov function, and that there are countably many equilibria in $L(X)$. Then $X(t)$ converges to an equilibrium as $t \to \infty$.*

The following corollary is particularly useful in applications since it provides a general convergence result for *stochastic gradient* algorithms.

Corollary 6.7 *Assume M is a smooth C^r Riemannian manifold of dimension $m \geq 1$, $V : M \to \mathbb{R}$ a C^r map and F the gradient vector field*

$$F(x) = -\nabla V(x).$$

Assume

(i) *F induces a global flow* Φ

(ii) *X is a precompact asymptotic pseudotrajectory of* Φ

(iii) $r \geq m$

Then $L(X)$ consists of equilibria and $V(X(t))$ converges as $t \to \infty$.

Proof Let $\Lambda = Eq(\Phi)$. By Sard's theorem (Hirsch, 1976; chapter 3) $V(\Lambda)$ has Lebesgue measure zero in $\mathbb{R}$ and the result follows from Proposition 6.4 applied with the strict Lyapounov function V. **QED**

6.3 Attractors

Let $X : \mathbb{R} \to M$ be an asymptotic pseudotrajectory of Φ. For any $T > 0$ define

$$d_X(T) = \sup_{k \in \mathbb{N}} d(\Phi_T(X(kT)), X(kT + T)). \tag{23}$$

If a point $x \in M$ belongs to the basin of attraction of an attractor $A \subset M$ then $\Phi_t(x) \to A$ as $t \to \infty$. The next lemma shows that the same is true for an asymptotic pseudotrajectory X provided that $d_X(T)$ is small enough and M is locally compact. This simple lemma will appear to be very useful in the next section.

Lemma 6.8 *Assume M is locally compact. Let $A \subset M$ be an attractor with basin $B(A)$ and let $K \subset B(A)$ be a nonempty compact set. There exist numbers $T > 0, \delta > 0$ depending only on K such that:*

If X is an asymptotic pseudotrajectory with $X(0) \in K$ and $d_X(T) < \delta$, then $L(X) \subset A$.

Proof Choose an open set W with compact closure such that $A \cup K \subset W \subset \overline{W} \subset B(A)$ and choose $\delta > 0$ such that $U_{2\delta}(A)$ (the 2δ neighborhood of A) is contained in W. Since A is an attractor there exists $T > 0$ such that $\Phi_T(W) \subset U_\delta(A)$. Now, if $X(0) \in K$ and $d_X(T) < \delta$ we have $\Phi_T(X(0)) \in U_\delta(A)$ and $d(X(T), \Phi_T(X(0))) < \delta$. Thus $X(T) \in U_{2\delta}(A) \subset W$. By induction it follows that $X(kT) \in W$ for all $k \in \mathbb{N}$. Thus, by compactness, $L(X) \cap \overline{W} \neq \emptyset$ and $L(X)$ is compact as a subset of $\Phi([0, T] \times \overline{W})$. Since points in $L(X) \cap \overline{W}$ are attracted by A and $L(X)$ is invariant, $L(X) \cap A \neq \emptyset$. The conclusion now follows from Proposition 5.3 and Theorem 5.7. **QED**

Below we assume that M is locally compact.

Theorem 6.9 *Let e be a asymptotically stable equilibrium with basin of attraction W and $K \subset W$ a compact set. If $X(t_k) \in K$ for some sequence $t_k \to \infty$, then $\lim_{t \to \infty} X(t) = e$.*

In the context of stochastic approximations this result was proved by Kushner and Clark (1978). It is an easy consequence of Theorem 5.7 because the only chain recurrent point in the basin of e is e.

More generally we have:

Theorem 6.10 *Let A be an attractor with basin W and $K \subset W$ a compact set. If $X(t_k) \in K$ for some sequence $t_k \to \infty$, then $L(X) \subset A$.*

Proof Follows from Theorem 5.7 and Lemma 6.8. **QED**

Corollary 6.11 *Suppose M is noncompact but locally compact and that Φ is dissipative meaning that there exists a global attractor for Φ. Let $M \cup \{\infty\}$ denote the one-point compactification of M. Then either $L(X)$ is an internally chain transitive subset of M or $\lim_{t \to \infty} X(t) = \infty$.*

When applied to stochastic approximation processes such as those described in section 4 Propositions 4.2 and 4.4 under the assumption that F is bounded and Lipschitz, Corollary 6.11 implies that with probability one either $X(t) \to \infty$ or $L(X)$ is internally chain transitive for the flow induced by F.

6.4 Planar Systems

The following result of Benaïm and Hirsch (1994) goes far towards describing the dynamics of internally chain recurrent sets for planar flows with isolated equilibria:

Theorem 6.12 *Assume Φ is a flow defined on $\mathbb{R}^2$ with isolated equilibria. Let L be an internally chain recurrent set. Then for any $p \in L$ one of the following holds:*

(i) p is an equilibrium.

(ii) p is periodic (i.e $\Phi_T(p) = p$ for some $T > 0$).

(iii) There exists a cyclic orbit chain $\Gamma \subset L$ which contains p.

Notice that this rules out trajectories in L which spiral toward a periodic orbit, or even toward a cyclic orbit chain.

In view of Theorem 5.7 we obtain:

Corollary 6.13 *Let Φ be a flow in $\mathbf{R}^2$ with isolated equilibria. If X is a bounded asymptotic pseudotrajectory of Φ then $L(X)$ is a connected union of equilibria, periodic orbits and cyclic orbit chains of Φ.*

The following corollary can be seen as a Poincaré-Bendixson result for asymptotic pseudotrajectories:

Corollary 6.14 *Let Φ be a flow defined on $\mathbf{R}^2$, $K \subset \mathbf{R}^2$ a compact subset without equilibria, X an asymptotic pseudotrajectory of Φ. If there exists $T > 0$ such that $X(t) \in K$ for $t \geq T$, then $L(X)$ is either a periodic orbit or a cylinder of periodic orbits.*

Of course if $X(t)$ is an actual trajectory of Φ, the Poincaré-Bendixson theorem precludes a cylinder of periodic orbits. But this can easily occur for an asymptotic pseudotrajectory.

The next result extends Dulac's criterion for convergence in planar flows having negative divergence:

Theorem 6.15 *Let Φ be a flow in an open set in the plane, and assume that Φ_t decreases area for $t > 0$. Then:*

(a) $L(X)$ *is a connected set of equilibria which is nowhere dense and which does not separate the plane.*

(b) *If Φ has at most countably many stationary points, than $L(X)$ consists of a single stationary point.*

Proof The proof is contained in that of Theorem 1.6 of (Benaïm and Hirsch 1994); here is a sketch. The assumption that Φ decreases area implies that no invariant continuum can separate the plane. A generalization of the Poincaré-Bendixson theorem (Hirsch and Pugh, 1988) shows that an internally chain recurrent continuum (such as $L(X)$) which does not separate the plane consists entirely of stationary points. Simple topological arguments complete the proof. **QED**

Example 6.16 Consider the learning process described in section 2.3. Assume that the probability law of ξ_n is such that functions h^1, h^2 are smooth. Then the divergence of the vector field (6) at every point (x^1, x^2) is

$$Trace(DF(x^1, x^2)) = -2.$$

This implies that Φ_t decreases area for $t > 0$. Since the interpolated process of $\{x_n\}$ is almost surely an asymptotic pseudotrajectory of Φ (use Proposition 4.4), the results of Theorem 6.15 apply almost surely to the limit set of the sequence $\{x_n\}$.

For more details and examples of nonconvergence with more that two players see (Benaïm and Hirsch, 1994; Fudenberg and Levine, 1998).

7 Convergence with positive probability toward an attractor

Throughout this section X is a continuous time stochastic process defined on some probability space $(\Omega, \mathcal{F}, P)$ with continuous (or *càd làg*) paths taking value in M.

We suppose that $X(.)$ is adapted to a non-decreasing sequence of sub-σ algebras $\{\mathcal{F}_t : t \geq 0\}$ and that for all $\delta > 0$ and $T > 0$

$$P(\sup_{s \geq t} [\sup_{0 \leq h \leq T} d(X(s+h), \Phi_h(X(s)))] \geq \delta | \mathcal{F}_t) \leq w(t, \delta, T) \qquad (24)$$

for some function $w : \mathbb{R}_+^3 \to \mathbb{R}_+$ such that

$$\lim_{t \to \infty} w(t, \delta, T) \downarrow 0.$$

A sufficient condition for (24) is that

$$P(\sup_{0 \le h \le T} d(X(t+h), \Phi_h(X(t))) \ge \delta | \mathcal{F}_t) \le \int_t^{t+T} r(t, \delta, T) dt \qquad (25)$$

for some function $r : \mathbb{R}_+^3 \to \mathbb{R}_+$ such that

$$\int_0^\infty r(t, \delta, T) dt < \infty.$$

This last condition is satisfied by most examples of stochastic approximation processes (see section (4) and section (7.2) below).

Our goal is to give simple conditions ensuring that X converges with positive probability toward a given attractor. We develop here some ideas which originally appeared in Benaïm (1997) and Duflo (1997).

7.1 Attainable Sets

A point $p \in M$ is said to be *attainable* by X if for each $t > 0$ and every open neighborhood U of p

$$P(\exists s \ge t : X(s) \in U) > 0.$$

Lemma 7.1 *The set $Att(X)$ of attainable points by X is closed, positively invariant under Φ and contains almost surely $L(X)$.*

Proof A point p lies in $M \setminus Att(X)$ if there exists a neighborhood U of p and $t > 0$ such that

$$P(\forall s \ge t : X(s) \notin U) = 1.$$

Hence it is clear that $M \setminus Att(X)$ is an open set almost surely disjoint from $L(X)$.

It remains to prove that given any $p \in Att(X)$ and $T > 0$, $\Phi_T(p) \in Att(X)$. Fix $\epsilon > 0$. By continuity of Φ_T there exists $\alpha > 0$ such that $\Phi_T(B_\alpha(p)) \subset B_{\epsilon/2}(\Phi_T(p))$. Since $p \in Att(X)$ and X is continuous there exists a sequence of rational numbers $\{s_k\}_{k \in \mathbb{N}}$ with $s_k \to \infty$ such that $P(X(s_k) \in B_\alpha(p)) > 0$.

Choose k large enough so that $P(d(\Phi_T(X(s_k)), X(s_k + T)) \ge \epsilon/2)|\mathcal{F}_{s_k}) < 1/2$. Hence $P(X(s_k + T) \in B_\epsilon(\Phi_T(p))) > 0$. This proves that $\Phi_T(p) \in Att(X)$. **QED**

Example 7.2 Let X be the interpolated process associated to the urn process $\{x_n\}$ described in section 2.2. Suppose that the urn function f maps Δ^m into $Int(\Delta^m)$. Then it is not hard to verify that every point of Δ^m is attainable. The proof is left to the reader.

Theorem 7.3 *Suppose M is locally compact. Let $A \subset M$ be an attractor for Φ with basin of attraction $B(A)$. If $Att(X) \cap B(A) \ne \emptyset$ then*

$$P(L(X) \subset A) > 0.$$

Furthermore, if $U \subset M$ is an open set relatively compact with $\overline{U} \subset B(A)$ there exist numbers $T, \delta > 0$ (depending on U) so that

$$P(L(X) \subset A) \geq (1 - w(t, \delta, T))P(\exists s \geq t : X(s) \in U).$$

Proof Let U be an open set such that $K = \overline{U}$ is a compact subset of $B(A)$. To the compact set K we can associate the numbers $T > 0, \delta > 0$ given by Lemma (6.8).

Let $t > 0$ sufficiently large so that $w(t, \delta, T) < 1$. For $n \in \mathbb{N}$ and $k \in \mathbb{N}$ set $t_n(k) = \frac{k}{2^n}$ and let

$$\tau_n = \inf_{k \in \mathbb{N}} \{t_n(k) : X(t_n(k)) \in U \text{ and } t_n(k) \geq t\}.$$

By Lemma (6.8)

$$\{\tau_n < \infty\} \cap \{\sup_{s \geq \tau_n} d(X(s+T), \Phi_T(X(s)) \leq \delta\} \subset \{L(X) \subset A\}.$$

Hence

$$P(L(X) \subset A) \geq \sum_{k \geq [2^n t]+1} E[P(\sup_{s \geq t_n(k)} d(X(s+T), \Phi_T(X(s))) \leq \delta | \mathcal{F}_{t_n(k)}) 1_{\tau_n = t_n(k)}]$$

$$\geq \sum_{k \geq [2^n t]+1} (1 - w(t_n(k), \delta, T))P(\tau_n = t_n(k)) \geq (1 - w(t, \delta, T))P(\tau_n < \infty).$$

Since $\lim_{n \to \infty} P(\tau_n < \infty) = P(\exists s \geq t : X(s) \in U)$ we obtain

$$P(L(X) \subset A) \geq (1 - w(t, \delta, T))P(\exists s \geq t : X(s) \in U).$$

Now, to prove that $P(L(X) \subset A) > 0$ it suffices to choose for U a neighborhood of a point $p \in Att(X) \cap B(A)$. **QED**

7.2 Examples

Proposition 7.4 *Let $F : \mathbb{R}^m \to \mathbb{R}^m$ be a Lipschitz vector field. Consider the diffusion process*

$$dX = F(X)dt + \sqrt{\epsilon(t)}dB_t$$

where ϵ is a positive decreasing function such that for all $c > 0$

$$\int_0^\infty exp(-\frac{c}{\epsilon(t)})dt < \infty.$$

Then

(i) For each attractor $A \subset \mathbb{R}^m$ of F the event

$$\Omega_A = \{\lim_{t \to \infty} d(X(t), A) = 0\} = \{L(X) \subset A\}$$

has positive probability and for each open set U relatively compact with $\overline{U} \subset B(A)$

$$P(\Omega_A) \geq P(\exists s \geq t : X(s) \in U)(1 - \int_t^\infty C \exp(-\delta^2 C(T)/\epsilon(s))ds$$

with δ and T given by Lemma 6.8 and $C, C(T)$ are positive constant (depending on F.)

(ii) On Ω_A $L(X)$ is almost surely internally chain transitive.

(iii) If F is a dissipative vector field with global attractor Λ

$$P(\Omega_\Lambda) = 1 - P(\lim_{t \to \infty} \|X(t)\| = \infty) > 0.$$

Proof (*i*) follows from the fact that the law of $X(t)$ has positive density with respect to the Lebesgue measure. Hence $Att(X) = \mathbb{R}^m$ and Theorem 7.3 applies. The lower bound for $P(\Omega_A)$ follows from Theorem 7.3 combined with Proposition 4.6, (*iii*). Statement (*iii*) follows from Theorems 7.3 and 6.11. **QED**

Similarly we have

Proposition 7.5 *Let* $F : \mathbb{R}^m \to \mathbb{R}^m$ *be a Lipschitz bounded vector field. Consider a Robbins-Monro algorithm (7) satisfying the assumptions of Proposition 4.2 or 4.4. Then*

(i) *For each attractor* $A \subset \mathbb{R}^m$ *whose basin has nonempty intersection with* $Att(X)$ *the event*

$$\Omega_A = \{\lim_{t \to \infty} d(X(t), A) = 0\} = \{L(X) \subset A\}$$

has positive probability and for each open set U *relatively compact such that* $\overline{U} \subset B(A)$

$$P(\Omega_A) \geq P(\exists s \geq t : X(s) \in U))(1 - \int_t^\infty r(\delta, T, s)ds)$$

where

$$r(\delta, T, s) = C exp(\frac{-\delta^2 C(T)}{\overline{\gamma}(s)})$$

if $\{U_n\}$ *is subgaussian (Proposition 4.4) and*

$$r(\delta, T, s) = C'(T, q)(\frac{\sqrt{\overline{\gamma}(s)}}{\delta})^q$$

under the weaker assumptions given by Proposition 4.2 with δ *and* T *given by Lemma 6.8. Here* $C, C(T), C'(T, q)$ *denote positive constants.*

(ii) *On* Ω_A $L(X)$ *is almost surely internally chain transitive.*

(iii) *If* F *is a dissipative vector field with global attractor* Λ

$$P(\Omega_\Lambda) = (1 - P(\lim_{t \to \infty} \|X(t)\| = \infty)) > 0.$$

7.3 Stabilization

Most of the results given in the preceding sections assume a precompact asymptotic pseudotrajectory X for a semiflow Φ. Actually when X is not precompact the long term behavior of X usually presents little interest (See Corollary 6.11).

For stochastic approximation processes there are several *stability* conditions which ensure that the paths of the process are almost surely bounded. Such conditions can be found in numerous places such as Nevelson and Khasminskii (1976), Benveniste et al (1990), Delyon (1996), Duflo (1996, 1997), Fort and Pages (1996), Kushner and Yin (1997), to name just a few. We present here a theorem due to Kushner and Yin (1997, Theorem 4.3).

Theorem 7.6 *Let*

$$x_{n+1} - x_n = \gamma_{n+1}(F(x_n) + U_{n+1})$$

be a Robbins-Monro algorithm (section 4.2). Suppose that there exists a C^2 function $V : \mathbb{R}^m \to \mathbb{R}_+$ with bounded second derivatives and a nonnegative function $k : \mathbb{R}^m \to \mathbb{R}_+$ such that:

(i) $\langle \nabla V(x), F(x) \rangle \leq -k(x)$.

(ii) $\lim_{\|x\| \to \infty} V(x) = +\infty$.

(iii) *There are positive constants K, R such that*

$$E(\|U_{n+1}\|^2 + \|F(x_n)\|^2 | \mathcal{F}_n) \leq K k(x_n)$$

when $\|x_n\| \geq R$ and

$$E(\sum_n \gamma_{n+1}^2(\|U_{n+1}\|^2 + \|F(x_n)\|^2) \mathbf{1}_{\|x_n\| \leq R}) < \infty.$$

(iv) $E(k(x_n)) < \infty$ if $V(x_n) < \infty$ and $E(V(x_0)) < \infty$.

Then $\limsup_{n \to \infty} \|x_n\| < \infty$ with probability one.

Proof A second order Taylor expansion and boundedness of the second derivative of V implies the existence of some constant $K_1 > 0$ such that

$$E(V(x_{n+1}) - V(x_n)|\mathcal{F}_n) \leq -\gamma_{n+1}k(x_n) + \gamma_{n+1}^2 K_1 E(\|U_{n+1}\|^2 + \|F(x_n)\|^2)|\mathcal{F}_n).$$

The hypotheses then imply that $E(V(x_n)) < \infty$ for all n. Let

$$W_n = K_1 E(\sum_{i \geq n} \gamma_{i+1}^2(\|U_{i+1}\|^2 + \|F(x_i)\|^2)\mathbf{1}_{\|x_i\| \leq R})|\mathcal{F}_n)$$

and $V_n = V(x_n) + W_n$. V_n is nonnegative and

$$E(V_{n+1} - V_n|\mathcal{F}_n) \leq -k(x_n)\gamma_{n+1} + K_1 K k(x_n)\gamma_{n+1}^2.$$

Since $\gamma_n \to 0$ there exists $n_0 \geq 0$ such that $E(V_{n+1} - V_n | \mathcal{F}_n) \leq 0$ for $n \geq n_0$. Since $V_n \geq 0$ and $E(V_{n_0}) < \infty$ the supermartingale convergence theorem implies that $\{V_n\}$ converges with probability one toward some nonnegative L^1 random variable V. Since assumption (iii) implies that $W_n \to 0$ with probability one, $V(x_n) \to V$ with probability one. By assumption (ii) we then must have $\limsup_{n \to \infty} V(x_n) < \infty$ **QED**

8 Shadowing Properties

In this section we consider the following question:

Given a stochastic approximation process such as (7) (or more generally an asymptotic pseudotrajectory for a flow Φ) does there exist a point x such that the omega limit set of the trajectory $\{\Phi_t(x) : t \geq 0\}$ is $L(X)$?

The answer is generally negative and $L(X)$ can be an arbitrary chain transitive set. However it is useful to understand what kind of conditions ensure a positive answer to this question. A case of particular interest in applications is given by the following problem: Assume that each $\Phi-$ trajectory converges toward an equilibrium. Does X converge also toward an equilibrium ?

The material presented in this section is based on the works of Hirsch (1994), Benaïm (1996), Benaïm and Hirsch (1996), Duflo (1996) and Schreiber (1997).

We begin by a illustrative example borrowed from Benaïm (1996) and Duflo (1996).

Example 8.1 Consider the Robbins Monro algorithm given in polar coordinates (ρ, θ) by the system

$$\rho_{n+1} - \rho_n = \gamma_{n+1}\rho_n(h(\rho_n^2) + g(\rho_n)\xi_{n+1}),$$

$$\theta_{n+1} - \theta_n = \gamma_{n+1}((\rho_n sin(\theta_n))^2 + \xi_{n+1}).$$

where $\{\xi_n\}$ is a sequence of $i.i.d$ random variables uniformly distributed on $[-1, 1]$ and $\{\gamma_n\}$ satisfies the condition of Proposition 4.4. The function h is a smooth function such that $h(u) = 1 - u$ for $0 \leq u \leq 4$ and $-3 \leq h(u) \leq -4$ for $u \geq 4$, $g(\rho) = 1_{\{1/2 \leq \rho \leq 2\}}$, and $\gamma_n < 1/4$ for all n. These choices ensure that the algorithm is well defined (i.e $\rho_0 \geq 0$ implies $\rho_n \geq 0$ for all $n \geq 0$).

We suppose given $\rho_0 > 0$. It is then not hard to verify there exist some constants $0 < k(\rho_0) < K(\rho_0)$ such that $k(\rho_0) \leq \rho_n \leq K(\rho_0)$ for all $n \geq 0$.

Let $F : \mathbb{R}^2 \to \mathbb{R}^2$ be the vectorfield defined by

$$F(x, y) = (xh(x^2 + y^2) - y^3, \; yh(x^2 + y^2) + xy^2). \qquad (26)$$

Then $X_n = (x_n, y_n) = (\rho_n cos(\theta_n), \rho_n sin(\theta_n))$ satisfies a recursion of the form

$$X_{n+1} - X_n = \gamma_{n+1}(F(X_n) + U_{n+1}) + 0(\gamma_{n+1}^2)$$

where $\{U_n\}$ is a sequence of bounded random variables such that $E(U_{n+1}|\mathcal{F}_n) = 0$.

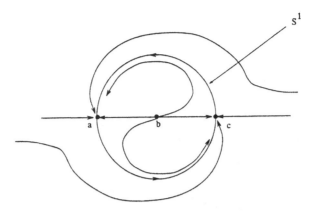

Figure 2: The phase portrait of $F(x, y) = (xh(x^2 + y^2) - y^3, yh(x^2 + y^2) + xy^2)$.

Let Φ be the flow induced by F (see figure 2). Equilibria of Φ are the points $a = (-1, 0)$, $b = (0, 0)$, $c = (0, 1)$, and every trajectory of Φ converges toward one of these equilibria. Internally chain transitive sets are the equilibria $\{a\}, \{b\}, \{c\}$ and the unit circle $S^1 = \{\rho = 1\}$ which is a cyclic orbit chain.

Since $\{(x_n, y_n)\}$ lives in some compact set disjoint from the origin, Theorem 5.7 combined with Proposition 4.4 and Remark 4.5 imply that the limit set of $\{(x_n, y_n)\}$ is almost surely one of the sets $\{a\}, \{c\}$ or S^1. We claim that if $\sum \gamma_n^2 = \infty$ then this limit set is almost surely S^1. Suppose on the contrary that $\{(x_n, y_n)\}$ converges toward one of the points a or c. Then $\lim_{n \to \infty} d(\theta_n, \pi\mathbb{Z}) = 0$ and since $|\theta_{n+1} - \theta_n| = 0(\gamma_{n+1})$ the sequence $\{\theta_n\}$ must converges. On the other hand by the law of iterated logarithm for martingales $\limsup_{n \to \infty} \sum_{i=1}^n \gamma_i \xi_i = \infty$. Thus

$$\limsup_{n \to \infty} \theta_n \geq \theta_0 + \limsup_{n \to \infty} \sum_{i=1}^n \gamma_i \xi_i = \infty.$$

A contradiction.

This example shows that the limiting behavior of a stochastic approximation process can be quite different from the limiting behavior of the associated ODE. We will show later (see Example 8.16) that $\{(x_n, y_n)\}$ actually converges toward one the points a or c provided that γ_n goes to zero "fast enough".

8.1 λ-Pseudotrajectories

Let X denote an asymptotic pseudotrajectory for a semiflow Φ on the metric space M. For $T > 0$ let

$$e(X, T) = \limsup_{t \to \infty} \frac{1}{t} \log(\sup_{0 \leq h \leq T} d(X(t + h), \Phi_h(X(t)))$$

and define the *asymptotic error rate of X* to be

$$e(X) = \sup_{T > 0} e(X, T).$$

If $e(X) \leq \lambda < 0$ we call X a $\lambda-pseudotrajectory$ of Φ.

The maps $\{\Phi_t\}$ are said to be Lipschitz, locally uniformly in $t > 0$ if for each $T > 0$ there exists $L(T) \geq 0$ such that

$$d(\Phi_h(x), \Phi_h(y)) \leq L(T)d(x, y)$$

for all $0 \leq h \leq T$, $x, y \in M$.

Lemma 8.2 *If the $\{\Phi_t\}$ are Lipschitz, locally uniformly in $t > 0$ then $e(X, T) = e(X)$ for all $T > 0$.*

Proof Let $T' > T > 0$. It is clear from the definition that $e(X, T) \leq e(X, T')$. Conversely, write $T' = kT + r$ with $k \in \mathbb{N}$ and $0 \leq r < T$ and set $D(X, R, t) = \sup_{0 \leq h \leq R} d(\Phi_h(X(t)), X(t + h))$. Then

$$D(X, T', t) \leq D(X, (k+1)T, t) \leq \sup\{D(X, kT, T), L(T)D(X, kT, t) + D(X, T, t+kT)\}.$$

Thus

$$e(X, T') \leq e(X, (k + 1)T) \leq \sup\{e(X, T), e(X, kT)\}.$$

Therefore $e(X, T') \leq e(X, (k + 1)T) \leq e(X, T)$. **QED**

Main Examples

Our main example of λ-pseudotrajectories is given by stochastic approximation processes whose step sizes go to zero at a "fast" rate:

Proposition 8.3 *Let $\{x_n\}$ given by (7) be a Robbins-Monro algorithm. Suppose*

$$l(\gamma) = \limsup_{n \to \infty} \frac{\log(\gamma_n)}{\tau_n} < 0.$$

Assume that F is a Lipschitz bounded vector field and that $\{U_n\}$ satisfies the assumptions of Proposition 4.2. Then X (the interpolated process) is almost surely a $\dfrac{l(\gamma)}{2}$-pseudotrajectory of Φ (the flow induced by F).

Proof Set $\lambda = l(\gamma)$ and let $0 < \epsilon < -\lambda$. For t large enough $\bar{\gamma}(t) \leq e^{(\lambda+\epsilon)t}$. Therefore, with $\Delta(t, T)$ given by (10) and $k \in \mathbb{N}$ large enough, equation (16) implies

$$E(\Delta(kT, T)^q) \leq TC(q, T) \exp \frac{kT(\lambda + \epsilon)q}{2}.$$

Let $\alpha > 0$ be such that $\frac{\lambda+\epsilon}{2} + \alpha < 0$. By Markov inequality

$$P(\Delta(kT, T) \geq e^{-kT\alpha}) \leq TC(q, T) \exp(qkT(\alpha + \frac{\lambda + \epsilon}{2})).$$

By Borel Cantelli Lemma this implies that

$$\limsup_{k \to \infty} \frac{1}{k} \log(\Delta(kT, T)) \leq -T\alpha$$

almost surely. Since α can be chosen arbitrary close to $-\lambda/2$ this proves that

$$\limsup_{k \to \infty} \frac{1}{kT} \log(\Delta(kT, T)) \leq \lambda/2$$

almost surely. Since $\Delta(t, T) \leq 2\Delta(kT, T) + \Delta((k+1)T, T)$ we get that

$$\limsup_{t \to \infty} \frac{1}{t} \log(\Delta(t, T)) \leq \lambda/2$$

almost surely and we conclude the proof by using inequality (11). **QED**

Remark 8.4 If $\gamma_n = f(n)$ for some positive decreasing function with $\int_1^\infty f(s)ds = \infty$, then

$$l(\gamma) = \limsup_{x \to \infty} \frac{\log(f(x))}{\int_1^x f(s)ds}.$$

For example, if

$$\gamma_n = \frac{A}{n^\alpha \log(n)^\beta}$$

then $l(\gamma) = 0$ for $0 < \alpha < 1$ and $\beta \geq 0$, $l(\gamma) = -1/A$ for $\alpha = 1$ and $\beta = 0$, and $l(\gamma) = -\infty$ for $\alpha = 1$ and $0 < \beta \leq 1$.

Similar to Proposition 8.3 is the next proposition whose proof is left to the reader:

Proposition 8.5 *Let X be the continuous time Markov process associated to the generator (21) and let*

$$\lambda = \limsup_{t \to \infty} \frac{\log(\epsilon(t))}{2t}.$$

If the vector field F given by (22) is Lipschitz continuous then X is almost surely a λ-pseudotrajectory of the flow induced by F.

Consider now the following situation. Suppose that X is a λ-pseudotrajectory whose limit set is contained in some compact positively invariant set K. Let $Y(t) \in K$ denote a point nearest to $X(t)$. It is not true in general that Y is a λ-pseudotrajectory for $\Phi|K$ but, for reasons that will be made clear later, it may be useful to know when this is true. The end of this section is devoted to this question.

Let $K \subset M$ be a compact positively invariant set for Φ and $B \subset M$ a set containing K. We say that K *attracts B exponentially at rate $\alpha < 0$* if there exists $C > 0$ such that

$$d(\phi_t(x), K) \leq Ce^{\alpha t}d(x, K)$$

for all $x \in B$ and $t \geq 0$.

Example 8.6 Suppose Φ is a C^1 flow on $\mathbb{R}^m$ and $\Gamma \subset \mathbb{R}^m$ is a periodic orbit of period $T > 0$. For any $p \in \Gamma$ let $\lambda_1 = e^{\mu_1}, \ldots, \lambda_m = e^{\mu_m}$ be the eigenvalues of $D\Phi_T(p)$ (counted with their multiplicities). The λ_i are are called the *characteristic (or Floquet) multipliers*. They are independent on p and the unity is always a Floquet multiplier (see Hartman (1964)).

Let $\alpha < 0$. If 1 has multiplicity 1 and the remaining $m-1$ Floquet multipliers are strictly inside the complex disk of center 0 and radius e^α then there exists a neighborhood B of Γ such that Γ attracts exponentially B at rate α/T. In this case Γ is called an *attracting hyperbolic periodic orbit*.

Lemma 8.7 Let $\alpha < 0, \lambda < 0$ and $\beta = \sup(\alpha, \lambda)$. Suppose $K \subset B$ attracts exponentially B at rate α. Let X be a λ-pseudotrajectory for Φ such that $X(t) \in B$ for all $t \geq 0$ and let $Y(t) \in K$ be a point nearest to $X(t)$. Then

(i)

$$\limsup_{t \to \infty} \frac{1}{t} \log d(X(t), Y(t)) \leq \beta.$$

(ii) *If the $\{\Phi_t\}$ are Lipschitz, locally uniformly in $t > 0$ then Y is a $\beta-$pseudotrajectory for $\Phi|K$*

Proof Choose $0 < \epsilon \leq -\beta$ and choose $T > 0$ large enough such that

$$d(\Phi_T(x), K) \leq e^{(\alpha+\epsilon)T} d(x, K)$$

for all $x \in B$. Thus there exists t_0 such that for $t \geq t_0$

$$d(X(t+T), K) \leq d(X(t+T), \Phi_T(X(t))) + d(\Phi_T(X(t)), K) \leq e^{(\lambda+\epsilon)t} + e^{(\alpha+\epsilon)T} d(X(t), K).$$

Let $v_k = d(X(kT), K)$, $\rho = e^{(\beta+\epsilon)T}$ and $k_0 = [t_0/T] + 1$. Then $v_{k+1} \leq \rho^k + \rho v_k$ for $k \geq k_0$. Hence

$$v_{k_0+m} \leq \rho^m (m\rho^{k_0-1} + v_{k_0})$$

for $m \geq 1$. It follows that

$$\limsup_{k \to \infty} \frac{\log(v_k)}{kT} \leq \frac{\log(\rho)}{T} = \beta + \epsilon.$$

Also for $kT \leq t \leq (k+1)T$ and $k \geq k_0$

$$d(X(t), K) \leq d(\Phi_{t-kT}(X(kT)), X(t)) + d(\Phi_{t-kT}(X(kT)), K)$$

$$\leq e^{(\lambda+\epsilon)kT} + Ce^{\alpha(t-kT)}v_k \leq e^{(\lambda+\epsilon)kT} + Cv_k.$$

Thus

$$\limsup_{t \to \infty} \frac{\log(d(X(t), K))}{t} \leq \beta + \epsilon$$

and since ϵ is arbitrary we get the desired result.

To obtain *(ii)* observe that

$$d(Y(t+h), \Phi_h(Y(t)) \leq d(\Phi_h(Y(t)), \Phi_h(X(t))) + d(\Phi_h(X(t)), X(t+h)) + d(X(t+h), Y(t+h))$$

Then for t large enough and $T > 0$

$$\sup_{0 \le h \le T} d(Y(t+h), \Phi_h(Y(t)) \le L(T)d(X(t), Y(t)) + e^{(\lambda+\epsilon)t} + \sup_{0 \le h \le T} d(X(t+h), Y(t+h)).$$

QED

8.2 Expansion Rate and Shadowing

From now on we assume that M is a Riemannian manifold and Φ a C^1 flow on M. The norm of a tangent vector v in the Riemannian metric is denoted by $||v||$. In our applications M will be a submanifold of $\mathbb{R}^m$ positively invariant under the flow generated by a smooth vector field.

Let $K \subset M$ denote a compact positively invariant set. The *expansion constant* of Φ_t at K (Hirsch, 1994) is the number

$$EC(\Phi_t, K) = \inf_{x \in K} m(D\Phi_t(x))$$

where

$$m(D\Phi_t(x)) = \inf\{||D\Phi_t(x)v|| : ||v|| = 1\}$$

denote the *minimal norm* of $D\Phi_t(x)$. Observe that since Φ is a flow then

$$m(D\Phi_t(x)) = ||D\Phi_t(x)^{-1}||^{-1} = ||D\Phi_{-t}(\Phi_t(x))||^{-1}.$$

The *expansion rate* of Φ at K is defined as

$$\mathcal{E}(\Phi, K) = \lim_{t \to \infty} \frac{1}{t} \log(EC(\Phi_t, K))$$

where the limit exists by a standard subadditivity argument whose verification is left to the reader.

Remark 8.8 It is important to understand that the expansion rate of Φ at K depends on the dynamics of Φ in M *and not only* in K. As a simple example illustrating this point, consider a smooth flow in $\mathbb{R}^m$ having a non-stationary periodic orbit Γ of period $T > 0$. Then it is not hard to see that $\mathcal{E}(\Phi, \Gamma)$ equals the smallest real part of the Floquet exponents of Γ divided by T. (this easily follows from Theorem 8.12.) If we now set $M = \Gamma$ and $\Psi = \Phi|\Gamma$ then $\mathcal{E}(\Psi, \Gamma) = 0$.

We now state a shadowing result due to Benaïm and Hirsch (1996) whose proof is an (easy) adaptation of Hirsch's shadowing theorem (Hirsch,1994).

Theorem 8.9 *Let $K \subset M$ be a compact positively invariant set. Let X be a λ-pseudotrajectory for Φ. Suppose*

(a) $L(X) \subset K$.

(b) $\lambda < \min\{0, \mathcal{E}(\Phi, K)\}$.

Then

(i) *There exists $r \geq 0$ and $x \in M$ such that*

$$\limsup_{t \to \infty} \frac{1}{t} \log d(X(t), \Phi_{t+r}(x)) \leq \lambda.$$

(ii) *Let x be as in (i). Suppose $l \geq 0$, $y \in M$ are such that*

$$\limsup_{t \to \infty} \frac{1}{t} \log d(X(t), \Phi_{t+l}(y)) \leq \lambda.$$

Then x and y are on the same orbit of Φ.

Proof Since $\lambda < \mathcal{E}(\Phi, K)$ we can choose $T > 0$ large enough so that

$$\min_{x \in K} m(D\Phi_T(x)) > e^{T\lambda}.$$

Set $f = \Phi_T$, $y_k = X(kT)$ and fix ν such that $e^{\lambda T} < \nu < \min_{x \in K} m(Df(x))$. Thus for k large enough

$$d(y_{k+1}, f(y_k)) \leq \nu^k. \tag{27}$$

By continuity of Df and compactness of K there exists a neighborhood U of K such that

$$\min_{x \in U} m(Df(x)) = \mu > \nu. \tag{28}$$

Claim: There exists a neighborhood $N \subset U$ of K and $\rho^* > 0$ such that

$$B(f(x), \rho\mu) \subset f(B(x, \rho))$$

for all $x \in N$ and $\rho \leq \rho^*$.

Proof of the claim: Choose a neighborhood U' of K, and $r > 0$, small enough such that for all $y \in U'$ and $d(y, z) \leq r$ there exists a C^1 curve $\gamma_{y,z} : [0, 1] \to U$ with the properties:

(i) $\gamma_{y,z}(0) = y$, $\gamma_{y,z}(1) = z$,

(ii) $\gamma_{y,z}([0, 1]) \subset U$,

(iii) $\int_0^1 \|\gamma'_{y,z}(s)\| ds = d(y, z)$.

Set $N = f^{-1}(U') \cap U$ and $\rho^* = \frac{r}{\mu}$. Let $x \in N, \rho \leq \rho^*$ and $d(z, f(x)) \leq \rho\mu$. Then

$$d(f^{-1}(z), x) = d(f^{-1}(z), f^{-1}(f(x))) \leq \int_0^1 \|Df^{-1}(\gamma_{f(x),z}(s))\gamma'_{f(x),z}(s)\| ds$$

$$\leq \frac{1}{\mu} \int_0^1 \|\gamma'_{f(x),z}(s)\| ds = \frac{1}{\mu} d(f(x), z) \leq \rho.$$

This proves the claim.

Since $L(X) \subset K$, $y_k \in N$ for k large enough. Let $\nu < \delta < \min\{1, \mu\}$. We claim that for k large enough

$$B(y_k, \delta^k) \subset f(B(y_{k-1}, \delta^{k-1})). \tag{29}$$

Indeed let $z \in B(y_k, \delta^k)$. Then (27) implies that for k large enough

$$d(z, f(y_{k-1})) \leq d(z, y_k) + d(f(y_{k-1}), y_k) \leq \delta^k + \nu^k$$

Since $\nu < \delta < \mu$, $\delta^k + \nu^{k-1} \leq \delta^{k-1}\mu$ for $k \geq \frac{\mu - \delta}{\log(\delta/\nu)} + 1$. Thus for k large enough, say $k \geq m$,

$$z \in B(f(y_{k-1}), \delta^{k-1}\mu) \subset f(B(y_{k-1}, \delta^{k-1})$$

where the last inclusion follows from the claim. This proves (29).

Set $B_k = B(y_k, \delta^k)$. For $n \geq m$ estimate (29) implies that

$$Q_n = \bigcap_{i \geq 0} f^{-i}(B_{n+i})$$

is a nonempty compact set. Also for $z \in Q_n$, $d(f^i(z), y_{n+i})) \leq \delta^{n+i}$. This proves statement (i) of the Theorem.

Since $\delta < \mu$ the claim shows that the diameter of $f^{-i}(B_{n+i})$ goes to zero as $i \to \infty$. This implies that $Q_n = \{z_n\}$ all $n \geq m$ where $z_n = \lim_{i\to\infty} f^{-i}(y_{n+i})$. This implies ($ii$). **QED**

Corollary 8.10 *Let Φ be a semiflow on M, $A \subset M$ a positively invariant submanifold of M and $K \subset A$ a compact positively invariant set. Let X be a λ-pseudotrajectory for Φ. Suppose that*

(a) $L(X) \subset K$.

(b) *There is a neighborhood of K (in M) which is attracted exponentially at rate $\alpha < 0$ by K.*

(c) *There is a C^1 flow Ψ on A such that $\Phi_t|A = \Psi_t$ for all $t \geq 0$.*

(d) $\beta = \sup(\alpha, \lambda) < \min\{0, \mathcal{E}(\Psi, K)\}$.

Then there exist $r \geq 0$ and $x \in A$ such that

$$\limsup_{t\to\infty} \frac{1}{t} \log d(X(t), \Phi_{t+r}(x)) \leq \beta.$$

Proof Let $Y(t) \in K$ be a point nearest to $X(t)$. By Lemma 8.7 (ii), Y is a β-pseudotrajectory for Ψ, and the result follows from Theorem 8.9 combined with estimate (i) of Lemma 8.7 **QED**

Example 8.11 As an illustration of Corollary 8.10 consider the diffusion on $\mathbb{R}^m$

$$dX = F(X)dt + \sqrt{\epsilon(t)}dB_t$$

where F and ϵ are as in Proposition 7.4.

Let $\Gamma \subset \mathbb{R}^m$ be an *attracting hyperbolic periodic orbit* of period $T > 0$. (see Example 8.6) such that the multipliers distinct of the unity have moduli $< e^\alpha$ for some $\alpha < 0$.

According to Proposition 7.4 the event $\Omega_\Gamma = \{L(X) \subset \Gamma\}$ has positive probability. If we furthermore assume that

$$\lambda = \limsup \frac{1}{2t} \log(\epsilon(t)) < 0$$

then Corollary 8.10 applied with $A = K = \Gamma$ and Proposition 8.5 imply that for almost every $\omega \in \Omega_\Gamma$ there exists $x(\omega) \in \Gamma$ such that

$$\limsup \frac{1}{t} \log(d(X(t), \Phi_t(x(\omega)))) \leq \sup(\lambda, \alpha/T).$$

8.3 Properties of the Expansion Rate

This section presents several useful estimates of the expansion rate. The key result is an ergodic characterization of the expansion rate due to Schreiber (1997).

We continue to assume that Φ is a C^1 flow on a Riemannian manifold M. Since we are concerned by the behavior of Φ restricted to a compact positively invariant set we furthermore assume without loss of generality that M is compact.

In order to present Schreiber's result we need to introduce a few notions of ergodic theory. We let $\mathcal{P}(M)$ denote the space of Borel probability measures on M with the topology of weak convergence[3]. A Borel probability measure $\mu \in \mathcal{P}(M)$ is said $\Phi-invariant$ or *invariant* under Φ if for every Borel set $A \subset M$ and every $t \in \mathbb{R}$

$$\mu(A) = \mu(\Phi_t(A)).$$

We let $\mathcal{M}(\Phi) \subset \mathcal{P}(M)$ denote the set of invariant measures. It is a nonempty compact convex subset of $\mathcal{P}(M)$ (Mañé, 1987, chapter 1). A measure $\mu \in \mathcal{M}(\Phi)$ is said $\Phi-ergodic$ if every Φ-invariant set has measure 0 or 1.

The *minimal center of attraction* of Φ is the set

$$MC(\Phi) = \overline{\bigcup_{\mu \in \mathcal{M}(\Phi)} supp(\mu)}.$$

The Birkhoff center of Φ is the set

$$BC(\Phi) = \overline{\{x \in M : x \in \omega(x)\}}.$$

By the Poincaré recurrence Theorem (Mañé, 1987, chapter 1)

$$MC(\Phi) \subset BC(\Phi). \tag{30}$$

---

[3] the one a functional analyst would call weak*

Let $\mu \in \mathcal{M}(\Phi)$. By the celebrated Oseledec's Theorem (see e.g Mañé 1987, chapter 11) there exists a Borel set $R \subset M$ of full measure ($\mu(R) = 1$) such that for all $x \in R$, there exist numbers $\lambda_1(x) < \ldots \lambda_{r(x)}(x)$ and a decomposition of $T_x M$ into

$$T_x M = E_1(x) \oplus \ldots \oplus E_{r(x)}(x)$$

such that for all $v \in E_i(x)$

$$\lim \frac{1}{t} \log \|D\Phi_t(x)v\| = \lambda_i(x).$$

The maps $x \to r(x), \lambda_i(x), E_i(x)$ are measurable.

When μ is ergodic there is an integer $r(\mu)$ and real numbers $\lambda_i(\mu)$, $i = 1, \ldots, r(\mu)$, such that $r(x) = r(\mu)$ and $\lambda_i(x) = \lambda_i(\mu)$ for μ almost every x. The $\lambda_i(\mu)$ are called the *Lyapounov exponents* of μ.

Schreiber (1997) proves the following result:

Theorem 8.12 *Let $K \subset M$ be a compact positively invariant set. Then*

$$\mathcal{E}(\Phi, K) = \inf_{\mu} \lambda_1(\mu)$$

where the infimum is taken over all ergodic measures with support in K.

Proof Let $f = \Phi_1$ and let $X = \{(x, v) : x \in K, v \in T_x M, \|v\| = 1.\}$ By compactness of X and definition of $\mathcal{E}(\Phi, K)$ there exists a sequence $(x_n, v_n) \in X$ such that

$$\mathcal{E}(\Phi, K) = \lim_{n \to \infty} \frac{1}{n} \log(EC(f^n, K)) = \lim_{n \to \infty} \frac{1}{n} \log(\|Df^n(x_n)v_n\|).$$

Define a map $G : X \to X$ by

$$G(x, v) = (f(x), \frac{Df(x)v}{\|Df(x)v\|}).$$

Let $\{\theta_n\}_{n \geq 0}$ be the sequence of probability measures defined on X by

$$\theta_n = \frac{1}{n} \sum_{i=0}^{n-1} \delta_{G^i(x_n, v_n)}.$$

By compactness of X we can always suppose (by replacing θ_n by some subsequence if necessary) that the sequence θ_n converges weakly toward some probability measure θ. Continuity of G easily implies that θ is G–invariant.

Let $h : X \to \mathbb{R}$ be the map defined by $h(x, v) = \log(\|Df(x)v\|)$. By the chain rule we get

$$\sum_{i=0}^{n-1} h(G^i(x, v)) = \log(\|Df^n(x)v\|).$$

Therefore

$$\int_X h d\theta = \lim_{n\to\infty} \int_X h d\theta_n = \lim_{n\to\infty} \frac{1}{n} \log(\|Df^n(x_n)v_n\|) = \mathcal{E}(\Phi, K).$$

Now, by the ergodic decomposition theorem (see Mañé, 1987 chapter 6 Theorem 6.4)

$$\int_X h d\theta = \int_X (\int_X h d\theta_{(x,v)}) d\theta$$

where $\theta_{(x,v)}$ are ergodic G-invariant probability measures. It follows that for each $\epsilon > 0$ there exists an ergodic G-invariant measure $\nu = \theta_{(x,v)}$ for some (x, v), such that $\int_X h d\nu \leq \mathcal{E}(\Phi, K) + \epsilon$. Birkhoff's ergodic theorem implies the existence of a Borel set $X' \subset X$ such that $\nu(X') = 1$ and

$$\lim_{n\to\infty} \frac{1}{n} \log(\|Df^n(x)v\|) = \lim_{n\to\infty} \frac{1}{n} \sum_{i=0}^{n-1} h(G^i(x, v)) = \int_X h d\nu \leq \mathcal{E}(\Phi, K) + \epsilon \quad (31)$$

for all $(x, v) \in X'$. Let μ be the marginal probability measure defined on K by

$$\mu(B) = \nu\{(x, v) \in X : x \in B\}.$$

Clearly μ is f-invariant and ergodic. On the other hand by Oseledec's Theorem for μ almost all $x \in K$ and all $v \in T_x M$ the limit $\lim_{n\to\infty} \frac{1}{n} \log(\|Df^n(x)v\|)$ exists and satisfies

$$\lambda_1(\mu) \leq \lim_{n\to\infty} \frac{1}{n} \log(\|Df^n(x)v\|) \quad (32)$$

Therefore from (31) and (32) we deduce

$$\inf_{\mu \, ergodic} \lambda_1(\mu) \leq \mathcal{E}(\Phi, K).$$

To prove the converse inequality let μ be any ergodic measure. By Oseledec's Theorem for μ almost all $x \in K$ and all $v \in E_1(x)$:

$$\lambda_1(\mu) = \lim_{n\to\infty} \frac{1}{n} \log(\|Df^n(x)v\|) \geq \mathcal{E}(\Phi, K)$$

QED

Corollary 8.13

$$\mathcal{E}(\Phi, K) = \mathcal{E}(\Phi, MC(\Phi|K)) = \mathcal{E}(\Phi, BC(\Phi|K)).$$

Proof The first equality follows from Theorem 8.12 and the second from Poincaré recurrence theorem (equation 30). **QED**

Theorem 8.12 and Corollary 8.13 can be used to estimate the Expansion rate. Here are two such estimates.

Corollary 8.14 *Assume that $M = \mathbb{R}^m$, Φ is generated by a smooth vector field F and $MC(\Phi|K) \subset Eq(\Phi)$ (the equilibria set). Then*

$$\mathcal{E}(\Phi, K) = \inf\{\lambda_1(p) : p \in Eq(\Phi) \cap K\}$$

where $\lambda_1(p)$ denote the smallest real part of the eigenvalues of the jacobian matrix $DF(p)$.

Proof Under the assumption that $MC(\Phi|K) \subset Eq(\Phi)$ every ergodic measure with support in K has to be a Dirac measure at an equilibrium point. Let $\mu = \delta_p$ be such a measure. Then $\lambda_1(\mu) = \lambda_1(p)$ and the result follows from Theorem 8.12. **QED**

The following result is based on Hirsch (1994).

Corollary 8.15 *Assume $M = \mathbb{R}^m$ and Φ is generated by a smooth vector field F. Let $\beta(x)$ equals the smallest eigenvalue of the matrix $\frac{1}{2}(DF(x) + DF(x)^T)$ where T denotes matrix transpose. Then*

$$\mathcal{E}(\Phi, K) \geq \inf_{x \in MC(\Phi|K)} \beta(x).$$

Proof Let $y \in MC(\Phi|K)$. By invariance of $MC(\Phi|K)$ $\Phi_t(y) \in MC(\Phi|K)$ for all $t \in \mathbb{R}$. The variational equation along orbits of the reversed time flow Φ_{-t} gives

$$\frac{d}{dt}D\Phi_{-t}(y) = -DF(\Phi_{-t}(y))D\Phi_{-t}(y)$$

Therefore for every nonzero vector $v \in \mathbb{R}^m$ and $t \geq 0$ we have,

$$\frac{d}{dt}||D\Phi_{-t}(y)v||^2 = -2\langle D\Phi_{-t}(y)v, DF(\Phi_{-t}(y))D\Phi_{-t}(y)v\rangle \leq -2\beta||D\Phi_{-t}(y)v||^2$$

with $\beta = \inf_{x \in MC(\Phi|K)} \beta(x)$. Therefore

$$||D\Phi_{-t}(y)|| \leq e^{-t\beta}$$

for all $y \in MC(\Phi|K)$ and $t \geq 0$. To conclude set $y = \Phi_t(x)$ for $x \in MC(\Phi|K)$ and the estimate follows from the definition of $\mathcal{E}(\Phi, K)$ combined with Corollary 8.13. **QED**

Example 8.16 Let $(x_n, y_n) \in \mathbb{R}^2$ be the Robbins Monro algorithm described in Example 8.1. It is convenient here to express the dynamics of the vector field (26) in polar coordinates. That is

$$\frac{d\rho}{dt} = \rho h(\rho^2), \; \frac{d\theta}{dt} = (\rho \sin \theta)^2.$$

Let $B_\epsilon = \{|1 - \rho| \leq \epsilon\}$. For $\epsilon << 1$ and $(\rho, \theta) \in B_\epsilon$

$$\frac{d(1 - \rho)^2}{dt} = -2(1 - \rho)\rho(1 - \rho^2) \leq -2(1 - \rho)^2(2 - \epsilon)(1 - \epsilon).$$

Thus $|1 - \rho(t)| \leq e^{-(2-\epsilon)(1-\epsilon)t}|1 - \rho|$. This shows that S^1 attract B_ϵ at rate $-2 + O(\epsilon)$.

To compute $\mathcal{E}(\Phi|S^1, S^1)$ we use Corollary 8.14. The dynamics on S^1 being given by $\dfrac{d\theta}{dt} = sin^2(\theta)$, 0 is the eigenvalue of the linearized ODE at equilibria points 0 and π. Thus $\mathcal{E}(\Phi|S^1, S^1) = 0$. Suppose now that

$$l(\gamma) = \limsup_{n \to \infty} \frac{\log(\gamma_n)}{\tau_n} < 0.$$

Then Corollary 8.10 and Proposition 8.3 imply that $\{x_n, y_n\}$ converges almost surely toward one of the points a or c of Figure 2.

9 Nonconvergence to Unstable points, Periodic orbits and Normally Hyperbolic Sets

Let $\{x_n\}$ be given by (7) and F a smooth vector field. Let $p \in \mathbb{R}^m$ be an equilibrium of F; that is $F(p) = 0$. As usual, if all eigenvalues of $DF(p)$ have nonzero real parts, p is called *hyperbolic*. If all eigenvalues of $DF(p)$ have negative real parts, p is *linearly stable*. If some eigenvalue has positive real part, p is *linearly unstable*.

Suppose p is a hyperbolic equilibrium of F which is linearly unstable. Then the set of initial values whose forward trajectories converge to p — the stable manifold $W_s(p)$ of p — is the image of an injective C^1 immersion $\mathbb{R}^k \to \mathbb{R}^m$ where $0 \leq k < m$. Consequently $W_s(p)$ has measure 0 in $\mathbb{R}^m$. This suggests that for the stochastic process (7), convergence of sample paths $\{x_n\}$ to p is a null event, provided the noise $\{U_n\}$ has sufficiently large components in the unstable directions at p. Such a result has been proved recently by Pemantle (1990) and Brandière and Duflo (1996) (under different sets of assumptions), provided the vector fields F is C^2 and the gain sequence is well behaved.

The ideas of Pemantle have been used in Benaïm and Hirsch (1995b) to tackle the case of hyperbolic unstable periodic orbits. This section is an extension of these results which covers a larger class of repelling sets. Recent works by Brandière (1996,1997) address similar questions and prove the nonconvergence of stochastic approximation processes toward certain types of repelling sets which are not considered here.

Throughout this section we assume given

- A smooth vector field $F : \mathbb{R}^m \to \mathbb{R}^m$ generating a flow $\Phi = \{\Phi_t\}$.

- A smooth $(m - d)$ dimensional (embedded) submanifold $S \subset \mathbb{R}^m$, where $d \in \{1, \ldots, m\}$.

- A nonempty compact set $\Gamma \subset S$ invariant under Φ.

We assume that S is *locally invariant*, meaning that there exists a neighborhood U of Γ (in $\mathbb{R}^m$) and a positive time t_0 such that

$$\Phi_t(U \cap S) \subset S$$

for all $|t| \leq t_0$.

We further assume that for every point $p \in \Gamma$,

$$\mathbb{R}^m = T_p S \oplus E_p^u$$

where

(i) $p \to E_p^u$ is a continuous map from Γ into the Grassman manifold $G(d, m)$ of d planes in $\mathbb{R}^m$.

(ii) $D\Phi_t(p)E_p^u = E_{\Phi_t(p)}^u$ for all $t \in \mathbb{R}, p \in E_p^u$.

(iii) There exist $\lambda > 0$ and $C > 0$ such that for all $p \in \Gamma$, $w \in E_p^u$ and $t \geq 0$

$$\|D\Phi_t(p)w\| \geq Ce^{\lambda t}\|w\|.$$

Examples

Linearly Unstable Equilibria: Suppose $\Gamma = \{p\}$ where $p \in \mathbb{R}^m$ is a linearly unstable equilibrium of F. Then $\mathbb{R}^m = E_p^s \oplus E_p^c \oplus E_p^u$ where E_p^s, E_p^c and E_p^u are the generalized eigenspaces of $DF(p)$ corresponding to eigenvalues with real parts < 0, equal to 0 and > 0. Because p is linearly unstable, the dimension of E_p^u is at least 1.

Using stable manifold theory (see e.g Shub, (1987) or Robinson, (1995)) there exists a locally invariant manifold S tangent to $E_p^s \oplus E_p^c$ - the *center stable manifold* of p - which is C^k when F is C^k.

Since $D\Phi_t(p) = e^{tDF(p)}$ there exist $\lambda > 0$ and $C > 0$ such that $\|D\Phi_t(p)w\| \geq Ce^{\lambda t}\|w\|$ for all $w \in E_p^u$.

Linearly Unstable Periodic Orbits: Let $\Gamma \subset \mathbb{R}^m$ be a periodic orbit. Γ is said to be hyperbolic if the unity is a multiplier with multiplicity one and the $m - 1$ other multipliers have moduli different from 1. Γ is said to be linearly unstable if some multiplier has modulus strictly greater than 1.

Suppose Γ is a hyperbolic linearly unstable periodic orbit for the C^k vector field F. By hyperbolicity (see for example Shub 1987) there exist positive constants C, λ and a decomposition of $T_\Gamma \mathbb{R}^m$ as the direct sum of three vector bundles:

$$T_\Gamma \mathbb{R}^m = E^s(\Gamma) \oplus E^u(\Gamma) \oplus E^\Phi(\Gamma)$$

which is invariant under $T_\Gamma \Phi$, and such that for all $p \in \Gamma$, $t \geq 0$:

$$\| D\Phi_t(p)|_{E_p^s} \| \leq Ce^{-\lambda t}, \tag{33}$$

$$\| D\Phi_{-t}(p)|_{E_p^u} \| \leq Ce^{-\lambda t},$$

$$E_p^\Phi = \text{span}(F(p))$$

where $p \times E_p^u$ denotes the fibre of $E^u(\Gamma)$ over p, and similar notation applies to E_p^s and E_p^Φ.

Because Γ is linearly unstable, the dimension of $E^u(p)$ is at least 1.

Each E_p^u is a linear subspace of $\mathbb{R}^m$. The map $p \mapsto E_p^u$ is a continuous map from Γ into the Grassmann manifold of linear subspaces of the appropriate dimension (it is actually C^k due to the fact that $T_\Gamma \Phi_t$ maps $p \times E_p^u$ to $\Phi_t(p) \times E_{\Phi_t(p)}^u$, and that $T_\Gamma \Phi$ is a C^k flow).

For $p \in \Gamma$ and sufficiently small $\epsilon > 0$, the *local stable manifold of p* is defined to be the set:

$$W_\epsilon^s(p) = \{x \in \mathbb{R}^m : \forall t \geq 0 \, ||\Phi_t(x) - \Phi_t(p)|| \leq \epsilon, \text{ and } \lim_{t \to \infty} ||\Phi_t(x) - \Phi_t(p))|| = 0\}.$$

Using stable manifold theory, we take ϵ small enough so that $W_\epsilon^s(p)$ is a C^k (embedded) submanifold of $\mathbb{R}^m$, whose tangent space at p is $T_p W_\epsilon^s(p) = E_p^s$.

We set $S = \bigcup_{p \in \Gamma} W_\epsilon^s(p)$. This is the *local stable manifold of Γ*. It follows from the above that S is a C^k locally invariant submanifold, with $T_p S = E_p^s \oplus E_p^\Phi$.

Let $0 < \alpha \leq 1$. We call a map $C^{1+\alpha}$ if it is C^1 and its derivative is αHölder. Note that C^{1+1} is weaker than C^2. A $C^{1+\alpha}$ manifold is a manifold whose transition functions may be chosen $C^{1+\alpha}$.

Theorem 9.1 *Let $\{x_n\}$ given by (7) be a Robbins Monro algorithm (section 4.2) Assume:*

(i) *There exists $K > 0$ such that $||U_n|| \leq K$ for all $n \geq 0$.*

(ii) *$\{\gamma_n\}$ is as in Proposition (4.4).*

(iii) *There exists a neighborhood $\mathcal{N}(\Gamma)$ of Γ and $b > 0$ such that for all unit vector $v \in \mathbb{R}^m$*

$$\mathbf{E}(\langle U_{n+1}, v \rangle^+ | \mathcal{F}_n) \geq b 1_{\{x_n \in \mathcal{N}(\Gamma)\}}.$$

(iv) *There exists $0 < \alpha \leq 1$ such that:*

    (a) *F and S are $C^{1+\alpha}$,*

    (b)

$$\lim_{n \to \infty} \frac{\gamma_{n+1}^\alpha}{\sqrt{\lim_{m \to \infty} \sum_{i=n+1}^m \gamma_i^2}} = 0.$$

Then

$$\mathbf{P}(\lim_{n \to \infty} d(x_n, \Gamma) = 0) = 0.$$

Remark 9.2 If $\frac{A}{n^\beta} \leq \gamma_n \leq \frac{B}{n^\beta}$ with $\beta > 1/2$, $0 < A \leq B$ then condition (iv), (b) of Theorem 9.1 is fulfilled provided that

$$\alpha > \frac{2\beta - 1}{2}.$$

If $\sum_n \gamma_n^2 = \infty$ condition (iv) of Theorem 9.1 is always satisfied for $\alpha > 0$.

9.1 Proof of Theorem 9.1

The proof of this result relies, on one hand, upon the construction of a suitable Lyapounov function, and on the other hand on probabilistic estimates due to Pemantle (1990).

Construction of a suitable Lyapounov function

The construction given here is very similar to the construction given in Benaïm and Hirsch (1995b), but instead of defining the Lyapounov function as the distance to S in the unstable direction for an adapted Riemann metric obtained by time averaging (as in Benaïm and Hirsch (1995b)), we consider the usual distance and we then define the Lyapounov function by averaging over time.

It appears that this construction leads to much easier estimates and allows us to handle the fact that the splitting $T_\Gamma \mathbb{R}^m = T_\Gamma S \oplus E^u$ is only continuous.

step1 The first step of the construction is to replace the continuous invariant splitting $T_\Gamma \mathbb{R}^m = T_\Gamma S \oplus E^u$ by a smooth (noninvariant) splitting $T_\Gamma \mathbb{R}^m = T_\Gamma S \oplus \tilde{E}^u$ close enough to the first one to control the expansion of $D\Phi_t$ along fibers of $\tilde{E}^u$.

Choose $T > 0$ large enough so that for $f = \Phi_T$, $p \in \Gamma$, and $w \in E_p^u$:

$$\|Df(p)w\| \geq 5\|w\|.$$

By Whitney embedding theorem (Hirsch, 1976, chapter 1) we can embed $G(d, m)$ into $\mathbb{R}^D$ for some $D \in \mathbb{N}$ large enough so that we can see $p \to E_p^u$ as a map from $\Gamma \to \mathbb{R}^D$. Thus by Tietze extension theorem (Munkres, 1975) we can extend this map to a continuous map from $\mathbb{R}^m$ into $\mathbb{R}^D$. Let ϱ denote a C^∞ retraction from a neighborhood of $G(d, m) \subset \mathbb{R}^D$ onto $G(d, m)$ whose existence follows from a classical result in differential topology (Hirsch, 1976, chapter 4). By composing the extension of $p \to E_p^u$ with ϱ we obtain a continuous map defined on a neighborhood N of Γ, taking values in $G(d, m)$ and which extends $p \to E_p^u$. To shorten notation, we keep the notation $p \in N \to E_p^u \in G(d, m)$ to denote this new map.

By replacing N by a smaller neighborhood if necessary, we can further assume that N is compact, $N \subset U$, $f(N) \subset U$ and

$$\|Df(p)w\| \geq 4\|w\|$$

for all $p \in N$, and $w \in E_p^u$.

Now, by a standard approximation procedure, we can approximate $p \in N \to E_p^u \in G(d, m)$ by a C^∞ function from N into $\mathbb{R}^D$. Then, by composing with ϱ, we obtain a C^∞ map $p \in N \to \tilde{E}_p^u \in G(d, m)$ which can be chosen arbitrary close to $p \in N \to E_p^u$ in the C^0 topology.

For $p \in N$, let

$$P_p : T_p S \oplus E_p^u \to T_p S,$$

$$u + v \to u$$

and let

$$\tilde{P}_p : T_pS \oplus \tilde{E}_p^u \to T_pS,$$

$$u + v \to u.$$

Fix $\epsilon > 0$ small enough so for all $p \in N$, $\epsilon(||Df(p)||+4) < 1$ and $\epsilon||Df(p)||(||P_p||+ ||\tilde{P}_p|| + 1)) < 1$ (this choice will be clarified in the next lemma). From now on, we will assume that the map $p \in N \to \tilde{E}^u{}_p \in G(d, m)$ is chosen such that for all $p \in N \cap S$:

(i) $\mathbb{R}^m = T_pS \oplus \tilde{E}^u{}_p$

(ii) The projector $\tilde{P}_p : T_pS \oplus \tilde{E}^u{}_p \to T_pS$, satisfies

$$||P_p - \tilde{P}_p|| \le \epsilon.$$

Let $\tilde{E}^u = \{(p, v) \in S \cap N \times \mathbb{R}^m : v \in \tilde{E}_p^u\}$. Since S is $C^{1+\alpha}$, $\tilde{E}^u$ is a $C^{1+\alpha}$ vector bundle over $S \cap N$. Let $H : \tilde{E}^u \to \mathbb{R}^m$ be the map defined by $H(p, v) = p + v$. It is easy to see that the tangent map of H at a point $(p, 0)$ is invertible. The inverse function theorem implies that H is a local $C^{1+\alpha}$ diffeomorphism at each point of the zero section of $\tilde{E}$. Since H maps the zero section to $S \cap N$ by the diffeomorphism $(p, 0) \to p$, it follows that H restricts to a $C^{1+\alpha}$ diffeomorphism $H : N_0' \to N_0$ between open neighborhoods N_0' or the zero section and $N_0 \subset N$ of $S \cap N$. We now define the maps

$$\Pi : N_0 \to S,$$

$$x \to \Pi(x) = p$$

for $H^{-1}(x) = (p, v)$ and

$$V : N_0 \to \mathbb{R}_+,$$

$$x \to ||x - \Pi(x)||.$$

Observe that for any $p \in N_0 \cap S$

$$D\Pi(p) = \tilde{P}_p.$$

Step 2 The second step consists in the construction of Lyapounov function which is zero on S and increases exponentially along trajectories outside S. This function (see Proposition 9.5) is obtained from V by some averaging procedure.

Lemma 9.3 *There exists a neighborhood of Γ, $N_1 \subset N_0$, and $\rho > 1$ such that for all $x \in N_1$*

$$V(f(x)) \ge \rho V(x).$$

To prove this Lemma we use the following estimates

Lemma 9.4 *Let $P, \tilde{P} : \mathbb{R}^m \to \mathbb{R}^m$ be two projectors and $A : \mathbb{R}^m \to \mathbb{R}^m$ a linear map. Assume there exist $\epsilon, \alpha > 0$ such that*

(i) $||P - \tilde{P}|| \le \epsilon$,

(ii) $||Au|| \ge \alpha||u||$ *for all* $u \in KerP$.

Then

(i) $||Av|| \ge (\alpha(1 - \epsilon) - \epsilon||A||)||v||$ *for all* $v \in Ker\tilde{P}$.

(ii) $||\tilde{P}A(Id - \tilde{P}) - PA(Id - P)|| \le \epsilon||A||(1 + ||P|| + ||\tilde{P}||)$.

Proof Let $v \in Ker\tilde{P}$ be a unit vector. Write $v = (v - Pv) + (P - \tilde{P})v$. Thus

$$||Av|| \ge \alpha||v - Pv|| - \epsilon||A|| \ge \alpha(1 - \epsilon) - \epsilon||A||.$$

This proves (i) while (ii) follows from

$$||\tilde{P}A(Id - \tilde{P}) - PA(Id - P)|| = ||(P - \tilde{P})A(Id - \tilde{P}) - PA(P - \tilde{P})||$$

$$\le \epsilon||A||(||Id - \tilde{P}|| + ||P||).$$

QED

Proof of Lemma 9.3

let $x \in N_0 \cap f^{-1}(N_0)$ and set $p = \Pi(x)$.

$$V(f(x)) \ge ||f(x) - f(p)|| - ||f(p) - \Pi(f(x))|| = ||Df(p)(x - p)||$$

$$- ||D\Pi(f(p))Df(p)(x - p)|| + o(||x - p||).$$

Lemma 9.4 (i) applied to $Df(p), P_p, \tilde{P}_p$ and our choice for ϵ imply

$$||Df(p)(x - p)|| \ge 3||x - p|| = 3V(x)$$

Also, by Lemma 9.4 (ii)

$$||\tilde{P}_{f(p)}Df(p)(Id - \tilde{P}_p) - P_{f(p)}Df(p)(Id - P_p)|| \le \epsilon||Df(p)||(1 + ||P_p|| + ||\tilde{P}_{f(p)}||) < 1.$$

Since for $p \in \Gamma$, $P_{f(p)}Df(p)(Id - P_p) = 0$ the preceding inequality implies the existence of a neighborhood of Γ, $N_1 \subset N_0$ such that for $x \in N_1$ and $p = \Pi(x)$: $||\tilde{P}_{f(p)}Df(p)(Id - \tilde{P}_p)|| < 1$. This implies

$$||D\Pi(f(p))Df(p)(x - p)|| < ||x - p|| = V(x).$$

It follows that $V(f(x)) \ge 3V(x) - V(x) + o(V(x))$. Replacing N_1 by a smaller neighborhood gives

$$V(f(x)) \ge \rho V(x)$$

for some $\rho > 1$. **QED**

Recall that a map $\eta : \mathbb{R}^m \to \mathbb{R}$ is said to have a right derivative at point x if for all $h \in \mathbb{R}^m$ the limit

$$D\eta(x).h = \lim_{t \to 0, t > 0} \frac{\eta(x + th) - \eta(x)}{t}$$

exists. If η is differentiable at x, then $D\eta(x).h = \langle \nabla\eta(x), h \rangle$ where $\nabla\eta(x) \in \mathbb{R}^m$ is the usual gradient.

Proposition 9.5 *There exists a compact neighborhood of* Γ, $\mathcal{N}(\Gamma) \subset N_1$ *and real numbers* $l > 0, \beta > 0$ *such that the map* $\eta : \mathcal{N}(\Gamma) \to \mathbb{R}$ *given by*

$$\eta(x) = \int_0^l V(\Phi_{-t}(x))dt$$

enjoys the following properties:

(i) η *is* C^r *on* $\mathcal{N}(\Gamma) \setminus S$.

(ii) *For all* $x \in \mathcal{N}(\Gamma) \cap S$, η *admits a right derivative* $D\eta(x) : \mathbb{R}^m \to \mathbb{R}^m$ *which is Lipschitz, convex and positively homogeneous.*

(iii) *If* $r \geq 1 + \alpha$ *for some* $0 < \alpha \leq 1$ *there exists* $k > 0$ *and a neighborhood* $U \subset \mathbb{R}^m$ *of* 0 *such that for all* $x \in \mathcal{N}(\Gamma)$ *and* $v \in U$

$$\eta(x + v) \geq \eta(x) + D\eta(x)v - k||v||^{1+\alpha}.$$

(iv) *There exists* $c_1 > 0$ *such that for all* $x \in \mathcal{N}(\Gamma) \setminus S$

$$||\nabla\eta(x)|| \geq c_1$$

and for all $x \in \mathcal{N}(\Gamma) \cap S$ *and* $v \in \mathbb{R}^m$

$$D\eta(x)v \geq c_1||v - D\Pi(x)v||.$$

(v) *For all* $x \in \mathcal{N}(\Gamma) \cap S$, $u \in T_x S$ *and* $v \in \mathbb{R}^m$

$$D\eta(x)(u + v) = D\eta(x)v$$

(vi) *For all* $x \in \mathcal{N}(\Gamma)$
$$D\eta(x).F(x) \geq \beta\eta(x),$$

Proof *Notation:* Given $A > 0$ we let $N^A \subset N_1$ denote a compact neighborhood of Γ such that for all $|t| \leq A$, $\Phi_t(N^A) \subset N_1$ and we let $C(A) > 0$ denote the Lipschitz constant of the map $t, x \to V(\Phi_t(x))$ restricted to $[-A, A] \times N^A$. Remark that for $|t| \leq A$ and $x \in N^A$ we have:

$$V(\Phi_t(x)) = |V(\Phi_t(x)) - V(\Phi_t(\Pi(x)))| \leq C(A)V(x). \tag{34}$$

We first fix $l > 2T$ and assume that $\mathcal{N}(\Gamma) \subset N^l$. We will see below (in proving (*vi*)) how to choose l.

    (*i*) is obvious.

    (*ii*) follows from the fact that Π is C^1, and $x \to ||x||$ admits a right derivative at the origin of $\mathbb{R}^m$ given as $h \to ||h||$.

    Before passing to the proof of (*iii*) let us compute $D\eta(x)$. For $x \in N^l$ let

$$G_t(x) = \Phi_{-t}(x) - \Pi(\Phi_{-t}(x)),$$

$$B(t, x) = DG_t(x) = [Id - D\Pi(\Phi_{-t}(x))]D\Phi_{-t}(x),$$

and for $x \in N^l \setminus S$ let

$$b(x) = \frac{x - \Pi(x)}{||x - \Pi(x)||}.$$

It is easy to verify that

$$D\eta(x).h = \int_0^l \langle B(t, x)h, b(\Phi_{-t}(x)) \rangle dt \tag{35}$$

for $x \in N^l \setminus S$ and

$$D\eta(x).h = \int_0^l ||B(t, x)h|| dt \tag{36}$$

for $x \in N^l \cap S$.

(iii) If $r \geq 1 + \alpha$, Φ and Π are C^1 with α Hölder derivatives. Hence there exits $k > 0$ such that

$$||G_t(x + u)|| - ||G_t(x)|| \geq ||G_t(x) + B(t, x)u|| - ||G_t(x)|| - k||u||^{1+\alpha}.$$

If $x \in N^l \cap S$ then $||G_t(x)|| = 0$ and the result follows from (36).
If $x \in N^l \setminus S$, convexity of the norm implies

$$||G_t(x) + B(t, x)u|| - ||G_t(x)|| \geq \frac{\langle G_t(x), B(t, x)u \rangle}{||G_t(x)||} = \langle B(t, x)u, b(\Phi_{-t}(x)) \rangle$$

and the result follows from (35).

(iv). *Claim:* There exists $c_0 > 0$ such that $||B(t, p)b|| \geq c_0$ for all $0 \leq t \leq l$, $p \in N^l \cap S$ and unit vector $b \in \tilde{E}_p^u$.

Proof of the claim: Suppose the contrary. Then by compactness of $\{(t, p, v) : 0 \leq t \leq l, p \in N^l \cap S, b \in \tilde{E}_p^u, ||b|| = 1\}$ there exists $0 \leq t \leq l, p \in N^l \cap S$ and a unit vector $b \in \tilde{E}_p^u$ such that $B(t, p)b = 0$. Therefore $D\Phi_{-t}(p)b \in Ker(Id - D\Pi(\Phi_{-t}(p)) = T_{\Phi_{-t}(p)}S$. Thus $b \in D\Phi_{-t}(p)^{-1}T_{\Phi_{-t}(p)}S = T_pS$. But this is impossible because b is a unit vector in $\tilde{E}_p^u$ and $\mathbb{R}^m = T_pS \oplus \tilde{E}_p^u$. This proves the claim.

For $x \in N^l \setminus S$

$$\Phi_{-t}(x) - \Pi(\Phi_{-t}(x)) = B(t, \Pi(x))(x - \Pi(x)) + o(||x - \Pi(x)||)$$

and for any $h \in \mathbb{R}^m$ with $||h|| = 1$,

$$\langle B(t, x)h, \Phi_{-t}(x) - \Pi(\Phi_{-t}(x)) \rangle = \langle B(t, \Pi(x))h, B(t, \Pi(x))(x - \Pi(x)) \rangle + o(||x - \Pi(x)||).$$

Thus, if we set $h = b(x)$ we get

$$\langle B(t, x)h, b(\Phi_{-t}(x)) \rangle = \frac{||B(t, \Pi(x)).b(x)||^2 + \epsilon(||x - \Pi(x)||)}{||B(t, \Pi(x)).b(x)|| + \epsilon_1(||x - \Pi(x)||)}$$

where $\lim_{u \to 0} \epsilon(u) = \lim_{u \to 0} \epsilon_1(u) = 0$. Since, according to the claim,

$$||B(t, \Pi(x)).b(x)|| >$$

c_0 this implies

$$\langle B(t, x)h, b(\Phi_{-t}(x)) \rangle \geq c_0/2$$

for $||x - \Pi(x)||$ small enough. Formulae (35) implies that $D\eta(x)b(x) \geq c_1$ with $c_1 = lc_0/2$. Hence $||\nabla \eta(x)|| \geq c_1$.

Suppose now $x \in \mathcal{N}(\Gamma) \cap S$. It follows from (iii) that $B(t, x)v = B(t, x)(v - D\Pi(p)v)$ for all $v \in \mathbb{R}^m$. Now the claim together with (36) imply that $D\eta(x)v \geq c_1||v - D\Pi(x)v||$.

(vi) For $x \in N^l$ and $0 \leq t \leq l$ we can write $t = kT + r$ for $k \in \mathbb{N}$ and $0 \leq r < T$. Thus, by Lemma 9.3 and equation (34),

$$V(\Phi_t(x)) = V(f^k(\Phi_r(x)) \geq \rho^k V(\Phi_r(x)) \geq \frac{\rho^k}{C(T)} V(x) \geq C_1(T)e^{at}V(x)$$

where $C_1(T) = \dfrac{1}{\rho C(T)}$ and $a = \dfrac{\log(\rho)}{T}$.

For $s > 0$,

$$\eta(\Phi_s(x)) - \eta(x) = -\int_{l-s}^{l} V(\Phi_{-t}(x))dt + \int_{-s}^{0} V(\Phi_{-t}(x))dt \geq -V(x)\frac{e^{-al}e^{as}}{C_1(T)}s$$

$$+ \int_0^s V(\Phi_{-t}(x))dt.$$

It follows that

$$\lim_{s \to 0, s > 0} \frac{\eta(\Phi_s(x)) - \eta(x)}{s} \geq V(x)(1 - \frac{e^{-al}}{C_1(T)})$$

It then suffices to choose l large enough so that $1 - \dfrac{e^{-al}}{C_1(T)} > 0$. With this choice of l we get that

$$D\eta(x).F(x) \geq \beta\eta(x)$$

with $\beta = \dfrac{1}{lC(l)}(1 - \dfrac{e^{-al}}{C_1(T)}) > 0$.   **QED**

Probabilistic Estimates

The following lemma adapted from Pemantle (1992, Lemma 5.5) is the probabilistic key of the proof of Theorem 9.1.

Lemma 9.6 *Let $\{S_n\}$ be a nonnegative stochastic process, $S_n = S_0 + \sum_{i=1}^{n} X_i$ where X_n is $\mathcal{F}_n$ measurable. Let $\{\gamma_n\}$ be a sequence of positive numbers such that $\sum_n \gamma_n^2 < \infty$ and let $\alpha_n = \sum_{i=n+1}^{\infty} \gamma_i^2$.*

Assume there exist a sequence $0 \leq \epsilon_n = o(\sqrt{\alpha_n})$, constants $a_1 > 0, a_2 > 0$ and an integer N_0 such that for all $n \geq N_0$:

(i) $| X_n | = o(\sqrt{\alpha_n})$.

(ii) $1_{\{S_n > \epsilon_n\}} E(X_{n+1}|\mathcal{F}_n) \geq 0$.

(iii) $\mathsf{E}(S_{n+1}^2 - S_n^2|\mathcal{F}_n) \geq a_1 \gamma_{n+1}^2$.

(iv) $E(X_{n+1}^2|\mathcal{F}_n) \leq a_2 \gamma_{n+1}^2$.

Then $\mathsf{P}(\lim_{n \to \infty} S_n = 0) = 0$.

This lemma is stated and proved in (Pemantle, 1992) for $\gamma_n = 1/n$ and but the proof adapts without difficulty to the present situation.

Proof Assume without loss of generality that $N_0 = 0$, $|X_n| \leq \sqrt{b_1 \alpha_n}$ and $\epsilon_n < \frac{1}{2}\sqrt{b_2 \alpha_n}$ where

$$2(b_1 + b_2) < a_1.$$

Given $n \in \mathbb{N}$ let T be the stopping time defined as

$$T = \inf\{i \geq n : S_i \geq \sqrt{b_2 \alpha_i}\}$$

Claim:

$$P(T < \infty|\mathcal{F}_n) \geq 1 - \frac{2(b_1 + b_2)}{a_1} \qquad (37)$$

Proof of (37): By assumption (iii), the process $Z_k = S_k^2 - a_1 \sum_{i=0}^{k} \gamma_i^2$ is a submartingale. Therefore $\{Z_{k \wedge T}\}_{k \geq n}$ is a submartingale and for all $m \geq n$ $E(Z_{m \wedge T} - Z_n|\mathcal{F}_n) \geq 0$. Hence

$$E(S_{m \wedge T}^2 - S_n^2|\mathcal{F}_n) \geq a_1 E\Big(\sum_{i=n+1}^{m \wedge T} \gamma_i^2|\mathcal{F}_n\Big) \geq a_1 \Big(\sum_{i=n+1}^{m} \gamma_i^2\Big) P(T > m|\mathcal{F}_n).$$

On the other hand

$$S_{m \wedge T}^2 - S_n^2 \leq 2(b_1 + b_2)\alpha_n$$

by definition of T and condition (i). It follows that

$$P(T > m|\mathcal{F}_n) \leq \frac{2(b_1 + b_2)\alpha_n}{a_1 \sum_{i=n+1}^{m} \gamma_i^2}.$$

Letting $m \to \infty$ proves the claim.

Now, let σ be the stopping time defined as

$$\sigma = \inf\{i \geq n : S_i < \frac{1}{2}\sqrt{b_2 \alpha_n}\}.$$

Claim: Let E_n be the event $E_n = \{S_n \geq \sqrt{b_2 \alpha_n}\}$. Then

$$P(\sigma = \infty|\mathcal{F}_n)1_{E_n} \geq \Big(\frac{b_2}{4a_2 + b_2}\Big)1_{E_n} \qquad (38)$$

Proof of (38): The process $\{S_{i \wedge \sigma}\}_{i \geq n}$ is a submartingale. Indeed

$$E(S_{i+1 \wedge \sigma} - S_{i \wedge \sigma}|\mathcal{F}_i) = 1_{\{\sigma > i\}} E(S_{i+1} - S_i|\mathcal{F}_i) \geq 1_{\{\sigma \geq i\}} 1_{\{S_i \geq \frac{1}{2}\sqrt{b_2 \alpha_i}\}} E(X_{i+1}|\mathcal{F}_i)$$

where the last term is nonnegative by condition (ii). Therefore by Doob's decomposition Lemma there exist a martingale $\{M_i\}_{i \geq n}$ and a previsible process $\{I_i\}_{i \geq n}$ such that $S_{i \wedge \sigma} = M_i + I_i$, $I_n = 0$ and $I_{i+1} \geq I_i$. The fact that $S_{i \wedge \sigma} \geq M_i$ implies

$$P(\sigma = \infty | \mathcal{F}_n) \geq P(\forall i \geq n : M_i \geq \frac{1}{2}\sqrt{b_2 \alpha_n} | \mathcal{F}_n).$$

Thus

$$P(\sigma = \infty | \mathcal{F}_n) 1_{E_n} \geq P(\forall i \geq n : M_i - M_n \geq -\frac{1}{2}\sqrt{b_2 \alpha_n} | \mathcal{F}_n) 1_{E_n} \qquad (39)$$

Our next goal is to estimate the right hand term of (39). Set $M'_i = M_i - M_n$. For $i \geq n$:

$$E(M'^2_i | \mathcal{F}_n) = \sum_{j=n}^{i-1} E((M_{j+1} - M_j)^2 | \mathcal{F}_n) \leq a_2 \alpha_n \qquad (40)$$

where we have used the fact that

$$E((M_{j+1} - M_j)^2 | \mathcal{F}_j) = E((S_{j+1} - S_j)^2 | \mathcal{F}_j) - (I_{j+1} - I_j)^2 \leq a_2 \gamma^2_{j+1}$$

by condition (iv). Therefore for $s > 0$, $m \geq n$ and $t > 0$

$$P(\inf_{n \leq i \leq m} M'_i < -s | \mathcal{F}_n) \leq P(\inf_{n \leq i \leq m}(M'_i - t) < -s - t | \mathcal{F}_n) \leq P(\sup_{n \leq i \leq m} |M'_i - t| \geq s + t | \mathcal{F}_n)$$

$$\leq \frac{E(M'^2_m | \mathcal{F}_n) + t^2}{(s+t)^2} \leq \frac{a_2 \alpha_n + t^2}{(s+t)^2}$$

where the last two inequalities follow from Doob's inequality combined with (40). With $s = \frac{1}{2}\sqrt{b_2 \alpha_n}$ and $t = \frac{a_2 \alpha_n}{s}$ we get that

$$P(\inf_{n \leq i \leq m} M'_i < -s | \mathcal{F}_n) \leq \frac{4a_2}{4a_2 + b_2}.$$

Thus

$$P(\forall i \geq n : M_i - M_n \geq -\frac{1}{2}\sqrt{b_2 \alpha_n} | \mathcal{F}_n) \geq 1 - \frac{4a_2}{4a_2 + b_2}.$$

This proves (38).

We can now finish the proof of the Lemma. Let G denote the event that $\{S_n\}$ does not converge to zero. By definition of T and inequality (38):

$$E(1_G | \mathcal{F}_i) 1_{T=i} = E(1_G | \mathcal{F}_i) 1_{E_i} 1_{T=i} \geq \frac{b_2}{4a_2 + b_2} 1_{E_i} 1_{T=i} = \frac{b_2}{4a_2 + b_2} 1_{T=i}$$

for all $i \geq n$. Therefore

$$E(1_G | \mathcal{F}_n) \geq \sum_{i \geq n} E(1_G 1_{T=i} | \mathcal{F}_n) = \sum_{i \geq n} E(E(1_G | \mathcal{F}_i) 1_{T=i} | \mathcal{F}_n)$$

$$\geq \frac{b_2}{4a_2 + b_2} P(T < \infty | \mathcal{F}_n) \geq \frac{b_2}{4a_2 + b_2}(1 - \frac{2(b_1 + b_2)}{a_1}) > 0$$

where the last inequality follows from (37). Since $\lim_{n\to\infty} E(1_G | \mathcal{F}_n) = 1_G$ almost surely this proves that $1_G = 1$ almost surely. **QED**

If $\sum_n \gamma_n^2 = \infty$ we use the next lemma:

Lemma 9.7 *Let $\{S_n\}$ be a stochastic process, $S_n = S_0 + \sum_{i=1}^n X_i$ where X_n is $\mathcal{F}_n$ measurable and $|X_n| \leq C$. Let $\{\gamma_n\}$ be such that $\sum_i \gamma_i^2 = \infty$. Assume there exists $a_1 > 0$ and some integer N_0 such that for all $n \geq N_0$ $E(S_{n+1}^2 - S_n^2 | \mathcal{F}_n) \geq a_1 \gamma_{n+1}^2$.*

Then $P(\lim_{n\to\infty} S_n = 0) = 0$.

Proof. As already noticed $Z_n = S_n^2 - \sum_{i=0}^n a_1 \gamma_i^2$ is a submartingale.

Suppose $P(\lim_{n\to\infty} S_n = 0) > 0$. Then for all $\epsilon > 0$ there exists $N \geq N_0$ such that $P(\bigcap_{n\geq N}\{| S_n | \leq \epsilon\}) > 0$.

Assume $| S_N | \leq \epsilon$ and define the stopping time $T = \inf\{k \geq N; | S_k | > \epsilon\}$. The sequence $\{(Z_{n\wedge T}, \mathcal{F}_n)\}_{n\geq N}$ is a submartingale and we have $Z_{n\wedge T} \leq (\epsilon + C)^2$. It follows from the submartingale convergence theorem that $\{Z_{n\wedge T}\}_{n\geq N}$ converges almost surely. Thus $\{\sum_{i=0}^{n\wedge T} a_1 \gamma_i^2\}_{n\geq N}$ is almost surely bounded. This implies $T < \infty$ almost surely. **QED**

We now prove Theorem 9.1.

Let $N \in \mathbb{N}$. Assume $x_N \in \mathcal{N}(\Gamma)$ where $\mathcal{N}(\Gamma)$ is the neighborhood given by Proposition 9.5. Let T be the stopping time defined by

$$T = \inf\{k \geq N; x_n \notin \mathcal{N}(\Gamma)\}.$$

We prove Theorem 9.1 by showing that $P(T < \infty) = 1$.

Without loss of generality we assume $N = 0$. (The proof is the same for any N).

Define two sequences of random variables $\{X_n\}_{n\geq 1}$ and $\{S_n\}$ as follows:

$$X_{n+1} = [\eta(x_{n+1}) - \eta(x_n)]1_{\{n\leq T\}} + \gamma_{n+1}1_{\{n>T\}},$$

$$S_0 = \eta(x_0), \quad S_n = S_0 + \sum_{i=1}^n X_i.$$

The process $\{S_n\}$ is clearly nonnegative. Notice that if $T = \infty$ then $X_{n+1} = \eta(x_{n+1}) - \eta(x_n)$ and S_n telescopes into $S_n = \eta(x_n)$. This will be used at the end of the proof.

We now suppose that $\sum \gamma_i^2 < \infty$ and verify that hypotheses (i) to (iv) of Lemma 9.6 are satisfied.

Conditions (i) and (iv). By Lipschitz continuity of η, and the boundedness of the sequences $\{F(x_n)\}$, $\{U_n\}$ we have $|X_{n+1}| = O(\gamma_{n+1}) = o(\sqrt{\alpha_n})$.

Condition (ii). Let $k' = k(\|F\| + K)$ where k is given by Proposition 9.5, (iii), $\|F\| = \sup\{F(x); x \in \mathcal{N}(\Gamma)\}$ and K is the uniform bound of the U_n. If $n \leq T$, using Proposition 9.5, $(ii), (iii), (v)$ and (vi) we have

$$\eta(x_{n+1}) - \eta(x_n) \geq \gamma_{n+1}\beta\eta(x_n) + \gamma_{n+1}D\eta(x_n)U_{n+1} - k'\gamma_{n+1}^{1+\alpha}. \tag{41}$$

Thus

$$1_{\{n \leq T\}}\mathsf{E}(X_{n+1}|\mathcal{F}_n) \geq 1_{\{n \leq T\}}\left[(\gamma_{n+1}\beta\eta(x_n) - k'\gamma_{n+1}^{1+\alpha}) + \gamma_{n+1}\mathsf{E}(D\eta(x_n)U_{n+1}|\mathcal{F}_n)\right].$$

By convexity of the right derivative of η (Proposition 9.5, (ii)) and the conditional Jensen inequality we have

$$\mathsf{E}(D\eta(x_n)U_{n+1}|\mathcal{F}_n) \geq D\eta(x_n)\mathsf{E}(U_{n+1}|\mathcal{F}_n) = 0.$$

Thus

$$1_{\{n \leq T\}}\mathsf{E}(X_{n+1}|\mathcal{F}_n) \geq 1_{\{n \leq T\}}(\gamma_{n+1}\beta\eta(x_n) - k'\gamma_{n+1}^{1+\alpha}) \tag{42}$$

If $n > T$, $X_{n+1} = \gamma_{n+1}$, so

$$1_{\{n > T\}}\mathsf{E}(X_{n+1}|\mathcal{F}_n) \geq 1_{\{n > T\}}\gamma_{n+1} \geq 0 \tag{43}$$

Putting (42) and (43) together and letting $\epsilon_n = \dfrac{k'}{\beta}\gamma_{n+1}^\alpha$ proves condition (ii) of Lemma 9.6.

For Condition (iii) of Lemma 9.6, we observe that

$$\mathsf{E}(S_{n+1}^2 - S_n^2|\mathcal{F}_n) = \mathsf{E}(X_{n+1}^2|\mathcal{F}_n) + 2S_n\mathsf{E}(X_{n+1}|\mathcal{F}_n).$$

If $S_n \geq \epsilon_n$, the right hand term is nonnegative by condition (ii), previously proved. If $S_n < \epsilon_n$, (42) and (43) imply $S_n\mathsf{E}(X_{n+1}|\mathcal{F}_n) \geq -\epsilon_n k'\gamma_{n+1}^{1+\alpha} = -O(\gamma_{n+1}^{1+2\alpha})$. Thus

$$\mathsf{E}(S_{n+1}^2 - S_n^2|\mathcal{F}_n) \geq \mathsf{E}(X_{n+1}^2|\mathcal{F}_n) - O(\gamma_{n+1}^{1+2\alpha}).$$

Therefore, to prove condition (iii) of Lemma 9.6, it suffices to show that

$$\mathsf{E}(X_{n+1}^2|\mathcal{F}_n) \geq b_1\gamma_{n+1}^2$$

for some $b_1 > 0$ and n large enough. From (41) we deduce

$$1_{\{n \leq T\}}\left(\mathsf{E}(X_{n+1}^+|\mathcal{F}_n) - [\gamma_{n+1}\mathsf{E}((D\eta(x_n)U_{n+1})^+|\mathcal{F}_n) - k'\gamma_{n+1}^{1+\alpha}]\right) \geq 0 \tag{44}$$

Using Proposition 9.5, (iv) and assumption (iii) of Theorem 9.1 we see that

$$1_{\{n \leq T\} \cap \{x_n \notin S\}}(\mathsf{E}((D\eta(x_n)U_{n+1})^+|\mathcal{F}_n) - c_1 b) \geq 0. \tag{45}$$

If $x_n \in S$, choose a unit vector $v_n \in \mathrm{Ker}(\mathrm{Id} - D\Pi(x_n))^\perp$. We have

$$< U_{n+1}, v_n > = < U_{n+1} - D\Pi(x_n)U_{n+1}, v_n > .$$

Let $\mathcal{A}$ denotes the event $\mathcal{A} = \{n \leq T\} \cap \{x_n \in S\}$. By using Proposition 9.5 (iv), the Cauchy-Schwartz inequality and assumption (iii) of Theorem 9.1 we obtain

$$
\begin{aligned}
1_{\mathcal{A}} \mathsf{E}((D\eta(x_n)U_{n+1})^+ | \mathcal{F}_n) &\geq \\
c_1 1_{\mathcal{A}} \mathsf{E}(\|U_{n+1} - D\Pi(x_n)U_{n+1}\| \, | \mathcal{F}_n) &\geq \\
c_1 1_{\mathcal{A}} \mathsf{E}(\langle U_{n+1} - D\Pi(x_n)U_{n+1}, v_n \rangle^+ | \mathcal{F}_n) &= \\
c_1 1_{\mathcal{A}} \mathsf{E}(\langle U_{n+1}, v_n \rangle^+ | \mathcal{F}_n) &\geq \quad c_1 b 1_{\mathcal{A}}.
\end{aligned}
\tag{46}
$$

Putting (44), (45), (46) together and (43) give

$$
\mathsf{E}(X_{n+1}^+ | \mathcal{F}_n) \geq \gamma_{n+1} c_1 b - k' \gamma_{n+1}^{1+\alpha}
$$

On the other hand $\mathsf{E}(X_{n+1}^2 | \mathcal{F}_n) \geq \mathsf{E}(X_{n+1}^+ | \mathcal{F}_n)^2$ by the Jensen inequality. It follows that $\mathsf{E}(X_{n+1}^2 | \mathcal{F}_n) \geq b_1 \gamma_{n+1}^2$ for $b_1 > 0$ and n large enough, as is desired.

Condition (i) through (iv) of Lemma 9.6 being satisfied, *the probability is zero that* $\{S_n\}$ *converges to zero*, according to Lemma 9.6. If $\sum \gamma_i^2 = \infty$ the proof given here also shows that conditions of Lemma 9.7 are satisfied.

Now suppose $T = \infty$. Then $\eta(x_n) = S_n$ and $\{x_n\}$ remains in $\mathcal{N}(\Gamma)$. Therefore (by Theorem 5.7) $L(\{x_n\})$ (the limit set of $\{x_n\}$) is a nonempty compact invariant subset of $\mathcal{N}(\Gamma)$, so that for all $y \in L(\{x_n\})$ and $t \in \mathbb{R}$ $\Phi_t(y) \in \mathcal{N}(\Gamma)$. By condition (vi) or Proposition (9.5) this implies that $\eta(\Phi_t(y)) \geq e^{\beta t} \eta(y)$ for all $t > 0$ forcing $\eta(y)$ to be zero. Thus $L(\{x_n\}) \subset S$. This implies $S_n = \eta(x_n) \to 0$. Since $\mathsf{P}(S_n \to 0) = 0$, T is almost surely finite. **QED**

10 Weak Asymptotic Pseudotrajectories

In the previous sections we have been mainly concerned with the asymptotic behavior of stochastic approximations processes with "fast" decreasing stepsizes, typically

$$
\gamma_n = o(\frac{1}{\log(n)})
$$

(Proposition 4.4) or

$$
\gamma_n = 0(n^{-\alpha}), \alpha \leq 1
$$

(Proposition 4.2).

If the step-sizes go to zero at a slower rate we cannot expect to characterize precisely the limit sets of the process[4]. However it is always possible to describe the "ergodic" or statistical behavior of the process in term of the corresponding behavior for the associated deterministic system. This is the goal of this section which is mainly based on Benaïm and Schreiber (1997). It is worth mentioning

---

[4] For instance, with a step-size of the order of $1/log(n)$ it is easy to construct examples for which the process never converges even though the chain recurrent set of the ODE consists of isolated equilibria.

that Fort and Pages (1997) in a recent paper largely generalize results of this section and address several interesting questions which are not considered here.

Let $(\Omega, \mathcal{F}, P)$ be a probability space and $\{\mathcal{F}_t : t \geq 0\}$ a nondecreasing family of sub-σ-algebras. Let (M, d) be a **separable** metric space equipped with its Borel σ-algebra.

A process

$$X : \mathbb{R}_+ \times \Omega \to M$$

$$(t, \omega) \to X(t, \omega)$$

is said to be a *weak asymptotic pseudotrajectory* of the semiflow Φ if

(i) It is *progressively measurable*: $X|[0,T] \times \Omega$ is $\mathcal{B}_{[0,T]} \times \mathcal{F}_T$ measurable for all $T > 0$ where $\mathcal{B}_{[0,T]}$ denotes the Borel $\sigma-$ field over $[0,T]$.

(ii)

$$\lim_{t \to \infty} P\{ \sup_{0 \leq h \leq T} d(X(t+h), \Phi_h(X(t))) \geq \alpha \,|\mathcal{F}_t\} = 0$$

almost surely for each $\alpha > 0$ and $T > 0$.

Recall (see section 8.3) that $\mathcal{P}(M)$ denotes the space of Borel probability measures on M with the topology of weak convergence and $\mathcal{M}(\Phi)(\subset \mathcal{P}(M))$ denotes the set of $\Phi-$invariant measures.

Let $\mu_t(\omega)$ denote the (random) occupation measure of the process :

$$\mu_t(\omega) = \frac{1}{t} \int_0^t \delta_{X(s,\omega)} ds$$

and let $\mathcal{M}(X, \omega)$ denote weak limit points of $\{\mu_t(\omega)\}$. The set $\mathcal{M}(X, \omega)$ is a (possibly empty) subset of $\mathcal{P}(M)$. However if $\{\mu_t(\omega)\}$ is tight (for example if $t \to X(t, \omega)$ is precompact) then by the Prohorov theorem $\mathcal{M}(X, \omega)$ is a nonempty compact subset of $\mathcal{P}(\mathcal{M})$.

Theorem 10.1 *Let X be a weak asymptotic pseudotrajectory of Φ. There exists a set $\tilde{\Omega} \subset \Omega$ of full measure $(P(\tilde{\Omega}) = 1)$ such that for all $\omega \in \tilde{\Omega}$*

$$\mathcal{M}(X, \omega) \subset \mathcal{M}(\Phi).$$

Proof

Let $f : M \to [0,1]$ be a uniformly continuous function and $T > 0$. For $n \geq 1$ set

$$U_n(f, T) = \int_{(n-1)T}^{nT} f(X(x(s))) ds,$$

$$M_n(f, T) = \sum_{i=1}^{n} \frac{1}{i} [U_i(f, T) - E(U_i(f, T)|\mathcal{F}_{(i-1)T})],$$

and

$$N_n(f, T) = \sum_{i=2}^{n+1} \frac{1}{i} [E(U_i(f, T)|\mathcal{F}_{(i-1)T}) - E(U_i(f, T)|\mathcal{F}_{(i-2)T})].$$

The processes $\{M_n(f,T)\}_{n\geq 1}$ and $\{N_n(f,T)\}_{n\geq 1}$ are martingales with respect to the filtration $\{\mathcal{F}_{nT} : n \geq 1\}$. Since

$$\sup_n E(M_n(f,T)^2) \leq 4T^2 \sum_i \frac{1}{i^2},$$

Doob's convergence theorem implies that $\{M_n(f,T)\}_{n\geq 1}$ converges almost surely. Hence, by Kronecker lemma,

$$\lim_{n\to\infty} \frac{1}{n} \sum_{i=1}^n [U_i(f,T) - E(U_i(f,T)|\mathcal{F}_{(i-1)T})] = 0 \tag{47}$$

almost surely. Similar reasoning with $\{N_n(f,T)\}$ leads to

$$\lim_{n\to\infty} \frac{1}{n} \sum_{i=2}^{n+1} [E(U_i(f,T)|\mathcal{F}_{(i-1)T}) - E(U_i(f,T)|\mathcal{F}_{(i-2)T})] = 0 \tag{48}$$

almost surely. Since $|U_i(f,T)| \leq T$, (47) implies

$$\lim_{n\to\infty} \frac{1}{n} \sum_{i=2}^{n+1} [U_i(f,T) - E(U_i(f,T)|\mathcal{F}_{(i-1)T})] = 0 \tag{49}$$

and by adding (48) and (49) we obtain

$$\lim_{n\to\infty} \frac{1}{n} \sum_{i=1}^n [U_{i+1}(f,T) - E(U_{i+1}(f,T)|\mathcal{F}_{(i-1)T})] = 0 \tag{50}$$

almost surely.

We claim that

$$\lim_{i\to\infty} E(U_{i+1}(f,T) - U_i(f \circ \Phi_T, T)|\mathcal{F}_{(i-1)T}) = 0 \tag{51}$$

almost surely. Let $\epsilon > 0$. By uniform continuity of f there exists $\alpha > 0$ such that $d(x,y) \leq \alpha$ implies $|f(x) - f(y)| \leq \epsilon$. Hence

$$|E(U_{i+1}(f,T) - U_i(f\circ\Phi_T, T)|\mathcal{F}_{(i-1)T})| \leq E\{\int_{(i-1)T}^{iT} |f(X(s+T)) - (f\circ\Phi_T)(X(s))|ds|\mathcal{F}_{(i-1)T}\}$$

$$\leq 2TP\{\sup_{(i-1)T\leq s\leq iT} |X(s+T) - \Phi_T(X(s))| \geq \alpha|\mathcal{F}_{(i-1)T}\} + T\epsilon.$$

Since X is a weak asymptotic pseudotrajectory of Φ the first term in the right of the inequality goes to zero almost surely as $i \to \infty$ and since ϵ is arbitrary, this proves the claim.

Now, write

$$U_{i+1}(f,T) - U_i(f \circ \Phi_T) = [U_{i+1}(f,T) - E(U_{i+1}(f,T)|\mathcal{F}_{(i-1)T})] +$$

$[E(U_{i+1}(f,T)|\mathcal{F}_{(i-1)T}) - E(U_i(f \circ \Phi_T, T)|\mathcal{F}_{(i-1)T})] + [E(U_i(f \circ \Phi_T, T)|\mathcal{F}_{(i-1)T})$
$- U_i(f \circ \Phi_T)].$

Then use equations (50), (51), and equation (47) with $f \circ \Phi_T$ in lieu of f. It follows that there exists a set $\Omega(f,T) \subset \Omega$ of full measure such that for all $\omega \in \Omega(f,T)$

$$\lim_{n \to \infty} \frac{1}{n} \sum_{i=1}^{n} U_{i+1}(f,T) - \frac{1}{n} \sum_{i=1}^{n} U_i(f \circ \Phi_T, T) = 0. \tag{52}$$

Since (M, d) is a separable metric space it admits a metric $\tilde{d}$ inducing the same topology as d such that

(a) $\tilde{d} \leq d$

(b) There exists a countable set $H = \{f^k \ k \in \mathbb{N}\}$ of uniformly continuous functions $f^k : (M, \tilde{d}) \to [0,1]$ such that the topology of $\mathcal{P}(M)$ is induced by the metric

$$R(\mu, \nu) = \sum_k \frac{|\int_M f^k d\mu - \int_M f^k d\nu|}{2^k}$$

Statement (a) follows for example from the construction given in Lemma 3.1.4 of Stroock (1993) while (b) follows from Theorem 3.1.5 of Stroock (1993).

Let

$$\tilde{\Omega} = \bigcap_{k \in \mathbb{N}, T \in \mathbb{Q}_+} \Omega(f^k, T)$$

Given $\omega \in \tilde{\Omega}$ and $\mu \in \mathcal{M}(X, \omega)$ there exists a sequence $t_j \to \infty$ (depending on ω) such that $\{\mu_{t_j}(\omega)\}$ converges weakly toward μ.

Let $n_j = [\frac{t_j}{T}]$ denote the integer part of $\frac{t_j}{T}$. Then

$$\lim_{j \to \infty} \frac{1}{n_j T} \sum_{i=0}^{n_j - 1} \int_{iT}^{(i+1)T} f(X_s) ds = \int_M f(x) \mu(dx) \tag{53}$$

for any continuous and bounded function $f : M \to \mathbb{R}$. Hence by combining (53) and (52) we get that

$$\int_M (f^k \circ \Phi_T)(x)\mu(dx) = \int_M f^k(x)\mu(dx)$$

for all $f^k \in H$ and $T \in \mathbb{Q}^+$. This proves that μ is Φ_T invariant for all $T \in \mathbb{Q}^+$. By continuity of Φ this implies that μ is Φ_T invariant for all $T > 0$ and since Φ is a semiflow, μ is Φ invariant. **QED**

Given $\mu \in \mathcal{P}(M)$ let $supp(\mu)$ denote the support of μ. Given a weak asymptotic pseudotrajectory X of Φ and $\omega \in \Omega$ we define the *minimal center of attraction* of $\{X(t, \omega) : t \geq 0\}$ as the (random) set

$$\overline{supp(X, \omega)} = \bigcup_{\mu \in \mathcal{M}(X, \omega)} supp(\mu)$$

Corollary 10.2 *Given a Borel set $A \subset M$ define*

$$\tau(\omega)(A) = \liminf_{t \to \infty} \mu_t(\omega)(A).$$

Suppose that for $P-$almost every ω $\{\mu_t(\omega)\}_{t \geq 0}$ is tight. Then for $P-$almost every ω

(i) $\tau(\omega)(supp(X, \omega)) = 1$ *and for any other closed set $A \subset M$ such that $\tau(\omega)(A) = 1$ it follows that $supp(X, \omega) \subset A$.*

(ii)

$$supp(X, \omega) \subset BC(\Phi) = \overline{\{x \in M : x \in \omega(x)\}}$$

Proof The proof of part (i) is an easy consequence of Theorem 10.1 and (ii) follows Theorem 10.1 and Poincaré recurrence Theorem (equation (30). **QED**

This last corollary has the interpretation that the fraction of time spends by a weak asymptotic pseudotrajectory in an arbitrary neighborhood of $BC(\Phi)$ goes to one with probability one.

10.1 Stochastic Approximation Processes with Slow Decreasing Step-Size

Consider a Robbins-Monro algorithm as described in section (4). Recall that $\overline{X} : \mathbb{R}_+ \to \mathbb{R}^m$ denotes the piecewise constant interpolated process given by $\overline{X}(t) = x_n$ for $\tau_n \leq t < \tau_{n+1}$. Set $\mathcal{F}_t = \mathcal{F}_n$ for $\tau_n \leq t < \tau_{n+1}$.

Proposition 10.3 *Let $\{x_n\}$ given by (7) be a Robbins-Monro algorithm. Assume*

(i) *F is Lipschitz on a neighborhood of $\{x_n : n \geq 0\}$,*

(ii) *x_0 is $\mathcal{F}_0$ measurable.*

(iii) *$\lim_{R \to \infty}\{\sup_n E(\|U_{n+1}\|\mathbf{1}_{\{\|U_{n+1}\| \geq R\}}|\mathcal{F}_n)\} = 0$.*

(iv) *$\lim_{n \to \infty} \gamma_n = 0$.*

Then $\overline{X}$ is a weak asymptotic pseudotrajectory of Φ. Hence $\overline{X}$ and X satisfy conclusion of Theorem 10.1 and Corollary 10.2.

Proof Given $R > 0$ let

$$U_{i+1}(R) = U_{i+1}\mathbf{1}_{\{|U_{i+1}| \leq R\}} - E(U_{i+1}\mathbf{1}_{\{\|U_{i+1}\| \leq R\}}|\mathcal{F}_i)$$

and

$$V_{i+1}(R) = U_{i+1} - U_{i+1}(R).$$

Then

$$P(\sup_{n\leq k\leq m(\tau_n+T)}\|\sum_{i=n}^{k}\gamma_{i+1}U_{i+1}\|\geq\alpha|\mathcal{F}_n)\leq$$

$$P(\sup_{n\leq k\leq m(\tau_n+T)}\|\sum_{i=n}^{k}\gamma_{i+1}U_{i+1}(R)\|\geq\alpha/2|\mathcal{F}_n)+P(\sup_{n\leq k\leq m(\tau_n+T)}$$

$$\|\sum_{i=n}^{k}\gamma_{i+1}V_{i+1}(R)\|\geq\alpha/2|\mathcal{F}_n)$$

$$\leq\frac{4}{\alpha^2}C_2R^2\sum_{i=n}^{m(\tau_n+T)}\gamma_{i+1}^2+\frac{4}{\alpha}E(\sum_{i=n}^{m(\tau_n+T)}\gamma_{i+1}E(\|U_{i+1}\|\mathbf{1}_{\{\|U_{i+1}\|>R\}}|\mathcal{F}_i)|\mathcal{F}_n)$$

where the first term in the right side of this inequality follows from inequality
(16) obtained with $q=2$ and the second term is an obvious estimate based on
Markov inequality. Let $\epsilon>0$. Assumption (iv) implies the existence of R large
enough so that

$$\sup_{i}E(\|U_{i+1}\|\mathbf{1}_{\{\|U_{i+1}\|>R\}}|\mathcal{F}_i)\leq\epsilon.$$

Hence

$$\limsup_{n\to\infty}P(\sup_{n\leq k\leq m(\tau_n+T)}\|\sum_{i=n}^{k}\gamma_{i+1}U_{i+1}\|\geq\alpha|\mathcal{F}_n)\leq\frac{4T}{\alpha}\epsilon. \qquad (54)$$

Inequality (54) combined with the estimate (11) proves the result.   **QED**

References

Akin, E. (1993). *The General Topology of Dynamical Systems*. American Mathematical Society, Providence.

Arthur, B., Ermol'ev, Y., and Kaniovskii, Y. (1983). A generalized urn problem and its applications. *Cybernetics*, 19:61–71.

Arthur, B. M. (1988). Self-reinforcing mechanisms in economics. In W, A. P., Arrow, K. J., and Pines, D., editors, *The Economy as an Evolving Complex System, SFI Studies in the Sciences of Complexity*. Addison-Wesley.

Benaïm, M. (1996). A dynamical systems approach to stochastic approximations. *SIAM Journal on Control and Optimization*, 34:141–176.

Benaïm, M. (1997). Vertex reinforced random walks and a conjecture of Pemantle. *The Annals of Probability*, 25:361–392.

Benaïm, M. and Hirsch, M. W. (1994). Learning processes, mixed equilibria and dynamical systems arising from repeated games. Submitted.

Benaïm, M. and Hirsch, M. W. (1995a). Chain recurrence in surface flows. *Discrete and Continuous Dynamical Systems*, 1(1):1–16.

Benaïm, M. and Hirsch, M. W. (1995b). Dynamics of morse-smale urn processes. *Ergodic Theory and Dynamical Systems*, 15:1005–1030.

Benaïm, M. and Hirsch, M. W. (1996). Asymptotic pseudotrajectories and chain recurrent flows, with applications. *J. Dynam. Differential Equations*, 8:141–176.

Benaïm, M. and Schreiber, S. J. (1997). Weak asymptotic pseudotrajectories for semiflows: Ergodic properties. Preprint.

Benveniste, A., Métivier, M., and Priouret, P. (1990). *Stochastic Approximation and Adaptive Algorithms*. Springer-Verlag, Berlin and New York.

Bowen, R. (1975). Omega limit sets of Axiom A diffeomorphisms. *J. Diff. Eq*, 18:333–339.

Brandière, O. (1996). Autour des pièges des algorithmes stochastiques. Thèse de Doctorat, Université de Marne-la-Vallée.

Brandière., O. (1997). Some pathological traps for stochastic approximation. *SIAM Journal on Control and Optimization*. To Appear.

Brandière, O. and Duflo., M. (1996). Les algorithmes stochastique contournent ils les pièges. *Annales de l'IHP*, 32:395–427.

Conley, C. C. (1978). *Isolated invariant sets and the Morse index*. CBMS Regional conference series in mathematics. American Mathematical Society, Providence.

Delyon, B. (1996). General convergence results on stochastic approximation. *IEEE trans. on automatic control*, 41:1245–1255.

Duflo, M. (1990). *Méthodes Récursives Aléatoires*. Masson. English Translation: Random Iterative Models, Springer Verlag 1997.

Duflo, M. (1996). *Algorithmes Stochastiques*. Mathématiques et Applications. Springer-Verlag.

Duflo, M. (1997). Cibles atteignables avec une probabilité positive d'après M. BENAIM. Unpublished manuscript.

Ethier, S. N. and Kurtz, T. G. (1986). *Markov Processes, Characterization and Convergence*. John Wiley and Sons, Inc.

Fort, J. C. and Pages, G. (1994). Réseaux de neurones: des méthodes connexionnistes d'apprentissage. *Matapli*, 37:31–48.

Fort, J. C. and Pages, G. (1996). Convergence of stochastic algorithms: From Kushner-Clark theorem to the lyapounov functional method. *Adv. Appl. Prob*, 28:1072–1094.

Fort, J. C. and Pages, G. (1997). Stochastic algorithm with non constant step: a.s. weak convergence of empirical measures. Preprint.

Fudenberg, D. and Kreps, K. (1993). Learning mixed equilibria. *Games and Econom. Behav.*, 5:320-367.

Fudenberg, F. and Levine, D. (1998). *Theory of Learning in Games*. MIT Press, Cambridge, MA. In Press.

Hartman, P. (1964). *Ordinary Differential Equationq*. Wiley, New York.

Hill, B. M., Lane, D., and Sudderth, W. (1980). A strong law for some generalized urn processes. *Annals of Probability*, 8:214-226.

Hirsch, M. W. (1976). *Differential Topology*. Springer-Verlag, Berlin, New York, Heidelberg.

Hirsch, M. W. (1994). Asymptotic phase, shadowing and reaction-diffusion systems. In *Differential equations, dynamical systems and control science*, volume 152 of *Lectures notes in pure and applied mathematics*, pages 87-99. Marcel Dekker, New-York.

Hirsch, M. W. and Pugh, C. C. (1988). Cohomology of chain recurrent sets. *Ergodic Theory and Dynamical Systems*, 8:73-80.

Kaniovski, Y. and Young, H. (1995). Learning dynamics in games with stochastic perturbations. *Games and Econom. Behav.*, 11:330-363.

Kiefer, J. and Wolfowitz, J. (1952). Stochastic estimation of the maximum of a regression function. *Ann. Math. Statis*, 23:462-466.

Kushner, H. J. and Clarck, C. C. (1978). *Stochastic Approximation for Constrained and Unconstrained Systems*. Springer-Verlag, Berlin and New York.

Kushner, H. J. and Yin, G. G. (1997). *Stochastic Approximation Algorithms and Applications*. Springer-Verlag, New York.

Ljung, L. (1977). Analysis of recursive stochastic algorithms. *IEEE Trans. Automat. Control.*, AC-22:551-575.

Ljung, L. (1986). *System Identification Theory for the User*. Prentice Hall, Englewood Cliffs, NJ.

Ljung, L. and Söderström, T. (1983). *Theory and Practice of Recursive Identification*. MIT Press, Cambridge, MA.

Mañé, R. (1987). *Ergodic Theory and Differentiable Dynamics*. Springer-Verlag, New York.

Métivier, M. and Priouret, P. (1987). Théorèmes de convergence presque sure pour une classe d'algorithmes stochastiques à pas décroissant. *Probability Theory and Related Fields*, 74:403-428.

Munkres, J. R. (1975). *Topology a first course.* Prentice Hall.

Nevelson, M. B. and Khasminskii, R. Z. (1976). *Stochastic Approximation and Recursive Estimation.* Translation of Math. Monographs. American Mathematical Society, Providence.

Pemantle, R. (1990). Nonconvergence to unstable points in urn models and stochastic approximations. *Annals of Probability*, 18:698–712.

Pemantle, R. (1992). Vertex reinforced random walk. *Probability Theory and Related Fields*, 92:117–136.

Robbins, H. and Monro, S. (1951). A stochastic approximation method. *Ann. Math. Statis*, 22:400–407.

Robinson, C. (1977). Stability theorems and hyperbolicity in dynamical systems. *Rocky Journal of Mathematics*, 7:425–434.

Robinson, C. (1995). *Introduction to the Theory of Dynamical Systems.* Studies in Advances Mathematics. CRC Press, Boca Raton.

Schreiber, S. J. (1997). Expansion rates and Lyapunov exponents. *Discrete and Conts. Dynam. Sys.*, 3:433–438.

Shub, M. (1987). *Global Stability of Dynamical Systems.* Springer-Verlag, Berlin, New York, Heidelberg.

Stroock, D. W. (1993). *Probability Theory. An analytic view.* Cambridge University Press.

White, H. (1992). *Artificial Neural Networks: Approximation and Learning Theory.* Blackwell, Cambridge, Massachussets.

SIMULATED ANNEALING ALGORITHMS AND MARKOV CHAINS WITH RARE TRANSITIONS

OLIVIER CATONI

ABSTRACT. In these notes, written for a D.E.A. course at University Paris XI during the first term of 1995, we prove the essentials about stochastic optimisation algorithms based on Markov chains with rare transitions, under the weak assumption that the transition matrix obeys a large deviation principle. We present a new simplified line of proofs based on the Freidlin and Wentzell graphical approach. The case of Markov chains with a periodic behaviour at null temperature is considered. We have also included some pages about the spectral gap approach where we follow Diaconis and Stroock [13] and Ingrassia [23] in a more conventional way, except for the application to non reversible Metropolis algorithms (subsection 6.2.2) where we present an original result.

ALGORITHMES DE RECUIT SIMULÉ ET CHAÎNES DE MARKOV À TRANSITIONS RARES: Dans ces notes, tirées d'un cours de D.E.A. donné au premier trimestre 1995, nous établissons les bases de la théorie des algorithmes d'optimisation stochastiques fondés sur des chaînes de Markov à transitions rares, sous l'hypothèse faible selon laquelle la matrice des transitions vérifie un principe de grandes déviations. Nous présentons un nouvel ensemble de preuves originales fondées sur l'approche graphique de Freidlin et Wentzell. Le cas des chaînes présentant un comportement périodique à température nulle est traité. De plus nous avons aussi inclus quelques pages sur les méthodes de trou spectral, dans lesquelles nous suivons Diaconis et Stroock [13] et Ingrassia [23] d'une façon plus conventionnelle, si ce n'est pour l'application aux algorithmes de Metropolis non réversibles de la section 6.2.2, qui est originale.

INTRODUCTION

These lecture notes were written on the occasion of a course of lectures which took place from January to April 1995. We seized the opportunity of the present English translation to add some proofs which were left to the reader and to correct some misprints and omissions. Sections 4.1, 4.2 and 4.3 contain standard material from [13] and [23]. The rest is more freely inspired by the existing literature. The presentation of the cycle decomposition is new, as well as lemma 1. We chose to make weak large deviation assumptions on the transition matrix p_β at inverse temperature β, and to give results which are accordingly concerned

Date: May 1995, English translation January 1997, in revised form November 1998.

only with equivalents for the logarithm of the probability of some events of interest. In the study of simulated annealing, we considered piecewise constant temperature sequences, in order to avoid introducing specifically non-homogeneous
techniques. Our aim was to give tools to study a wide variety of stochastic optimisation algorithms with discrete time and finite state space. For related results
directed towards applications to statistical mechanics, we refer to [8].

1. EXAMPLES OF HOMOGENEOUS MARKOV CHAINS

We are going to study in this section homogeneous Markov chains related to
stochastic optimisation algorithms.

1.1. The Metropolis Algorithm.

This algorithm can be applied to any finite
state space E on which an energy function $U : E \to \mathbb{R}$ is defined (U can be any
arbitrary real valued function). Its purpose can be either:

- to simulate the equilibrium distribution of a system from statistical mechanics with state space E and energy U interacting with a heat bath at
  temperature T,
- or to find a state $x \in E$ for which $U(x)$ is close to $\min_{y \in E} U(y)$.

We will mainly be interested in the second application in these notes.

Description of the algorithm

Let us consider a Markov matrix $q : E \times E \to [0,1]$ which is irreducible and
reversible with respect to its invariant measure. In other words let us assume
that

- $\displaystyle\sum_{y \in E} q(x,y) = 1, \quad x \in E,$
- $\displaystyle\sup_{m} q^m(x,y) > 0, \quad x, y \in E.$

(This last equation means that there is a path $x_0 = x, x_1, \ldots, x_m = y$ leading
from x to y such that $q(x_i, x_{i+1}) > 0$, $i = 0, \ldots, l-1$.)

- the invariant probability distribution μ of q (which is unique under the
preceding assumptions) is such that

$$\mu(x)q(x,y) = \mu(y)q(y,x).$$

Let us consider also an inverse temperature $\beta > 0$, $\beta \in \mathbb{R}$. To this temperature
corresponds the Gibbs distribution $G(E, \mu, U, \beta)$, defined by

$$G(E, \mu, U, \beta)(x) = \frac{\mu(x)}{Z} \exp(-\beta U(x))$$

where Z (the "partition function") is

$$Z = \sum_{x \in E} \mu(x) \exp(-\beta U(x)).$$

The distribution $G(E, \mu, U, \beta)$ describes the thermal equilibrium of the thermo-
dynamic system (E, μ, U, β). We then define the transition matrix at inverse
temperature β. This is the Markov matrix $p_\beta : E \times E \to [0,1]$ defined by

$$p_\beta(x,y) = q(x,y) \exp -\beta(U(y) - U(x))^+, \quad x \neq y \in E,$$

where $r^+ = \max\{0, r\}$.

Proposition 1.1. *The matrix p_β is irreducible. It is aperiodic as soon as U is not constant, and therefore*

$$\forall \zeta, \nu \in \mathcal{M}_1^+(E), \quad \lim_{n \to +\infty} (\zeta - \nu) p_\beta^n = 0,$$

where $\mathcal{M}_1^+(E)$ is the set of probability measures on E. Moreover p_β is reversible with respect to $\mu_\beta = G(E, \mu, U, \beta)$.

Proof: It is irreducible because $p_\beta(x, y) > 0$ as soon as $q(x, y) > 0$. If U is not constant there are $x, y \in E$ such that $q(x, y) > 0$ and $U(x) < U(y)$, which implies that $p_\beta(x, x) > 0$ and therefore that p_β is aperiodic. Moreover

$$\begin{aligned} \mu_\beta(x) p_\beta(x, y) &= \frac{1}{Z} \mu(x) q(x, y) \exp\left(-\beta(U(x) \vee U(y))\right) \\ &= \mu_\beta(y) p_\beta(y, x), \quad x, y \in E, x \neq y. \end{aligned}$$

1.1.1. *Construction of the Metropolis algorithm.* On the canonical space $(E^{\mathbb{N}}, \mathcal{B})$ where $\mathcal{B}$ is the sigma field generated by the events depending on a finite number of coordinates, we consider the canonical process $(X_n)_{n \in \mathbb{N}}$ defined by

$$X_n(x) = x_n, \quad x \in E^{\mathbb{N}},$$

and the family of probability distributions $(P_\beta^x)_{x \in E}$ on $(E^{\mathbb{N}}, \mathcal{B})$ defined by

$$P_\beta^x \circ X_0^{-1} = \delta_x,$$

$$P_\beta^x(X_n = y \mid (X_0, \ldots, X_{n-1}) = (x_0, \ldots, x_{n-1})) = p_\beta(x_{n-1}, y).$$

The homogeneous Markov chain $(E^{\mathbb{N}}, (X_n)_{n \in \mathbb{N}}, \mathcal{B}, (P_\beta^x)_{x \in E})$ is the canonical realization of the Metropolis algorithm with state space E, Markov matrix q, energy function U and inverse temperature β. We will use the notation $M(E, q, U, \beta)$.

1.1.2. *Computer implementation.* Assuming that $X_{n-1} = x \in E$, choose a state y according to the distribution $q(x, y)$, compute $U(y) - U(x)$, if $U(y) \leq U(x)$, put $X_n = y$, if $U(y) > U(x)$, put $X_n = y$ with probability $\exp -\beta(U(y) - U(x))$ and $X_n = x$ otherwise.

1.1.3. *Behaviour at temperature zero ($\beta = +\infty$).* Letting β tend to $+\infty$ in the definition of $M(E, q, U, \beta)$, we define the infinite inverse temperature algorithm $M(E, q, U, +\infty)$ by

$$P_{+\infty}(X_n = y \mid X_{n-1} = x) = q(x, y) \mathbf{1}(U(y) \leq U(x)), \quad x \neq y \in E.$$

This is a relaxation algorithm: $U(X_n)$ is almost surely non increasing. It is still homogeneous, but no more ergodic in general (if U is not constant on E, E has at least one transient component).

When β tends to infinity, $M(E, q, U, \beta)$ weakly tends to $M(E, q, U, +\infty)$, in the sense that for any function $f : E^{\mathbb{N}} \to \mathbb{R}$ depending on a finite number of coordinates we have

$$\lim_{\beta \to +\infty} \int_{E^{\mathbb{N}}} f(y) P_\beta(dy) = \int_{E^{\mathbb{N}}} f(y) P_{+\infty}(dy).$$

(Note that it implies that the same holds for any continuous function f, $E^{\mathbb{N}}$ being equipped with the product topology, because any such function is a uniform limit of functions depending on a finite number of coordinates.)

When it is observed during a fixed interval of time, $M(E, q, U, \beta)$ is a small perturbation of $M(E, q, U, +\infty)$ at low temperature.

We can see now that the Metropolis algorithm is suitable for the two purposes we announced at the beginning:

- *Simulation of the thermal equilibrium distribution $G(E, \mu, U, \beta)$*: As p_β is irreducible and aperiodic and as E is finite, $(P_\beta \circ X_0^{-1})p_\beta^n = P_\beta \circ X_n^{-1}$ tends to $G(E, \mu, U, \beta)$ when n tends to infinity (at exponential rate, as will be seen in the following).

- *Minimisation of U*: The Gibbs distributions $\mu_\beta = G(E, \mu, U, \beta)$ get concentrated around $\arg\min U$ when β tends to $+\infty$.

  Indeed, for any $\eta > 0$,

  $$\mu_\beta (U(x) < \min U + \eta) \geq 1 - \frac{1}{Z} \exp\left(-\beta(\eta + \min U)\right),$$
  $$Z \geq \mu(\arg\min U) \exp\left(-\beta \min U\right),$$

  therefore we have the following rough estimate

  $$\mu_\beta(U(x) < \min_E U + \eta) \geq 1 - \mu(\arg\min U)^{-1} e^{-\beta\eta}.$$

  Taking $\eta = \min\{U(y), y \in E \setminus \arg\min U\} - \min_E U$, we see that, as a consequence,

  $$\lim_{\beta \to +\infty} G(E, \mu, U, \beta)(\arg\min U) = 1.$$

Thus

Proposition 1.2. *For any $\epsilon > 0$ there are $N \in \mathbb{N}$ and $\beta \in \mathbb{R}_+$ such that for any $n > N$*

$$P_\beta(U(X_n) = \min U) \geq 1 - \epsilon.$$

1.2. The Gibbs sampler. This algorithm is meant for a product state space $E = \prod_{i=1}^r F_i$, where the components F_i are finite sets. The purpose is the same as for the Metropolis algorithm (simulate the Gibbs distribution or minimise the energy).

Description: Let us consider

- An energy function $U : E \to \mathbb{R}$, which can in fact be any real valued function.
- An "infinite temperature" probability distribution $\mu \in \mathcal{M}_1^+$.
- An inverse temperature $\beta \in \mathbb{R}_+^*$.
- The Gibbs distribution

  $$G(E, \mu, U, \beta)(x) = \frac{\mu(x)}{Z} \exp(-\beta U(x)).$$

- A permutation $\sigma \in \mathfrak{S}_r$ of $\{1, \ldots, r\}$.

Let us define

- For any $i \in \{1, \ldots, r\}$ the transition matrix $p_\beta^i : E \times E \to [0, 1]$ at site i and inverse temperature β

$$p_\beta^i(x, y) = \mathbf{1}(\overline{y}^i = \overline{x}^i)\, G(E, \mu, U, \beta)(y \mid \overline{y}^i = \overline{x}^i), \quad x, y \in E,$$

where we have used the notations $x = (x^j)_{j=1}^r$, $x^j \in F_j$ and $\overline{x}^i = (x^j)_{j, j \neq i}$.

- The global transition matrix at temperature β

$$p_\beta = \prod_{i=1}^r p_\beta^{\sigma(i)} = p_\beta^{\sigma(1)} \cdots p_\beta^{\sigma(r)},$$

which corresponds to the scan of the sites defined by the permutation σ.

Properties of p_β:

- It is a full matrix, $(p_\beta(x, y) > 0, x, y \in E)$, thus it is irreducible and aperiodic.
- The Gibbs distribution G is p_β^i invariant for any $i \in \{1, \ldots, r\}$, therefore G is also the (unique) invariant probability measure of p_β.

We consider then the Markov chain with canonical realization $(E^{\mathbb{N}}, (X_n)_{n \in \mathbb{N}}, \mathcal{B}, P_\beta)$ where P_β is the probability measure on $(E^{\mathbb{N}}, \mathcal{B})$ of the Markov chain defined by $P_\beta \circ X_0^{-1}$ and

$$P(X_n = y \mid X_{n-1} = x) = p_\beta(x, y), \quad x, y \in E.$$

The homogeneous Markov chain (X, P_β) is called a Gibbs sampler with state space E, energy function U, reference measure μ, scan function σ, inverse temperature β and initial distribution $P_\beta \circ X_0^{-1} = \mathcal{L}_0$. The notation $GS(E, \mu, \sigma, U, \beta, \mathcal{L}_0)$ will denote this process in the following. Let us describe its computer implementation with more details.

Computer implementation:

Each step of the chain corresponds to one scan of all the sites, in the order defined by σ. It includes thus r sub-steps.

To perform the ith sub-step, $i = 1, \ldots, r$, if x is the starting configuration, we have to draw at random $f \in F_{\sigma(i)}$ according to the conditional thermal equilibrium distribution at site $\sigma(i)$ knowing that the configuration should coincide with x on the other sites.

This computation is easy if

- The number of elements of $F_{\sigma(i)}$ is small,
- The conditional distribution $G(X^{\sigma(i)} = f \mid X^j = x^j, j \neq \sigma(i))$ depends on few coordinates, as it is the case for a Markov random field. The new state at the end of the ith sub-step is $y \in E$, given by $y^{\sigma(i)} = f$ and $y^j = x^j$, $j \neq \sigma(i)$.

Behaviour at "zero temperature": Here again $\lim_{\beta \to +\infty} p_\beta^i$ exists, therefore $\lim_{\beta \to +\infty} p_\beta$ exists and defines a Markov chain at temperature zero. This zero temperature dynamic is a relaxation algorithm: the energy is almost surely non-increasing. It is not in general an ergodic process, and P_β converges weakly to $P_{+\infty}$, as in the case of the Metropolis dynamic. Moreover the purposes of simulation of the equilibrium distribution and of minimisation of the energy are fulfilled in the same way, and, as for the Metropolis algorithm, proposition 1.2 holds also for the Gibbs sampler.

2. MARKOV CHAINS WITH RARE TRANSITIONS

2.1. Construction. We are going to put the two previous examples into a more general framework. Let us consider

- An arbitrary finite state space E,
- A rate function $V : E \times E \to \mathbb{R}_+ \cup \{+\infty\}$. Assume that V is irreducible in the sense that the matrix $\exp(-V(x, y))$ is irreducible.
- A family $\mathcal{F} = (E^{\mathbb{N}}, (X_n)_{n \in \mathbb{N}}, \mathcal{B}, P_\beta)_{\beta \in \mathbb{R}_+}$ of homogeneous Markov chains indexed by a real positive parameter β.

Definition 2.1. The family of homogeneous Markov chains $\mathcal{F}$ is said to have rare transitions with rate function V if for any $x, y \in E$

$$\lim_{\beta \to +\infty} \frac{-\log P_\beta(X_n = y \mid X_{n-1} = x)}{\beta} = V(x, y),$$

(with the convention that $\log 0 = -\infty$).

Remarks about this definition:

- This is a large deviation assumption with speed β and rate function V about the transition matrix. We will see that it implies large deviation estimates for the exit time and point from any subdomain of E.
- The two examples of algorithms given previously fit into this framework. Indeed the rate function of the Metropolis algorithm $M(E, q, U, \beta, \mathcal{L}_0)$ is

$$V(x, y) = \begin{cases} (U(y) - U(x))_+ & \text{if } p_\beta(x, y) > 0 \text{ for } \beta > 0 \\ +\infty & \text{otherwise.} \end{cases}$$

As for the Gibbs Sampler $GS(E, \mu, \sigma, U, \beta, \mathcal{L}_0)$ with $E = \prod_{i=1}^{r} F_i$, the rate function V is built in the following way:
For any $x, y \in E$, any $i \in \{1, \ldots, r\}$, let us put

$$V^i(x, y) = \begin{cases} U(y) - \inf\{U(z) \mid \bar{z}^i = \bar{x}^i\}, & \text{if } \bar{x}^i = \bar{y}^i \\ +\infty & \text{otherwise,} \end{cases}$$

and let us consider the path $\gamma = (\gamma_k)_{k=0}^{r}$ defined by

$$\gamma_k^{\sigma(i)} = \begin{cases} y^{\sigma(i)} & \text{if } i \leq k, \\ x^{\sigma(i)} & \text{otherwise.} \end{cases}$$

The rate function of the Gibbs sampler is

$$V(x, y) = \sum_{k=1}^{r} V^{\sigma(k)}(\gamma_{k-1}, \gamma_k).$$

2.2. Rate function induced by a potential.

Definition 2.2. We will say that the rate function $V : E \times E \to \mathbb{R}_+ \cup \{+\infty\}$ is induced by the potential $U : E \to \mathbb{R}$ if for all $x, y \in E$

$$U(x) + V(x, y) = U(y) + V(y, x),$$

with the convention that $+\infty + r = +\infty$ for any $r \in \mathbb{R}$.

Proposition 2.1. *The rate function of the Metropolis algorithm $M(E, \mu, U, \beta, \mathcal{L}_0)$ is induced by U.*

Proof:

As q is irreducible, $\mu(x) > 0$ for any $x \in E$. Indeed there is x_0 such that $\mu(x_0) > 0$ and there is n such that $q^n(x_0, x) > 0$, therefore $\mu(x) = \mu q^n(x) \geq \mu(x_0) q^n(x_0, x) > 0$. Thus $q(x, y) > 0$ if and only if $q(y, x) > 0$, from the μ reversibility of q. Therefore $V(x, y) = +\infty$ if and only if $V(y, x) = +\infty$. In the case when $q(x, y) > 0$, $x \neq y$,

$$V(x, y) - V(y, x) = (U(y) - U(x))^+ - (U(x) - U(y))^+ = U(y) - U(x). \qquad \square$$

3. Lemmas on irreducible Markov chains

Let E be a finite state space, $p : E \times E \to [0, 1]$ an irreducible Markov matrix, $(E^{\mathbb{N}}, (X_n)_{n \in \mathbb{N}}, \mathcal{B}, P)$ an homogeneous Markov chain with transition matrix p, $W \subset E$ a given subset of E and $\overline{W} = E \setminus W$ its complement. For any oriented graph $g \subset E \times E$ and any $x \in E$, we write $g(x) = \{y \mid (x, y) \in g\}$ and more generally $g^n(x) = \bigcup_{y \in g^{n-1}(x)} g(y)$.

Definition 3.1. We let $G(W)$ be the set of oriented graphs $g \subset E \times E$ satisfying

1. For any $x \in E$, $|g(x)| = 1_{\overline{W}}$ (no arrow starts from W, exactly one arrow starts from each state outside W).

2. For any $x \in E$, $x \notin O_g(x)$, where $O_g(x) = \bigcup_{n=1}^{+\infty} g^n(x)$ is the orbit of x under g, (g is without loop).

Equivalently, the second condition can be replaced by: For any $x \in E \setminus W, O_g(x) \cap W \neq \emptyset$ (any point in $\overline{W}$ leads to W).

Definition 3.2. For any $x \in E$, $y \in W$, we will write

$$G_{x,y}(W) = \begin{cases} \{g \in G(W) \mid y \in O_g(x)\} & \text{if } x \in \overline{W} \\ G(W) & \text{if } x = y \\ \emptyset & \text{if } x \in W \setminus \{y\}. \end{cases}$$

Thus $G_{x,y}(W)$ is the set of graphs $g \in G(W)$ linking x to y. We will also write

$$G_{A,B}(W) = \{g \mid \forall x \in A, \exists y \in B \text{ such that } g \in G_{x,y}(W)\}.$$

We will give three formulas which express the equilibrium distribution of p, the probability distribution of the hitting point of W, and the expectation of the corresponding hitting time, as the ratio of two finite sums of positive terms.

They have been introduced in the large deviation theory of random dynamical systems by Freidlin and Wentzell [16]. The idea of using graphs to compute determinants has been known since the nineteenth century and presumably goes back to Kirchhoff [24]. The proofs which we propose are based on a preliminary lemma:

Lemma 3.1. *For any* $W \subset E$, $W \neq \emptyset$, *let* $p_{|\overline{W} \times \overline{W}}$ *be the matrix* p *restricted to* $\overline{W} \times \overline{W}$:

$$p_{|\overline{W} \times \overline{W}}(x, y) = p(x, y)\mathbf{1}(x \notin W)\mathbf{1}(y \notin W).$$

Let $\tau(W)$ *be the first hitting time of* W: $\tau(W) = \inf\{n \geq 0 | X_n \in W\}$. *For any* $x, y \in \overline{W}$ *we have*

$$(\mathrm{id}_{|\overline{W}} - p_{|\overline{W} \times \overline{W}})^{-1}(x, y) = \left(\sum_{n=0}^{+\infty} p_{|\overline{W} \times \overline{W}}^n \right)(x, y)$$

$$= E_\beta \left(\sum_{n=0}^{\tau(W)} \mathbf{1}(X_n = y) \mid X_0 = x \right)$$

$$= \left(\sum_{g \in G_{x,y}(W \cup \{y\})} p(g) \right) \left(\sum_{g \in G(W)} p(g) \right)^{-1},$$

where $p(g) = \displaystyle\prod_{(z,t) \in g} p(z, t).$

Remark: The fact that $\mathrm{id}_{\overline{W}} - p_{\overline{W} \times \overline{W}}$ is non singular is a consequence of the fact that p is irreducible ($\lim_n p_{|\overline{W} \times \overline{W}}^n = 0$ and therefore all the eigenvalues of $p_{|\overline{W} \times \overline{W}}$ are of module lower than one).

Lemma 3.2. *The (unique) invariant probability distribution of* p *is given by*

$$\mu(x) = \left(\sum_{g \in G(\{x\})} p(g) \right) \left(\sum_{y \in E} \sum_{g \in G(\{y\})} p(g) \right)^{-1}, \quad x \in E.$$

Lemma 3.3. *The distribution of the first hitting point can be expressed as*

$$P(X_{\tau(W)} = y \mid X_0 = x) = \left(\sum_{g \in G_{x,y}(W)} p(g) \right) \left(\sum_{g \in G(W)} p(g) \right)^{-1},$$

for any $W \neq \emptyset$, $x \in \overline{W}$, $y \in W$.

Lemma 3.4. *For any* $W \neq \emptyset$, *any* $x \in \overline{W}$,

$$E(\tau(W) \mid X_0 = x) = \left(\sum_{y \in \overline{W}} \sum_{g \in G_{x,y}(W \cup \{y\})} p(g) \right) \left(\sum_{g \in G(W)} p(g) \right)^{-1}.$$

Proof of lemma 3.1:

As p is irreducible, for any $W \neq \emptyset$, there is $g \in G(W)$ such that $p(g) > 0$ (the proof of this is left to the reader).

Let us write for any $x, y \in \overline{W}$

$$m(x, y) = \left(\sum_{g \in G_{x,y}(W \cup \{y\})} p(g) \right) \left(\sum_{g \in G(W)} p(g) \right)^{-1}.$$

We want to check that for any $x, y \in \overline{W}$

(1)
$$\sum_{z \in \overline{W}} (\mathrm{id}(x, z) - p(x, z))\, m(z, y) = \mathrm{id}(x, y).$$

Using the equality

$$p(x, x) = 1 - \sum_{z \in E \setminus \{x\}} p(x, z),$$

we can equivalently check that

(2)
$$\sum_{z \in \overline{\{x\}}} p(x, z) m(x, y) = \mathrm{id}(x, y) + \sum_{z \in \overline{W \cup \{x\}}} p(x, z) m(z, y).$$

The left hand side of this equation is equal to

$$\left(\sum_{(z,g) \in C_1} p(x, z) p(g) \right) \left(\sum_{g \in G(W)} p(g) \right)^{-1}.$$

where $C_1 = \{(z, g) \in \overline{\{x\}} \times G(W \cup \{y\}) \ : \ g \in G_{x,y}(W \cup \{y\})\}$, the right hand side is equal to

$$\mathrm{id}(x, y) + \left(\sum_{(z,g) \in C_2} p(x, z) p(g) \right) \left(\sum_{g \in G(W)} p(g) \right)^{-1},$$

where

$$C_2 = \{(z, g) \in \overline{W \cup \{x\}} \times G(W \cup \{y\}) \mid g \in G_{z,y}(W \cup \{y\})\}.$$

Let us consider first the case when $x \neq y$. Then we can define a one to one mapping $\varphi : C_1 \to C_2$ by

$$\varphi(z, g) = \begin{cases} (z, g) & \text{if } g \in G_{z,y}(W \cup \{y\}), \\ (g(x), (g \cup \{(x, z)\}) \setminus \{(x, g(x))\}) & \text{if } g \notin G_{z,y}(W \cup \{y\}). \end{cases}$$

The easiest way to check that φ is one to one is to check that

$$\varphi^{-1}(z, g) = \begin{cases} (z, g) & \text{if } g \in G_{x,y}(W \cup \{y\}), \\ (g(x), (g \cup \{(x, z)\}) \setminus \{(x, g(x))\}) & \text{if } g \notin G_{x,y}(W \cup \{y\}). \end{cases}$$

Let us write $\varphi = (\varphi_1, \varphi_2)$ to show the two components of φ. The following change of variable

$$\sum_{(z,g)\in C_2} p(x,z)p(g) = \sum_{(z,g)\in C_1} p(x,\varphi_1(z,g))p(\varphi_2(z,g))$$

$$= \sum_{(z,g)\in C_1} p(x,z)p(g)$$

shows that

$$\sum_{z\in\overline{\{x\}}} p(x,z)m(x,y) = \sum_{z\in\overline{W\cup\{x\}}} p(x,z)m(z,y).$$

We have now to check the case when $x = y$. In this case $C_2 \subset C_1$. Let us consider the one to one mapping $\varphi : C_1 \setminus C_2 \to G(W)$ defined by $\varphi(z,g) = g \cup \{(x,z)\}$, with inverse $\varphi^{-1}(g) = (g(x), g \setminus \{(x,g(x))\})$.

We have

$$\sum_{(z,g)\in C_1\setminus C_2} p(x,z)p(g) = \sum_{g\in G(W)} p(g),$$

and therefore

$$\left(\sum_{(z,g)\in C_1} p(x,z)p(g)\right)\left(\sum_{g\in G(W)} p(g)\right)^{-1}$$

$$= 1 + \left(\sum_{(z,g)\in C_2} p(x,z)p(g)\right)\left(\sum_{g\in G(W)} p(g)\right)^{-1}. \qquad \square$$

Proof of lemma 3.3:

$$P(X_{\tau(W)} = y \,|\, X_0 = x) = \sum_{z\in\overline{W}}\sum_{n=0}^{+\infty} P(X_n = z, \tau(W) > n \,|\, X_0 = x)p(z,y)$$

$$= \sum_{z\in\overline{W}}\left(\sum_{g\in G_{x,z}(W\cup\{z\})} p(g)p(z,y)\right)\left(\sum_{g\in G(W)} p(g)\right)^{-1}$$

$$= \left(\sum_{g\in G_{x,y}(W)} p(g)\right)\left(\sum_{g\in G(W)} p(g)\right)^{-1}.$$

Proof of lemma 3.4 :

$$
E(\tau(W)\,|\,X_0 = x) \;=\; E\left(\sum_{n=0}^{\tau(W)-1} 1(X_n \in \overline{W})\,|\,X_0 = x\right)
$$

$$
=\; E\left(\sum_{y \in \overline{W}}\sum_{n=0}^{+\infty} 1(X_n = y, \tau(W) > n)\,|\,X_0 = x\right)
$$

$$
=\; \frac{\displaystyle\sum_{y \in \overline{W}}\sum_{g \in G_{x,y}(W \cup \{y\})} p(g)}{\displaystyle\sum_{g \in G(W)} p(g)}.
$$

Proof of lemma 3.2 :
Let $\nu(x) = \inf\{n \ge 1 \,|\, X_n = x\}$.

$$
\mu(x) \;=\; E(\nu(x)\,|\,X_0 = x)^{-1}
$$

$$
=\; \left(\sum_{y,y\neq x} p(x,y)E(\tau(\{x\})\,|\,X_0 = y) + 1\right)^{-1}
$$

$$
=\; \left(\sum_{y,y\neq x}\sum_{z,z\neq x}\sum_{g \in G_{y,z}(\{x,z\})} p(x,y)p(g) + \sum_{g \in G(\{x\})} p(g)\right)^{-1}\left(\sum_{g \in G(\{x\})} p(g)\right)
$$

$$
=\; \left(\sum_{g \in G(\{x\})} p(g)\right)\left(\sum_{z \in E}\sum_{g \in G(\{z\})} p(g)\right)^{-1},
$$

because for any $z \neq x$ $\varphi_z : \{(y,g)\,|\,y \neq x, g \in G_{y,z}(\{x,z\})\} \to G(\{z\})$ defined by $\varphi_z(y,g) = g \cup \{(x,y)\}$ is one to one.

4. Cycle decomposition of a family of Markov chains with rare transitions

4.1. Behaviour of the invariant distribution, virtual energy.

Definition 4.1. The rate function $V : E \times E \longrightarrow \mathbb{R}_+ \cup \{+\infty\}$ is said to be irreducible when the matrix $(\exp -V(x,y))_{(x,y)\in E^2}$ is irreducible. This means namely that for any $x, y \in E$ there is a path $z_0 = x, \dots, z_r = y$ such that

$$
V(z_{i-1}, z_i) < +\infty, \quad i = 1, \cdots, r.
$$

Proposition 4.1. *Let $\mathcal{F} = (E^{\mathbb{N}}, (X_n)_{n\in\mathbb{N}}, \mathcal{B}, P_\beta)_{\beta\in\mathbb{R}_+}$ be a family of homogeneous Markov chains with rare transitions, with irreducible rate function V. Then for β large enough (X, P_β) is irreducible and its invariant probability distribution μ_β is such that for any $x \in E$*

$$
\lim_{\beta\to+\infty} -\beta^{-1}\log\mu_\beta(x) = \tilde{U}(x) \in \mathbb{R}_+.
$$

The "virtual energy" function $\tilde{U} : E \to \mathbb{R}$ can be expressed as

$$\tilde{U}(x) = \min_{g \in G(\{x\})} V(g) - \min_{y \in E} \min_{g \in G(\{y\})} V(g),$$

where $V(g) = \sum_{(z,t) \in g} V(z,t)$. In the case when V is induced by a potential function U, we have for any $x \in E$ that $\tilde{U}(x) = U(x) - \min_{y \in E} U(y)$.

Corollary 4.1. *The family $\mathcal{F}$ describes an optimisation algorithm for the minimisation of the virtual energy $\tilde{U}$: For any $\epsilon > 0$, there are $N \in \mathbb{N}$ and $\beta \in \mathbb{R}_+$ such that, for any $n > N$,*

$$\min_{x \in E} P_\beta(\tilde{U}(X_n) = 0 \mid X_0 = x) \geq 1 - \epsilon.$$

This algorithm is called a "generalised Metropolis algorithm".

Proof: The first part of the proposition is a straightforward consequence of lemma 2. In the case when V is induced by U, consider the one to one mapping

$$\varphi : G(\{y\}) \longrightarrow G(\{x\}),$$

defined by

$$\varphi(g) = \{(z,t) \in g, t \notin O_g(x)\} \cup \{(t,z), (z,t) \in g, t \in O_g(x)\}.$$

It is obtained by reversing in $g \in G(\{y\})$ the path leading from x to y. We have

$$\tilde{U}(y) + U(x) + \min_{z \in E} \min_{g \in G(\{z\})} V(g) = \min_{g \in G(\{y\})} (V(g) + U(x))$$

$$= \min_{g \in G(\{y\})} \left(\sum_{\substack{z \notin O_g(x) \cup \{x\} \\ (z,t) \in g}} V(z,t) + U(x) + \sum_{\substack{z \in O_g(x) \cup \{x\} \\ (z,t) \in g}} V(z,t) \right)$$

$$= \min_{g \in G(\{y\})} \left(\sum_{\substack{z \notin O_g(x) \cup \{x\} \\ (z,t) \in g}} V(z,t) + U(y) + \sum_{\substack{z \in O_g(x) \cup \{x\} \\ (z,t) \in g}} V(t,z) \right)$$

$$= \min_{g \in G(\{y\})} (U(y) + V(\varphi(y)))$$

$$= U(y) + \min_{g \in G(\{x\})} V(g)$$

$$= U(y) + \tilde{U}(x) + \min_{z \in E} \min_{g \in G(\{z\})} V(g),$$

The proof of the corollary is the same as in the case of the classical Metropolis algorithm when the chain is aperiodic. When the chain has period d, then each chain $(X_{nd+k})_{n \in \mathbb{N}}$ is aperiodic for $k \in \{0, \dots, d-1\}$, and the combination of the inequalities obtained for these d processes gives the result for $(X_n)_{n \in \mathbb{N}}$. $\quad\square$

4.2. Large deviation estimates for the exit time and exit point from a subdomain.

In this paragraph we will study the limiting behaviour of the law of the exit time and exit point from an arbitrary subdomain D of E. Let us recall some notations introduced in section 3:

$$\overline{D} = E \setminus D,$$

$$\tau(D) = \inf\{n \in \mathbb{N} : X_n \in D\}.$$

Proposition 4.2. *For any $D \subset E, D \neq \emptyset$, for any $x \in D$, under the same hypotheses as previously,*

$$\lim_{\beta \to +\infty} \frac{\log E_\beta(\tau(\overline{D}) \mid X_0 = x)}{\beta} = \min_{g \in G(\overline{D})} V(g) - \min_{y \in \overline{D}} \min_{g \in G_{x,y}(\overline{D} \cup \{y\})} V(g),$$

moreover, for any $y \in \overline{D}$

$$\lim_{\beta \to +\infty} -\frac{1}{\beta} P_\beta(X_{\tau(\overline{D})} = y \mid X_0 = x) = \min_{g \in G_{x,y}(\overline{D})} V(g) - \min_{g \in G(\overline{D})} V(g).$$

We will use the following notations for these new rate functions:

$$\lim_{\beta \to +\infty} -\beta^{-1} \log P_\beta(X_{\tau(\overline{D})} = y \mid X_0 = x) \overset{\text{def}}{=} V_D(x, y)$$

$$\lim_{\beta \to +\infty} \beta^{-1} \log E_\beta(\tau(\overline{D}) \mid X_0 = x) \overset{\text{def}}{=} H_D(x).$$

In the next paragraph, we will link the rate functions appearing in these two large deviation estimates with the virtual energy $\tilde{U}$. For this purpose, we will introduce the decomposition of the state space into cycles due to Freidlin and Wentzell.

4.3. Definition of cycles.

Definition 4.2. Under the preceding hypotheses, a subdomain $C \subset E$ is said to be a cycle if it is a one point set or if for any $x, y \in C$, $x \neq y$, the probability, starting from x, to leave C without visiting y is exponentially small, by which we mean that

$$\lim_{\beta \to +\infty} -\frac{1}{\beta} \log P_\beta(X_{\tau(\overline{C} \cup \{y\})} \neq y \mid X_0 = x) > 0.$$

As a consequence we have of course

$$\lim_{\beta \to +\infty} P_\beta(X_{\tau(\overline{C} \cup \{y\})} = y \mid X_0 = x) = 1.$$

4.4. Some properties of cycles.

Proposition 4.3. *The subdomain C of E is a cycle if and only if it is a one point set or for any $x, y \in C$, $x \neq y$ the number $N_C(x, y)$ of round trips including x and y performed by the chain starting from x before it leaves C satisfies*

$$\lim_{\beta \to +\infty} \frac{1}{\beta} \log E_\beta(N_C(x, y) \mid X_0 = x) > 0.$$

Remark: This property justifies the name "cycle".

Proof: Let us give a more formal mathematical definition of $N_C(x, y)$. For this, let us introduce the sequences of stopping times $(\mu_k(x, y), \nu_k(x, y))_{n \in \mathbb{N}}$ defined by the following induction

$$
\begin{aligned}
\nu_{-1}(x, y) &= 0 \\
\mu_k(x, y) &= \inf\{n > \nu_{k-1}(x, y) : X_n \in \{y\} \cup \overline{C}\} \\
\nu_k(x, y) &= \inf\{n > \mu_k(x, y) : X_n \in \{x\} \cup \overline{C}\},
\end{aligned}
$$

then $N_C(x, y) = \inf\{k : X_{\mu_k(x,y)} \notin C \text{ or } X_{\nu_k(x,y)} \notin C\}$.
We have

$$
\begin{aligned}
E_\beta(N_C(x, y) \mid X_0 = x) &= \sum_{n=0}^{+\infty} P_\beta(N_C(x, y) > n \mid X_0 = x) \\
&= \sum_{n=0}^{+\infty} \left(P_\beta(X_{\mu_0(x,y)} = y \text{ and } X_{\nu_0(x,y)} = x \mid X_0 = x)\right)^n \\
&= \left(1 - P_\beta(X_{\mu_0(x,y)} = y \text{ and } X_{\nu_0(x,y)} = x \mid X_0 = x)\right)^{-1}.
\end{aligned}
$$

Moreover

$$
\begin{aligned}
&P_\beta(X_{\mu_0(x,y)} = y \text{ and } X_{\nu_0(x,y)} = x \mid X_0 = x) \\
&= P_\beta(X_{\tau(\overline{C} \cup \{y\})} = y \mid X_0 = x) P_\beta(X_{\tau(\overline{C} \cup \{x\})} = x \mid X_0 = y) \\
&= \left(1 - P_\beta(X_{\tau(\overline{C} \cup \{y\})} \neq y \mid X_0 = x)\right) \left(1 - P_\beta(X_{\tau(\overline{C} \cup \{x\})} \neq x \mid X_0 = y)\right).
\end{aligned}
$$

Therefore

$$
\lim_{\beta \to +\infty} \frac{1}{\beta} \log E_\beta(N_C(x, y) \mid X_0 = x)
$$
$$
= \inf\{V_{C \setminus \{z\}}(t, u) : (z, t) \in \{(x, y), (y, x)\} \text{ and } u \in \overline{C}\},
$$

which proves that

$$
\lim_{\beta \to +\infty} \frac{1}{\beta} \log E_\beta(N_C(x, y) \mid X_0 = x) > 0
$$

for all $x, y \in C$, $x \neq y$, if and only if $V_{C \setminus \{y\}}(x, z) > 0$ for all $(x, y) \in C^2$, $z \in \overline{C}$. $\square$

Proposition 4.4. *Let $\mathcal{C}(E, V)$ be the set of cycles of (E, V). It has a tree structure for the inclusion relation, with root E and leaves the one point sets. This means that if C_1 and C_2 are cycles, either $C_1 \subset C_2$ or $C_2 \subset C_1$ or $C_1 \cap C_2 = \emptyset$.*

Proof: If it were the case that $x \in C_1 \cap C_2$, $y \in C_1 \setminus C_2$ and $z \in C_2 \setminus C_1$, we would obtain a contradiction: we would have

$$
0 = \lim_{\beta \to +\infty} \frac{1}{\beta} \log P_\beta(X_{\tau(\overline{C}_2 \cup \{z\})} = z \mid X_0 = x)
$$

$$
\leq \lim_{\beta \to +\infty} \frac{1}{\beta} \log P_\beta(X_{\tau(\overline{C}_1 \cup \{y\})} \neq y \mid X_0 = x)
$$

$$
< 0. \qquad \square
$$

Proposition 4.5. *For any subdomain D of E, we define the principal boundary $B(D)$ of D by*

$$
B(D) = \{y \notin D : V_D(x, y) = 0 \text{ for some } x \in D\}
$$

Then for any cycle $C \in \mathcal{C}(E, V)$, any subdomain $D \subset C$, $D \neq \emptyset$, $D \neq C$, $B(D) \subset C$.

Proof:
If $y \in B(D) \setminus C$, $x \in D$, $z \in C \setminus D$, then

$$
P_\beta(X_{\tau(\overline{D})} = y \mid X_0 = x) \leq P(X_{\tau(\overline{C} \cup \{z\})} = y \mid X_0 = x),
$$

because in this case $y \in \overline{C} \cup \{z\}$ and $\overline{C} \cup \{z\} \subset \overline{D}$. This is in contradiction with the fact that

$$
\lim_{\beta \to +\infty} -\frac{1}{\beta} \log P_\beta(X_{\tau(\overline{D})} = y \mid X_0 = x) = 0
$$

and

$$
\lim_{\beta \to +\infty} -\frac{1}{\beta} \log P_\beta(X_{\tau(\overline{C} \cup \{z\})} = y \mid X_0 = x) > 0.
$$

Therefore $B(D) \subset C$. $\qquad \square$

An important property of a cycle is that, at low temperature, the exit time and exit point become independent from the starting point when it belongs to the cycle.

Proposition 4.6 (Independence from the starting point). *For any cycle $C \in \mathcal{C}(E, V)$, any $x \in C$, $y \in C$, $z \notin C$,*

$$
V_C(x, z) = V_C(y, z) \overset{\text{def}}{=} V(C, z)
$$

and

$$
H_C(x) = H_C(y) \overset{\text{def}}{=} H(C).
$$

The quantity $H(C)$ is called the depth of the cycle C.

Proof:

$$
\begin{aligned}
P_\beta(X_{\tau(\overline{C})} = z \mid X_0 = y) &= P_\beta(X_{\tau(\overline{C})} = z \mid X_0 = x) P_\beta(X_{\tau(\overline{C} \cup \{x\})} = x \mid X_0 = y) \\
&\quad + P_\beta(X_{\tau(\overline{C} \cup \{x\})} = z \mid X_0 = y) \\
&\geq P_\beta(X_{\tau(\overline{C})} = z \mid X_0 = x) P_\beta(X_{\tau(\overline{C} \cup \{x\})} = x \mid X_0 = y).
\end{aligned}
$$

Therefore $V_C(y, z) \leq V_C(x, z) + V_{C \setminus \{x\}}(y, x)$ by the definition of cycles $V_{C \setminus \{x\}}(y, x) = 0$, therefore $V_C(y, z) \leq V_C(x, z)$ and, exchanging x and y, $V_C(y, z) = V_C(x, z)$. Similarly we have

$$E_\beta(\tau(\overline{C}) \mid X_0 = x) = \sum_{u \in C} E_\beta \left(\sum_{n=0}^{\tau(\overline{C})} \mathbf{1}(X_n = u) \mid X_0 = x \right)$$

and

$$E_\beta \left(\sum_{n=0}^{\tau(\overline{C})} \mathbf{1}(X_n = u) \mid X_0 = x \right) = E_\beta \left(\sum_{n=0}^{\tau(\overline{C})} \mathbf{1}(X_n = u) \mid X_0 = u \right)$$
$$\times P_\beta(X_{\tau(\overline{C} \cup \{u\})} = u \mid X_0 = x).$$

Therefore $H_C(x)$ is independent of $x \in C$. $\quad \Box$

Now we will give some properties of cycles linked with W-graph computations:

Proposition 4.7 (characterisation of cycles in terms of W-graphs). *A subset C of E is a cycle if and only if it is either a one point set or satisfies: for any $y \in C$, any $\hat{g} \in \arg \min\limits_{g \in G(\overline{C} \cup \{y\})} V(g)$, $\hat{g}(C \setminus \{y\}) \subset C$.*

Proof: For any subset C of E, $|C| > 1$, any $y \in C$,

$$\min_{x \in C \setminus \{y\}} \lim_{\beta \to +\infty} -\frac{1}{\beta} \log P_\beta(X_{\tau(\overline{C} \cup \{y\})} \neq y \mid X_0 = x)$$
$$= \min_{x \in C \setminus \{y\}} \min_{z \notin C} \min_{g \in G_{x,z}(\overline{C} \cup \{y\})} V(g) - \min_{g \in G(\overline{C} \cup \{y\})} V(g)$$
$$= \min_{g \in G(\overline{C} \cup \{y\}), g(C \setminus \{y\}) \not\subset C} V(g) - \min_{g \in G(\overline{C} \cup \{y\})} V(g),$$

therefore $\min\limits_{x \in C \setminus \{y\}} \lim\limits_{\beta \to +\infty} -\frac{1}{\beta} \log P_\beta(X_{\tau(\overline{C} \cup \{y\})} \neq y \mid X_0 = x) > 0$ if and only if $\arg \min\limits_{g \in G(\overline{C} \cup \{y\})} V(g) \subset \{g \in G(\overline{C} \cup \{y\}) : g(C \setminus \{y\}) \subset C\}$. $\quad \Box$

Proposition 4.8 (leading terms in a $\overline{C}$-graph). *For any cycle $C \in \mathcal{C}(V)$, any $x \in C$, any $y \notin C$ such that $V(C, y) < +\infty$, any graph $\hat{g} \in \arg \min\limits_{g \in G_{x,y}(\overline{C})} V(g)$, $\hat{g}(C) \subset C \cup \{y\}$.*

Proof: Let us consider the state $z \in O_{\hat{g}}(x) \cup \{x\}$ such that $(z, y) \in \hat{g}$. Let $\tilde{g} \in \arg \min\limits_{g \in G(\overline{C} \cup \{z\})} V(g)$, then according to the preceding proposition $\tilde{g}(C \setminus \{z\}) \subset C$, therefore $\tilde{g} \cup \{(z, y)\}$ belongs to $G_{x,y}(\overline{C})$. Thus

$$V(z, y) + V(\tilde{g}) = V(\tilde{g} \cup \{(z, y)\}) \geq V(\hat{g}) = V(\hat{g} \setminus \{(z, y)\}) + V(z, y).$$

This shows that $V(\hat{g} \setminus \{(z, y)\}) = \arg \min\limits_{g \in G(\overline{C} \cup \{z\})} V(g)$, and therefore that $\hat{g}(C \setminus \{z\}) \subset C$. $\quad \Box$

Proposition 4.9 (local computation of the virtual energy). *For any cycle* $C \in \mathcal{C}(V)$, *any* $x, y \in C$,

$$\tilde{U}(x) - \tilde{U}(y) = \min_{g \in G(\overline{C} \cup \{x\})} V(g) - \min_{g \in G(\overline{C} \cup \{y\})} V(g).$$

This shows that the computation of the virtual energy within a cycle up to an additive constant depends only on the restriction of the rate function V to this cycle.

Proof: For any graph $g \in E \times E$, any subset $A \subset E$, let us put $g_{|A} = \{(u, v) \in g : u \in A\}$. With the notations of the proposition, let $g \in G(\{x\})$, then $g_{|C} \in G(\overline{C} \cup \{x\})$, $g_{|\overline{C}} \in G(C)$ and

$$V(g) = V(g_{|C}) + V(g_{|\overline{C}}).$$

Therefore

$$\min_{g \in G(\{x\})} V(g) \geq \min_{g \in G(\overline{C} \cup \{x\})} V(g) + \min_{g \in G(C)} V(g).$$

On the other hand, if $\hat{g} \in \arg\min_{g \in G(\overline{C} \cup \{x\})} V(g)$ and $g \in G(C)$, then $g \cup \hat{g}$ is without loop, because $\hat{g}(C) \subset C$, and thus $g \cup \hat{g} \in G(\{x\})$, and

$$\min_{g \in G(C)} V(g) + V(\hat{g}) = \min_{g \in G(C)} V(g \cup \hat{g}) \geq \min_{g \in G(\{x\})} V(g).$$

We have proved that $\min_{g \in G(C)} V(g) + \min_{g \in G(\overline{C} \cup \{x\})} V(g) = \min_{g \in G(\{x\})} V(g)$ and the proposition follows from the fact that

$$\tilde{U}(x) - \tilde{U}(y) = \min_{g \in G(\{x\})} V(g) - \min_{g \in G(\{y\})} V(g). \qquad \square$$

4.5. Iterative construction of cycles and the virtual energy function.

For any subset $D \subset E$, let us put $\tilde{U}(D) = \min_{x \in D} \tilde{U}(x)$.

Proposition 4.10. *Let $E = \bigcup_{i \in I} C_i$ be a partition of E into disjoint cycles. Assume that it is not trivial, namely that $|I| \geq 2$. Let us consider on $\mathcal{C}_I = \{C_i \mid i \in I\}$ the graph s of the typical jumps, defined by*

$$s = \{(C_i, C_i) \mid i \in I\} \cup \{(C_i, C_j) \mid B(C_i) \cap C_j \neq \emptyset\}.$$

Let $\mathcal{C}_J = \{C_j \mid j \in J\}$ be an irreducible and stable component of s, that is a component of $\mathcal{C}_I$ for the equivalence relation

$$\mathcal{R}_s = \{(C_i, C_j) : i, j \in I, C_i \in O_s(C_j) \text{ and } C_j \in O_s(C_i)\} \cup \{(C_i, C_i) : i \in I\},$$

such that $s(\mathcal{C}_J) \subset \mathcal{C}_J$. There exists at least one such component, because s induces on $\mathcal{C}_J/\mathcal{R}_s$ a graph without loop, which has therefore at least one leaf (or terminal node). Moreover J is not reduced to one point, because this would mean that the principal boundary of the would be unique cycle in $\mathcal{C}_J$ would be empty, which is impossible.

Then $C = \bigcup_{j \in J} C_j$ is a cycle, and $C_j, j \in J$ are the maximal strict subcycles of C for the inclusion relation. Moreover, for any $i, j \in J$

$$\tilde{U}(C_i) + H(C_i) = \tilde{U}(C_j) + H(C_j),$$
$$H(C) = \min_{j \in J, y \notin C} V(C_j, y) + \max_{j \in J} H(C_j),$$

and for any $y \notin C$

$$V(C, y) = \min_{j \in J} V(C_j, y) - \min_{z \notin C, j \in J} V(C_j, z).$$

Remark: This proposition allows to build iteratively all the cycles, starting from the trivial partition of E into one point sets, computing in the same time the quantities $\tilde{U}(x) - \tilde{U}(C)$, $x \in C$, $H(C)$ and $V(C, y), y \notin C$.

Proof: Let $y \in C$. We will prove that for any $\hat{g} \in \arg\min_{g \in G(\overline{C} \cup \{y\})} V(g)$, $\hat{g}(C) \subset C$.

Let us assume that $y \in C_{j_0}$. As $\mathcal{C}_J$ is a component of $\mathcal{C}_I/\mathcal{R}_s$, it is possible to extract from $s/\mathcal{C}_J$ an oriented tree α with root C_{j_0} (we mean by this that a graph without loop connecting each point of $\mathcal{C}_J$ to C_{j_0}). Let $g^{j_0} \in \arg\min_{g \in G(\overline{C}_{j_0} \cup \{y\})} V(g)$ and for any $j \in J \setminus \{j_0\}$, let $g^j \in \arg\min_{g \in G(\overline{C_j})} V(g)$ be such that $g^j(C_j) \subset C_j \cup \alpha(C_j)$. Such a g^j exists, according to proposition 4.8, because $B(C_j) \cap \alpha(C_j) \neq \emptyset$. The graph $\bigcup_{j \in J} g^j$ is without loop, and therefore belongs to $G(\overline{C} \cup \{y\})$, thus

$$\sum_{j \in J} V(g^j) = V\left(\bigcup_{j \in J} g^j\right) \geq V(\hat{g}) = \sum_{j \in J} V(\hat{g}_{|C_j}).$$

This proves that

$$V(\hat{g}_{|C_j}) = \min_{g \in G(\overline{C}_j)} V(g), \qquad j \in J \setminus \{j_0\}$$

and

$$V(\hat{g}_{|C_{j_0}}) = \min_{g \in G(\overline{C}_{j_0} \cup \{y\})} V(g),$$

and therefore that

$$\hat{g}(C_{j_0}) = C_{j_0}$$

and

$$\hat{g}(C_j) \subset C_j \cup B(C_j), \qquad j \in J \setminus \{j_0\}.$$

Thus $\hat{g}(C) = C$.

This shows that C is a cycle. Let us prove now that $C_j \subset C$ are maximal among the subcycles of C distinct from C itself.

Assume that for some $j_0 \in J$ and some cycle $C' \in \mathcal{C}(V)$, $C_{j_0} \subset C' \subset C$, $C_{j_0} \neq C'$. As $\mathcal{C}(V)$ is a tree (proposition 4.4), there is $J' \subset J$ such that $C' = \bigcup_{j \in J'} C_j$ and $\{j_0\} \neq J'$. From a preceding proposition, for any $j \in J'$, $B(C_j) \subset C'$, since $C_j \neq C'$. Therefore, $s(C_j, j \in J') \subset \{C_j, j \in J'\}$, which implies that $J' = J$ and therefore that $C' = C$.

From the local computation of the virtual energy into cycles, we see that

$$\tilde{U}(C_i) - \tilde{U}(C_j) = \min_{x \in C_i} \min\{V(g) : g \in G(\overline{C} \cup \{x\})$$
$$- \min_{y \in C_j} \min\{V(g) : g \in G(\overline{C} \cup \{y\})\}.$$

From the preceding computation, for any $x \in C_i$

$$\min\{V(g) : g \in G(\overline{C} \cup \{x\})\} = \sum_{k \in J \setminus \{i\}} \min\{V(g) : g \in G(\overline{C}_k)\}$$
$$+ \min\{V(g) : g \in G(\overline{C}_i \cup \{x\})\}.$$

Therefore

$$\tilde{U}(C_i) - \tilde{U}(C_j) = \min_{x \in C_i} \min\{V(g) : g \in G(\overline{C}_i \cup \{x\})\}$$
$$- \min_{y \in C_j} \min\{V(g) : g \in G(\overline{C}_j \cup \{y\})\}$$
$$+ \min\{V(g) : g \in G(\overline{C}_j)\} - \min\{V(g) : g \in G(\overline{C}_i)\}$$
$$= H(C_j) - H(C_i).$$

Similarly

$$H(C) = \min\{V(g) : g \in G(\overline{C})\} - \min_x \min\{V(g) : g \in G(\overline{C} \cup \{x\})\}$$
$$= \min_{z \in \overline{C}} \min_{j \in J} \Big\{ \min\{V(g) : g \in G_{C_j,z}(\overline{C}_j)\}$$
$$+ \sum_{k \in J \setminus \{j\}} \min\{V(g) : g \in G(\overline{C}_k)\} \Big\}$$
$$- \min\{V(g) : x \in C, g \in G(\overline{C} \cup \{x\})\}$$
$$= \min_{z \in \overline{C}} \min_{j \in J} V(C_j, z) + \sum_{k \in J} \min\{V(g) : g \in G(\overline{C}_k)\}$$
$$- \min\{V(g) : g \in G(\overline{C} \cup \{x\}), x \in C\}$$
$$= \min_{z \in \overline{C}} \min_{j \in J} V(C_j, z) - \max_{j \in J} H(C_j).$$

We have also for any $x \in C$,

$$V(C, y) = \min_{g \in G_{C,y}(\overline{C})} V(g) - \min_{g \in G(\overline{C})} V(g)$$
$$= \min_j \min_{g \in G_{C_j,y}(\overline{C}_j)} V(g) + \sum_{k \in J \setminus \{j\}} \min_{g \in G(\overline{C}_j)} V(g)$$
$$- \min_z \min_j \left(\min_{g \in G_{C_j,z}(\overline{C}_j)} V(g) + \sum_{k \in J \setminus \{j\}} \min_{g \in \overline{C}_j} V(g) \right)$$
$$= \min_{j \in J} V(C_j, y) - \min_z \min_{j \in J} V(C_j, z). \qquad \square$$

4.6. **Maximal depth and maximal partition of a domain.** In this subsection we will compute the maximal depth $H(D) \overset{\text{def}}{=} \max_{x \in D} H_D(x)$, of a domain $D \subset E$ in terms of the maximal partition of D defined below:

Definition 4.3. For any domain $D \subset E$, we let $\mathcal{M}(D)$ be the set of maximal elements of $\{C \in \mathcal{C}(E) : C \subset D\}$ for the inclusion relation. Due to the tree structure of $\mathcal{C}(E)$, this is a partition of D. We call it the maximal partition of D.

From the graph point of view, the maximal partition has the important following property:

Lemma 4.1. *For any domain $D \subsetneq E$*

(3)
$$\min_{g \in G(\overline{D})} V(g) = \sum_{C \in \mathcal{M}(D)} \min_{g \in G(\overline{C})} V(g).$$

Proof. A first remark is that for any $g \in G(\overline{D})$

$$V(g) = \sum_{C \in \mathcal{M}(D)} V(g_{|C}),$$

This proves that the left hand side of equation (3) is not smaller than the right hand side. To prove the reverse inequality, consider the graph s on $\mathcal{M}(D) \cup \{\overline{D}\}$ defined by

$$(C_1, C_2) \in s \text{ iff } C_1 \in \mathcal{M}(D) \text{ and } C_2 \cap B(C_1) \neq \emptyset.$$

Then according to proposition 4.10, s is without any stable irreducible component and connects every cycle of $\mathcal{M}(D)$ to $\overline{D}$. Therefore it can be spanned by a disjoint union of oriented trees leading to $\overline{D}$, from which we can build as in the proof of proposition 4.10 a graph $\hat{g} \in G(\overline{D})$ such that $\hat{g}_{|C} = \min\{V(g) ; g \in G(\overline{C})\}$ for any $C \in \mathcal{M}(D)$, proving that the right hand side is not smaller than the left hand side of equation (3). $\square$

We are ready now to compute the maximal depth of a domain:

Proposition 4.11. *For any domain $D \subsetneq E$ let us define*

$$H(D) = \max_{x \in D} H_D(x).$$

Then

$$H(D) = \max\{H(C) ; C \in \mathcal{M}(D)\}.$$

Proof. Let us put for short for any set of graphs G

$$V(G) = \min_{g \in G} V(g).$$

By definition we have

$$H(D) = V(G(\overline{D})) - \min_{y \in D} V(G(\overline{D} \cup \{y\})).$$

For any fixed $y \in D$, let $C \in \mathcal{M}(D)$ be such that $y \in C$. Remarking that

$$\mathcal{M}(D \setminus \{y\}) = \mathcal{M}(D \setminus C) \cup \mathcal{M}(C \setminus \{y\}),$$

and using the previous lemma we get that

$$H(D) = \max_{C \in \mathcal{M}(D)} \max_{y \in C} V(G(\overline{C})) - V(G(\overline{C} \cup \{y\}))$$

$$= \max_{C \in \mathcal{M}(D)} H(C).$$

$\square$

4.7. Computing the cycles in term of path elevations.

In order to give a description of cycles and therefore of the behaviour of the trajectories which recalls what happens in the case when the rate function V derives from an energy function U, we will introduce a characterisation of the energy based on paths instead of graphs.

Energy barrier between two points

For any two states $x, y \in E$, let $\Gamma_{x,y}$ be the set of paths joining x to y:

$$\Gamma_{x,y} = \{(x_0, \ldots, x_r) : r > 0, x_0 = x, x_r = y\} \subset \bigcup_r E^r.$$

For any path $\gamma = (x_0, \ldots, x_r)$, let

$$H(\gamma) = \max_{i=1,\ldots,r} \tilde{U}(x_{i-1}) + V(x_{i-1}, x_i),$$

with the convention that when $r = 0$ we put

$$H((x_0)) = \tilde{U}(x_0).$$

The energy barrier between x and y is defined to be

$$H(x, y) = \min_{\gamma \in \Gamma_{x,y}} H(\gamma).$$

Proposition 4.12 (energy barrier of a cycle). *For any cycle $C \in \mathcal{C}(E, V)$, any $y \notin C$, we have*

$$\min_{x \in C} \tilde{U}(x) + V(x, y) = \tilde{U}(C) + H(C) + V(C, y).$$

Proof:

$$\min_{x \in C} \tilde{U}(x) + V(x, y) - \tilde{U}(C)$$

$$= \min_{x \in C} \left\{ \min_{g \in G(\overline{C} \cup \{x\})} V(g) + V(x, y) \right\} - \min_{z \in C} \min_{g \in G(\overline{C} \cup \{z\})} V(g)$$

$$= \min_{g \in G_{C,y}(\overline{C})} V(g) - \min_{z \in C} \min_{g \in G(\overline{C} \cup \{z\})} V(g)$$

$$= V(C, y) + H(C). \qquad \square$$

Proposition 4.13 (elevation of paths within a cycle). *For any cycle $C \in \mathcal{C}(V)$, any $x \in C$, any $y \notin C$, there is a path $\varphi \in \Gamma_{x,y}$, $\varphi = (\varphi_0, \ldots, \varphi_s)$ such that $\varphi_i \in C$, $i = 0, \ldots, s - 1$ and $H(\varphi) = \tilde{U}(C) + H(C) + V(C, y)$. For any x, $y \in C$, there is a path $\varphi = (\varphi_0, \ldots, \varphi_s) \in \Gamma_{x,y}$ such that $\varphi_i \in C$, $i = 0, \ldots, s$ and $H(\varphi) \le \tilde{U}(C) + \sup\{H(\tilde{C}) \,|\, \tilde{C} \in \mathcal{C}, \tilde{C} \subset C, \tilde{C} \ne C\}$ (with the convention that $\sup \emptyset = 0$).*

Proof. Let us proceed by induction on the size of cycles. For any $x, y \in C$, there are $C_0, \ldots, C_k \in \mathcal{C}(V)$ such that $C_i \subset C$, $C_i \neq C$, C_i are maximal, $B(C_{i-1}) \cap C_i \neq \emptyset$, $i = 1, \ldots, k$, $x \in C_0$, $y \in C_k$. This is a consequence of proposition 4.10 on the iterative construction of cycles. Let y_i, $i = 1, \ldots, k$ be a point in $B(C_{i-1}) \cap C_i$ and let $y_0 = x$. According to our induction hypothesis that proposition 4.13 is true for the strict subcycles of C, we can find paths $\varphi^i \in \Gamma_{y_{i-1}, y_i}$, $i = 1, \ldots, k$, such that $\varphi^i \subset C$ and $H(\varphi^i) = \tilde{U}(C_{i-1}) + H(C_{i-1})$. We can also find $\varphi^{k+1} \in \Gamma_{y_k, y}$ such that $\varphi^{k+1} \subset C_k \subset C$ and $H(\varphi^{k+1}) \leq \tilde{U}(C_k) + H(C_k)$. The concatenated path $\varphi_{x,y} = (\varphi^1, \ldots, \varphi^{k+1}) \in \Gamma_{x,y}$ belongs to C and has an elevation lower than $\tilde{U}(C) + \max\{H(\tilde{C}) : \tilde{C} \in \mathcal{C}(V), \tilde{C} \subset C, \tilde{C} \neq C\}$. Let us now consider $x \in C$ and $z \notin C$, we can find according to proposition 4.12 a point y such that $\tilde{U}(y) + V(y, z) = \tilde{U}(C) + H(C) + V(C, z)$. Let $\varphi_{x,y}$ be constructed as above. The path $(\varphi_{x,y}, z)$ is included in C except its end point z and has an elevation equal to $\tilde{U}(C) + H(C) + V(C, z)$. Proposition 4.13 being easily seen to be true for one point cycles is therefore proved by induction. $\qquad\square$

Proposition 4.14. *The elevation function is symmetric:*

$$H(x, y) = H(y, x), \quad x, y \in E.$$

Proof: Let $C_1 \in \mathcal{C}(V)$ be the largest cycle such that $x \in C_1$, $y \notin C_1$. Let $C_2 \in \mathcal{C}(V)$ be the largest cycle such that $x \notin C_2$, $y \in C_2$. Let $C_3 \in \mathcal{C}(V)$ be the smallest cycle such that $\{x, y\} \in C_3$. The cycles C_1 and C_2 are maximal strict subcycles of C_3, therefore $H(x, y) = H(C_1) + \tilde{U}(C_1) = H(C_2) + \tilde{U}(C_2) = H(y, x)$. $\square$

Proposition 4.15. *For any cycle $C \in \mathcal{C}$,*

$$H(C) = \max_{x \in C} \min_{y \notin C} H(x, y) - \tilde{U}(x),$$

and more generally for any $D \subset E$, $D \neq E$, $D \neq \emptyset$,

$$H(D) = \max_{x \in D} \min_{y \notin D} H(x, y) - \tilde{U}(x).$$

Proof. The case of a cycle is a direct consequence of propositions 4.12 and 4.13 . In the case of a general domain D, one has to consider the maximal partition $\mathcal{M}(D)$ of D and apply proposition 4.11, to see that if C_0 is one of the deepest cycles in $\mathcal{M}(D)$ then $H(D) = H(C_0)$. Taking x in the bottom of C_0, and remarking that

$$\min_{y \notin D} H(x, y) \geq \min_{y \notin C_0} H(x, y),$$

we get that

$$H(D) = H(C_0) \leq \max_{x \in D} \min_{y \notin D} H(x, y) - \tilde{U}(x).$$

Now, for the converse, let x be any point in D and let C_0 be the maximal cycle of $\mathcal{M}(D)$ to which x belongs. As seen in the proof of equation (3), there is a sequence of cycles $C_0, \ldots, C_r$ such that $B(C_i) \cap C_{i+1} \neq \emptyset$, $i = 0, \ldots, r-1$ and $B(C_r) \cap \overline{D} \neq \emptyset$. Remark that $\tilde{U}(C_i) + H(C_i)$ is decreasing: indeed, taking

$u \in C_i$ and $v \in B(C_i) \cap C_{i+1}$, we see that $\tilde{U}(C_i) + H(C_i) = H(u, v) = H(v, u) \geq \tilde{U}(C_{i+1}) + H(C_{i+1})$. With the help of proposition 4.13 we build a path γ, starting at x, going through this sequence of cycles and ending in $\overline{D}$ such that

$$\min_{y \notin D} H(x, y) \leq H(\gamma)$$

$$\leq \max_{0 \leq i \leq r} \tilde{U}(C_i) + H(C_i)$$

$$= \tilde{U}(C_0) + H(C_0)$$

$$\leq \tilde{U}(x) + H(D).$$

$\square$

Proposition 4.16 (Weak reversibility condition of Hajek and Trouvé).
Let $U : E \longrightarrow \mathbb{R}$ be an arbitrary real valued function defined on E. Let the elevation $H_U(\gamma)$ of a path $\gamma = (z_0, \ldots, z_r) \in E^{r+1}$ with respect to U and V be defined by

$$H_U(\gamma) = \max_{i=1,\ldots,r} U(z_{i-1}) + V(z_{i-1}, z_i).$$

Let $H_U(x, y) = \min_{\gamma \in \Gamma_{x,y}} H_U(\gamma)$, $x, y \in E$, $x \neq y$.
Then $U(x) = \tilde{U}(x) + \min_{y \in E} U(y)$, $x \in E$ if and only if H_U is symmetric.

Proof: See question 10.3 of the appendix for some hints about the proof.

Proposition 4.17. *For any $x, y \in E$,*

$$\tilde{U}(y) \leq \tilde{U}(x) + V(x, y).$$

Consequently for any path $\gamma = (\gamma_0, \ldots, \gamma_r)$

$$H(\gamma) \leq \tilde{U}(\gamma_0) + \sum_{k=1}^{r} V(\gamma_{k-1}, \gamma_k).$$

Proof:

$$\tilde{U}(y) \leq H(y, x) = H(x, y) \leq \tilde{U}(x) + V(x, y). \qquad \square$$

4.8. Another construction of cycles. For any $\lambda \in \mathbb{R}$, let us introduce the equivalence relation

$$\mathcal{R}_\lambda = \{(x, y) \in E^2 \mid x \neq y, H(x, y) < \lambda\} \cup \{(x, x) \mid x \in E\}.$$

Proposition 4.18. *The components of $E/\mathcal{R}_\lambda$ are the one point sets $\{x\}$ such that $\tilde{U}(x) \geq \lambda$ and the cycles $C \in \mathcal{C}(V)$ such that*

(4) $\max\{\tilde{U}(\tilde{C}) + H(\tilde{C}) \mid \tilde{C} \in \mathcal{C}, \tilde{C} \subset C, \tilde{C} \neq C\} < \lambda \leq \tilde{U}(C) + H(C)$.

Thus $\mathcal{C}(V) = \bigcup_{\lambda \in \mathbb{R}_+} E/\mathcal{R}_\lambda$.

Proof: If $C \in \mathcal{C}(V)$ satisfies equation (4), then $C \in E/\mathcal{R}_\lambda$ according to previous propositions. On the other hand, let us compute for any $x \in E$ the component of x in $E/\mathcal{R}_\lambda$. Let us consider the maximal sequence of distinct cycles $C_0 = \{x\} \subset C_1 \subset C_2 \subset \cdots \subset E$ containing x (C_i is the smallest cycle strictly containing

C_{i-1}). If $H(C_0) + \tilde{U}(C_0) > \lambda$, then $\{x\} \in E/\mathcal{R}_\lambda$, otherwise let us consider $i_0 = \min\{i \mid H(C_i) + \tilde{U}(C_i) \geq \lambda\}$, then according to the first part of the proof $x \in C_{i_0} \in E/\mathcal{R}_\lambda$. $\quad\square$

4.9. Exit time from a subdomain.

From Freidlin and Wentzell's lemma we deduce that

Proposition 4.19. *For any subdomain $D \subset E$, $D \neq E$, any $x \in D$, any $\epsilon > 0$,*

$$\lim_{\beta \to +\infty} -\frac{1}{\beta} \log P_\beta(\tau(\overline{D}) > e^{\beta(H(D)+\epsilon)} \mid X_0 = x) = +\infty,$$

and

$$\liminf_{\beta \to +\infty} -\frac{1}{\beta} \log(\min_{y \in D} P_\beta(\tau(\overline{D}) < e^{\beta(H(D)-\epsilon)} \mid X_0 = y) \geq \epsilon,$$

where $H(D) = \max_{y \in D} H_D(y)$.

Proof: Applying the Markov property, we see that:

$$P(\tau(\overline{D}) > e^{\beta(H(D)+\epsilon)} \mid X_0 = x)$$
$$\leq \left(\max_{y \in D} P(\tau(\overline{D}) > e^{\beta(H(D)+\epsilon/2)} \mid X_0 = y) \right)^{\lfloor e^{\beta\epsilon/2} \rfloor}$$
$$\leq \left(\max_{y \in D} E(\tau(\overline{D}) \mid X_0 = y) e^{-\beta(H(D)+\epsilon/2)} \right)^{\lfloor e^{\beta\epsilon/2} \rfloor}$$
$$\leq \exp -\beta \left(\frac{\epsilon}{4} \left(e^{\beta\epsilon/2} - 1 \right) \right),$$

To prove the second equation, let us notice that

$$\sum_{k=0}^{+\infty} P(\tau(\overline{D}) > ke^{\gamma\beta} \mid X_0 = x) \geq E(e^{-\gamma\beta}\tau(\overline{D}) \mid X_0 = x),$$

and that

$$\sum_{k=0}^{+\infty} P(\tau(\overline{D}) > ke^{\gamma\beta} \mid X_0 = x) \leq \sum_{k=0}^{+\infty} \left(\max_{y \in D} P(\tau(\overline{D}) > e^{\gamma\beta} \mid X_0 = y) \right)^k$$
$$= \left(\min_{y \in D} P(\tau(\overline{D}) \leq e^{\gamma\beta} \mid X_0 = y) \right)^{-1},$$

thus

$$\min_{y \in D} P(\tau(\overline{D}) \leq e^{\gamma\beta} \mid X_0 = y) \leq e^{\gamma\beta} E(\tau(\overline{D}) \mid X_0 = x)^{-1}. \quad\square$$

Proposition 4.20. *For any cycle $C \in \mathcal{C}(V)$, any sufficiently small $\epsilon > 0$, any $x \in C$,*

$$\liminf_{\beta \to +\infty} -\frac{1}{\beta} P_\beta(\tau(\overline{C}) < e^{\beta(H(C)-\epsilon)} \mid X_0 = x) \geq \epsilon.$$

Proof:

For any $x, y \in C$, $\gamma > 0$,

$$P(\tau(\overline{C}) < e^{\gamma\beta} \mid X_0 = x) \leq P(\tau(\overline{C}) < e^{\beta\gamma} \mid X_0 = y)P(X_{\tau(\overline{C}\cup\{y\})} = y \mid X_0 = x)$$
$$+ P(X_{\tau(\overline{C}\cup\{y\})} \neq y \mid X_0 = x).$$

Let

$$\epsilon_0 = \min_{x,y \in C, x \neq y} \lim_{\beta \to +\infty} -\frac{1}{\beta} \log P(X_{\tau(\overline{C}\cup\{y\})} \neq y \mid X_0 = x)$$

$$= \min\{V_{C\setminus\{y\}}(x, z) : x, y \in C, x \neq y, z \in \overline{C}\} > 0,$$

then for all $\epsilon < \epsilon_0$ and β large enough

$$P(\tau(\overline{C}) < e^{\beta\gamma} \mid X_0 = x) \leq \min_{y \in \overline{C}} P(\tau(\overline{C}) < e^{\beta\gamma} \mid X_0 = y) + e^{-\beta\epsilon}.$$

We end the proof by taking $\gamma = H(C) - \epsilon$ and applying the preceding proposition.
□

5. CONVERGENCE TOWARDS EQUILIBRIUM

Proposition 5.1. *For any cycle $C \in \mathcal{C}(V)$, any $\gamma > 0$ such that $H(\{t \in C \mid \tilde{U}(t) > \tilde{U}(C)\}) < \gamma < H(C)$, any $x, y \in C$,*

$$\liminf_{\beta \to +\infty} -\frac{1}{\beta} \log P(X_{\lfloor e^{\beta\gamma} \rfloor} = y, \tau(\overline{C}) > e^{\gamma\beta} \mid X_0 = x) \geq \tilde{U}(y) - \tilde{U}(C).$$

Corollary 5.1.

$$\liminf_{\beta \to +\infty} -\frac{1}{\beta} \log P(\tilde{U}(X_{\lfloor e^{\beta\gamma} \rfloor}) \neq \tilde{U}(C) \mid X_0 = x) > 0$$

Proof: Let us put $N = \lfloor e^{\gamma\beta} \rfloor$. Let $A = \arg\min_{x \in C} \tilde{U}(x)$. For any $x, y \in C$,

$$P_\beta(X_n = y, \tau(\overline{C}) > e^{\gamma\beta} \mid X_0 = x) \leq P_\beta(\tau(\overline{C} \cup A) > e^{\gamma\beta} \mid X_0 = x)$$
$$+ \sup_{k \in \mathbb{N}, z \in A} P(X_k = y \mid X_0 = z).$$

Let $f_k(x) = P_\beta(X_k = x \mid X_0 = z)\mu_\beta(x)^{-1}$. We have $\sum_{x \in E} f_k(x)p_\beta(x, y)\dfrac{\mu_\beta(x)}{\mu_\beta(y)} = f_{k+1}(y)$, and $\sum_{x \in E} p_\beta(x, y)\dfrac{\mu_\beta(x)}{\mu_\beta(y)} = 1$, therefore

$$\max_{y \in E} f_k(x) \leq \max_{y \in E} f_0(x) = \frac{1}{\mu_\beta(z)},$$

and $\sup_{k \in \mathbb{N}} P(X_k = y \mid X_0 = z) \leq \dfrac{\mu_\beta(y)}{\mu_\beta(z)}.$ □

Proposition 5.2. *Let us assume that $C \in \mathcal{C}(V)$ is such that for some $z \in \arg\min_{x \in C} \tilde{U}(x)$, considering the graph s of the null cost jumps,*

$$s = \{(x, y) \in E^2 : V(x, y) = 0\},$$

the orbit $O_s(z)$ is aperiodic. Then for any $x, y \in C$, any γ such that $H(C \setminus \{z\}) < \gamma < H(C)$,

$$\lim_{\beta \to +\infty} -\frac{1}{\beta} \log P_\beta(X_{\lfloor e^{\gamma\beta}\rfloor} = y, \tau(\overline{C}) > e^{\gamma\beta} \mid X_0 = x) = \tilde{U}(y) - \tilde{U}(C).$$

Proof:

Let us consider the Markov chain $(Y_n)_{n \in \mathbb{N}}$ on C with transitions

$$P_\beta(Y_n = y \mid Y_{n-1} = x) = \lim_{M \to +\infty} P_\beta(X_n = y \mid X_{n-1} = x, \tau(\overline{C}) > M).$$

The existence of this limit is a consequence of the Perron-Frobenius theorem applied to the (non stochastic) aperiodic irreducible non negative matrix $p_{\beta|C \times C}$. This theorem says that

$$p_{\beta|C \times C} = \rho \, \pi_1 + R \circ \pi_2,$$

where (π_1, π_2) forms a system of projectors (i.e. $\pi_1 \circ \pi_2 = \pi_2 \circ \pi_1 = 0$ and $\pi_1 + \pi_2 = \mathrm{Id}$), where π_1 is the projection on the one dimensional vector space generated by a positive eigenvector, where $\rho > 0$ is the spectral radius of $p_{\beta|C \times C}$ and where the spectral radius of R is strictly lower than ρ. This implies that

$$\lim_{M \to +\infty} \frac{\delta_z (p_{\beta|C \times C})^M \mathbf{1}}{\delta_y (p_{\beta|C \times C})^M \mathbf{1}} = \lim_{M \to +\infty} \frac{P_\beta(\tau(\overline{C}) > M \mid X_0 = z)}{P_\beta(\tau(\overline{C}) > M \mid X_0 = y)}$$

exists for any $y, z \in C$ and is equal to

$$\frac{\delta_z \pi_1 \mathbf{1}}{\delta_y \pi_1 \mathbf{1}}.$$

Therefore as soon as $p_\beta(x, y) > 0$,

$$\frac{P_\beta(X_n = z \mid X_{n-1} = x, \tau(\overline{C}) > M)}{P_\beta(X_n = y \mid X_{n-1} = x, \tau(\overline{C}) > M)} = \frac{p_\beta(x, z)}{p_\beta(x, y)} \frac{P_\beta(\tau(\overline{C}) > M - n \mid X_0 = z)}{P_\beta(\tau(\overline{C}) > M - n \mid X_0 = y)}$$

has a limit when M tends to infinity, which proves in turn the existence of the limit defining the transitions of Y at temperature β.

Now that the definition of Y is justified, let us return to the main stream of our proof. We have

$$P_\beta(X_{\lfloor e^{\gamma\beta}\rfloor} = y, \tau(\overline{C}) > \lfloor e^{\gamma\beta}\rfloor \mid X_0 = x) P_\beta(\tau(\overline{C}) > M - \lfloor e^{\gamma\beta}\rfloor \mid X_0 = y)$$
$$= P_\beta(X_{\lfloor e^{\gamma\beta}\rfloor} = y, \tau(\overline{C}) > M \mid X_0 = x),$$

and therefore

$$P_\beta(X_{\lfloor e^{\gamma\beta}\rfloor} = y, \tau(\overline{C}) > e^{\gamma\beta} \mid X_0 = x)$$
$$= P_\beta(X_{\lfloor e^{\gamma\beta}\rfloor} = y \mid \tau(\overline{C}) > M, X_0 = x) \frac{P_\beta(\tau(\overline{C}) > M \mid X_0 = x)}{P_\beta(\tau(\overline{C}) > M - \lfloor e^{\gamma\beta}\rfloor \mid X_0 = y)}.$$

Moreover

$$P_\beta(\tau(\overline{C}) > M \mid X_0 = x) =$$
$$\sum_{z \in C} P_\beta(X_{\lfloor e^{\gamma\beta}\rfloor} = z, \tau(\overline{C}) > e^{\gamma\beta} \mid X_0 = x) P_\beta(\tau(\overline{C}) > M - \lfloor e^{\gamma\beta}\rfloor \mid X_0 = z).$$

Let $K = M - \lfloor e^{\gamma\beta} \rfloor$,

$$P_\beta(\tau(\overline{C}) > K \mid X_0 = z) \geq P_\beta(X_{\tau(\overline{C} \cup \{y\})} = y \mid X_0 = z)P_\beta(\tau(\overline{C}) > K \mid X_0 = y),$$

therefore

$$\limsup_{\beta \to +\infty} \sup_{K \in \mathbb{N}} \left| \frac{P_\beta(\tau(\overline{C}) > K \mid X_0 = z)}{P_\beta(\tau(\overline{C}) > K \mid X_0 = y)} - 1 \right| = 0.$$

Thus

$$\lim_{\beta \to +\infty} \sup_{M > e^{\gamma\beta}} \left| \frac{P_\beta(\tau(\overline{C}) > M \mid X_0 = x)}{P_\beta(\tau(\overline{C}) > M - \lfloor e^{\gamma\beta} \rfloor \mid X_0 = y)} - 1 \right|$$

$$= \lim_{\beta \to +\infty} \left| \sum_{z \in C} P_\beta(X_{\lfloor e^{\gamma\beta} \rfloor} = z, \tau(\overline{C}) > e^{\gamma\beta} \mid X_0 = x) - 1 \right|$$

$$= 0,$$

and, letting $M \to +\infty$,

$$\lim_{\beta \to +\infty} \frac{P_\beta(X_{\lfloor e^{\gamma\beta} \rfloor} = y, \tau(\overline{C}) > e^{\gamma\beta} \mid X_0 = x)}{P(Y_{\lfloor e^{\gamma\beta} \rfloor} = y \mid Y_0 = x)} = 1.$$

In the same way, we can prove that for any $x, y \in C$,

$$\lim_{\beta \to +\infty} \frac{p_\beta(x, y)}{P_\beta(Y_1 = y \mid Y_0 = x)} = 1.$$

Therefore Y is a Markov chain with rare transitions and rate function $V_{C \times C}$. According to proposition 4.9, the virtual energy of Y is $(\tilde{U}(x) - \tilde{U}(C))_{x \in C}$. Therefore it is enough to prove the proposition in the special case when $C = E$. We will assume in the following of the proof that we are in this case. Let us consider the family of product Markov chains $\left((E \times E)^{\mathbb{N}}, (X^1, X^2)_{n \in \mathbb{N}}, \mathcal{B} \otimes \mathcal{B}, P_\beta^1 \otimes P_\beta^2 \right)_{\beta \in \mathbb{R}_+}$, where P_β^1 and P_β^2 have the same transitions as P_β and have the following initial distributions:

$$\begin{aligned} P_\beta^1 \circ (X_0^1)^{-1} &= \delta_x, \\ P_\beta^2 \circ (X_0^2)^{-1} &= \mu_\beta, \end{aligned}$$

(here μ_β is as usual the invariant distribution at inverse temperature β). It is a family of Markov chains with rare transitions with rate function

$$V^2(x, y) = V(x^1, y^1) + V(x^2, y^2).$$

Moreover $H^2((E \times E) \setminus \{(z, z)\}) = H(E \setminus \{z\})$. Indeed there is n_0 such that for any $n \geq n_0$, there is a path $(\varphi_1, \ldots, \varphi_n)$ such that $\varphi_1 = \varphi_n = z$ and $V(\varphi_{i-1}, \varphi_i) = 0$. For any $x \in E$ there is an infinite path $(\psi_n)_{n \in \mathbb{N}}$. such that $\psi_1 = x$ and $\tilde{U}(\psi_i) \leq \tilde{U}(x)$, (take a path such that $V(\psi_{i-1}, \psi_i) = 0$). Moreover, for any $i \in \mathbb{N}$, $H(\psi_i, z) \leq H(x, z)$, indeed

$$
\begin{aligned}
H(\psi_i, z) &\leq \max(H(\psi_i, x), H(x, z)) \\
&= \max(H(x, \psi_i), H(x, z) \\
&= \max(\tilde{U}(x), H(x, z)) \\
&= H(x, z).
\end{aligned}
$$

With these two types of paths, it is easy to build in $E \times E$ a path $\psi \in \Gamma_{(x,y),(z,z)}$ such that

$$
H(\psi) \leq (\tilde{U}(x) + H(y, z)) \vee H(x, z)
$$

(Let the first component follow ψ while the second component is led to z via a path of minimal elevation $H(y, z)$, then let the first component follow a path of minimum elevation, while the second component follows a path φ of suitable length.) This proves that $H^2(E \times E \setminus \{(z, z)\}) = H(E \setminus \{z\})$, because it cannot obviously be lower.

Now for any $y \in C$, putting $N = \lfloor e^{\gamma \beta} \rfloor$, applying the Markov property at time $\tau^2(\{(z, z)\})$, and remarking that X^1 and X^2 conditioned by the same initial condition have the same distribution, we have

$$
\begin{aligned}
P_\beta^1 \otimes P_\beta^2(X_N^1 = y) &\geq P_\beta^1 \otimes P_\beta^2((X_N^1 = y) \text{ and } \tau^2(\{(z, z)\}) \leq N) \\
&= P_\beta^1 \otimes P_\beta^2((X_N^2 = y) \text{ and } \tau^2(\{(z, z)\}) \leq N) \\
&\geq P_\beta^2(X_N^2 = y) - P_\beta^1 \otimes P_\beta^2(\tau^2(\{(z, z)\}) > N).
\end{aligned}
$$

(This argument is equivalent to considering a "coupled" Markov chain where X^1 and X^2 are glued together once they meet.) As

$$
\lim_{\beta \to +\infty} -\frac{1}{\beta} \log P_\beta^1 \otimes P_\beta^2(\tau^2(\{(z, z)\}) > N) = +\infty,
$$

we get the desired result. □

Theorem 5.1 (convergence rate). *Let us put*

$$
\begin{aligned}
H_1 &= H(E \setminus \arg\min \tilde{U}) \\
H_2 &= H(E \setminus \{z\}), \ z \in \arg\min \tilde{U}, \\
H_3 &= H^2((E \times E) \setminus \Delta),
\end{aligned}
$$

where the value of H_2 is independent from the choice of $z \in \arg\min_{x \in E} \tilde{U}(x)$ and where $\Delta = \{(x, x) : x \in E\}$. For any $\gamma > H_1$, any $x \in E$, $y \in E$,

$$
\liminf_{\beta \to +\infty} -\frac{1}{\beta} \log P_\beta(X_{\lfloor e^{\gamma \beta} \rfloor} = y \mid X_0 = x) \geq \tilde{U}(y).
$$

For any $\gamma > H_2$

$$
\lim_{\beta \to +\infty} -\frac{1}{\beta} \log P_\beta(\tau(\{z\}) > e^{\gamma \beta} \mid X_0 = x) = +\infty, \ x \in E, z \in \arg\min_{x \in E} \tilde{U}(x).
$$

For any $\gamma > H_3$, any $x, y \in E$

$$
\lim_{\beta \to +\infty} -\frac{1}{\beta} \log P_\beta(X_{\lfloor e^{\gamma \beta} \rfloor} = y \mid X_0 = x) = \tilde{U}(y).
$$

In general the constants $H_1 \leq H_2 \leq H_3$ are distinct. However, when the null cost graph $s = \{(x, y) \in E^2 \mid V(x, y) = 0\} \cup \Delta$ has an aperiodic component in $\arg\min \tilde{U}$, we have $H_2 = H_3$. Moreover if $\arg\min \tilde{U}$ is a one point set, then $H_1 = H_2 = H_3$.

Eventually the following non-convergence results holds: for any $\gamma < H_1$, there is $x \in E$ such that

$$\liminf_{\beta \to +\infty} -\frac{1}{\beta} \log P_\beta(\tilde{U}(X_{\lfloor e^{\gamma\beta} \rfloor}) = 0 \mid X_0 = x) > 0;$$

for any $\gamma < H_2$, any $z \in \arg\min \tilde{U}$, there is $x \in E$ such that

$$\liminf_{\beta \to +\infty} -\frac{1}{\beta} \log P_\beta(\tau(\{z\})) \leq e^{\gamma\beta} \mid X_0 = x) > 0,$$

for any $\gamma < H_3$, any $z \in \arg\min \tilde{U}$, there is $x \in E$ such that

$$\limsup_{\beta \to +\infty} -\frac{1}{\beta} \log P_\beta(X_{\lfloor e^{\gamma\beta} \rfloor} = z \mid X_0 = x) > 0.$$

Remark 5.1. The second and the third critical depths are distinct when the chain is "almost" periodic on the set on ground states, that is when it behaves as a periodic chain on a time scale larger than $e^{H_2\beta}$. The non convergence results show that H_1, H_2 and H_3 are sharp.

Proof. The first convergence result is a consequence of proposition 5.1, the second one is a consequence of proposition 4.19, and the third one is proved exactly as the end of the proof of proposition 5.2. The first and second non convergence results are easy corollaries of proposition 4.20. The third non convergence result is proved in the following way: take $(x, y) \in E^2$ in the bottom of the deepest cycle of $E \times E \setminus \{(z, z) ; z \in E\}$. By definition, the depth of this cycle is the third critical depth H_3, therefore for any $\gamma < H_3$, any $z \in \arg\min \tilde{U}$,

$$\lim_{\beta \to +\infty} -\frac{1}{\beta} \log P_\beta(\tau^2(\{(z, z)\})) \leq e^{\gamma\beta} \mid (X^1, X^2)_0 = (x, y)) > 0.$$

But

$$\min\{P_\beta(X_{\lfloor e^{\gamma\beta} \rfloor} = z \mid X_0 = x), P_\beta(X_{\lfloor e^{\gamma\beta} \rfloor} = z \mid X_0 = y)\}$$
$$\leq \sqrt{P_\beta((X^1, X^2)_{\lfloor e^{\gamma\beta} \rfloor} = (z, z) \mid (X^1, X^2)_0 = (x, y))}$$
$$\leq \sqrt{P_\beta(\tau(\{(z, z)\})) \leq e^{\gamma\beta} \mid (X^1, X^2)_0 = (x, y))}.$$

This proves that either

$$\limsup_{\beta \to +\infty} -\frac{1}{\beta} \log P_\beta(X_{\lfloor e^{\gamma\beta} \rfloor} = z \mid X_0 = x) > 0$$

or

$$\limsup_{\beta \to +\infty} -\frac{1}{\beta} \log P_\beta(X_{\lfloor e^{\gamma\beta} \rfloor} = z \mid X_0 = y) > 0$$

$\square$

Corollary 5.2 (choice of β as a function of N). *For any $\eta > 0$ and any $\gamma > H(E \setminus \arg \min \tilde{U}) = H_1$, we have*

$$\liminf_{N \to +\infty} -\frac{1}{\log N} \log P_{(\log N)/\gamma}(\tilde{U}(X_N) \geq \eta \mid X_0 = x) \geq \frac{\eta}{\gamma}.$$

(The probability of failure of the algorithm with N steps has an upper bound of order $\left(\dfrac{1}{N}\right)^{\eta/H_1}$.) On the contrary for any $\gamma < H_1$, there is $x \in E$ such that

$$\liminf_{N \to +\infty} -\frac{1}{\log N} \log P_{(\log N)/\gamma}(\tilde{U}(X_N) < \eta \mid X_0 = x) > 0,$$

(the probability of failure consequently tends to one.)

Remarks:

- The inverse temperature parameter β has to be chosen as a function of the number of iterations N.
- To get an approximate solution y such that $\tilde{U}(y) < \eta$ with probability $1 - \epsilon$, the number of iterations needed is of order $\epsilon^{-H_1/\eta}$.
- To get an exact solution with probability $1 - \epsilon$, it is necessary to set in the previous estimate the value of the constant η to $\eta = \min\{\tilde{U}(z) \mid z \in E, \tilde{U}(z) > 0\}$, which may be very close to zero, in which case the number of iterations needed is very large. Therefore, in some situations, the Metropolis algorithm is very slow and speed-up methods are required.
- Another weakness of the Metropolis algorithm is that it is as a rule impossible to compute explicitly the value of H_1, whereas this value is needed to set the temperature parameter in an efficient way.

6. GEOMETRIC INEQUALITIES FOR EIGENVALUES OF MARKOV CHAINS

6.1. Reversible Markov chains.

6.1.1. *Spectral gap estimates.*

Theorem 6.1. *Let E be a finite set and $(E^{\mathbb{N}}, (X_n)_{n \in \mathbb{N}}, \mathcal{B}, P)$ be the canonical realization of a Markov chain with irreducible and reversible transition matrix p and invariant probability distribution π. Let us define the operator $p : L^2(\pi) \to L^2(\pi)$ by $pf(x) = \sum_{y \in E} p(x, y) f(y)$. This operator is self-adjoint, therefore it can be put in diagonal form and its eigenvalues $\lambda_0 \geq \cdots \geq \lambda_{m-1}$ (where $m = |E|$), counted with their multiplicities, satisfy:*

$$1 = \lambda_0 > \lambda_1 \geq \lambda_2 \geq \cdots \geq \lambda_{m-1} \geq -1.$$

For any probability distribution $\mu \in \mathcal{M}_1^+(E)$, any integer $n \in \mathbb{N}$,

$$\|\mu p^n - \pi\|_{2,\pi} \leq (\max(\lambda_1, -\lambda_{m-1}))^n \|\mu - \pi\|_{2,\pi}.$$

Moreover, for any subset D of E, any $n \in \mathbb{N}$,

$$|P(X_n \in D \mid X_0 = x) - \pi(D)|$$

$$\leq \left(\frac{1 - \pi(x)}{\pi(x)}\right)^{1/2} \min\left(\pi(D)^{1/2}, \frac{1}{2}\right) (\max(\lambda_1, -\lambda_{m-1}))^n.$$

Proof: For any functions $f, g \in L^2(\pi)$,

$$(f, pg) = \sum_{x \in E, y \in E} f(x)p(x, y)g(y)\pi(x)$$

$$(pf, g) = \sum_{x \in E, y \in E} p(x, y)f(y)g(x)\pi(x).$$

As $\pi(x)p(x, y) = \pi(y)p(y, x)$, we have that $(f, pg) = (pf, g)$. The strict inequality between $\beta_0 > \beta_1$ is part of the Perron-Frobenius theorem which we will not prove here. We have $\lambda_{m-1} = -1$ when p is 2-periodic.

The matrix p being irreducible, its invariant measure π is everywhere strictly positive. Therefore we can define a representation

$$i : \mathcal{M}_1^+(E) \longrightarrow L^2(\pi)$$

by $i(\mu) = \dfrac{d\mu}{d\pi}$ and put on $\mathcal{M}_1^+(E)$ the corresponding Euclidean norm $\|\mu\|_{2,\pi} = \|i(\mu)\|_{L^2(\pi)}$. The adjoint operator $\tilde{p} = i^{-1} \circ p \circ i : \mathcal{M}_1^+(E) \longrightarrow \mathcal{M}_1^+(E)$ is nothing but the right action of p: $\tilde{p}(\mu)(y) = (\mu p)(y) = \sum_x \mu(x)p(x, y)$. Note that it is self adjoint for $\| \ \|_{2,\pi}$, with the same spectrum as p.

Let $\rho = \max_{k \geq 1} |\lambda_k| = \max(\lambda_1, -\lambda_{m-1})$. We have $\|\mu p^n - \pi\|_{2,\pi} = \|(\mu - \pi)p^n\|_{2,\pi}$. Moreover

$$(\mu - \pi, \pi)_{2,\pi} = \int \left(\frac{d\mu}{d\pi} - 1\right) d\pi = 0,$$

therefore $\mu - \pi$ is in the space generated by the eigenvectors of $\tilde{p}$ corresponding to the eigenvalues $\lambda_1, \ldots, \lambda_{m-1}$. Let $\nu_1, \ldots, \nu_{m-1}$ be some choice of these eigenvectors

$$(\mu - \pi) = \sum_{k=1}^{m-1} \alpha_k \nu_k,$$

$$(\mu - \pi)p^n = \sum_{k=1}^{m-1} \alpha_k \lambda_k^n \nu_k,$$

$$\|(\mu - \pi)p^n\|_{2,\pi}^2 = \sum_{k=1}^{m-1} |\alpha_k|^2 |\lambda_k|^{2n} \|\nu_k\|_{2,\pi}^2$$

$$\leq \rho^{2n} \sum_{k=1}^{m-1} |\alpha_k|^2 \|\nu_k\|_{2,\pi}^2$$

$$= \rho^{2n} \|\mu - \pi\|_{2,\pi}^2.$$

Moreover

$$|P(X_n \in D \mid X_0 = x) - \pi(D)| = \left| \int_D (f_n(y) - 1) \, d\pi(y) \right|$$

where $f_n(y) = \dfrac{P(X_n = y \mid X_0 = x)}{\pi(y)} = \dfrac{\delta_x p^n(y)}{\pi(y)}$. Applying the Cauchy-Schwartz inequality, we obtain that

$$
\begin{aligned}
|P(X_n \in D \mid X_0 = x) - \pi(D)| &\le \left(\int_D (f_n(y) - 1)^2 d\pi(y) \right)^{1/2} \pi(D)^{1/2} \\
&\le \|\delta_x p^n - \pi\|_{2,\pi} \pi(D)^{1/2} \\
&\le \rho^n \|\delta_x - \pi\|_{2,\pi} \pi(D)^{1/2}.
\end{aligned}
$$

Moreover

$$\|\delta_x - \pi\|_{2,\pi} = \left(\frac{(1 - \pi(x))^2}{\pi(x)} + 1 - \pi(x) \right)^{1/2} = \left(\frac{1 - \pi(x)}{\pi(x)} \right)^{1/2}.$$

In the same way

$$
\begin{aligned}
|P(X_n \in D \mid X_0 = x) - \pi(D)| &\le \int_{y, f_n(y) > 1} (f_n(y) - 1) d\pi(y) \\
&\le \frac{1}{2} \int_E |f_n(y) - 1| d\pi(y) \\
&\le \frac{1}{2} \|\delta_x p^n - \pi\|_{2,\pi} \\
&\le \frac{1}{2} \rho^n \|\delta_x - \pi\|_{2,\pi}. \qquad \square
\end{aligned}
$$

6.1.2. *Poincaré inequalities.* Let us call a "routing function" any function $\gamma : E^2 \longrightarrow \bigcup_{n=2}^{+\infty} E^n$ such that $\gamma(x, y) = (z_0 = x, z_1, \ldots, z_{r(x,y)} = y)$ is a path (of arbitrary length $r(x, y)$) going from x to y, with the supplementary condition that $r(x, y)$ be odd when $x = y$. Let Γ be the set of all routing functions.

For any Markov matrix p, irreducible and reversible with respect to its invariant probability distribution π, we define the length of $(z_0, \ldots, z_r)$ with respect to p by

$$|(z_0, \ldots, z_r)|_p = \sum_{i=1}^{r} (\pi(z_{i-1}) p(z_{i-1}, z_i))^{-1},$$

with the convention that $0^{-1} = +\infty$.

Let us introduce the constants

$$\kappa = \min_{\gamma \in \Gamma} \max_{(z,t) \in E^2 \setminus \Delta} \sum_{\substack{(x,y) \in E^2 \setminus \Delta, \\ (z,t) \in \gamma(x,y)}} |\gamma(x, y)|_p \pi(x) \pi(y)$$

$$\iota = \min_{\gamma \in \Gamma} \max_{(z,t) \in E^2} \sum_{\substack{x \in E, \\ (z,t) \in \gamma(x,x)}} |\gamma(x, x)|_p \pi(x).$$

Theorem 6.2. *With the previous notations, the spectrum* $\lambda_0 = 1 > \lambda_1 \geq \cdots \geq \lambda_{m-1} \geq -1$ *of* p *satisfies*

$$\lambda_1 \leq 1 - \frac{1}{\kappa}$$
$$-\lambda_{m-1} \leq 1 - \frac{2}{\iota}.$$

Proof:

Let us write λ_1 as

$$\lambda_1 = \sup_{\substack{\varphi \in L^2(\pi), \\ E_\pi(\varphi) = 0}} \frac{(\varphi, p\varphi)_\pi}{(\varphi, \varphi)_\pi},$$

this gives

$$1 - \lambda_1 = \inf_{\varphi, E_\pi(\varphi)=0} \frac{\mathcal{E}(\varphi, \varphi)}{(\varphi, \varphi)_\pi},$$

where

$$
\begin{aligned}
\mathcal{E}(\varphi, \varphi) &= (\varphi, \varphi - p\varphi)_\pi \\
&= \sum_{x,y \in E} \varphi(x)(\varphi(x) - p(x,y)\varphi(y))\pi(x) \\
&= \sum_{x,y \in E} \varphi(x)(\varphi(x)p(x,y) - p(x,y)\varphi(y))\pi(x).
\end{aligned}
$$

Let us put $\pi(x)p(x,y) = Q(x,y)$. We have

$$
\begin{aligned}
\mathcal{E}(\varphi, \varphi) &= \sum_{x,y \in E} \varphi(x)(\varphi(x) - \varphi(y))Q(x,y) \\
&= \frac{1}{2} \sum_{x,y \in E} (\varphi(x) - \varphi(y))^2 Q(x,y)
\end{aligned}
$$

(*Remark:* The quadratic form $\mathcal{E}$ is called the Dirichlet form of p.)

When $E_\pi(\varphi) = 0$, we have for any routing function $\gamma \in \Gamma$

$$
\begin{aligned}
(\varphi, \varphi)_\pi &= \frac{1}{2} \sum_{x,y \in E^2 \setminus \Delta} (\varphi(x) - \varphi(y))^2 \pi(x)\pi(y) \\
&= \frac{1}{2} \sum_{x,y \in E^2 \setminus \Delta} \left(\sum_{(z,t) \in \gamma(x,y)} \varphi(t) - \varphi(z) \right)^2 \pi(x)\pi(y) \\
&\leq \frac{1}{2} \sum_{(x,y) \in E^2 \setminus \Delta} \pi(x)\pi(y) \left(\sum_{(z,t) \in \gamma(x,y)} \frac{1}{Q(z,t)} \right) \\
&\qquad \times \sum_{(z,t) \in \gamma(x,y), z \neq t} Q(z,t) \left(\varphi(t) - \varphi(z) \right)^2 \\
&\leq \frac{1}{2} \sum_{(z,t) \in E^2 \setminus \Delta} Q(z,t) \left(\varphi(t) - \varphi(z) \right)^2 \\
&\qquad \times \sum_{\substack{(x,y) \in E^2 \setminus \Delta, \\ (z,t) \in \gamma(x,y)}} \pi(x)\pi(y)|\gamma(x,y)|_p \\
&\leq \mathcal{E}(\varphi, \varphi) \max_{(z,t) \in E^2 \setminus \Delta} \sum_{\substack{(x,y) \in E^2 \setminus \Delta, \\ (z,t) \in \gamma(x,y)}} \pi(x)\pi(y)|\gamma(x,y)|_p.
\end{aligned}
$$

This being true for any choice of $\gamma \in \Gamma$, we have

$$(\varphi, \varphi)_\pi \leq \kappa \mathcal{E}(\varphi, \varphi),$$

whence $1 - \lambda_1 \geq \dfrac{1}{\kappa}$.

Let us come now to the second inequality. We have

$$1 + \lambda_{m-1} = \inf_{\varphi \in L^2(\pi)} \frac{(\varphi, p\varphi + \varphi)_\pi}{(\varphi, \varphi)_\pi}.$$

Moreover

$$
\begin{aligned}
(\varphi, p\varphi + \varphi)_\pi &= \sum_{(x,y) \in E} \varphi(x)(p(x,y)\varphi(y) + \varphi(x))\pi(x) \\
&= \sum_{(x,y) \in E} \varphi(x)(\varphi(x) + \varphi(y))Q(x,y) \\
&= \frac{1}{2} \sum_{(x,y) \in E} (\varphi(x) + \varphi(y))^2 Q(x,y).
\end{aligned}
$$

$$(\varphi, \varphi)_\pi = \sum_{x \in E, \gamma(x,x)=(z_0,\ldots,z_r)} \pi(x) \frac{1}{4} \left(\sum_{i=0}^{r-1} (-1)^i \left(\varphi(z_i) + \varphi(z_{i+1}) \right) \right)^2$$

$$\leq \frac{1}{4} \sum_{x \in E, \gamma(x,x)=(z_0,\ldots,z_r)} \pi(x) \left(\sum_{i=0}^{r-1} \left(\varphi(z_i) + \varphi(z_{i+1}) \right)^2 Q(z_i, z_{i+1}) \right)$$

$$\times \left(\sum_{i=0}^{r-1} \frac{1}{Q(z_i, z_{i+1})} \right)$$

$$\leq \frac{1}{4} \sum_{x \in E} \pi(x) |\gamma(x,x)|_p \sum_{(z,t) \in \gamma(x,x)} \left(\varphi(z) + \varphi(t) \right)^2 Q(z,t)$$

$$\leq (\varphi, p\varphi + \varphi)_\pi \frac{1}{2} \max_{(z,t) \in E^2} \sum_{x \in E, (z,t) \in \gamma(x,x)} \pi(x) |\gamma(x,x)|_p.$$

Thus

$$(\varphi, \varphi)_\pi \leq \frac{\iota}{2} (\varphi, p\varphi + \varphi)_\pi,$$

and $1 + \lambda_{m-1} \geq \dfrac{2}{\iota}$. $\square$

6.2. Application to the generalised Metropolis algorithm.

6.2.1. *Reversible case.*

Theorem 6.3. *Let us consider a family* $(E^{\mathbb{N}}, (X_n)_{n \in \mathbb{N}}, \mathcal{B}, P_\beta)_{\beta \in \mathbb{R}_+}$ *of homogeneous Markov chains with rate function* V *and transition matrix* p_β. *Let us assume that* V *is irreducible and that* p_β *is reversible with respect to its invariant probability distribution* μ_β. *Let us assume moreover that for some strictly positive constants* β_0, a, b, c, d, *for any* $\beta \geq \beta_0$, *any* $x, y \in E$,

$$ae^{-\beta U(x)} \leq \mu_\beta(x) \leq be^{-\beta U(x)}$$
$$ce^{-\beta V(x,y)} \leq p_\beta(x,y).$$

Let us assume that $V(x,x) = 0$, *for any* $x \in E$. *Let* $\gamma \in \Gamma$ *be a routing function such that for any* $(x,y) \in E^2 \setminus \Delta$

$$H(\gamma(x,y)) = H(x,y),$$

and let

$$L(\gamma) = \max_{(x,y) \in E^2 \setminus \Delta} |\gamma(x,y)|, \quad \text{(nb. of edges)}$$

$$D(\gamma) = \max_{(z,t) \in E^2 \setminus \Delta} \left| \{(x,y) \in E^2 \setminus \Delta \mid (z,t) \in \gamma(x,y)\} \right|.$$

Then the eigenvalues of p_β, $\lambda_0 = 1 > \lambda_1 \geq \cdots \geq \lambda_{m-1} \geq -1$, *satisfy, for any* $\beta \geq \beta_0$,

$$\lambda_1 \leq 1 - \frac{a \, c}{b^2 \, L(\gamma) D(\gamma)} e^{-\beta H_2}$$

$$-\lambda_{m-1} \leq 1 - 2 \, c.$$

Consequently for any $\beta \geq \beta_0$, any $D \subset E$, any $x \in E$,

$$|P_\beta(X_n \in D \mid X_0 = x) - \mu_\beta(D)| \leq a^{-1/2} e^{\beta U(x)/2} (\mu_\beta(D))^{1/2}$$
$$\times \left(1 - \min\left(2c, \frac{ac \exp(-\beta H_2)}{b^2 L(\gamma) D(\gamma)}\right)\right)^n.$$

Proof: The upper bound for λ_1 is a consequence of the expression for κ. To get the lower bound for λ_{m-1}, consider the routing function $\gamma(x, x) = (x, x)$.

6.2.2. *The non-reversible case.* Let us consider a family $(E^{\mathbb{N}}, (X_n)_{n \in \mathbb{N}}, \mathcal{B}, P_\beta)_{\beta \in \mathbb{R}_+}$ of Markov chains with rare transitions with irreducible rate function V and transition matrix p_β.

Given some real number $\lambda \in]0, 1[$, let us consider the Markov matrices

$$q_\beta(x, y) = \lambda\delta(x, y) + (1 - \lambda)p_\beta(x, y)$$
$$\overline{q}_\beta(x, y) = \sum_{z \in E} q_\beta(x, z)q_\beta(y, z)\frac{\mu_\beta(y)}{\mu_\beta(z)}.$$

The matrices q_β and $\overline{q}_\beta$ are irreducible and μ_β is their common invariant distribution. Moreover $\overline{q}_\beta$ is reversible, it is a non negative self-adjoint operator in $L^2_{\mu_\beta}$, since it is the product of q_β and of its adjoint. Let ρ_β be the spectral gap of $\overline{q}_\beta$,

$$\rho_\beta = 1 - \max\{|\xi| \mid \xi \in Sp(\overline{q}_\beta), \xi \neq 1\}$$
$$= 1 - \max\{\xi \mid \xi \in Sp(\overline{q}_\beta), \xi \neq 1\}.$$

Theorem 6.4. *We have*

$$\limsup_{\beta \to +\infty} -\frac{1}{\beta} \log \rho_\beta \leq H_2 = H(E \setminus \{z\})$$

for any $z \in \arg\min \tilde{U}$. Moreover for any $D \subset E$, any $x \in E$, any $n \in \mathbb{N}$,

$$P_\beta(\tau(D) \leq n \mid X_0 = x) \geq \mu_\beta(D) - \left(\frac{1 - \mu_\beta(x)}{\mu_\beta(x)}\right)^{1/2} \mu_\beta(D)^{1/2}(1 - \rho_\beta)^{n/2}.$$

Proof. Let $(E^{\mathbb{N}}, (X_n)_{n \in \mathbb{N}}, \mathcal{B}, Q_\beta)_{\beta \in \mathbb{R}_+}$ be the canonical realization of a family of Markov chains with transition matrix q_β (and some irrelevant arbitrary initial distribution). We have

$$Q_\beta(X_n \in D \mid X_0 = x) = \sum_{k=0}^{n} \binom{n}{k} \lambda^{n-k}(1 - \lambda)^k P_\beta(X_k \in D \mid X_0 = x)$$
$$\leq \max_{k=0,\ldots,n} P_\beta(X_k \in D \mid X_0 = x)$$
$$\leq P_\beta(\tau(D) \leq n \mid X_0 = x).$$

Moreover

$$|Q_\beta(X_n \in D \mid X_0 = x) - \mu_\beta(D)| \leq \mu_\beta(D)^{1/2}\|\delta_x q_\beta^n - \mu_\beta\|_{2,\mu_\beta}.$$

For any probability distribution ν

$$\left\| \nu \frac{q_\beta}{\mu_\beta} - 1 \right\|^2_{2,\mu_\beta} = \left(\frac{(\nu - \mu_\beta)\bar{q}_\beta}{\mu_\beta}, \frac{(\nu - \mu_\beta)}{\mu_\beta} \right)_{\mu_\beta}$$

$$\leq \left\| \frac{\nu - \mu_\beta}{\mu_\beta} \right\|^2_{2,\mu_\beta} (1 - \rho_\beta),$$

therefore

$$P_\beta(\tau(D) \leq n \mid X_0 = x) \geq \mu_\beta(D) - \left(\frac{1 - \mu_\beta(x)}{\mu_\beta(x)} \right)^{1/2} \mu_\beta(D)^{1/2} (1 - \rho_\beta)^{n/2}.$$

Let $\overline{V}$ be the rate function corresponding to $(\bar{q}_\beta)_{\beta \in \mathbb{R}_+}$. It is easy to check that for any $x, y \in E$,

$$\overline{V}(x, y) \leq V(x, y),$$

thus $H_2(\overline{V}) = \max_{x \in E} H_{\overline{V}}(x, y) - \tilde{U}(x) \leq \max_{x \in E} H_V(x, y) - \tilde{U}(x) = H_2$, where y is an arbitrary point in $\arg \min \tilde{U}$. Thus, considering a routing function satisfying $H_{\overline{V}}(\gamma(x, y)) = H_{\overline{V}}(x, y)$, we see that

$$\limsup_{\beta \to +\infty} -\frac{1}{\beta} \log \rho_\beta \leq H_2.$$

$\square$

Remark 6.1. We have in fact more precisely that

$$\lim_{\beta \to +\infty} -\frac{\log \rho_\beta}{\beta} = H_2.$$

This can be seen from [29] where the reader will find an alternative approach to the non reversible case, or from the fact that a lower limit would contradict the optimal rate of convergence of q_β^n towards its invariant distribution proved in theorem 5.1. Another interesting reference for the "multiplicative reversiblization" method is [15].

Remark 6.2. Theorem 6.4 can be used as an alternative tool to prove convergence results, such as proposition 5.1 and 5.2. Indeed these propositions rely on upper bounds for the tail distribution of the exit times from domains. As an illustration, let us see how we can prove the first bound in proposition 4.19 from theorem 6.4.

Let $D \subsetneq E$ be a domain and $\tau(\overline{D})$ its exit time. The estimate we are looking for does not depend on the behaviour of the Markov chain outside from D, so we can modify the state space, creating a unique outer state Δ standing for $\overline{D}$. Let $E' = D \cup \{\Delta\}$ be the new state space. Let the modified transition matrix p'_β be

$$p'_\beta = \begin{cases} p_\beta(x, y) & \text{if } (x, y) \in D^2, \\ \sum_{z \notin D} p_\beta(x, z) & \text{if } x \in D, y = \Delta, \\ \dfrac{1}{|D|} \exp(-\beta H_\Delta) & \text{if } x = \Delta, y \in D. \end{cases}$$

By taking $H_\Delta > H(D)$, we get a modified virtual energy $\tilde{U}'$ such that $\tilde{U}'(\Delta) = 0$ and

$$H_\Delta - H(D) \le \tilde{U}'(x) \le H_\Delta, \quad x \in D.$$

Indeed

$$\begin{aligned}
H_\Delta - \tilde{U}'(x) &= H'(\Delta, x) - \tilde{U}'(x) \\
&= H'(x, \Delta) - \tilde{U}'(x) \\
&\le H(D),
\end{aligned}$$

and

$$\tilde{U}'(x) \le H'(x, \Delta) = H_\Delta.$$

Moreover the critical depths of the modified landscape are $H_1' = H_2' = H_3' = H(D)$. We see immediately that we have

$$\lim_{\beta \to +\infty} \mu_\beta'(\Delta) = 1,$$

$$\lim_{\beta \to +\infty} \mu_\beta'(x) \exp\big((H_\Delta + \epsilon)\beta\big) = +\infty, \qquad x \in D, \epsilon > 0,$$

$$\lim_{\beta \to +\infty} \rho_\beta' \exp\big((H(D) + \epsilon)\beta\big) = +\infty, \qquad \epsilon > 0.$$

Plugging this altogether into theorem 6.4 applied to $\{\Delta\}$ gives that for any $\epsilon > 0$

$$\lim_{\beta \to +\infty} \max_{x \in D} P_\beta(\tau(\overline{D}) > \exp\big(\beta(H(D) + \epsilon)\big) \mid X_0 = x) = 0,$$

(taking $H_\Delta = H(D) + \epsilon/2$). This can be immediately strengthened to

$$\lim_{\beta \to +\infty} -\frac{1}{\beta} \log \max_{x \in D} P_\beta(\tau(\overline{D}) > \exp\big(\beta(H(D) + \epsilon)\big) \mid X_0 = x) = +\infty$$

using the Markov property as in the beginning of the proof of proposition 4.19.

We have sketched the link between theorem 6.4 and proposition 4.19 to show that semigroup methods can be extended to the same generality as the Freidlin and Wentzell approach, however the reader should keep in mind that their main interest is to provide more explicit bounds and constants when stronger assumptions are made on the transition matrix p_β than what is assumed in these notes.

7. SIMULATED ANNEALING ALGORITHMS

7.1. Description. Let us consider a finite state space E and a family $(p_\beta)_{\beta \in \mathbb{R}_+}$ of Markov matrices with rare transitions and irreducible rate function V.

For any increasing inverse temperature sequence $(\beta_n)_{n \in \mathbb{N}^*}$ (of real positive numbers), we can construct a non-homogeneous Markov chain $(E^{\mathbb{N}}, (X_n)_{n \in \mathbb{N}}, \mathcal{B}, P_{(\beta_n)_{n \in \mathbb{N}^*}})$ with transitions

$$P_{(\beta.)}(X_n = y \mid X_{n-1} = x) = p_{\beta_n}(x, y), \quad x, y \in E.$$

This chain describes the generalised simulated annealing algorithm. It is used to minimise the virtual energy $\tilde{U}$ corresponding to (E, V).

7.2. Convergence results. These results make use of two important constants of (E, V). We have already used the first, it is the first critical depth

$$H_1 = \max\{H(C) \mid C \in \mathcal{C}(V), \tilde{U}(C) > 0\}$$

The second will be called the difficulty of (E, V), and is defined to be

$$D = \max\{\frac{H(C)}{\tilde{U}(C)} \mid C \in \mathcal{C}(V), \tilde{U}(C) > 0\}.$$

Theorem 7.1. *With the preceding hypotheses and notations, for any bounds $\overline{H}$ and $\underline{D}$ such that $\overline{H} > H_1$, and $0 < \underline{D} < D$, for any η such that $0 < \eta < \overline{H}/\underline{D}$, any integer $r > 0$, the triangular sequence of inverse temperatures*

$$\beta_n^N = \frac{1}{\overline{H}} \log \left(\frac{N}{r}\right) \left(\frac{\overline{H}}{\underline{D}\eta}\right)^{\frac{1}{r}\left\lfloor \frac{(n-1)r}{N} \right\rfloor}, \qquad 1 \le n \le N,$$

satisfies for any $x \in E$

$$\liminf_{N \to +\infty} -\frac{1}{\log N} \log P_{(\beta^N)}(\tilde{U}(X_N) \ge \eta \mid X_0 = x) \ge \frac{1}{D}\left(\frac{\underline{D}\eta}{\overline{H}}\right)^{1/r}.$$

Remarks:

- For r large, the order of magnitude of the upper bound for $P_{(\beta^N)}(\tilde{U}(X_N) \ge \eta \mid X_0 = x)$ is close to $N^{-1/D}$. More precisely, for any $\epsilon > 0$,

  any $r \ge \dfrac{\log\left(\overline{H}/(\underline{D}\eta)\right)}{\log(1 + \epsilon)}$, any $x \in E$,

  $$\liminf_{N \to +\infty} -\frac{1}{\log N} \log P_{(\beta^N)}(\tilde{U}(X_N) \ge \eta \mid X_0 = x) \ge \frac{1}{(1 + \epsilon)D}.$$

  The number of iterations needed to bring down the probability of failure to a given order of magnitude is therefore independent of the precision $\eta > 0$. The upper bound for the probability of failure is at best of order $N^{-1/D}$. One can show that this is the best one can achieve using non-decreasing inverse temperature sequences (see [6, 31, 33]).

- The choice of parameters is robust: it is not necessary to know the exact value of D to choose the values of the parameters. We get a probability

  of failure of order $\left(\dfrac{1}{N}\right)^{\frac{1}{D}\left(\frac{\eta\underline{D}}{\overline{H}}\right)^{1/r}}$ uniformly for any rate function V

  such that $\overline{H} > H_1(E, V)$ and $\underline{D} < D(E, V)$. This is not the case with the Metropolis algorithm in which the choice of β requires a precise knowledge of $H_1(V)$ (namely the proved exponent of convergence of simulated an-

  nealing $\dfrac{1}{D}\left(\dfrac{\eta\underline{D}}{\overline{H}}\right)^{1/r}$ is uniformly close to the optimal exponent $\dfrac{1}{D}$ when

  $\left(\dfrac{\eta\underline{D}}{\overline{H}}\right)^{1/r}$ is close to one, which can be obtained by taking a large value for

r, even when the gaps $\overline{H} - H_1$ and $D - \underline{D}$ are large, whereas the exponent of convergence of the Metropolis algorithm $\dfrac{\eta}{\gamma}$ is close to optimal only when $\dfrac{\gamma - H_1}{H_1}$ is small.)

- Triangular sequences of inverse temperatures are absolutely needed: one can show that for any infinite (unique) non-decreasing temperature sequence (β_n)

$$\min_{x \in E} \limsup_{N \to +\infty} -\frac{1}{\log N} \log P_{(\beta.)}(\tilde{U}(X_N) \geq \eta \mid X_0 = x) \leq \frac{\eta}{H_1(E, V)}.$$

(See [5].) When a non triangular sequence is used, the convergence speed is in first approximation of the same order as for the Metropolis algorithm. This means that triangular sequences are crucial to get a significative speed-up with respect to the Metropolis algorithm.

Proof:

Let us put, to simplify notations,

$$\gamma_k^N = \beta_n^N = \frac{1}{\overline{H}} \log \frac{N}{r} \left(\frac{\overline{H}}{\underline{D}\eta}\right)^{k/r} \quad , k\frac{N}{r} < n \leq (k+1)\frac{N}{r}.$$

Let us also put $P_{(\beta^N.)} = P_N$ and assume that $N/r \in \mathbb{N}$ (the modifications needed to handle the general case are left to the reader).

Let $\xi > 0$ be fixed and let

$$\eta_k = \frac{\overline{H}}{(1 + \xi)D} \left(\frac{\overline{H}}{\underline{D}\eta}\right)^{-\frac{k+1}{r}} , \quad k = 0, \ldots, r - 1.$$

$$\begin{cases} \lambda_0 = +\infty \\ \lambda_k = \frac{(1+\frac{1}{\underline{D}})\overline{H}}{(1+\xi)} \left(\frac{\overline{H}}{\underline{D}\eta}\right)^{-k/r} , & k = 1, \ldots, r - 1. \end{cases}$$

Let us consider the events

$$B_k = \{\tilde{U}(X_n) + V(X_n, X_{n+1}) \leq \lambda_k, k\frac{N}{r} \leq n < (k+1)\frac{N}{r}\},$$

$$A_k = B_k \cap \{\tilde{U}(X_{(k+1)N/r}) < \eta_k\}.$$

We have

$$\exp \overline{H}\gamma_0^N = \frac{N}{r},$$

$$\exp\left((1 + \xi)\left(1 + \frac{1}{D}\right)^{-1} \lambda_k \gamma_k^N\right) = \frac{N}{r}, \quad k > 0,$$

and

$$\eta_{r-1} = \frac{1}{\xi + 1}\frac{D}{\underline{D}}\eta \leq \eta.$$

Therefore

$$P_N(\tilde{U}(X_N) \geq \eta \mid X_0 = x) \leq P_N(\tilde{U}(X_N) \geq \eta_{r-1} \mid X_0 = x)$$
$$\leq 1 - P_N\left(\bigcap_{k=0}^{r-1} A_k \mid X_0 = x\right)$$
$$\leq \sum_{k=0}^{r-1} P_N\left(\overline{A}_k \cap \bigcap_{l=0}^{k-1} A_l \mid X_0 = x\right).$$

Moreover

$$P_N\left(\overline{A}_k \cap \bigcap_{l=0}^{k-1} A_l \mid X_0 = x\right) \leq P_N\left(\overline{B}_k \cap \bigcap_{l=0}^{k-1} A_l \mid X_0 = x\right)$$
$$+ P_N\left((\tilde{U}(X_{(k+1)N/r}) \geq \eta_k) \cap \bigcap_{l=0}^{k-1} A_l \cap B_k \mid X_0 = x\right).$$

Let us remark first that for any cycle C such that $\tilde{U}(C) > 0$ and $H(C) + \tilde{U}(C) \leq \lambda_k$ we have

$$H(C) \leq \left(1 + \frac{1}{D}\right)^{-1} \lambda_k.$$

For any $z \in E$ such that $\tilde{U}(z) < \eta_{k-1}$, let us consider the smallest cycle $C_z \in \mathcal{C}(V)$ containing z such that $\tilde{U}(C_z) + H(C_z) > \lambda_k$. We have $\tilde{U}(C_z) = 0$. Indeed, if we had $\tilde{U}(C_z) > 0$, we would have also $\tilde{U}(C_z) + H(C_z) \leq \tilde{U}(z)(1 + D) \leq \eta_{k-1}(1 + D) = \lambda_k$, which is a contradiction.

From the preceding remarks, we deduce that

$$H_1(C_z, V_{|C_z \times C_z}) \leq \left(1 + \frac{1}{D}\right)^{-1} \lambda_k.$$

Indeed any cycle $C \subset C_z$ such that $C \neq C_z$ and $\tilde{U}(C) > 0$ satisfies $H(C) \leq \left(1 + \frac{1}{C}\right)^{-1} \lambda_k$, and

$$H_1(C_z, V_{|C_z \times C_z}) = \max\{H(C) : C \subset C_z, C \in \mathcal{C}(V), \tilde{U}(C) > 0\}$$
$$= H(C_z \setminus \arg\min_{y \in C_z} \tilde{U}(y)).$$

Let us remark now that

$$P(X_{(k+1)N/r} = y, B_k \mid X_{kN/r} = z)$$
$$\leq P(X_{(k+1)N/r} = y, X_n \in C_z, k\frac{N}{r} < n \leq (k+1)\frac{N}{r} \mid X_{kN/r} = z).$$

Let us consider on C_z the Markov chain $(Y_n)_{n\in\mathbb{N}}$ with transitions $P(Y_n = y \mid Y_{n-1} = x) = q(x,y)$ defined by

$$
q(x,y) = \begin{cases} p_{\gamma_k^N}(x,y) & \text{if } x \neq y \in C_z, \\ 1 - \displaystyle\sum_{w \in (C_z \backslash \{x\})} q(x,w) & \text{otherwise.} \end{cases}
$$

(We obtain this new chain by reflecting $(X_n)_{n\in\mathbb{N}}$ on the boundary of C_z.)

As for any $x, y \in C_z$, $p_{\gamma_k^N}(x,y) \leq q(x,y)$, we have

$$
P_N(X_{(k+1)N/r} = y, X_n \in C_z, k\frac{N}{r} < n \leq (k+1)\frac{N}{r} \mid X_{kN/r} = z)
$$
$$
\leq P(Y_{N/r} = y \mid Y_0 = z).
$$

Applying to Y the theorem on the convergence speed of the Metropolis algorithm, we see that for any $\epsilon > 0$, there is N_0 such that for any $N \geq N_0$,

$$
P_N\left((X_{(k+1)N/r} = y) \cap \bigcap_{l=0}^{k-1} A_l \cap B_k \mid X_0 = x\right) \leq \exp\left(-\gamma_k^N(\tilde{U}(y) - \epsilon)\right).
$$

We have now to find an upper bound for

$$
P_N\left(\overline{B}_k \cap \bigcap_{l=0}^{k-1} A_l \mid X_0 = x\right).
$$

For any $z \in E$ such that $\tilde{U}(z) < \eta_{k-1}$, we have

$$
\begin{aligned}
P_N(\overline{B}_k \mid X_{kN/r} = z) &\leq \sum_{n=kN/r+1}^{(k+1)N/r} P_N(\tilde{U}(X_{n-1}) + V(X_{n-1}, X_n) > \lambda_k \mid X_{kN/r} = z) \\
&= \sum_{\substack{kN/r < n \leq (k+1)N/r, \\ (u,v) \in E^2, \\ \tilde{U}(u) + V(u,v) > \lambda_k}} P_N(X_{n-1} = u \mid X_{kN/r} = z) p_{\gamma_k^N}(u,v) \\
&\leq \sum_{\substack{kN/r < n \leq (k+1)N/r, \\ (u,v) \in E^2, \\ \tilde{U}(u) + V(u,v) > \lambda_k}} \frac{\mu_{\gamma_k^N}(u)}{\mu_{\gamma_k^N}(z)} p_{\gamma_k^N}(u,v) \\
&\leq \frac{N}{r} \exp\left(-\gamma_k^N(\lambda_k - \tilde{U}(z) - \epsilon)\right),
\end{aligned}
$$

for any $\epsilon > 0$ and N large enough. Thus for any $\epsilon > 0$, for N large enough,

$$P_N\left(\overline{B}_k \cap \bigcap_{l=0}^{k-1} A_l \mid X_0 = x\right) = \sum_{z,\tilde{U}(z)<\eta_{k-1}} P_N(\overline{B}_k \mid X_{kN/r} = z)$$

$$\times P_N\left((X_{kN/r} = z) \cap \bigcap_{l=0}^{k-2} A_l \cap B_{k-1} \mid X_0 = x\right)$$

$$\leq \sum_{z,\tilde{U}(z)<\eta_{k-1}} \frac{N}{r} \exp\left(-\gamma_k^N(\lambda_k - \tilde{U}(z) - \epsilon)\right)$$

$$\times \exp\left(-\gamma_{k-1}^N(\tilde{U}(z) - \epsilon)\right)$$

$$\leq \frac{N}{r} \exp\left(-\gamma_k^N(\lambda_k - \eta_{k-1} - \epsilon)\right)$$

$$\times \exp\left(-\gamma_{k-1}^N(\eta_{k-1} - 2\epsilon)\right).$$

Therefore, for any $\epsilon > 0$, there is N_0 such that, for any $N \geq N_0$,

$$P\left(\overline{A}_k \cap \bigcap_{l=0}^{k-1} A_l \mid X_0 = x\right)$$

$$\leq \frac{N}{r} \exp\left(-(\lambda_k - \eta_{k-1})\gamma_k^N - \eta_{k-1}\gamma_{k-1}^N + \epsilon(\gamma_k^N + \gamma_{k-1}^N)\right)$$

$$+ \exp\left(-(\eta_k - \epsilon)\gamma_k^N\right).$$

Coming back to the definitions, we see that

$$\eta_k \gamma_k^N = \log\left(\frac{N}{r}\right)\frac{1}{(1+\xi)}\frac{1}{D}\left(\frac{D\eta}{\overline{\overline{H}}}\right)^{1/r},$$

$$\lambda_k \gamma_k^N + \eta_{k-1}(\gamma_{k-1}^N - \gamma_k^N) = \left(1 + \frac{1}{D}\right)(1+\xi)^{-1}\log\frac{N}{r}$$

$$+ \frac{1}{(1+\xi)D}\left(\left(\frac{\eta D}{\overline{\overline{H}}}\right)^{1/r} - 1\right)\log\frac{N}{r}$$

$$= \log\frac{N}{r}\left((1+\xi)^{-1}\left(1 + \frac{1}{D}\left(\frac{\eta D}{\overline{\overline{H}}}\right)^{1/r}\right)\right),$$

$$\epsilon\gamma_k^N \leq \frac{\epsilon}{\eta \underline{D}}\log\frac{N}{r}.$$

Thus

$$P\left(\overline{A}_k \cap \bigcap_{l=0}^{k-1} A_l \mid X_0 = x\right) \leq \left(\frac{N}{r}\right)^{-(1+\xi)^{-1}\left(\frac{1}{D}\left(\frac{\eta D}{\overline{\overline{H}}}\right)^{1/r} - \xi\right) + \frac{2\epsilon}{\eta \underline{D}}}$$

$$+ \left(\frac{N}{r}\right)^{-(1+\xi)^{-1}\frac{1}{D}\left(\frac{D\eta}{\overline{\overline{H}}}\right)^{1/r} + \frac{\epsilon}{\eta \underline{D}}}.$$

Letting ξ and ϵ tend to zero, we get eventually that

$$\liminf_{N \to +\infty} -\frac{1}{\log N} \log P_N(\tilde{U}(X_N) \geq \eta \mid X_0 = x) \geq \frac{1}{D} \left(\frac{D\eta}{\overline{\overline{H}}} \right)^{1/r}.$$

For more precise computations under the stronger hypothesis that for some constant $a > 0$

$$a^{-1} e^{-\beta V(x,y)} \leq p_\beta(x, y) \leq a e^{-\beta V(x,y)},$$

we refer to [11].

8. THE ENERGY TRANSFORMATION ALGORITHM

8.1. The energy transformation method. The purpose of this algorithm is to minimise a function $U : E \longrightarrow \mathbb{R}$ defined on a finite set E, using to explore the states of E an irreducible Markov matrix $q : E \times E \longrightarrow [0, 1]$ with a fixed symmetric support. The method is to use a rate function of the form

$$V(x, y) = (F \circ U(y) - F \circ U(x))^+, \quad q(x, y) > 0,$$

where $F : \mathbb{R} \longrightarrow \mathbb{R} \cup \{-\infty\}$ is a suitable increasing function.

8.2. Convergence result for a single transformation.

Proposition 8.1. *Let* $q : E \times E \longrightarrow [0, 1]$ *be an irreducible Markov matrix with symmetric support. Let* $(E^{\mathbb{N}}, (X_n)_{n \in \mathbb{N}}, \mathcal{B}, P_{\beta,\eta})_{\beta \in \mathbb{R}_+, \eta \in \mathbb{R}_+^*}$ *be the canonical realization of a family of Markov chains with transitions*

$$p_{\beta,\eta}(x, y) = q(x, y) \exp\left(-\beta \left(F_\eta \circ U(y) - F_\eta \circ U(x) \right)^+ \right), \quad x \neq y,$$

where $F_\eta(u) = \log(u + \eta)$, *where* $\eta + U_{\min} > 0$.
Let us introduce the two rate functions:

$$V(x, y) = \begin{cases} (U(y) - U(x))^+, & p_{\beta,\eta}(x, y) > 0, \\ +\infty & otherwise \end{cases}$$

$$W_\eta(x, y) = \begin{cases} (F_\eta \circ U(y) - F_\eta \circ U(x))^+, & p_{\beta,\eta}(x, y) > 0 \\ +\infty & otherwise. \end{cases}$$

Then W_η *is the rate function describing the rare transitions of the sub-family* $(P_{\beta,\eta})_{\beta \in \mathbb{R}_+}$, *and for any* $\eta > -U_{\min}, \rho > 0, \epsilon > 0$ *and any* $x \in E$,

$$\liminf_{N \to +\infty} -\frac{1}{\log N} \log P_{\beta_{N,\eta},\eta}(U(X_N) - U_{\min} \geq \rho(\eta + U_{\min}) \mid X_0 = x) \geq$$
$$\frac{\log(1 + \rho)}{\log(1 + D_{(\eta+U_{\min})})} (1 + \epsilon)^{-1},$$

with

$$\beta_{N,\eta} = \frac{\log N}{(1 + \epsilon) \log(1 + D_{(\eta+U_{\min})})},$$

$$D_\alpha = \max\{ \frac{H_V(C)}{U(C) - U_{\min} + \alpha} \mid C \in \mathcal{C}(V), U(C) > U_{\min} \},$$

where $H_V(C)$ *is the depth of* C *with respect to the rate function* V, *induced by* U.

Remark: If it is known in advance that $a < U_{\min} < b$, it is possible to take $F(u) = \log(u - a)$. This ensures a probability of failure bounded by

$$\left(\frac{1}{N}\right)^{\frac{\log(1+\rho)}{\log(1+D_{(U_{\min}-a)})}(1+\epsilon)^{-1}}$$ when failure means $U(X_N) \geq U_{\min} + \delta$ with $\delta =$

$\rho(b - a)$. The interesting thing is that the exponent

$$\alpha = (1 + \epsilon)^{-1} \frac{\log(1 + \rho)}{\log(1 + D_{(U_{\min}-a)})} = (1 + \epsilon)^{-1} \frac{\log\left(1 + \frac{\delta}{b-a}\right)}{\log\left(1 + D_{(U_{\min}-a)}\right)}$$ describing the

convergence speed depends on the precision $(b - a)$ with which $U_{\min}$ is known in advance, and that, for a fixed value of δ, α tends to $+\infty$ when the precision $b - a$ tends to 0.

Proof:

As F_η is increasing, it is easy to see that $\mathcal{C}(V) = \mathcal{C}(W_\eta)$. In the case when $H_1(V) = H_1(W_\eta) = 0$, there are no local minimum, $D_{(\eta+U_{\min})} = 0$, and the proposition is true with the convention that $1/0 = +\infty$, since the convergence of the probability of error to zero is in this case easily seen to be exponential and not polynomial in N. Therefore we will assume in this proof that $H_1(W_\eta) > 0$. For any cycle $C \in \mathcal{C}(V) = \mathcal{C}(W_\eta)$ such that $U(C) > U_{\min}$,

$$\begin{aligned}
H_{W_\eta}(C) &= F_\eta(U(C) + H_V(C)) - F_\eta(U(C)) \\
&= \log\left(1 + \frac{H_V(C)}{U(C) + \eta}\right) \\
&\leq \log(1 + D_{(\eta+U_{\min})}),
\end{aligned}$$

Moreover

$$\begin{aligned}
F_\eta(\rho(\eta + U_{\min}) + U_{\min}) - F_\eta(U_{\min}) &= \log(1 + \rho), \\
\text{and} \quad \exp\left(\beta_{N,\eta} H_1(W_\eta)(1 + \epsilon)\right) &\leq N,
\end{aligned}$$

therefore

$$\liminf_{N \to +\infty} -\frac{1}{\beta_{N,\eta}} \log P_{\beta_{N,\eta},\eta}(U(X_N) - U_{\min} \geq \rho(\eta + U_{\min}) \mid X_0 = x)$$
$$\geq \log(1 + \rho). \qquad \square$$

In the following paragraph, we will use the energy transformation method repeatedly to improve a rough initial lower bound for $U_{\min}$.

8.3. The Iterated Energy Transformation algorithm.

Theorem 8.1. *Let $\gamma < U_{\min}$ be a lower bound for $U_{\min}$ which is assumed to be known beforehand. Let $\eta_0 \geq 0$ be a non negative parameter, and let us consider the (non-Markovian) stochastic process $(E^{\mathbb{N}}, (X_n)_{n=1,\ldots,N}, \mathcal{B}, P_N)$ with transitions*

$$P_N(X_n = y \mid (X_0, \ldots, X_{n-1}) = (x_0, \ldots, x_{n-1})) = p_{\beta_N, \tau_k}(x_{n-1}, y),$$
$$k\frac{N}{r} < n \leq (k+1)\frac{N}{r},$$

where

$$\beta_N = \frac{\log(N/r)}{(1+\epsilon)\log(1+D_{\eta_0})}$$

$$\tau_k = \tau_{k-1} - \frac{1}{(1+\rho)}\left(\tau_{k-1} + U(x_{kN/r})\right) + \eta_0,$$

$$\tau_0 = \eta_0 - \gamma,$$

and where r is the number of steps of the algorithm. Then for any $x \in E$

$$\liminf_{N\to+\infty} -\frac{1}{\log N}$$

$$\log P_N\left(U(X_N) - U_{\min} \geq \rho\left(\frac{\rho}{1+\rho}\right)^{r-1}(U_{\min} - \gamma + \eta_0) + \eta_0\rho(1+\rho) \mid X_0 = x\right)$$

$$\geq \frac{\log(1+\rho)}{(1+\epsilon)\log(1+D_{\eta_0})}.$$

Remarks:

- The probability of failure can be reduced to order N^ξ with ξ arbitrarily large by increasing ρ and r and decreasing η_0. A more precise study of the algorithm (see [7]) would allow us to choose r and ρ as functions of N and to get a convergence speed better than polynomial.
- The I.E.T. algorithm is well suited when D is large and $|E|$ is moderate. In order to fight against the number of states in E, it is possible to use an energy transform of the form $\alpha u + \beta \log(u + \eta)$.
- The energy transformation method can also be used for the simulated annealing algorithm: any concave increasing energy transformation will decrease the difficulty (see [2]).

Proof: Let us introduce the events

$$A_k = \{U(X_{(k+1)N/r}) - U_{\min} < (\tau_k + U_{\min})\rho\}.$$

We have

$$P_N(U(X_N) \geq U_{\min} + \rho(U_{\min} + \tau_{r-1}) \mid X_0 = x) = P_N(\overline{A}_{r-1} \mid X_0)$$

$$\leq P_N\left(\bigcap_{k=0}^{r-1} A_k \mid X_0 = x\right)$$

$$\leq \sum_{k=0}^{r-1} P_N\left(\overline{A}_k \cap \bigcap_{l=0}^{k-1} A_l \mid X_0 = x\right)$$

$$\leq \sum_{k=0}^{r-1} P_N\left(\overline{A}_k \mid X_0 = x, \bigcap_{l=0}^{k-1} A_l\right).$$

When $\bigcap_{l=0}^{k-1} A_l$ holds,

$$U(X_{kN/r}) + \tau_{k-1} < (U_{\min} + \tau_{k-1})(\rho + 1),$$

therefore $U_{\min} + \tau_k > \eta_0$ and, applying the previous proposition,

$$\liminf_{N \to +\infty} -\frac{1}{\log N} \log P_N(\overline{A}_k \mid X_0 = x, \bigcap_{l=0}^{k-1} A_l) \geq \frac{\log(1+\rho)}{(1+\epsilon)\log(1+D_{\eta_0})},$$

thus

$$\liminf_{N \to +\infty} -\frac{1}{N} \log P_N(U(X_N) \geq U_{\min} + \rho(U_{\min} + \tau_{r-1}) \mid X_0 = x)$$

$$\geq \frac{\log(1+\rho)}{(1+\epsilon)\log(1+D_{\eta_0})}.$$

Moreover

$$\tau_k + U_{\min} \leq (\tau_{k-1} + U_{\min})\frac{\rho}{1+\rho} + \eta_0$$

$$\leq \eta_0 \sum_{l=0}^{k-1} \left(\frac{\rho}{1+\rho}\right)^l + \left(\frac{\rho}{1+\rho}\right)^k (\tau_0 + U_{\min}),$$

whence

$$(\tau_{r-1} + U_{\min}) \leq \eta_0(1+\rho) + \left(\frac{\rho}{1+\rho}\right)^{r-1} (U_{\min} + \eta_0 - \gamma). \qquad \square$$

9. A GENERAL REMARK ABOUT THE INTEREST OF REPEATED OPTIMISATION SCHEMES

All the algorithms we have encountered in these notes have a probability of failure bounded by $\epsilon(N)$, where N is their number of iterations and where $\lim_{N \to +\infty} N^{-1} \log \epsilon(N) = 0$. Due to this slow convergence speed, these algorithms should be used repeatedly. Indeed performing $N/\hat{M}$ repetitions of the algorithm with $\hat{M}$ iterations, where $\hat{M} \in \arg\min_{M \in \mathbb{N}} M^{-1} \log \epsilon(M)$, and keeping in the end the best solution among the $N/\hat{M}$ computed solutions, gives a probability of failure bounded from above by ξ^N with $\xi = \epsilon(\hat{M})^{1/\hat{M}}$ (when $N/\hat{M} \in \mathbb{N}$). The fact that $\lim_{M \to +\infty} M^{-1} \log \epsilon(M) = 0$ ensures that $\arg\min_{M \in \mathbb{N}} M^{-1} \log \epsilon(M)$ is not void and is bounded. See [2] and [7, 10] for more details.

10. PROBLEM

The different questions are independent. The integer part of r is noted $\lfloor r \rfloor = \max\{n \in \mathbb{Z} \mid n \leq r\}$.

10.1. Question 1. Let us consider the state space $E = \{1, 2, 3, 4, 5\}$ and the rate function $V : E \times E \to \mathbb{R}_+ \cup \{+\infty\}$ defined by the following matrix

$$\begin{pmatrix} 0 & 1 & 3 & 0 & 2 \\ 8 & 0 & 2 & 2 & 3 \\ 9 & 5 & 0 & 7 & 4 \\ 0 & 2 & +\infty & 0 & +\infty \\ 8 & 5 & 4 & 11 & 0 \end{pmatrix}$$

(for instance $V(3,4) = 7$.)

1.1. Compute the virtual energy of each state and construct all the cycles by induction.

1.2. Compute $H_1(V)$, $H_2(V)$ et $H_3(V)$.

1.3. Let us consider a family $(E^\mathbb{N}, (X_n)_{n \in \mathbb{N}}, \mathcal{B}, P_\beta)_{\beta \in \mathbb{R}_+}$ of homogeneous Markov chains with rare transitions with rate function V. For any subset D of E, we put $\tau(D) = \inf\{n \in \mathbb{N} \mid X_n \in D\}$. Compute

$$\lim_{\beta \to +\infty} -\frac{1}{\beta} \log P_\beta(X_{\tau(\{2,3,5\})} = 3 \mid X_0 = 4),$$

and

$$\lim_{\beta \to +\infty} \frac{1}{\beta} \log E_\beta(\tau(\{2,5\}) \mid X_0 = 3).$$

10.2. **Question 2.** Let us consider a family $(E^\mathbb{N}, (X_n)_{n \in \mathbb{N}}, \mathcal{B}, P_\beta)_{\beta \in \mathbb{R}_+}$ of homogeneous Markov chains with rare transitions defined on a finite state space E, with an irreducible rate function $V : E \times E \to \mathbb{R}_+ \cup \{+\infty\}$. Let $\tilde{U}$ be its virtual energy.

Let us assume that for some real positive constants a and b and for any $(x, y) \in E^2$

$$a \exp(-\beta V(x, y)) \le p_\beta(x, y) \le b \exp(-\beta V(x, y)),$$

where $p_\beta : E \times E \to [0, 1]$ is the transition matrix of the chain P_β: For any $n \in \mathbb{N}$, $n > 0$,

$$p_\beta(x, y) = P_\beta(X_n = y \mid X_{n-1} = x).$$

2.1. Show that there is a positive real constant c such that for any subset D of E, $D \ne E$, $D \ne \emptyset$, any $x \in E \setminus D$, any $n \in \mathbb{N}$, any $\beta \in \mathbb{R}_+$,

$$P_\beta(\tau(D) > n \mid X_0 = x) \le \exp\left(-\left\lfloor cne^{-\beta H(E \setminus D)}\right\rfloor\right),$$

where $\tau(D)$ is the first hitting time of D:

$$\tau(D) = \inf\{n \in \mathbb{N} \mid X_n \in D\}.$$

2.2. Deduce from this that there is a positive real constant d such that for any real positive $\eta \in \mathbb{R}_+$, any $x \in E$, any $\beta \in \mathbb{R}_+$,

$$P_\beta(\tilde{U}(X_n) \ge \eta \mid X_0 = x) \le \exp\left(-\left\lfloor \frac{n}{d}e^{-\beta H_1(V)}\right\rfloor\right) + de^{-\eta\beta}.$$

2.3. Using the preceding inequalities, state a convergence theorem concerning $P_\beta(\tilde{U}(X_{N(\beta)}) \ge \eta \mid X_0 = x)$ for a suitable function $N(\beta)$.

10.3. Question 3: Weak reversibility condition of Hajek and Trouvé.

On a finite state space E, let us consider an irreducible rate function $V : E \times E \to \mathbb{R}_+ \cup \{+\infty\}$ and a real valued function $U : E \to \mathbb{R}$. Let us define the elevation $H_U(\gamma)$ of a path $\gamma = (z_0, \cdots, z_r) \in E^{r+1}$ with respect to U by the formula

$$H_U(\gamma) = \max_{i=1,\cdots,r} U(z_{i-1}) + V(z_{i-1}, z_i).$$

For any $(x, y) \in E^2$, let $\Gamma_{x,y}$ be the set of paths joining x to y:

$$\Gamma_{x,y} = \bigcup_{r=1}^{+\infty} \{(z_0, \cdots, z_r) \in E^{r+1} \mid z_0 = x, z_r = y\}.$$

Let us define the minimum elevation between two states $x \in E$ and $y \in E$ by

$$H_U(x, y) = \min\{H_U(\gamma) \mid \gamma \in \Gamma_{x,y}\}.$$

3.1. Let us assume that the function $H_U(x, y)$ is symmetric. Namely, let us assume that for any $(x, y) \in E^2$

$$H_U(x, y) = H_U(y, x).$$

(This is a "weak reversibility condition", due to Hajek in the case when $p_\beta(x, y) = q(x, y) \exp\left(-\beta(U(y) - U(x))_+\right)$ with a non reversible kernel q and to Trouvé in the general case). Let $\tilde{U}$ be the virtual energy corresponding to (E, V). For any cycle $C \in \mathcal{C}(V)$, consider the following property $\mathcal{H}(C)$:

$$\forall (x, y) \in C^2, \quad U(x) - \tilde{U}(x) = U(y) - \tilde{U}(y).$$

Show by induction on $|C|$ that $\mathcal{H}(C)$ is true for any cycle $C \in \mathcal{C}(E, V)$. Hints:

- Consider the partition $(C_i)_{i \in I}$ of C in strict maximal subcycles. Introduce the constants $c_i \in \mathbb{R}$, $i \in I$, defined by

$$U(x) = \tilde{U}(x) + c_i, \quad x \in C_i.$$

- Show that if $B(C_i) \cap C_j \neq \emptyset$, (where $B(C_i)$ is the principal boundary of C_i), then $c_i \geq c_j$. (For $x \in C_i$ and $y \in B(C_i) \cap C_j$ compare $H_{\tilde{U}}(x, y)$, $H_U(x, y)$, $H_U(y, x)$ and $H_{\tilde{U}}(y, x)$.)
- Draw from this the conclusion that $c_i = c_j$ for any $(i, j) \in I^2$.

This shows that $(H_U(x, y))_{(x,y) \in E^2}$ is symmetric if and only if for any $x \in E$

$$U(x) = \min_{y \in E} U(y) + \tilde{U}(x).$$

10.4. Question 4.

4.1. Give an example of a finite state space E and of an irreducible rate function $V : E \times E \to \mathbb{R}_+ \cup \{+\infty\}$ such that

$$
\begin{aligned}
H_1(E, V) &= 1, \\
H_2(E, V) &= 2, \\
H_3(E, V) &= 3.
\end{aligned}
$$

4.2. Could-you give such an example in which $|E| = 4$?

4.3. Could-you give such an example in which $|E| = 5$?

Acknowledgement: I would like to thank Alain Trouvé and Cécile Cot for their useful remarks on a first French draft of these lecture notes. My thanks also go to Robert Azencott, who encouraged my interest in stochastic optimisation since the early times of my Phd dissertation. I am grateful to the referee for his careful reading and numerous suggestions that were very precious to help me improve the presentation of many results.

REFERENCES

[1] Azencott Robert (1988) Simulated Annealing, *Séminaire Bourbaki 40ième année, 1987-1988* **697.**

[2] Azencott Robert (1992) Sequential Simulated Annealing: Speed of Convergence and Acceleration Techniques, in *Simulated Annealing: Parallelization Techniques*, R. Azencott Ed., Wiley Interscience.

[3] Azencott Robert (1992) A Common Large Deviations Mathematical Framework for Sequential Annealing and Parallel Annealing, in *Simulated Annealing: Parallelization Techniques*, R. Azencott Ed., Wiley Interscience.

[4] Azencott Robert and Graffigne Christine (1992) Parallel Annealing by Periodically Interacting Multiple Searches: Acceleration Rates, in *Simulated Annealing: Parallelization Techniques*, R. Azencott Ed., Wiley Interscience.

[5] Catoni Olivier (1991) Exponential Triangular Cooling Schedules for Simulated Annealing Algorithms: a case study, *Applied Stochastic Analysis, Proceedings of a US-French Workshop, Rutgers University, April 29 - May 2, 1991*, Karatzas I. and Ocone D. eds., Lecture Notes in Control and Information Sciences No 177, Springer Verlag, 1992.

[6] Catoni Olivier (1992) Rough Large Deviation Estimates for Simulated Annealing: Application to Exponential Schedules, *The Annals of Probability*, Vol. 20, nb. 3, pp. 1109 - 1146.

[7] Catoni Olivier, (1998) The Energy Transformation Method for the Metropolis Algorithm Compared with Simulated Annealing. *Probab. Theory Related Fields 110 (1998), no. 1.*, pages 69–89.

[8] Catoni Olivier and Cerf Raphael (1997) The Exit Path of a Markov Chain with Rare Transitions, *ESAIM:P&S*, vol 1, pp. 95-144, http://www.emath.fr/Maths/Ps/ps.html.

[9] Catoni Olivier (1998) Solving Scheduling Problems by Simulated Annealing. *SIAM J. Control Optim. 36, no. 5, (electronic)*, pages 1539–1575.

[10] Catoni Olivier (1996) Metropolis, Simulated Annealing and I.E.T. Algorithms: Theory and Experiments. *Journal of Complexity 12, special issue on the conference Foundation of Computational Mathematics, January 5-12 1997, Rio de Janeiro*, pages 595–623, December 1996.

[11] Cot Cécile and Catoni Olivier (1998) Piecewise constant triangular cooling schedules for generalized simulated annealing algorithms. *Ann. Appl. Probab. 8, no. 2,*, pages 375–396.

[12] Deuschel J.D. and Mazza C. (1994) L^2 convergence of time nonhomogeneous Markov processes: I. Spectral Estimates, *The annals of Applied Probability*, vol. 4, no. 4, 1012-1056.

[13] Diaconis Persi and Stroock Daniel (1991) Geometric Bounds for Eigenvalues of Markov Chains, *The Annals of Applied Probability*, Vol. 1, No 1, 36 - 61.

[14] Duflo M. (1996) *Algorithmes Stochastiques*, Mathématiques & Applications (Paris), Springer Verlag.

[15] Fill J. A. (1991) Eigenvalue bounds on the convergence to stationarity for nonreversible Markov chains, with an application to the exclusion process, *Ann. Applied Probab.*, **1.**

[16] Freidlin, M. I. and Wentzell, A. D. (1984). *Random Perturbations of Dynamical Systems.* Springer, New York.

[17] Geman S., Geman D., *Stochastic relaxation, Gibbs distribution, and the Bayesian restoration of images*, I.E.E.E. Transactions on Pattern Analysis and Machine Intelligence, 6, 721- 741, 1984.

[18] Götze F. (1991) Rate of Convergence of Simulated Annealing Processes, *preprint*.

[19] Graffigne Christine (1992) Parallel Annealing by Periodically Interacting Multiple Searches: An Experimental Study, in *Simulated Annealing: Parallelization Techniques*, R. Azencott Ed., Wiley Interscience.

[20] Holley R. and Stroock D. (1988) Annealing via Sobolev inequalities, *Comm. Math. Phys.*, 115:553-559.

[21] Holley, R. A., Kusuoka, S. and Stroock, D. W. (1989), Asymptotics of the spectral gap with applications to the theory of simulated annealing, *Journal of functional analysis*, **83**, 333-347.

[22] Hwang, C. R. and Sheu, S. J. (1992) Singular perturbed Markov chains and exact behaviour of simulated annealing processes. *J. Theoret. Prob.*, **5**, 2, 223-249.

[23] Ingrassia S. (1994) On the rate of convergence of the Metropolis algorithm and Gibbs sampler by geometric bounds, *Ann. Appl. Probab.* **4**, no.2, 347-389.

[24] Kirchhoff G. (1847) Über die Auflösung der Gleichungen, auf welche man beider Untersuchung der linearen Verteilung galvanischer Ströme gefuhrt wird, *Ann. Phys. Chem.*, 72, pp. 497-508.(English transl. IRE Trans. Circuit Theory CT-5 (1958) 4-7).

[25] Kirkpatrick S., Gelatt C. D. and Vecchi M. P., (1983) Optimization by simulated annealing, *Science*, 220, 621-680, 1983.

[26] Miclo Laurent (1991) Evolution de l'énergie libre. Application à l'étude de la convergence des algorithmes du recuit simulé. *Doctoral Dissertation*, Université d'Orsay, February 1991.

[27] Miclo Laurent (1996) Sur les problèmes de sortie discrets inhomogènes *Ann. Appl. Probab.* 6, no 4, 1112-1156.

[28] Miclo Laurent (1995) Sur les temps d'occupations des processus de Markov finis inhomogènes à basse température, *submitted to Stochastics and Stochastics Reports*.

[29] Miclo Laurent (1997) Remarques sur l'hypercontractivité et l'évolution de l'entropie pour des chaînes de Markov finies, *Séminaire de Probabilités XXXI*, Lecture Notes in Mathematics 1655, Springer.

[30] Saloff-Coste, Laurent (1997) Lectures on finite Markov chains *Lectures on probability theory and statistics (Saint-Flour, 1996)*, 301–413, Lecture Notes in Math., 1665, Springer, Berlin.

[31] Trouvé Alain (1993) Parallélisation massive du recuit simulé, *Doctoral Dissertation*, Université Paris 11, January 5 1993.

[32] Trouvé Alain (1994) Cycle Decomposition and Simulated Annealing, *S.I.A.M. J. Control Optim., 34(3)*, 1996.

[33] Trouvé Alain (1995) Rough Large Deviation Estimates for the Optimal Convergence Speed Exponent of Generalized Simulated Annealing Algorithms, *Ann. Inst. H. Poincaré, Probab. Statist.*, 32(2), 1996.

CONCENTRATION OF MEASURE
AND LOGARITHMIC SOBOLEV INEQUALITIES

MICHEL LEDOUX

TABLE OF CONTENTS

INTRODUCTION   123

1. ISOPERIMETRIC AND CONCENTRATION INEQUALITIES   126
1.1 Introduction   126
1.2 Isoperimetric inequalities for Gaussian and Boltzmann measures   127
1.3 Some general facts about concentration   134

2. SPECTRAL GAP AND LOGARITHMIC SOBOLEV INEQUALITIES   139
2.1 Abstract functional inequalities   139
2.2 Examples of logarithmic Sobolev inequalities   145
2.3 Herbst's argument   148
2.4 Entropy-energy inequalities and non-Gaussian tails   154
2.5 Poincaré inequalities and concentration   159

3. DEVIATION INEQUALITIES FOR PRODUCT MEASURES   161
3.1 Concentration with respect to the Hamming metric   161
3.2 Deviation inequalities for convex functions   163
3.3 Information inequalities and concentration   166
3.4 Applications to bounds on empirical processes   171

4. MODIFIED LOGARITHMIC SOBOLEV INEQUALITIES FOR
   LOCAL GRADIENTS   173
4.1 The exponential measure   173
4.2 Modified logarithmic Sobolev inequalities   178
4.3 Poincaré inequalities and modified logarithmic Sobolev inequalities   179

5. MODIFIED LOGARITHMIC SOBOLEV INEQUALITIES IN
   DISCRETE SETTINGS   182
5.1 Logarithmic Sobolev inequality for Bernoulli and Poisson measures   182
5.2 Modified logarithmic Sobolev inequalities and Poisson tails   188
5.3 Sharp bounds   190

6. SOME APPLICATIONS TO LARGE DEVIATIONS AND TO
   BROWNIAN MOTION ON A MANIFOLD   193
6.1 Logarithmic Sobolev inequalities and large deviation upper bounds   193
6.2 Some tail estimate for Brownian motion on a manifold   194

7. ON REVERSED HERBST'S INEQUALITIES AND BOUNDS ON
   THE LOGARITHMIC·SOBOLEV CONSTANT   199
7.1 Reversed Herbst's inequality   199
7.2 Dimension free lower bounds   204
7.3 Upper bounds on the logarithmic Sobolev constant   205
7.4 Diameter and the logarithmic Sobolev constant for Markov chains   209

   REFERENCES   214

INTRODUCTION

The concentration of measure phenomenon was put forward in the seventies by V. D. Milman in the local theory of Banach spaces. Of isoperimetric inspiration, it is of powerful interest in applications, in particular in probability theory (probability in Banach spaces, empirical processes, geometric probabilities, statistical mechanics...) One main example is the Gaussian concentration property which expresses that, whenever A is a Borel set in $\mathbb{R}^n$ of canonical Gaussian measure $\gamma(A) \geq \frac{1}{2}$, for every $r \geq 0$,

$$\gamma(A_r) \geq 1 - e^{-r^2/2}$$

where A_r is the r-th Euclidean neighborhood of A. As r increases, the enlargement A_r thus gets very rapidly a measure close to one. This Gaussian concentration property can be described equivalently on functions. If F is a Lipschitz map on $\mathbb{R}^n$ with $\|F\|_{\mathrm{Lip}} \leq 1$, for every $r \geq 0$,

$$\gamma(F \geq \int F d\gamma + r) \leq e^{-r^2/2}.$$

Together with the same inequality for $-F$, the Lipschitz function F is seen to be concentrated around some mean value with very high probability. These quantitative estimates are dimension free and extend to arbitrary infinite dimensional Gaussian measures. As such, they are a main tool in the study of Gaussian processes and measures.

Simultaneously, hypercontractive estimates and logarithmic Sobolev inequalities came up in quantum field theory with the contributions of E. Nelson and L. Gross. In particular, L. Gross proved in 1975 a Sobolev inequality for Gaussian measures of logarithmic type. Namely, for all smooth functions f on $\mathbb{R}^n$,

$$\int f^2 \log f^2 d\gamma - \int f^2 d\gamma \log \int f^2 d\gamma \leq 2 \int |\nabla f|^2 d\gamma.$$

This inequality is again independent of the dimension and proved to be a substitute of the classical Sobolev inequalities in infinite dimensional settings. Logarithmic Sobolev inequalities have been used extensively in the recent years as a way to measure the smoothing properties (hypercontractivity) of Markov semigroups. In particular, they are a basic ingredient in the investigation of the time to equilibrium.

One of the early questions on logarithmic Sobolev inequalities was to determine which measures, on $\mathbb{R}^n$, satisfy an inequality similar to the one for Gaussian measures. To this question, raised by L. Gross, I. Herbst (in an unpublished letter to L.

Gross) found the following necessary condition: if μ is a probability measure such that for some $C > 0$ and every smooth function f on $\mathbb{R}^n$,

$$\int f^2 \log f^2 \, d\mu - \int f^2 \, d\mu \log \int f^2 \, d\mu \leq C \int |\nabla f|^2 \, d\mu,$$

then,

$$\int e^{\alpha |x|^2} \, d\mu(x) < \infty$$

for every $\alpha < \frac{1}{C}$. Furthermore, for any Lipschitz function F on $\mathbb{R}^n$ with $\|F\|_{\text{Lip}} \leq 1$, and every real λ,

$$\int e^{\lambda F} \, d\mu \leq e^{\lambda \int F \, d\mu + C\lambda^2/4}.$$

By a simple use of Chebyshev's inequality, the preceding thus relates in an essential way to the Gaussian concentration phenomenon.

Herbst's result was mentioned in the early eighties by E. Davies and B. Simon, and has been revived recently by S. Aida, T. Masuda and I. Shigekawa. It was further developed and refined by S. Aida, S. Bobkov, F. Götze, L. Gross, O. Rothaus, D. Stroock and the author. Following these authors and their contributions, the aim of these notes is to present a complete account on the applications of logarithmic Sobolev inequalities to the concentration of measure phenomenon. We exploit Herbst's original argument to deduce from the logarithmic Sobolev inequalities some differential inequalities on the Laplace transforms of Lipschitz functions. According to the family of entropy-energy inequalities we are dealing with, these differential inequalities yield various behaviors of the Laplace transforms of Lipschitz functions and of their concentration properties. In particular, the basic product property of entropy allows us to investigate with this tool concentration properties in product spaces. The principle is rather simple minded, and as such convenient for applications.

The first part of this set of notes includes a introduction to isoperimetry and concentration for Gaussian and Boltzmann measures. The second part then presents spectral gap and logarithmic Sobolev inequalities, and describes Herbst's basic Laplace transform argument. In the third part, we investigate by this method deviation and concentration inequalities for product measures. While concentration inequalities do not necessarily tensorize, we show that they actually follow from stronger logarithmic Sobolev inequalities. We thus recover most of M. Talagrand's recent results on isoperimetric and concentration inequalities in product spaces. We briefly mention there the information theoretic inequalities by K. Marton which provide an alternate approach to concentration also based on entropy, and which seems to be well suited to dependent structures. We then develop the subject of modified logarithmic Sobolev inequalities investigated recently in joint works with S. Bobkov. We examine in this way concentration properties for the product measure of the exponential distribution, as well as, more generally, of measures satisfying a Poincaré inequality. In the next section, the analogous questions for discrete gradients are addressed, with particular emphasis on Bernoulli and Poisson measures. We then present some applications to large deviation upper bounds and to tail estimates for

Brownian motion on a manifold. In the final part, we discuss some recent results on the logarithmic Sobolev constant in Riemannian manifolds with non-negative Ricci curvature. The last section is an addition of L. Saloff-Coste on the logarithmic Sobolev constant and the diameter for Markov chains. We sincerely thank him for this contribution.

It is a pleasure to thank the organizers (in particular M. Scheutzow) and the participants of the "Graduierten- kolleg" course which was held in Berlin in November 1997 for the opportunity to present, and to prepare, these notes. These notes would not exist without the collaboration with S. Bobkov which led to the concept of modified logarithmic Sobolev inequality and whose joint work form most of Parts 4 and 5. Thanks are also due to S. Kwapień for numerous exchanges over the years on the topic of these notes. D. Piau and D. Steinsaltz were very helpful with their comments and corrections on the manuscript.

1. ISOPERIMETRIC AND CONCENTRATION INEQUALITIES

In this first part, we present the Gaussian isoperimetric inequality as well as a Gaussian type isoperimetric inequality for a class of Boltzmann measures with a sufficiently convex potential. Isoperimetry is a natural way to introduce to the concentration of measure phenomenon. For completness, we propose a rather short, self-contained proof of these isoperimetric inequalities following the recent contributions [Bob4], [Ba-L]. Let us mention however that our first goal in these notes is to produce simpler, more functional arguments to derive concentration properties. We then present the concentration of measure phenomenon, and discuss a few of its first properties.

1.1 Introduction

The classical isoperimetric inequality in Euclidean space states that among all subsets with fixed finite volume, balls achieve minimal surface area. In probabilistic, and also geometric, applications one is often interested in finite measure space, such as the unit sphere S^n in $\mathbb{R}^{n+1}$ equipped with its normalized invariant measure σ^n. On S^n, (geodesic) balls, or caps, are again the extremal sets, that is achieve minimal surface measure among sets with fixed measure.

The isoperimetric inequality on the sphere was used by V. D. Milman in the early seventies as a tool to prove the famous Dvoretzky theorem on Euclidean sections of convex bodies (cf. [Mi], [M-S]). Actually, V. D. Milman is using the isoperimetric property as a concentration property. Namely, in its integrated version, the isoperimetry inequality states that whenever $\sigma^n(A) = \sigma^n(B)$ where B is a ball on S^n, for every $r \geq 0$,

$$\sigma^n(A_r) \geq \sigma^n(B_r) \tag{1.1}$$

where A_r (resp. B_r) is the neighborhood of order r of A (resp. B) for the geodesic metric on the sphere. Since, for a set A on S^n with smooth boundary ∂A, the surface measure σ_s^n of ∂A can be described by the Minkowski content formula as

$$\sigma_s^n(\partial A) = \liminf_{r \to 0} \frac{1}{r} \left[\sigma^n(A_r) - \sigma^n(A) \right],$$

(1.1) is easily seen to be equivalent to the isoperimetric statement. Now, the measure of a cap may be estimated explicitely. For example, if $\sigma^n(A) \geq \frac{1}{2}$, it follows from

(1.1) that

$$\sigma^n(A_r) \geq 1 - \sqrt{\tfrac{\pi}{8}} e^{-(n-1)r^2/2} \tag{1.2}$$

for every $r \geq 0$. Therefore, if the dimension is large, only a small increase of r (of the order of $\frac{1}{\sqrt{n}}$) makes the measure of A_r close to 1. In a sense. the measure σ^n is concentrated around the equator, and (1.2) describes the so-called concentration of measure phenomenon of σ^n. One significant aspect of this concentration phenomenon is that the enlargements are not infinitesimal as for isoperimetry, and that emphasis is not on extremal sets. These notes will provide a sample of concentration properties with the functional tool of logarithmic Sobolev inequalities.

1.2 Isoperimetric inequalities for Gaussian and Boltzmann measures

It is well known that uniform measures on n-dimensional spheres with radius $\sqrt{n}$ approximate (when projected on a finite number of coordinates) Gaussian measures (Poincaré's lemma). In this sense, the isoperimetric inequality on spheres gives rise to an isoperimetric inequality for Gaussian measures (cf. [Le3]). Extremal sets are then half-spaces (which may be considered as balls with centers at infinity). Let, more precisely, $\gamma = \gamma^n$ be the canonical Gaussian measure on $\mathbb{R}^n$ with density

$$(2\pi)^{-n/2} \exp(-|x|^2/2)$$

with respect to Lebesgue measure. Define the Gaussian surface measure of a Borel set A in $\mathbb{R}^n$ as

$$\gamma_s(\partial A) = \liminf_{r \to 0} \frac{1}{r} [\gamma(A_r) - \gamma(A)] \tag{1.3}$$

where $A_r = \{x \in \mathbb{R}^n; d_2(x, A) < r\}$ is the r-Euclidean open neighborhood of A. Then, if H is a half-space in $\mathbb{R}^n$, that is $H = \{x \in \mathbb{R}^n; \langle x, u \rangle < a\}$, where $|u| = 1$ and $a \in [-\infty, +\infty]$, and if $\gamma(A) = \gamma(H)$, then

$$\gamma_s(\partial A) \geq \gamma_s(\partial H).$$

Let $\Phi(t) = (2\pi)^{-1/2} \int_{-\infty}^{t} e^{-x^2/2} dx$, $t \in [-\infty, +\infty]$, be the distribution function of the canonical Gaussian measure in dimension one and let $\varphi = \Phi'$. Then $\gamma(H) = \Phi(a)$ and $\gamma_s(\partial H) = \varphi(a)$ so that,

$$\gamma_s(\partial A) \geq \varphi(a) = \varphi \circ \Phi^{-1}(\gamma(A)). \tag{1.4}$$

Moreover, half-spaces are the extremal sets in this inequality. In this form, the Gaussian isoperimetric inequality is dimension free.

In applications, the Gaussian isoperimetric inequality is often used in its integrated version. Namely, if $\gamma(A) = \gamma(H) = \Phi(a)$ (or only $\gamma(A) \geq \Phi(a)$), then, for every $r \geq 0$,

$$\gamma(A_r) \geq \gamma(H_r) = \Phi(a + r). \tag{1.5}$$

In particular, if $\gamma(A) \geq \tfrac{1}{2} (= \Phi(0))$,

$$\gamma(A_r) \geq \Phi(r) \geq 1 - e^{-r^2/2}. \tag{1.6}$$

To see that (1.4) implies (1.5), we may assume, by a simple approximation, that A is given by a finite union of open balls. The family of such sets A is closed under the operation $A \mapsto A_r$, $r \geq 0$. Then, the lim inf in (1.3) is a true limit. Actually, the boundary ∂A of A is a finite union of piecewise smooth $(n-1)$-dimensional surfaces in $\mathbb{R}^n$ and $\gamma_s(\partial A)$ is given by the integral of the Gaussian density along ∂A with respect to Lebesgue measure on ∂A. Now, by (1.4), the function $v(r) = \Phi^{-1} \circ \gamma(A_r)$, $r \geq 0$, satisfies

$$v'(r) = \frac{\gamma_s(\partial A_r)}{\varphi \circ \Phi^{-1}(\gamma(A_r))} \geq 1$$

so that $v(r) = v(0) + \int_0^r v'(s)ds \geq v(0) + r$. which is (1.5). (Alternatively, see [Bob3].)

The Euclidean neighborhood A_r of a Borel set A can be viewed as the Minkowski sum $A + rB_2 = \{a + rb ; a \in A, b \in B_2\}$ with B_2 the Euclidean open unit ball. If γ is any (centered) Gaussian measure on $\mathbb{R}^n$, B_2 has to be replaced by the ellipsoid associated to the covariance structure of γ. More precisely, denote by $\Gamma = M\,^t M$ the covariance matrix of the Gaussian measure γ on $\mathbb{R}^n$. Then γ is the image of the canonical Gaussian measure by the linear map $M = (M_{ij})_{1 \leq i,j \leq n}$. Set $\mathcal{K} = M(B_2)$. Then, if $\gamma(A) \geq \Phi(a)$, for every $r \geq 0$,

$$\gamma(A + r\mathcal{K}) \geq \Phi(a + r). \tag{1.7}$$

In this formulation, the Gaussian isoperimetric inequality extends to infinite dimensional (centered) Gaussian measures. the set $\mathcal{K}$ being the unit ball of the reproducing kernel Hilbert space $\mathcal{H}$ (the Cameron-Martin space for Wiener measure for example). Cf. [Bor], [Led3].

To see moreover how (1.6) or (1.7) may be used in applications. let for example $X = (X_t)_{t \in T}$ be a centered Gaussian process indexed by some, for simplicity, countable parameter set T. Assume that $\sup_{t \in T} X_t < \infty$ almost surely. Fix $t_1, \ldots, t_n$ in T and consider the distribution γ of the sample $(X_{t_1}, \ldots, X_{t_n})$. Choose m finite such that $\mathbb{P}\{\sup_{t \in T} X_t \leq m\} \geq \frac{1}{2}$. In particular, if

$$A = \big\{ \max_{1 \leq i \leq n} X_{t_i} \leq m \big\},$$

then $\gamma(A) \geq \frac{1}{2}$. Therefore, by (1.7) (with $a = 0$), for every $r \geq 0$,

$$\gamma(A + r\mathcal{K}) \geq \Phi(r) \geq 1 - e^{-r^2/2}.$$

Now, for any h in $\mathcal{K} = M(B_2)$,

$$\max_{1 \leq i \leq n} h_i \leq \max_{1 \leq i \leq n} \Big(\sum_{j=1}^n M_{ij}^2 \Big)^{1/2} = \max_{1 \leq i \leq n} \big(\mathbb{E}(X_{t_i}^2) \big)^{1/2}$$

by the Cauchy-Schwarz inequality, so that

$$A + r\mathcal{K} \subset \big\{ \max_{1 \leq i \leq n} X_{t_i} \leq m + r \max_{1 \leq i \leq n} \big(\mathbb{E}(X_{t_i}^2) \big)^{1/2} \big\}.$$

Set $\sigma = \sup_{t \in T}(\mathbb{E}(X_t^2))^{1/2}$. (It is easily seen that σ is always finite under the assumption $\sup_{t \in T} X_t < \infty$. Let indeed m' be such that $\mathbb{P}\{\sup_{t \in T} X_t \leq m'\} \geq \frac{3}{4}$. Then, if $\sigma_t = (\mathbb{E}(X_t^2))^{1/2}$, $\frac{m'}{\sigma_t} \geq \Phi^{-1}(\frac{3}{4}) > 0$.) It follows from the preceding that

$$\mathbb{P}\{\max_{1 \leq i \leq n} X_{t_i} \leq m + \sigma r\} \geq 1 - e^{-r^2/2}.$$

By monotone convergence, and taking complements, for every $r \geq 0$,

$$\mathbb{P}\{\sup_{t \in T} X_t \geq m + \sigma r\} \leq e^{-r^2/2}. \tag{1.8}$$

This inequality describes the strong integrability properties of almost surely bounded Gaussian processes. It namely implies in particular (cf. Proposition 1.2 below) that for every $\alpha < \frac{1}{2\sigma^2}$,

$$\mathbb{E}\left(\exp\left(\alpha(\sup_{t \in T} X_t)^2\right)\right) < \infty. \tag{1.9}$$

Equivalently, in a large deviation formulation,

$$\lim_{r \to \infty} \frac{1}{r^2} \log \mathbb{P}\{\sup_{t \in T} X_t \geq r\} = -\frac{1}{2\sigma^2}. \tag{1.10}$$

(The lower bound in (1.10) is just that

$$\mathbb{P}\{\sup_{t \in T} X_t \geq r\} \geq \mathbb{P}\{X_t \geq r\} = 1 - \Phi\left(\frac{r}{\sigma_t}\right) \geq \frac{e^{-r^2/2\sigma_t^2}}{\sqrt{2\pi}(1 + (r/\sigma_t))}$$

for every $t \in T$ and $r \geq 0$.) But inequality (1.8) actually contains more information than just this integrability result. (For example, if X^n is a sequence of Gaussian processes as before, and if we let $\|X^n\| = \sup_{t \in T} X_t^n$, $n \in \mathbb{N}$, then $\|X^n\| \to 0$ almost surely as soon as $\mathbb{E}(\|X^n\|) \to 0$ and $\sigma^n \sqrt{\log n} \to 0$ where $\sigma^n = \sup_{t \in T}(\mathbb{E}((X_t^n)^2))^{1/2}$.) (1.8) describes a sharp deviation inequality in terms of two parameters, m and σ. In this sense, it belongs to the concentration of measure phenomenon which will be investigated in these notes (cf. Section 1.3). Note that (1.8), (1.9), (1.10) hold similarly with $\sup_{t \in T} X_t$ replaced by $\sup_{t \in T} |X_t|$ (under the assumption $\sup_{t \in T} |X_t| < \infty$ almost surely).

The Gaussian isoperimetric inequality was established in 1974 independently by C. Borell [Bor] and V. N. Sudakov and B. S. Tsirel'son [S-T] on the basis of the isoperimetric inequality on the sphere and Poincaré's lemma. A proof using Gaussian symmetrizations was developed by A. Ehrhard in 1983 [Eh]. We present here a short and self-contained proof of this inequality. Our approach will be functional. Denote by $\mathcal{U} = \varphi \circ \Phi^{-1}$ the Gaussian isoperimetric function in (1.4). In a recent striking paper, S. Bobkov [Bob4] showed that for every smooth enough function f with values in the unit interval $[0, 1]$,

$$\mathcal{U}\left(\int f d\gamma\right) \leq \int \sqrt{\mathcal{U}^2(f) + |\nabla f|^2}\, d\gamma \tag{1.11}$$

where $|\nabla f|$ denotes the Euclidean length of the gradient ∇f of f. It is easily seen that (1.11) is a functional version of the Gaussian isoperimetric inequality (1.4). Namely, if (1.11) holds for all smooth functions, it holds for all Lipschitz functions with values in $[0, 1]$. Assume again that the set A in (1.4) is a finite union of non-empty open balls. In particular, $\gamma(\partial A) = 0$. Apply then (1.11) to $f_r(x) = (1 - \frac{1}{r} d_2(x, A))^+$ (where d_2 is the Euclidean distance function). Then $f_r \to I_A$ and $\mathcal{U}(f_r) \to 0$ almost everywhere since $\gamma(\partial A) = 0$ and $\mathcal{U}(0) = \mathcal{U}(1) = 0$. Moreover, $|\nabla f_r| = 0$ on A and on the complement of the closure of A_r, and $|\nabla f_r| \leq \frac{1}{r}$ everywhere. Note that the sets $\partial(A_r)$ are of measure zero for every $r \geq 0$. Therefore

$$\mathcal{U}(\gamma(A)) \leq \liminf_{r \to 0} \int |\nabla f_r| d\gamma \leq \liminf_{r \to 0} \frac{1}{r} \left[\gamma(A_r) - \gamma(A) \right] = \gamma_s(\partial A).$$

To prove (1.11), S. Bobkov first establishes the analogous inequality on the two-point space and then uses the central limit theorem, very much as L. Gross in his proof of the Gaussian logarithmic Sobolev inequality [Gr1] (cf. Section 2.2). The proof below is direct. Our main tool will be the so-called Ornstein-Uhlenbeck or Hermite semigroup with invariant measure the canonical Gaussian measure γ. For every f, in $L^1(\gamma)$ say, set

$$P_t f(x) = \int_{\mathbb{R}^n} f\big(e^{-t/2} x + (1 - e^{-t})^{1/2} y\big) d\gamma(y), \quad x \in \mathbb{R}^n, \quad t \geq 0. \qquad (1.12)$$

The operators P_t are contractions on all $L^p(\gamma)$-spaces, and are symmetric and invariant with respect to γ. That is, for any sufficiently integrable functions f and g, and every $t \geq 0$, $\int f P_t g \, d\gamma = \int g P_t f \, d\gamma$. The family $(P_t)_{t \geq 0}$ is a semigroup $(P_s \circ P_t = P_{s+t})$. P_0 is the identity operator whereas $P_t f$ converges in $L^2(\gamma)$ towards $\int f d\gamma$ as t tends to infinity. All these properties are immediately checked on the preceding integral representation of P_t together with the elementary properties of Gaussian measures. The infinitesimal generator of the semigroup $(P_t)_{t \geq 0}$, that is the operator L such that

$$\frac{d}{dt} P_t f = P_t L f = L P_t f,$$

acts on all smooth functions f on $\mathbb{R}^n$ by

$$L f(x) = \tfrac{1}{2} \Delta f(x) - \tfrac{1}{2} \langle x, \nabla f(x) \rangle.$$

In other words, L is the generator of the Ornstein-Uhlenbeck diffusion process $(X_t)_{t \geq 0}$, the solution of the stochastic differential equation $dX_t = dB_t - \frac{1}{2} X_t dt$ where $(B_t)_{t \geq 0}$ is standard Brownian motion in $\mathbb{R}^n$. Moreover, the integration by parts formula for L indicates that, for f and g smooth enough on $\mathbb{R}^n$,

$$\int f(-Lg) d\gamma = \frac{1}{2} \int \langle \nabla f, \nabla g \rangle d\gamma. \qquad (1.13)$$

Let now f be a fixed smooth function on $\mathbb{R}^n$ with values in $[0, 1]$. It might actually be convenient to assume throughout the argument that $0 < \varepsilon \leq f \leq 1 - \varepsilon$

and let then ε tend to zero. Recall $\mathcal{U} = \varphi \circ \Phi^{-1}$. To prove (1.11) it will be enough to show that the function

$$J(t) = \int \sqrt{\mathcal{U}^2(P_t f) + |\nabla P_t f|^2}\, d\gamma$$

is non-increasing in $t \geq 0$. Indeed, if this is the case, $J(\infty) \leq J(0)$, which, together with the elementary properties of P_t recalled above, amounts to (1.11). Towards this goal, we first emphasize the basic property of the Gaussian isoperimetric function $\mathcal{U}$ that will be used in the argument, namely that $\mathcal{U}$ satisfies the fundamental differential equality $\mathcal{U}\mathcal{U}'' = -1$ (exercise). We now have

$$\frac{dJ}{dt} = \int \frac{1}{\sqrt{\mathcal{U}^2(P_t f) + |\nabla P_t f|^2}}\left[\mathcal{U}\mathcal{U}'(P_t f)\mathrm{L}P_t f + \langle \nabla(P_t f), \nabla(\mathrm{L}P_t f)\rangle\right] d\gamma.$$

To ease the notation, write f for $P_t f$. We also set $K(f) = \mathcal{U}^2(f) + |\nabla f|^2$. Therefore,

$$\frac{dJ}{dt} = \int \frac{1}{\sqrt{K(f)}}\left[\mathcal{U}\mathcal{U}'(f)\mathrm{L}f + \langle \nabla f, \nabla(\mathrm{L}f)\rangle\right] d\gamma. \tag{1.14}$$

For simplicity in the exposition, let us assume that the dimension n is one, the general case being entirely similar, though notationally a little bit heavier. By the integration by parts formula (1.13),

$$\int \frac{1}{\sqrt{K(f)}}\, \mathcal{U}\mathcal{U}'(f)\,\mathrm{L}f d\gamma = -\frac{1}{2}\int \left(\frac{\mathcal{U}\mathcal{U}'(f)}{\sqrt{K(f)}}\right)' f' d\gamma$$

$$= -\frac{1}{2}\int \frac{1}{\sqrt{K(f)}}\left[\mathcal{U}'^2(f) - 1\right]f'^2 d\gamma$$

$$+ \frac{1}{2}\int \frac{\mathcal{U}\mathcal{U}'(f)f'}{K(f)^{3/2}}\left[\mathcal{U}\mathcal{U}'(f)f' + f'f''\right]d\gamma$$

where we used that $\mathcal{U}\mathcal{U}'' = -1$ and that

$$K(f)' = 2\mathcal{U}\mathcal{U}'(f)f' + (f'^2)' = 2\mathcal{U}\mathcal{U}'(f)f' + 2f'f''. \tag{1.15}$$

In order to handle the second term on the right-hand side of (1.14), let us note that

$$\langle \nabla f, \nabla(\mathrm{L}f)\rangle = \tfrac{1}{2} f'(f'' - xf')' = -\tfrac{1}{2} f'^2 + f'\mathrm{L}f'.$$

Hence, again by the integration by parts formula (1.13), and by (1.15),

$$\int \frac{1}{\sqrt{K(f)}}\langle \nabla f, \nabla(\mathrm{L}f)\rangle d\gamma = -\frac{1}{2}\int \frac{f'^2}{\sqrt{K(f)}}\, d\gamma + \int \frac{f'}{\sqrt{K(f)}}\,\mathrm{L}f' d\gamma$$

$$= -\frac{1}{2}\int \frac{f'^2}{\sqrt{K(f)}}\, d\gamma - \frac{1}{2}\int \frac{f''^2}{\sqrt{K(f)}}\, d\gamma$$

$$+ \frac{1}{2}\int \frac{f'f''}{K(f)^{3/2}}\left[\mathcal{U}\mathcal{U}'(f)f' + f'f''\right]d\gamma.$$

Putting these equations together, we get, after some algebra,

$$\frac{dJ}{dt} = -\frac{1}{2} \int \frac{1}{K(f)^{3/2}} \left[\mathcal{U}'^2(f)f'^4 - 2\mathcal{U}\mathcal{U}'(f)f'^2 f'' + \mathcal{U}^2(f)f''^2 \right] d\gamma$$

and the result follows since

$$\mathcal{U}'^2(f)f'^4 - 2\mathcal{U}\mathcal{U}'(f)f'^2 f'' + \mathcal{U}^2(f)f''^2 = \left(\mathcal{U}'(f)f'^2 - \mathcal{U}(f)f'' \right)^2 \geq 0.$$

The preceding proof of the Gaussian isoperimetric inequality came up in the joint work [Ba-L] with D. Bakry. The argument is developed there in an abstract framework of Markov diffusion generators and semigroups and applies to a large class of invariant measures of diffusion generators satisfying a curvature assumption. We present here this result for some concrete class of Boltzmann measures for which a Gaussian-like isoperimetric inequality holds.

Let us consider a smooth (C^2 say) function W on $\mathbb{R}^n$ such that e^{-W} is integrable with respect to Lebesgue measure. Define the so-called Boltzmann measure as the probability measure

$$d\mu(x) = Z^{-1} e^{-W(x)} dx$$

where Z is the normalization factor. As is well-known, μ may be described as the invariant measure of the generator $L = \frac{1}{2}\Delta - \frac{1}{2}\nabla W \cdot \nabla$. Alternatively, L is the generator of the Markov semigroup $(P_t)_{t \geq 0}$ of the Kolmogorov process $X = (X_t)_{t \geq 0}$ solution of the stochastic differential Langevin equation

$$dX_t = dB_t - \frac{1}{2}\nabla W(X_t) dt.$$

The choice of $W(x) = \frac{1}{2}|x|^2$ with invariant measure the canonical Gaussian measure corresponds to the Ornstein-Uhlenbeck process. Denote by $W''(x)$ the Hessian of W at $x \in \mathbb{R}^n$.

Theorem 1.1. *Assume that, for some $c > 0$, $W''(x) \geq c\,\mathrm{Id}$ as symmetric matrices, uniformly in $x \in \mathbb{R}^n$. Then, whenever A is a Borel set in $\mathbb{R}^n$ with $\mu(A) \geq \Phi(a)$, for any $r \geq 0$,*

$$\mu(A_r) \geq \Phi(a + \sqrt{c}\, r).$$

As in the Gaussian case, the inequality of Theorem 1.1 is equivalent to its infinitesimal version

$$\mu_s(\partial A) \geq \sqrt{c}\,\mathcal{U}(\mu(A))$$

with the corresponding notion of surface measure and to the functional inequality

$$\mathcal{U}\left(\int f\,d\gamma \right) \leq \int \sqrt{\mathcal{U}^2(f) + \frac{1}{c}|\nabla f|^2}\, d\gamma$$

which is the result we established in the proof as before. Before turning to this proof, let us comment on the Gaussian aspect of the theorem. Let F be a Lipschitz map on $\mathbb{R}^n$ with Lipschitz coefficient $\|F\|_{\mathrm{Lip}} \le \sqrt{c}$. Then, the image measure ν of μ by F is a contraction of the canonical Gaussian measure on $\mathbb{R}$. Indeed, we may assume by some standard regularization procedure that ν is absolutely continuous with respect to Lebesgue measure on $\mathbb{R}$ with a strictly positive density. Set $\nu(r) = \nu((-\infty, r])$ so that the measure ν has density ν'. For $r \in \mathbb{R}$, apply Theorem 1.1, or rather its infinitesimal version, to $A = \{F \le r\}$ to get $\mathcal{U}(\nu(r)) \le \nu'(r)$. Then, setting $k = \nu^{-1} \circ \Phi$ and $x = \Phi^{-1} \circ \nu(r)$, $k'(x) \le 1$ so that ν is the image of the canonical Gaussian measure on $\mathbb{R}$ by the contraction k. In particular, in dimension one, every measure satisfying the hypothesis of Theorem 1.1 is a Lipschitz image of the canonical Gaussian measure.

Proof of Theorem 1.1. It is entirely similar to the proof of the Gaussian isoperimetric inequality in Section 1.1. Denote thus by $(P_t)_{t \ge 0}$ the Markov semigroup with generator $L = \frac{1}{2}\Delta - \frac{1}{2}\nabla W \cdot \nabla$. The integration by parts formula for L reads

$$\int f(-Lg)d\mu = \frac{1}{2}\int \langle \nabla f, \nabla g \rangle d\mu$$

for smooth functions f and g. Fix a smooth function f on $\mathbb{R}^n$ with $0 \le f \le 1$. As in the Gaussian case, we aim to show that, under the assumption on W,

$$J(t) = \int \sqrt{\mathcal{U}^2(P_t f) + \frac{1}{c}|\nabla P_t f|^2}\, d\mu$$

is non-increasing in $t \ge 0$. Remaining as before in dimension one for notational simplicity, the argument is the same than in the Gaussian case with now $K(f) = \mathcal{U}^2(f) + \frac{1}{c}|\nabla f|^2$ so that

$$K(f)' = 2\mathcal{U}\mathcal{U}'(f)f' + \frac{2}{c}f'f''.$$

Similarly,

$$\langle \nabla f, \nabla(Lf) \rangle = f'\left(\tfrac{1}{2}f'' - \tfrac{1}{2}W'f'\right)' = -\tfrac{1}{2}W''f'^2 + f'Lf'.$$

Hence, again by the integration by parts formula,

$$\int \frac{1}{\sqrt{K(f)}}\langle \nabla f, \nabla(Lf)\rangle d\gamma = -\frac{1}{2}\int \frac{W''f'^2}{\sqrt{K(f)}}d\mu + \int \frac{f'}{\sqrt{K(f)}}Lf'd\mu$$

$$= -\frac{1}{2}\int \frac{W''f'^2}{\sqrt{K(f)}}d\mu - \frac{1}{2}\int \frac{f''^2}{\sqrt{K(f)}}d\mu$$

$$+ \frac{1}{2}\int \frac{f'f''}{K(f)^{3/2}}\left[\mathcal{U}\mathcal{U}'(f)f' + \frac{1}{c}f'f''\right]d\mu.$$

In the same way, we then get

$$\frac{dJ}{dt} = -\frac{1}{2c}\int \frac{1}{K(f)^{3/2}}\left[\mathcal{U}'^2(f)f'^4 - 2\mathcal{U}\mathcal{U}'(f)f'^2 f'' + \mathcal{U}^2(f)f''^2\right]d\mu$$

$$- \frac{1}{2}\int \frac{f'^2}{K(f)^{3/2}}\left(\frac{W''}{c} - 1\right)\left[\mathcal{U}^2(f) + \frac{1}{c}f'^2\right]d\mu.$$

Since $W'' \ge c$, the conclusion follows. The proof of Theorem 1.1 is complete. $\qquad\square$

1.3 Some general facts about concentration

As we have seen in (1.6), one corollary of Gaussian isoperimetry is that whenever A is a Borel set in $\mathbb{R}^n$ with $\gamma(A) \geq \frac{1}{2}$ for the canonical Gaussian measure γ, then, for every $r \geq 0$,

$$\gamma(A_r) \geq 1 - e^{-r^2/2}. \tag{1.16}$$

In other words, starting with a set with positive measure ($\frac{1}{2}$ here), its (Euclidean) enlargement or neighborhood gets very rapidly a mass close to one (think for example of $r = 5$ or 10). We described with (1.2) a similar property on spheres. While true isoperimetric inequalities are usually quite difficult to establish, in particular identification of extremal sets, concentration properties like (1.2) or (1.16) are milder, and may be established by a variety of arguments, as will be illustrated in these notes.

The concentration of measure phenomenon, put forward most vigorously by V. D. Milman in the local theory of Banach spaces (cf. [Mi], [M-S]), may be described for example on a metric space (X, d) equipped with a probability measure μ on the Borel sets of (X, d). One is then interested in the concentration function

$$\alpha(r) = \sup\left\{1 - \mu(A_r); \mu(A) \geq \tfrac{1}{2}\right\}, \quad r \geq 0,$$

where $A_r = \{x \in X; d(x, A) < r\}$. As a consequence of (1.16), $\alpha(r) \leq e^{-r^2/2}$ in case of the canonical Gaussian measure γ on $\mathbb{R}^n$ with respect to the Euclidean metric. The important feature of this definition is that several measures, as we will see, do have very small concentration functions $\alpha(r)$ as r becomes "large". We will mainly be interested in Gaussian (or at least exponential) concentration functions throughout these notes. Besides Gaussian measures, Haar measures on spheres were part of the first examples (1.2). Martingale inequalities also yield family of examples (cf. [Mau1], [M-S], [Ta7]). In this work, we will encounter further examples, in particular in the context of product measures.

The concentration of measure phenomenon may also be described on functions. Let F be a Lipschitz map on X with $\|F\|_{\mathrm{Lip}} \leq 1$ (by homogeneity) and let m be a median of F for μ. Then, since $\mu(F \leq m) \geq \frac{1}{2}$, and $\{F \leq m\}_r \subset \{F \leq m + r\}$, we see that for every $r \geq 0$,

$$\mu(F \geq m + r) \leq \alpha(r). \tag{1.17}$$

When such an inequality holds, we will speak of a deviation inequality for F. Together with the same inequality for $-F$,

$$\mu(|F - m| \geq r) \leq 2\alpha(r). \tag{1.18}$$

We then speak of a concentration inequality for F. In particular, the Lipschitz map F concentrates around some fixed mean value m with a probability estimated by α. According to the smallness of α as r increases, F may be considered as almost constant on almost all the space. Note that these deviation or concentration inequalities on (Lipschitz) functions are actually equivalent to the corresponding statement on sets. Let A be a Borel set in (X, d) with $\mu(A) \geq \frac{1}{2}$. Set $F(x) = d(x, A)$ where $r > 0$. Clearly $\|F\|_{\mathrm{Lip}} \leq 1$ while

$$\mu(F > 0) = \mu(x; d(x, A) > 0) \leq 1 - \mu(A) \leq \tfrac{1}{2}.$$

Hence, there is a median m of F which is ≤ 0 and thus, by (1.17),

$$1 - \mu(A_r) \leq \mu(F \geq r) \leq \alpha(r). \tag{1.19}$$

In the Gaussian case, for every $r \geq 0$,

$$\gamma(F \geq m + r) \leq e^{-r^2/2} \tag{1.20}$$

when $\|F\|_{\text{Lip}} \leq 1$ and

$$\gamma(F \geq m + r) \leq e^{-r^2/2\|F\|_{\text{Lip}}^2}$$

for arbitrary Lipschitz functions, extending thus the simple case of linear functions. These inequalities emphasize the two main parameters in a concentration property, namely some deviation or concentration value m, mean or median, and the Lipschitz coefficient $\|F\|_{\text{Lip}}$ of F. An example of this type already occured in (1.8) which may be shown to follow equivalently from (1.20) (consider $F(x) = \max_{1 \leq i \leq n}(Mx)_i$). As a consequence of Theorem 1.1, if μ is a Boltzmann measure with $W''(x) \geq c\,\text{Id}$ for every $x \in \mathbb{R}^n$, and if F is Lipschitz with $\|F\|_{\text{Lip}} \leq 1$, we get similarly that for every $r \geq 0$,

$$\mu(F \geq m + r) \leq e^{-r^2/2c}. \tag{1.21}$$

Although this last bound covers an interesting class of measures, it is clear that its application is fairly limited. It is therefore of interest to investigate new tools, other than isoperimetric inequalities, to derive concentration inequalities for large families of measures. This is the task of the next chapters.

It might be worthwhile to note that while we deduced the preceding concentration inequalities from isoperimetry, one may also adapt the semigroup arguments to give a direct, simpler, proof of these inequalities. To outline the argument in case of (1.20), let F on $\mathbb{R}^n$ be smooth and such that $\int F d\gamma = 0$ and $\|F\|_{\text{Lip}} \leq 1$. For fixed $\lambda \in \mathbb{R}$, set $H(t) = \int e^{\lambda P_t F} d\gamma$ where $(P_t)_{t \geq 0}$ is the Ornstein-Uhlenbeck semigroup (1.12). Since $H(\infty) = 1$, we may write, for every $t \geq 0$,

$$\begin{aligned}
H(t) &= 1 - \int_t^\infty H'(s)ds \\
&= 1 - \lambda \int_t^\infty \left(\int LP_s F e^{\lambda P_s F} d\gamma \right) ds \\
&= 1 + \frac{\lambda^2}{2} \int_t^\infty \left(\int |\nabla P_s F|^2 e^{\lambda P_s F} d\gamma \right) ds
\end{aligned}$$

by the integration by parts formula (1.13). Since $\|F\|_{\text{Lip}} \leq 1$, $|\nabla F| \leq 1$ almost everywhere, so that

$$|\nabla P_s F|^2 = \left| e^{-s/2} P_s(\nabla F) \right|^2 \leq e^{-s} P_s(|\nabla F|^2) \leq e^{-s}$$

almost everywhere. Hence, for $t \geq 0$,

$$H(t) \leq 1 + \frac{\lambda^2}{2} \int_t^\infty e^{-s} H(s)ds.$$

By Gronwall's lemma,

$$H(0) = \int e^{\lambda F} d\gamma \leq e^{\lambda^2/2}.$$

To deduce the deviation inequality (1.20) from this result, simply apply Chebyshev's inequality: for every $\lambda \in \mathbb{R}$ and $r \geq 0$,

$$\gamma(F \geq r) \leq e^{-\lambda r + \lambda^2/2}.$$

Minimizing in λ ($\lambda = r$) yields

$$\gamma(F \geq r) \leq e^{-r^2/2},$$

where we recall that F is smooth and such that $\int F d\gamma = 0$ and $\|F\|_{\text{Lip}} \leq 1$. By a simple approximation procedure, we therefore get that, for every Lipschitz function F on $\mathbb{R}^n$ such that $\|F\|_{\text{Lip}} \leq 1$ and all $r \geq 0$,

$$\gamma(F \geq \int F d\gamma + r) \leq e^{-r^2/2}. \tag{1.22}$$

The same argument would apply for the Boltzmann measures of Theorem 1.1 to produce (1.21) with the mean instead of a median. We note that this direct proof of (1.22) is shorter than the proof of the full isoperimetric inequality.

Inequality (1.22) may be used to investigate supremum of Gaussian processes as (1.7) or (1.20). As before, let $(X_t)_{t \in T}$ be a centered Gaussian process indexed by some countable set T, and assume that $\sup_{t \in T} X_t < \infty$ almost surely. Fix $t_1, \ldots, t_n$ and denote by $\Gamma = M\,^tM$ the covariance matrix of the centered Gaussian sample $(X_{t_1}, \ldots, X_{t_n})$. This sample thus has distribution Mx under $\gamma(dx)$. Let $F(x) = \max_{1 \leq i \leq n}(Mx)_i$, $x \in \mathbb{R}^n$. Then F is Lipschitz with

$$\|F\|_{\text{Lip}} = \max_{1 \leq i \leq n} \left(\mathbb{E}(X_{t_i}^2) \right)^{1/2} \leq \sigma$$

where $\sigma = \sup_{t \in T}(\mathbb{E}(X_t^2))^{1/2}$. Therefore, by (1.22), for every $r \geq 0$,

$$\mathbb{P}\Big\{ \max_{1 \leq i \leq n} X_{t_i} \geq \mathbb{E}\big(\max_{1 \leq i \leq n} X_{t_i} \big) + \sigma r \Big\} \leq e^{-r^2/2}. \tag{1.23}$$

Similarly for $-F$,

$$\mathbb{P}\Big\{ \max_{1 \leq i \leq n} X_{t_i} \leq \mathbb{E}\big(\max_{1 \leq i \leq n} X_{t_i} \big) - \sigma r \Big\} \leq e^{-r^2/2}.$$

Choose now m such that $\mathbb{P}\{\sup_{t \in T} X_t \leq m\} \geq \frac{1}{2}$ and r_0 such that $e^{-r_0^2/2} < \frac{1}{2}$. Then

$$\mathbb{P}\Big\{ \max_{1 \leq i \leq n} X_{t_i} \leq m \Big\} \geq \frac{1}{2}.$$

Intersecting with the preceding probability, we get

$$\mathbb{E}\big(\max_{1 \leq i \leq n} X_{t_i} \big) \leq m + \sigma r_0$$

independently of $t_1, \ldots, t_n$ in T. In particular, $\mathbb{E}(\sup_{t \in T} X_t) < \infty$, and by monotone convergence in (1.23),

$$\mathbb{P}\left\{\sup_{t \in T} X_t \geq \mathbb{E}\left(\sup_{t \in T} X_t\right) + \sigma r\right\} \leq e^{-r^2/2}. \tag{1.24}$$

This inequality is the analogue of (1.8) with the mean instead of the median. Note that the condition $\mathbb{E}(\sup_{t \in T} X_t) < \infty$ came for free in the argument. It thus also implies (1.9) and (1.10).

This approximation argument may be used in the same way on infinite dimensional Gaussian measures γ with respect to their reproducing kernel Hilbert space $\mathcal{H}$. If F is Lipschitz with respect to $\mathcal{H}$ in the sense that

$$|F(x) - F(y)| \leq |x - y|_{\mathcal{H}},$$

then

$$\gamma(F \geq m + r) \leq e^{-r^2/2} \tag{1.25}$$

for all $r \geq 0$ with m either the mean or a median of F for γ. See [Le3].

The inequalities (1.20) and (1.22) yield deviation inequalities for either a median or the mean of a Lipschitz function. Up to numerical constants, these are actually equivalent ([M-S], p. 142). One example was the inequalities (1.8) and (1.24) for supremum of Gaussian processes, and also (1.25). Let us describe the argument in some generality for exponential concentration functions. The argument clearly extends to sufficiently small concentration functions. (We will use this remark in the sequel.)

Let F be a measurable function on some probability space $(X, \mathcal{B}, \mu)$ such that, for some $0 < p < \infty$, some $a \in \mathbb{R}$ and some constants $c, d > 0$,

$$\mu(|F - a| \geq r) \leq 2c\, e^{-r^p/d} \tag{1.26}$$

for all $r \geq 0$. Then, first of all,

$$\int |F - a| d\mu = \int_0^\infty \mu(|F - a| \geq r) dr \leq \int_0^\infty 2c\, e^{-r^p/d} dr \leq C_p cd^{1/p}$$

where $C_p > 0$ only depends on p. In particular, $|\int F d\mu - a| \leq C_p cd^{1/p}$. Therefore, for $r \geq 0$,

$$\mu(F \geq \textstyle\int F d\mu + r) \leq \mu(F \geq a - C_p cd^{1/p} + r).$$

According as $r \leq 2C_p cd^{1/p}$ or $r \geq 2C_p cd^{1/p}$ we easily get that

$$\mu(F \geq \textstyle\int F d\mu + r) \leq c'e^{-r^p/d'}$$

where $c' = \max(c, e^{C_p^p c^p})$ and $d' = 2^p d$. Together with the same inequality for $-F$, (1.26) thus holds with a the mean of F (and c' and d'). Similary, if we choose in (1.26) $r = r_0$ so that

$$2c\, e^{-r^p/d} < \tfrac{1}{2},$$

for example $r_0^p = d \log(8c)$, we see that $\mu(|F - a| \geq r_0) < \frac{1}{2}$. Therefore a median m of F for μ will satisfy

$$a - r_0 \leq m \leq a + r_0.$$

It is then easy to conclude as previously that, for every $r \geq 0$,

$$\mu(F \geq m + r) \leq c' e^{-r^p/d'}$$

where $c' = 8c$ and $d' = 2^p d$. We can therefore also choose for a in (1.26) a median of F.

An alternate argument may be given on the concentration function. For a probability measure μ on the Borel sets of a metric space (X, d), assume that for some non-increasing function α on $\mathbb{R}_+$,

$$\mu(F \geq \mathrm{E}_\mu(F) + r) \leq \alpha(r) \tag{1.27}$$

for every F with $\|F\|_{\mathrm{Lip}} \leq 1$ and every $r \geq 0$. Let A with $\mu(A) > 0$ and fix $r > 0$. Set $F(x) = \min(d(x, A), r)$. Clearly $\|F\|_{\mathrm{Lip}} \leq 1$ and

$$\mathrm{E}_\mu(F) \leq (1 - \mu(A))r.$$

Applying (1.27),

$$1 - \mu(A_r) = \mu(F \geq r) \leq \mu(F \geq \mathrm{E}_\mu(F) + \mu(A)r) \leq \alpha(\mu(A)r). \tag{1.28}$$

In particular, if $\mu(A) \geq \frac{1}{2}$,

$$\mu(A_r) \geq 1 - \alpha(\tfrac{r}{2}).$$

We conclude this section by emphasizing that a concentration inequality of such as (1.26) of course implies strong integrability properties of the Lipschitz function F. This is the content of the simple proposition which immediately follows by integration in $r \geq 0$.

Proposition 1.2. *Let F be a measurable function on $(X, \mathcal{B}, \mu)$ such that for some $0 < p < \infty$, some $a \in \mathbb{R}$ and some constants $c, d > 0$,*

$$\mu(|F - a| \geq r) \leq 2c\, e^{-r^p/d}$$

for every $r \geq 0$. Then

$$\int e^{\alpha|F|^p} d\mu < \infty$$

for every $\alpha < \frac{1}{d}$.

Proof. From the hypothesis, for every $r \geq |a|$,

$$\mu(|F| \geq r) \leq \mu(|F - a| \geq r - |a|) \leq 2c\, e^{-(r-|a|)^p/d}.$$

Now, by Fubini's theorem,

$$\int e^{\alpha|F|^p} d\mu = 1 + \int_0^\infty p\alpha\, r^{p-1} \mu(|F| \geq r) e^{\alpha r^p} dr$$

$$\leq e^{\alpha|a|^p} + \int_{|a|}^\infty p\alpha\, r^{p-1} \mu(|F| \geq r) e^{\alpha r^p} dr$$

$$\leq e^{\alpha|a|^p} + \int_{|a|}^\infty p\alpha\, r^{p-1} 2c\, e^{-(r-|a|)^p/d} e^{\alpha r^p} dr$$

from which the conclusion follows. $\qquad\square$

2. SPECTRAL GAP AND
LOGARITHMIC SOBOLEV INEQUALITIES

We present in this section the basic simple argument that produces Gaussian concentration under a logarithmic Sobolev inequality. We try to deal with a rather general framework in order to include several variations developed in the literature. Herbst's original argument, mentioned in [D-S], has been revived recently by S. Aida, T. Masuda and I. Shigekawa [A-M-S]. Since then, related papers by S. Aida and D. Stroock [A-S], S. Bobkov and F. Götze [B-G], L. Gross and O. Rothaus [G-R], O. Rothaus [Ro3] and the author [Le1] further developed the methods and results. Most of the results presented in these notes are taken from these works. We will mainly be concerned with Herbst's original differential argument on the Laplace transform. The papers [A-S], [Ro3] and [G-R] also deal with moment growth.

We present in the first paragraph a general setting dealing with logarithmic Sobolev and Poincaré inequalities. We then turn to Herbst's basic argument which yields Gaussian concentration under a logarithmic Sobolev inequality. We discuss next more general entropy-energy inequalities and exponential integrability under spectral gap inequalities.

2.1 Abstract functional inequalities

In order to develop the functional approach to concentration, we need to introduce a convenient setting in which most of the known results may be considered. We will go from a rather abstract and informal framework to more concrete cases and examples.

Let $(X, \mathcal{B}, \mu)$ be a probability space. We denote by E_μ integration with respect to μ, and by $(L^p(\mu), \| \cdot \|_\infty)$ the Lebesgue spaces over $(X, \mathcal{B}. \mu)$. For any function f in $L^2(\mu)$, we further denote by

$$\text{Var}_\mu(f) = E_\mu(f^2) - \left(E_\mu(f)\right)^2$$

the variance of f. If f is a non-negative function on E such that $E_\mu(f \log^+ f) < \infty$, we introduce the entropy of f with respect to μ as

$$\text{Ent}_\mu(f) = E_\mu(f \log f) - E_\mu(f) \log E_\mu(f).$$

(Actually, since the function $x \log x$ is bounded below, $\text{Ent}_\mu(f) < \infty$ if and only if $E_\mu(f \log^+ f) < \infty$.) Note that $\text{Ent}_\mu(f) \geq 0$ and that $\text{Ent}_\mu(\alpha f) = \alpha \text{Ent}_\mu(f)$ for $\alpha \geq 0$. We write E, Var, Ent when there is no confusion with respect to the measure.

On some subset $\mathcal{A}$ of measurable functions f on X, consider now a map, or energy, $\mathcal{E} : \mathcal{A} \to \mathbb{R}_+$. We say that μ satisfies a spectral gap or Poincaré inequality with respect to $\mathcal{E}$ (on $\mathcal{A}$) if there exists $C > 0$ such that

$$\mathrm{Var}_\mu(f) \le C\mathcal{E}(f) \tag{2.1}$$

for every function $f \in \mathcal{A}$ in $\mathrm{L}^2(\mu)$. We say that μ satisfies a logarithmic Sobolev inequality with respect to $\mathcal{E}$ (on $\mathcal{A}$) if there exists $C > 0$ such that

$$\mathrm{Ent}_\mu(f^2) \le 2C\mathcal{E}(f) \tag{2.2}$$

for every function $f \in \mathcal{A}$ with $\mathrm{E}_\mu(f^2 \log^+ f^2) < \infty$. (The choice of the normalization in (2.2) will become clear with Proposition 2.1 below.) By extension, the integrability properties on f will be understood when speaking of inequalities (2.1) and (2.2) for all f in $\mathcal{A}$.

These abstract definitions include a number of cases of interest. For example, if (X, d) is a metric space equipped with its Borel σ-field $\mathcal{B}$, one may consider the natural generalization of the modulus of the usual gradient

$$|\nabla f(x)| = \limsup_{d(x,y) \to 0} \frac{|f(x) - f(y)|}{d(x, y)} \tag{2.3}$$

(with $|\nabla f(x)| = 0$ for isolated points x in X). In this case, one may define, for a probability measure μ on $(X, \mathcal{B})$,

$$\mathcal{E}(f) = \mathrm{E}_\mu(|\nabla f|^2) \tag{2.4}$$

on the class $\mathcal{A}$ of all, say, (bounded) Lipschitz functions on X. One important feature of this situation is that the operator ∇ is a derivation in the sense that for a C^∞ function ψ on $\mathbb{R}$, and $f \in \mathcal{A}$, $\psi(f) \in \mathcal{A}$ and

$$|\nabla(\psi(f))| = |\nabla f||\psi'(f)|. \tag{2.5}$$

In particular,

$$\mathcal{E}(\psi(f)) \le \|\nabla f\|_\infty^2 \mathrm{E}_\mu(\psi'(f)^2). \tag{2.6}$$

For example,

$$\mathcal{E}(e^{f/2}) \le \frac{1}{4} \|\nabla f\|_\infty^2 \mathrm{E}_\mu(e^f).$$

Another setting of interest, following [A-S] and [G-R], consists of the gradients and Dirichlet forms associated to (symmetric) Markov semigroups. On a probability space $(X, \mathcal{B}, \mu)$, let $p_t(x, \cdot)$ be a Markov transition probability function on $(X, \mathcal{B})$. Assume that $p_t(x, dy)\mu(dx)$ is symmetric in x and y and that, for each bounded measurable function f on X,

$$P_t f(x) = \int f(y) p_t(x, dy)$$

converges to f in $L^2(\mu)$ as t goes to 0. Denote also by P_t the unique bounded extension of P_t to $L^2(\mu)$. Then $(P_t)_{t>0}$ defines a strongly continuous semigroup on $L^2(\mu)$ with Dirichlet form the quadratic form

$$\mathcal{E}(f,f) = \lim_{t \to 0} \frac{1}{2t} \iint (f(x) - f(y))^2 p_t(x, dy)\mu(dx). \qquad (2.7)$$

Let $\mathcal{D}(\mathcal{E})$ be the domain of $\mathcal{E}$ (the space of $f \in L^2(\mu)$ for which $\mathcal{E}(f,f) < \infty$). On the algebra $\mathcal{A}$ of bounded measurable functions f of $\mathcal{D}(\mathcal{E})$, one may then consider $\mathcal{E}(f) = \mathcal{E}(f,f)$. This energy functional does not necessarily satisfy a chain rule formula of the type of (2.6). However, as was emphasized in [A-S], we still have that, for every f in $\mathcal{A}$,

$$\mathcal{E}(e^{f/2}) \leq \frac{1}{2} |||f|||_\infty^2 E_\mu(e^f). \qquad (2.8)$$

Here

$$|||f|||_\infty^2 = \sup\{\mathcal{E}(gf, f) - \tfrac{1}{2}\mathcal{E}(g, f^2); g \in \mathcal{A}. \|g\|_1 \leq 1\}$$

that may be considered as a generalized norm of a gradient. To establish (2.8), note that, by symmetry,

$$\iint \left(e^{f(x)/2} - e^{f(y)/2}\right)^2 p_t(x, dy)\mu(dx)$$

$$= 2 \iint_{\{f(x)<f(y)\}} \left(e^{f(x)/2} - e^{f(y)/2}\right)^2 p_t(x, dy)\mu(dx)$$

$$\leq \frac{1}{2} \iint (f(x) - f(y))^2 e^{f(y)} p_t(x, dy)\mu(dx).$$

Now, for every g in $\mathcal{A}$,

$$\lim_{t \to 0} \frac{1}{2t} \iint g(x)(f(x) - f(y))^2 p_t(x, dy)\mu(dx) = \mathcal{E}(gf, f) - \tfrac{1}{2}\mathcal{E}(g, f^2)$$

from which (2.8) follows.

Examples fitting this general framework are numerous. Let $X = \mathbb{R}^n$ and write ∇f for the usual gradient of a smooth function f on $\mathbb{R}^n$. Let

$$M : \mathbb{R}^n \to \text{invertible matrices } \{n \times n\}$$

be measurable and locally bounded and let $d\mu(x) = w(x)dx$ be a probability measure on $\mathbb{R}^n$ with $w > 0$. For every C^∞ compactly supported function f on $\mathbb{R}^n$, set

$$\mathcal{E}(f,f) = \int_{\mathbb{R}^n} \langle M(x)\nabla f(x), M(x)\nabla f(x)\rangle d\mu(x).$$

We need not be really concerned here with the semigroup induced by this Dirichlet form. Ignoring questions on the closure of $\mathcal{E}$, it readily follows that in this case

$$|||f|||_\infty = \sup\{|M(x)\nabla f(x)|; x \in \mathbb{R}^n\}$$

where $|\cdot|$ is Euclidean length. More generally, if μ is a probability measure on a Riemannian manifold X, and if $\mathcal{E}(f,f) = \int_M |\nabla f|^2 d\mu$, then one has $|||f|||_\infty = \|\nabla f\|_\infty$.

With this class of examples, we of course rejoin the generalized moduli of gradients (2.3). In this case, the Dirichlet form $\mathcal{E}$ is actually local, that is, it satisfies the chain rule formula (2.6). In particular, (2.8) holds in this case with constant $\frac{1}{4}$ (and $|||f|||_\infty = \|\nabla f\|_\infty$). We freely use this observation throughout these notes.

Covering in another way the two preceding settings, one may also consider the abstract Markov semigroup framework of [Ba1] in which, given a Markov generator L on some nice algebra $\mathcal{A}$ of functions, one defines the carré du champ operator as

$$\Gamma(f,g) = \tfrac{1}{2} L(fg) - fLg - gLf.$$

For example, if L is the Laplace-Beltrami operator on a manifold M, then $\Gamma(f,g) = \nabla f \cdot \nabla g$. One may then define

$$\mathcal{E}(f) = E_\mu(\Gamma(f,f))$$

on the class $\mathcal{A}$. If L is symmetric, one shows that $|||f|||_\infty = \|\Gamma(f,f)\|_\infty$. Provided L is a diffusion (that is, it satisfies the change of variables formula $L\psi(f) = \psi'(f)Lf + \psi''(f)\Gamma(f,f)$) $\mathcal{E}$ will satisfy (2.6). A further discussion may be found in [Ba1].

We turn to discrete examples. Let X be a finite or countable set. Let $K(x,y) \geq 0$ satisfy

$$\sum_{y \in X} K(x,y) = 1$$

for every $x \in X$. Asssume furthermore that there is a symmetric invariant probability measure μ on X, that is $K(x,y)\mu(x)$ is symmetric in x and y and $\sum_x K(x,y)\mu(x) = \mu(y)$ for every $y \in X$. In other words, (K,μ) is a symmetric Markov chain. Define

$$\mathcal{E}(f,f) = \frac{1}{2} \sum_{x,y \in X} (f(x) - f(y))^2 K(x,y)\mu(\{x\}).$$

In this case,

$$|||f|||_\infty^2 = \frac{1}{2} \sup_{x \in X} \sum_{y \in X} (f(x) - f(y))^2 K(x,y).$$

It might be worthwhile noting that if we let

$$\|\nabla f\|_\infty = \sup\{|f(x) - f(y)|; K(x,y) > 0\},$$

then, since $\sum_y K(x,y) = 1$,

$$|||f|||_\infty^2 \leq \frac{1}{2} \|\nabla f\|_\infty^2.$$

It should be clear that the definition of the $|||\cdot|||_\infty$-norm tries to be as close as possible to the sup-norm of a gradient in a continuous setting. As such however, it does not always reflect accurately discrete situations. Discrete gradients may actually be examined in another way. If f is a function on $\mathbb{Z}$, set

$$Df(x) = f(x+1) - f(x), \quad x \in \mathbb{Z}. \tag{2.9}$$

One may then consider

$$\mathcal{E}(f) = \mathrm{E}_\mu(|Df|^2) \tag{2.10}$$

for a measure μ on $\mathbb{Z}$. This energy will not satisfy (2.6) but may satisfy (2.8). For reals $m(x)$, $x \in \mathbb{Z}$, let

$$\mathcal{E}(f,f) = \sum_{x \in \mathbb{Z}} Df(x)^2 m(x)^2 \mu(\{x\}).$$

One can check that for this Dirichlet form

$$|||f|||_\infty^2 = \sup_{x \in \mathbb{Z}} \frac{1}{2}\left(m(x)^2 Df(x)^2 + m(x-1)^2 \frac{\mu(\{x-1\})}{\mu(\{x\})} Df(x-1)^2 \right)^2. \tag{2.11}$$

As will be seen in Part 5, this uniform norm of the gradient is actually of little use in specific examples, such as Poisson measures. It will be more fruitful to consider $\sup_{x \in \mathbb{Z}^d} |Df(x)|$. The lack of chain rule (for example, $|D(e^f)| \leq |Df| e^{|Df|} e^f$ only in general) will then have to be handled by other means. The norm $||| \cdot |||_\infty$ is in fact only well adapted to produce Gaussian bounds as we will see in Section 2.3. It is actually defined in such a way to produce results similar to those which follows from a chain rule formula. As such, this norm is not suited to a number of discrete examples (see also [G-R]).

The preceding example may be further generalized to $\mathbb{Z}^d$. Similarly, in the context of statistical mechanics, set $X = \{-1, +1\}^{\mathbb{Z}^d}$ and let

$$|Df(\omega)| = \left(\sum_{k \in \mathbb{Z}^d} |\partial_k f(\omega)|^2 \right)^{1/2} \tag{2.12}$$

where $\partial_k f(\omega) = f(\omega^k) - f(\omega)$ where ω^k is the element of X obtained from ω by replacing the k-th coordinate with $-\omega_k$.

Logarithmic Sobolev inequalities were introduced to describe smoothing properties of Markov semigroups, especially in infinite dimensional settings. The key argument was isolated by L. Gross [Gr1] who showed how a logarithmic Sobolev inequality is actually equivalent to hypercontractivity of a Markov generator. Precisely, if $(P_t)_{t>0}$ is a Markov semigroup with invariant measure μ and Dirichlet form $\mathcal{E}$, then the logarithmic Sobolev inequality

$$\mathrm{Ent}(f^2) \leq 2C\,\mathcal{E}(f,f), \quad f \in \mathcal{A},$$

is equivalent to saying that, whenever $1 < p < q < \infty$ and $t > 0$ are such that $e^{2t/C} \geq (q-1)/(p-1)$, we have

$$\|P_t f\|_q \leq \|f\|_p$$

for every $f \in \mathcal{A}$ in $L^p(\mu)$ (cf. [Gr1], [Ba1] for the precise statement). Hypercontractivity is an important tool in deriving sharp estimates on the time to equilibrium of P_t [S-Z], [St], [D-S] etc.

Now, we mention a simple comparison between spectral and logarithmic Sobolev inequalities. The hypothesis on $\mathcal{E}$ is straightforward in all the previous examples.

Proposition 2.1. *Assume that μ satisfies the logarithmic Sobolev inequality*

$$\mathrm{Ent}_\mu(f^2) \leq 2\,\mathcal{E}(f), \quad f \in \mathcal{A},$$

and that $af + b \in \mathcal{A}$ and $\mathcal{E}(af + b) = a^2\mathcal{E}(f)$ for every $f \in \mathcal{A}$ and $a, b \in \mathbb{R}$. Then μ satisfies the spectral gap inequality

$$\mathrm{Var}_\mu(f^2) \leq \mathcal{E}(f), \quad f \in \mathcal{A}.$$

Proof. Fix f with $\mathrm{E}_\mu(f) = 0$ and $\mathrm{E}_\mu(f^2) = 1$ and apply the logarithmic Sobolev inequality to $1 + \varepsilon f$. As ε goes to 0, a Taylor expansion of $\log(1 + \varepsilon f)$ yields the conclusion. $\square$

It might be worthwhile mentioning that the converse to Proposition 2.1 is not true in general, even within constants. We will have the opportunity to encounter a number of such cases throughout these notes (cf. Sections 4.1, 5.1 and 7.3).

One important feature of both variance and entropy is their product property. Assume we are given probability spaces $(X_i, \mathcal{B}_i, \mu_i)$, $1 \leq i \leq n$. Denote by P the product probability measure $P = \mu_1 \otimes \cdots \otimes \mu_n$ on the product space $X = X_1 \times \cdots \times X_n$ equipped with the product σ-field $\mathcal{B}$. Given f on the product space, we write furthermore f_i, $1 \leq i \leq n$, for the function on X_i defined by

$$f_i(x_i) = f(x_1, \ldots, x_{i-1}, x_i, x_{i+1}, \ldots, x_n).$$

with $x_1, \ldots, x_{i-1}, x_{i+1}, \ldots, x_n$ fixed.

Proposition 2.2. *Under appropriate integrability conditions,*

$$\mathrm{Var}_P(f) \leq \sum_{i=1}^n \mathrm{E}_P\big(\mathrm{Var}_{\mu_i}(f_i)\big)$$

and

$$\mathrm{Ent}_P(f) \leq \sum_{i=1}^n \mathrm{E}_P\big(\mathrm{Ent}_{\mu_i}(f_i)\big).$$

Proof. Let us prove the assertion concerning entropy, the one for variance being (simpler and) similar. Recall first that for a non-negative function f on $(X, \mathcal{B}, \mu)$,

$$\mathrm{Ent}_\mu(f) = \sup\{\mathrm{E}_\mu(fg); \mathrm{E}_\mu(e^g) \leq 1\}. \tag{2.13}$$

Indeed, assume by homogeneity that $\mathrm{E}_\mu(f) = 1$. By Young's inequality

$$uv \leq u \log u - u + e^v, \quad u \geq 0, \quad v \in \mathbb{R},$$

we get, for $E_\mu(e^g) \leq 1$,

$$E_\mu(fg) \leq E_\mu(f \log f) - 1 + E_\mu(e^g) \leq E_\mu(f \log f).$$

The converse is obvious.

To prove Proposition 2.2, given g on $(X, \mathcal{B}, P)$ such that $E_P(e^g) \leq 1$, set, for every $i = 1 \ldots, n$,

$$g^i(x_i, \ldots, x_n) = \log\left(\frac{\int e^{g(x_1,\ldots,x_n)} d\mu_1(x_1) \cdots d\mu_{i-1}(x_{i-1})}{\int e^{g(x_1,\ldots,x_n)} d\mu_1(x_1) \cdots d\mu_i(x_i)}\right).$$

Then $g \leq \sum_{i=1}^n g^i$ and $E_{\mu_i}(e^{(g^i)_i}) = 1$. Therefore,

$$E_P(fg) \leq \sum_{i=1}^n E_P(fg^i) = \sum_{i=1}^n E_P\big(E_{\mu_i}(f_i(g^i)_i)\big) \leq \sum_{i=1}^n E_P\big(\text{Ent}_{\mu_i}(f_i)\big)$$

which is the result. Proposition 2.2 is established. □

What Proposition 2.2 will tell us in applications is that, whenever the energy on the product space is the sum of the energies on each coordinates, in order to establish a Poincaré or logarithmic Sobolev inequality in product spaces, it will be enough to deal with the dimension one. In particular, these inequalities will be independent of the dimension of the product space. This is why logarithmic Sobolev inequalities are such powerful tools in infinite dimensional analysis.

2.2 Examples of logarithmic Sobolev inequalities

The first examples of logarithmic Sobolev inequalities were discovered by L. Gross in 1975 [Gr1]. They concerned the two-point space and the canonical Gaussian measure. For the two point space $\{0, 1\}$ with uniform (Bernoulli) measure $\mu = \frac{1}{2}\delta_0 + \frac{1}{2}\delta_1$, L. Gross showed that for every f on $\{0, 1\}$,

$$\text{Ent}_\mu(f^2) \leq \frac{1}{2} E_\mu(|Df|^2) \tag{2.14}$$

where $Df(x) = f(1) - f(0)$, $x \in \{0, 1\}$. The constant is optimal. In its equivalent hypercontractive form, this inequality actually goes back to A. Bonami [Bon]. Due to Proposition 2.2, if μ^n is the n-fold product measure of μ on $\{0, 1\}^n$, for every f on $\{0, 1\}^n$,

$$\text{Ent}_{\mu^n}(f^2) \leq \frac{1}{2} E_{\mu^n}\left(\sum_{i=1}^n |D_i f|^2\right)$$

where, for $x = (x_1, \ldots, x_n) \in \{0, 1\}^n$ and $i = 1, \ldots, n$, $D_i f(x) = Df_i(x_i)$. Applying this inequality to

$$f(x_1, \ldots, x_n) = \varphi\left(\frac{x_1 + \cdots + x_n - \frac{n}{2}}{\sqrt{\frac{n}{4}}}\right)$$

for some smooth φ on $\mathbb{R}$, L. Gross deduced, with the classical central limit theorem, a logarithmic Sobolev inequality for the canonical Gaussian measure γ on $\mathbb{R}$ in the form of

$$\mathrm{Ent}_\gamma(\varphi^2) \leq 2\,\mathrm{E}_\gamma(\varphi'^2).$$

By the product property of entropy, if γ is the canonical Gaussian measure on $\mathbb{R}^n$, for every f on $\mathbb{R}^n$ with gradient in $L^2(\gamma)$,

$$\mathrm{Ent}_\gamma(f^2) \leq 2\,\mathrm{E}_\gamma(|\nabla f|^2). \tag{2.15}$$

Inequality (2.15) may be considered as the prototype of logarithmic Sobolev inequalities. The constant in (2.15) is optimal as can be checked for example on exponential functions $e^{\lambda x}$, which actually saturate this inequality. This observation is a first indication on the Laplace transform approach we will develop next. Several simple, alternative proofs of this inequality have been developed in the literature. For our purposes, it might be worthwhile noting that it may be seen as consequence of the Gaussian isoperimetric inequality itself. This has been noticed first in [Le1] but recently, W. Beckner [Be] kindly communicated to the author a simple direct argument on the basis of the functional inequality (1.11). Namely, let g be smooth with $\int g^2 d\gamma = 1$ and apply (1.11) to $f = \varepsilon g^2$ with $\varepsilon \to 0$. We get that

$$1 \leq \int \sqrt{\frac{\mathcal{U}^2(\varepsilon g^2)}{\mathcal{U}^2(\varepsilon)} + \frac{4\varepsilon^2}{\mathcal{U}^2(\varepsilon)}\, g^2 |\nabla g|^2}\, d\gamma.$$

Noticing that $\mathcal{U}^2(\varepsilon) \sim \varepsilon^2 \log(\frac{1}{\varepsilon^2})$ as $\varepsilon \to 0$, we see that

$$1 \leq \int g^2 \sqrt{1 - \frac{1}{M}\log g^2 + \frac{2}{M}\frac{|\nabla g|^2}{g^2} + o\left(\frac{1}{M}\right)}\, d\gamma$$

where $M = M(\varepsilon) = \log(\frac{1}{\varepsilon}) \to \infty$ as $\varepsilon \to 0$. Hence

$$1 \leq \int g^2\left(1 - \frac{1}{2M}\log g^2 \frac{1}{M}\frac{|\nabla g|^2}{g^2}\right)d\gamma + o\left(\frac{1}{M}\right)$$

from which the Gaussian logartihmic Sobolev inequality (2.15) follows. The same argument works for the Boltzmann measures of Theorem 1.1. On the other hand, the semigroup arguments leading to the Gaussian isoperimetric inequality may also be adapted to give a direct, simpler proof of the logarithmic Sobolev inequality (2.15) [Ba1], [Le3]. To briefly sketch the argument (following the same notation), let f be smooth and non-negative on $\mathbb{R}$. Then write

$$\mathrm{Ent}_\gamma(f) = -\int_0^\infty \frac{d}{dt}\,\mathrm{E}_\gamma(P_t f \log P_t f)dt$$

(with $(P_t)_{t \geq 0}$ the Ornstein-Uhlenbeck semigroup (1.12)). By the chain rule formula,

$$\frac{d}{dt}\,\mathrm{E}_\gamma(P_t f \log P_t f) = \mathrm{E}_\gamma(LP_t f \log P_t f) + \mathrm{E}_\gamma(LP_t f) = -\frac{1}{2}\,\mathrm{E}_\gamma\left(\frac{(P_t f)'^2}{P_t f}\right)$$

since γ is invariant under the action of P_t and thus $\mathrm{E}_\gamma(LP_tf)$. Now, $(P_tf)' = e^{-t/2}P_tf'$ so that, by the Cauchy-Schwarz inequality for P_t,

$$(P_tf')^2 \leq P_tf\, P_t\left(\frac{f'^2}{f}\right).$$

Summarizing,

$$\mathrm{Ent}_\gamma(f) \leq \frac{1}{2}\int_0^\infty e^{-t}\, \mathrm{E}_\gamma\left(P_t\left(\frac{f'^2}{f}\right)\right)dt = \frac{1}{2}\mathrm{E}_\gamma\left(\frac{f'^2}{f}\right)$$

which, by the change of f into f^2, is (2.15) in dimension one.

The preceding proof may be shown to imply in the same way the Poincaré inequality for Gaussian measures

$$\mathrm{Var}_\gamma(f) \leq \mathrm{E}_\gamma(|\nabla f|^2). \tag{2.16}$$

(Write, in dimension one for simplicity,

$$\begin{aligned}
\mathrm{Var}_\gamma(f) &= -\int_0^\infty \frac{d}{dt}\,\mathrm{E}_\gamma\big((P_tf)^2\big)dt = -2\int_0^\infty \mathrm{E}(P_tf L P_tf)dt \\
&= \int_0^\infty \mathrm{E}_\gamma\big((P_tf)'^2\big)dt \\
&= \int_0^\infty e^{-t}\mathrm{E}_\gamma\big((P_tf')^2\big)dt \\
&\leq \int_0^\infty e^{-t}\mathrm{E}_\gamma(f'^2)dt = \mathrm{E}_\gamma(f'^2).)
\end{aligned}$$

It may also be seen as a consequence of the logarithmic Sobolev inequality (2.15) by Proposition 2.1. Actually, (2.16) is a straigthforward consequence of a series expansion in Hermite polynomials, and may be found, in this form, in the physics literature of the thirties.

Both the (dimension free) logarithmic Sobolev and Poincaré inequalities (2.15) and (2.16) extend to infinite dimensional Gaussian measures replacing the gradient by the Gross-Malliavin derivatives along the directions of the reproducing kernel Hilbert space. This is easily seen by a finite dimensional approximation (cf. [Le3]).

The preceding semigroup proofs also apply to Boltzmann measures as studied in Section 1.2. In particular, under the curvature assumption of Theorem 1.1, these measures satisfy the logarithmic Sobolev inequality

$$\mathrm{Ent}_\mu(f^2) \leq \frac{2}{c}\mathrm{E}_\mu(|\nabla f|^2). \tag{2.17}$$

As wa have seen, this inequality may also be shown to follow from Theorem 1.1 (cf. also [Ba-L]). We discuss in Section 7.1 logarithmic Sobolev inequalities for a more general class of potentials.

Further logarithmic Sobolev inequalities have been established and studied throughout the literature, mainly for their hypercontractive content. We refer to the survey [Gr2] for more information. We investigate here logarithmic Sobolev inequalities for their applications to the concentration of measure phenomenon.

2.3 Herbst's argument

In this section, we illustrate how concentration properties may follow from a logarithmic Sobolev inequality. Although rather elementary, this observation is a powerful scheme which allows us to establish some new concentration inequalities. Indeed, as illustrated in particular in the next chapter, convexity of entropy allows to tensorize one-dimensional inequalities to produce concentration properties in product spaces whereas concentration itself does not usually tensorize.

To clarify the further developments, we first present Herbst's argument (or what we believe Herbst's argument was) in the original simple case. Let thus μ be a probability measure on $\mathbb{R}^n$ such that for some $C > 0$ and all smooth f on $\mathbb{R}^n$,

$$\text{Ent}(f^2) \leq 2C \, \text{E}(|\nabla f|^2) \tag{2.18}$$

(where ∇f is the usual gradient of f). Let now F be smooth (and bounded) such that $\|F\|_{\text{Lip}} \leq 1$. In particular, since we assume F to be regular enough, we can have that $|\nabla F| \leq 1$ at every point. Apply now (2.18) to $f^2 = e^{\lambda F}$ for every $\lambda \in \mathbb{R}$. We have

$$\text{E}(|\nabla f|^2) = \frac{\lambda^2}{4} \, \text{E}(|\nabla F|^2 e^{\lambda F}) \leq \frac{\lambda^2}{4} \, \text{E}(e^{\lambda F}).$$

Setting $H(\lambda) = \text{E}_\mu(e^{\lambda F})$, $\lambda \in \mathbb{R}$, we get by the definition of entropy,

$$\lambda H'(\lambda) - H(\lambda) \log H(\lambda) \leq \frac{C\lambda^2}{2} H(\lambda).$$

In other words, if $K(\lambda) = \frac{1}{\lambda} \log H(\lambda)$ (with $K(0) = F'(0)/F(0) = \text{E}_\mu(F)$),

$$K'(\lambda) \leq \frac{C}{2}$$

for every λ. Therefore,

$$K(\lambda) = K(0) + \int_0^\lambda K'(u)du \leq \text{E}_\mu(F) + \frac{C\lambda}{2}$$

and hence, for every λ,

$$H(\lambda) = \text{E}_\mu(e^{\lambda F}) \leq e^{\lambda \text{E}_\mu(F) + C\lambda^2/2}. \tag{2.19}$$

Replacing F by a smooth convolution, (2.19) extends to all Lipschitz functions with $\|F\|_{\text{Lip}} \leq 1$ (see below). By Chebyshev's inequality, for every $\lambda, r \geq 0$,

$$\mu(F \geq \text{E}_\mu(F) + r) \leq e^{-\lambda r + Cr^2/2}$$

and optimizing in λ, for every $r \geq 0$,

$$\mu\big(F \geq E_\mu(F) + r\big) \leq e^{-r^2/2C}.$$

The same inequality holds for $-F$.

The next proposition is some abstract formulation on the preceding argument. It aims to cover several situations at once so that it may look akward at first. The subsequent results will take a simpler form. At this point, they all yield Gaussian concentration under logarithmic Sobolev inequalities. In the next section, we study non-Gaussian tails which arise from more general entropy-energy inequalities, or from the lack of chain rule for discrete gradients (cf. Part 5).

Let $(X, \mathcal{B}, \mu)$ be a probability space. We write E for E_μ, and similarly Var, Ent. Let $\mathcal{A}$ be a subset of $L^1(\mu)$. For every f in $\mathcal{A}$, let $N(f) \geq 0$. Typically $N(f)$ will be our Lipschitz norm or generalized sup-norm of the gradient. For example, $N(f) = \|\nabla f\|_\infty$ in (2.4), or $\||f\||_\infty$ in (2.8), or $\sup_{x \in \mathbb{Z}} |Df(x)|$ in (2.10).

Proposition 2.3. *Let $\mathcal{A}$ and N be such that, for every $f \in \mathcal{A}$ and $\lambda \in \mathbb{R}$, $\lambda f \in \mathcal{A}$, $E(e^{\lambda f}) < \infty$ and $N(\lambda f) = |\lambda| N(f)$. Assume that for every $f \in \mathcal{A}$,*

$$\mathrm{Ent}(e^f) \leq \frac{1}{2} N(f)^2 \, E(e^f).$$

Then, whenever F in $\mathcal{A}$ is such that $N(F) \leq 1$, then

$$E(e^{\lambda F}) \leq e^{\lambda E(F) + \lambda^2/2} \tag{2.20}$$

for every $\lambda \in \mathbb{R}$. Furthermore, for every $r \geq 0$,

$$\mu\big(F \geq E(F) + r\big) \leq e^{-r^2/2}, \tag{2.21}$$

and similarly for $-F$.

Proof. It just reproduces the proof of (2.19). Fix $F \in \mathcal{A}$ with $N(F) \leq 1$ and write $H(\lambda) = E(e^{\lambda F})$, $\lambda \geq 0$. Similarly, set $K(\lambda) = \frac{1}{\lambda} \log H(\lambda)$, $K(0) = E(F)$. Applying the logarithmic Sobolev inequality of the statement to λF, $\lambda \geq 0$, we get $K'(\lambda) \leq \frac{1}{2}$ for $\lambda \geq 0$. Therefore,

$$K(\lambda) = K(0) + \int_0^\lambda K'(u)du \leq E(F) + \frac{\lambda}{2}$$

and hence, for every $\lambda \geq 0$,

$$H(\lambda) \leq e^{\lambda E(F) + \lambda^2/2}.$$

Changing F into $-F$ yields (2.20). The proof is completed similarly. $\square$

We begin by adding several comments to Proposition 2.3.
If $N(F) \leq c$ in Proposition 2.3, then, by homogeneity,

$$\mu\big(F \geq E(F) + r\big) \leq e^{-r^2/2c^2}, \quad r \geq 0.$$

Sometimes the class $\mathcal{A}$ in Proposition 2.3 only includes λf when $f \in \mathcal{A}$ and $\lambda \geq 0$. The proof above was written so as to show that (2.20) then only holds for all $\lambda \geq 0$. Such a modification can be proposed similarly on the subsequent statements. We use these remarks freely throughout this work.

Very often, the logarithmic Sobolev inequality is only available on a class $\mathcal{A}$ densely defined in some larger, more convenient, class. The class of cylindrical functions on an abstract Wiener space is one typical and important example. In particular, this class might consist of bounded functions, so that the integrability assumptions in Proposition 2.3 are immediate. The conclusions however are only of interest for unbounded functions. Rather than extend the logarithmic Sobolev inequality itself, one may note that the corresponding concentration inequality easily extends. Let us agree that a function f on X satisfies $N(f) \leq 1$ if there is a sequence of functions $(f_n)_{n \in \mathbb{N}}$ in $\mathcal{A}$ with $N(f_n) \leq 1$ (or, more generally, $N(f_n) \leq 1 + \frac{1}{n}$) that converge μ-almost everywhere to f. For example, under some stability properties of $\mathcal{A}$, f_n could be $f_n = \max(-n, \min(f, n))$ which thus define a sequence of bounded functions converging to f. Dirichlet forms associated to Markov semigroups are stable by Lipschitz functions and $\mathcal{E}(f_n, f_n) \leq \mathcal{E}(f, f)$, thus falling into this case. Energies given by generalized moduli of gradients (2.4) may also be considered. Then, if F on X such that $N(F) \leq 1$, F is integrable and the conclusions of Proposition 2.3 holds. To see this, let $(F_n)_{n \in \mathbb{N}}$ be a sequence in $\mathcal{A}$ with $N(F_n) \leq 1$ such that $F_n \to F$ almost everywhere. By Proposition 2.3, for every n and $r \geq 0$,

$$\mu\big(|F_n - \mathrm{E}(F_n)| \geq r\big) \leq 2\,e^{-r^2/2}. \qquad (2.22)$$

Let m be large enough that $\mu(|F| \leq m) \geq \frac{3}{4}$. Then, for some n_0 and every $n \geq n_0$, $\mu(|F_n| \leq m + 1) \geq \frac{1}{2}$. Choose furthermore $r_0 > 0$ with $2\,e^{-r_0^2/2} < \frac{1}{2}$. Therefore, intersecting the sets $\{|F_n| \leq m + 1\}$ and $\{|F_n - \mathrm{E}(F_n)| \geq r_0\}$, we see that

$$|\mathrm{E}(F_n)| \leq r_0 + m + 1$$

for every $n \geq n_0$ thus. Hence, by (2.22) again,

$$\mu\big(|F_n| \geq r + r_0 + m + 1\big) \leq 2\,e^{-r^2/2}$$

for every $r \geq 0$ and $n \geq n_0$. In particular $\sup_n \mathrm{E}(F_n^2) < \infty$ so that, by uniform integrability, $\mathrm{E}(|F|) < \infty$ and $\mathrm{E}(F_n) \to \mathrm{E}(F)$. Then, by Fatou's lemma, for every $\lambda \in \mathbb{R}$,

$$\mathrm{E}(e^{\lambda F}) \leq \liminf_{n \to \infty} \mathrm{E}(e^{\lambda F_n}) \leq \liminf_{n \to \infty} e^{\lambda \mathrm{E}(F_n) + \lambda^2/2} = e^{\lambda \mathrm{E}(F) + \lambda^2/2}.$$

One then concludes as in Proposition 2.3. We emphasize that the integrability of F came for free. A similar reasoning was used in (1.24)

Note furthermore that, in the preceding setting, if $N(F) \leq 1$, then

$$\mathrm{E}\big(e^{\alpha F^2}\big) < \infty \qquad (2.23)$$

for every $\alpha < \frac{1}{2}$. As will be seen below, this condition is optimal. (2.23) is a consequence of Proposition 1.2. A beautiful alternate argument in this case was suggested by L. Gross (cf. [A-M-S]) on the basis of (2.20). If γ is the canonical Gaussian measure on $\mathbb{R}$, by Fubini's theorem,

$$\begin{aligned}
\mathrm{E}\left(e^{\alpha F^2}\right) &= \mathrm{E}\left(\int_{\mathbb{R}} e^{\sqrt{2\alpha}xF} d\gamma(x)\right) \\
&\leq \int_{\mathbb{R}} e^{\sqrt{2\alpha}x\mathrm{E}(F)+\alpha x^2} d\gamma(x) \\
&= \frac{1}{\sqrt{1-2\alpha}} e^{\alpha\mathrm{E}(F)^2/(1-2\alpha)}.
\end{aligned}$$

The bound is optimal as can be seen from the example $F(x) = x$ (with respect to γ).

We now show how the preceding statement may be applied to the settings presented in Section 2.1 for logarithmic Sobolev inequalities in their more classical form. The results below are taken from [A-M-S], [A-S], [G-R], [Le1], [Ro3].

In the context of Dirichlet forms (2.7) associated to Markov semigroup. let $\mathcal{A}$ be the algebra of bounded functions on $(X, \mathcal{B})$ in the domain $\mathcal{D}(\mathcal{E})$ of the Dirichlet form. Take $N(f) = ||| \cdot |||_\infty$, and let us agree, as above, that a measurable function f on X is such that $||| \cdot |||_\infty \leq 1$ if there is a sequence $(f_n)_{n\in\mathbb{N}}$ in $\mathcal{A}$ with $|||f_n|||_\infty \leq 1$ that converge μ-almost everywhere to f.

Corollary 2.4. *Assume that for some $C > 0$ and every f in $\mathcal{A}$*

$$\mathrm{Ent}(f^2) \leq 2C\mathcal{E}(f).$$

Then, whenever F is such that $|||F|||_\infty \leq 1$, we have $\mathrm{E}(|F|) < \infty$ and, for every $r \geq 0$,

$$\mu\left(F \geq \mathrm{E}(F) + r\right) \leq e^{-r^2/4C}.$$

Proof. Apply the logarithmic Sobolev inequality to $e^{f/2}$ to get, according to (2.8),

$$\mathrm{Ent}(e^f) \leq C|||f|||_\infty^2 \mathrm{E}(e^f).$$

The conclusion then follows from Proposition 2.3 (and homogeneity). $\qquad\square$

As a second set of examples, consider an operator Γ on some class $\mathcal{A}$ such that $\Gamma(f) \geq 0$ and $\Gamma(\lambda f) = \lambda^2 \Gamma(f)$ for every f in $\mathcal{A}$. As a typical example, $\Gamma(f) = |\nabla f|^2$ for a generalized modulus of gradient, or $\Gamma(f) = \Gamma(f, f)$ for a more general carré du champ. One may also choose $\Gamma(f) = |Df|^2$ for a discrete gradient such as (2.9). Keeping with the preceding comments, we agree that a function f on X is such that $N(f) = \|\Gamma(f)\|_\infty \leq 1$ if there is a sequence $(f_n)_{n\in\mathbb{N}}$ in $\mathcal{A}$ converging to f such that $\|\Gamma(f)\|_\infty \leq 1$ for every n. The following corollary to Proposition 2.3 is immediate.

Corollary 2.5. *Let $\mathcal{A}$ be such that, for every $f \in \mathcal{A}$ and $\lambda \in \mathbb{R}$, $\lambda f \in \mathcal{A}$, $E(e^{\lambda f}) < \infty$. Assume that for some $C > 0$ and every $f \in \mathcal{A}$,*

$$\mathrm{Ent}(e^f) \leq \frac{C}{2} \mathrm{E}(\Gamma(f) e^f).$$

Then, whenever F is such that $\|\Gamma(f)\|_\infty \leq 1$, we have $\mathrm{E}(|F|) < \infty$ and

$$\mu(F \geq \mathrm{E}(F) + r) \leq e^{-r^2/2C}$$

for every $r \geq 0$.

In case of a local gradient operator $\Gamma(f) = |\nabla f|^2$ (2.3) on a metric space $X, d)$ satisfying the chain rule formula (2.5), a logarithmic Sobolev inequality of the type

$$\mathrm{Ent}(f^2) \leq 2C\,\mathrm{E}(|\nabla f|^2)$$

is actually equivalent to the logarithmic Sobolev inequality

$$\mathrm{Ent}(e^f) \leq \frac{C}{2}\,\mathrm{E}(|\nabla f|^2 e^f) \tag{2.24}$$

of Corollary 2.5 (on some appropriate class of functions $\mathcal{A}$ stable by the operations required for this equivalence to hold). As we will see, this is no more true for non-local gradients. Even in case of a local gradient, it may also happen that (2.20) holds for some class of functions for which the classical logarithmic Sobolev inequality is not satisfied. In the next statement, we do not specify the stability properties on $\mathcal{A}$.

Corollary 2.6. *Assume that for some $C > 0$ and all f in $\mathcal{A}$*

$$\mathrm{Ent}(f^2) \leq 2C\,\mathrm{E}(|\nabla f|^2).$$

Then, whenever F is such that $\|\nabla F\|_\infty \leq 1$, we have $\mathrm{E}(|F|) < \infty$ and, for every $r \geq 0$,

$$\mu(F \geq \mathrm{E}(F) + r) \leq e^{-r^2/2C}.$$

Together with (1.28), for every set A with $\mu(A) > 0$,

$$\mu(A_r) \geq 1 - e^{-\mu(A)^2 r^2/2C} \tag{2.25}$$

for every $r \geq 0$.

Let us consider, for example, in Corollary 2.6, the Gaussian measure γ on $\mathbb{R}^n$. The logarithmic Sobolev inequality (2.15) holds for all almost everywhere differentiable functions with gradients in $L^2(\gamma)$. Let $\mathcal{A}$ be the class of bounded Lipschitz functions on $\mathbb{R}^n$. Let F be a Lipschitz function on $\mathbb{R}^n$. For any $n \in \mathbb{N}$, set $F_n = \max(-n, \min(F, n))$. Then F_n is bounded Lipschitz and converges almost everywhere to F. Moreover, if $\|F\|_{\mathrm{Lip}} \leq 1$, $\|F_n\|_{\mathrm{Lip}} \leq 1$ for every n. By Rademacher's theorem, F_n is almost everywhere differentiable with $|\nabla F_n| \leq 1$ almost everywhere. Therefore, as an application of Corollary 2.6, we thus recover that for any Lipschitz F with $\|F\|_{\mathrm{Lip}} \leq 1$,

$$\gamma(F \geq \mathrm{E}_\gamma(F) + r) \leq e^{-r^2/2}, \quad r \geq 0,$$

which is the concentration property (1.22). In particular, the optimal constant in the exponent has been preserved throughout this procedure. We thus see how a logarithmic Sobolev inequality always determines a Gaussian concentration of isoperimetric nature.

The previous comment applies exactly similarly for the class of Boltzmann measures investigated in Theorem 1.1 (see also (2.17)). Moreover, the approximation procedure just described may be performed similarly for generalized gradients, on manifolds for example. Similarly, a cylindrical approximation would yield (1.25) for an infinite dimensional Gaussian measure from the Gaussian logarithmic Sobolev inequality. (1.25) would also follow from the logarithmic Sobolev inequality for infinite dimensional Gaussian measures, although the extension scheme is much simpler at the level of concentration inequalities.

We present next an application in a non-local setting following [A-S]. Recall the "gradient" (2.12) for a function f on $X = \{-1, +1\}^{\mathbb{Z}^d}$. Let μ be a Gibbs state on X corresponding to a finite range potential $\mathcal{J}$. It was shown by D. Stroock and B. Zegarlinski [S-Z] that the Dobrushin-Shlosman mixing condition ensures a logarithmic Sobolev inequality for μ

$$\text{Ent}_\mu(f^2) \leq 2C \, \text{E}(|Df|^2)$$

for some $C > 0$. Assume moreover that $\mathcal{J}$ is shift-invariant. Let ψ be a continuous function on X for which $\text{E}_\mu(\psi) = 0$ and

$$\beta = \sum_{k \in \mathbb{Z}^d} \|\partial_k \psi\|_\infty < \infty.$$

Let finally $(a_k)_{k \in \mathbb{Z}^d}$ be a sequence of real numbers with

$$\alpha^2 = \sum_{k \in \mathbb{Z}^d} a_k^2 < \infty.$$

For S^j the natural shift on $\mathbb{Z}^d$ (defined by $S^j(\omega) = \omega_{j+k}$), consider then a function F of the form

$$F = \sum_{j \in \mathbb{Z}^d} a_j \psi \circ S^j.$$

Such a function is actually defined as the limit in quadratic mean of the partial sums. As such, it is easily seen that

$$|||F|||_\infty \leq \alpha\beta.$$

The preceding results (Corollary 2.4) apply to yield concentration and integrability properties of such functions F. In particular, for every $r \geq 0$,

$$\mu(F \geq \text{E}_\mu(F) + r) \leq e^{-r^2/2C\alpha^2\beta^2}.$$

These results are thus very similar to the ones one gets in the non-interacting case (that is when μ is a product measure on $\{-1, +1\}^{\mathbb{Z}^d}$).

Before turning to variations of the previous basic argument to non-Gaussian tails in the next section, we present a recent result of S. Bobkov and F. Götze [B-G] which bounds, in this context, the Laplace transform of a function f in terms of some integral of its gradient. Up to numerical constants, this is an improvement upon the preceding statements. The proof however relies on the same ideas.

Let us consider, as in Corollary 2.5, an operator Γ on some class $\mathcal{A}$ in $L^1(\mu)$ such that $\Gamma(\lambda f) = \lambda^2 \Gamma(f) \geq 0$ for every $f \in \mathcal{A}$ and $\lambda \in \mathbb{R}$.

Theorem 2.7. *Let $\mathcal{A}$ be such that, for every $f \in \mathcal{A}$ and $\lambda \in \mathbb{R}$ (or only $\lambda \in [-1, +1]$), $\lambda f \in \mathcal{A}$, $E(e^{\lambda f}) < \infty$ and $E(e^{\lambda \Gamma(f)}) < \infty$. Assume that for every $f \in \mathcal{A}$,*

$$\mathrm{Ent}(e^f) \leq \frac{1}{2}\, E(\Gamma(f) e^f).$$

Then, for every $f \in \mathcal{A}$

$$E(e^{f - E(f)}) \leq E(e^{\Gamma(f)}).$$

Proof. Let, for every f, $g = \Gamma(f) - \log E(e^{\Gamma(f)})$, so that $E(e^g) = 1$. By (2.13),

$$E(\Gamma(f)\, e^f) - E(e^f) \log E(e^{\Gamma(f)}) \leq \mathrm{Ent}(e^f).$$

Together with the hypothesis $E(\Gamma(f)e^f) \geq 2\,\mathrm{Ent}(e^f)$, we get, for every f in $\mathcal{A}$,

$$\mathrm{Ent}(e^f) \leq E(e^f) \log E(e^{\Gamma(f)}).$$

Apply this inequality to λf for every λ. With the notation of the proof of Proposition 2.3, for every $\lambda \in \mathbb{R}$,

$$K'(\lambda) \leq \frac{1}{\lambda^2}\, \psi(\lambda^2)$$

where $\psi(\lambda) = \log E(e^{\lambda \Gamma(f)})$. Now, ψ is non-negative, non-decreasing and convex, and $\psi(0) = 0$. Therefore $\psi(\lambda)/\lambda$ is non-decreasing in $\lambda \geq 0$. Recalling that $K(0) = E(F)$, it follows that

$$K(1) \leq K(0) + \int_0^1 \frac{1}{\lambda^2}\, \psi(\lambda^2)\, d\lambda \leq E(F) + \psi(1)$$

which is the result. Theorem 2.7 is established. $\qquad\square$

2.4 Entropy-energy inequalities and non-Gaussian tails

The preceding basic argument admits a number of variations, some of which will be developed in the next chapters. We investigate first the case of defective logarithmic Sobolev inequality.

A defective logarithmic Sobolev inequality is of the type

$$\mathrm{Ent}_\mu(f^2) \leq a E_\mu(f^2) + 2\mathcal{E}(f), \quad f \in \mathcal{A} \tag{2.26}$$

where $a \geq 0$. Of course, if $a = 0$, this is just a classical logarithmic Sobolev inequality. We would like to know if the preceding concentration inequalities of Gaussian type still hold under such a defective inequality, and whether the latter again determines the best exponential integrability in (2.23). According to the discussion in the preceding section, it will be enough to deal with the setting of Proposition 2.3.

Proposition 2.8. *In the framework of Proposition 2.3, assume that for some $a > 0$ and for every $f \in \mathcal{A}$,*

$$\mathrm{Ent}(e^f) \leq a\mathrm{E}(e^f) + \frac{1}{2} N(f)^2 \, \mathrm{E}(e^f).$$

Then, whenever $N(F) \leq 1$,

$$\mathrm{E}(e^{\alpha F^2}) < \infty$$

for every $\alpha < \frac{1}{2}$.

Proof. Working first with a sequence $(F_n)_{n \in \mathbb{N}}$ in $\mathcal{A}$ such that $F_n \to F$, we may and do assume that $F \in \mathcal{A}$. Apply the defective logarithmic Sobolev inequality to λF for every $\lambda \in \mathbb{R}$. Letting as before $H(\lambda) = \mathrm{E}(e^{\lambda F})$, we get

$$\lambda H'(\lambda) - H(\lambda) \log H(\lambda) \leq \left(a + \frac{\lambda^2}{2}\right) H(\lambda).$$

If $K(\lambda) = \frac{1}{\lambda} \log H(\lambda)$, we see that, for every $\lambda > 0$,

$$K'(\lambda) \leq \frac{a}{\lambda^2} + \frac{1}{2}.$$

Hence, for every $\lambda \geq 1$,

$$K(\lambda) = K(1) + \int_1^\lambda K'(u)du \leq K(1) + a + \frac{\lambda}{2}.$$

It follows that, for $\lambda \geq 1$,

$$\mathrm{E}(e^{\lambda F}) \leq \left(\mathrm{E}(e^F)\right)^\lambda e^{a\lambda + \lambda^2/2}. \tag{2.27}$$

Let us choose first $\lambda = 2$. Then $\mathrm{E}(e^{2F}) \leq A\mathrm{E}(e^F)^2$ with $A = e^{2(a+1)}$. Let m be large enough so that $\mu(|F| \geq m) \leq 1/4A$. Then $\mu(e^F \geq e^m) < 1/4A$ and

$$\begin{aligned}
\mathrm{E}(e^F) &\leq e^m + \mu(e^F \geq e^m)^{1/2} \left(\mathrm{E}(e^{2F})\right)^{1/2} \\
&\leq e^m + \sqrt{A}\,\mu(e^F \geq e^m)^{1/2} \mathrm{E}(e^F) \\
&\leq 2\,e^m.
\end{aligned}$$

Coming back to (2.27), for every $\lambda \geq 1$,

$$\mathrm{E}(e^{\lambda F}) \leq 2^\lambda e^{(m+a)\lambda + \lambda^2/2} = e^{B\lambda + \lambda^2/2}$$

where $B = m + a + \log 2$. By Chebyshev's inequality,

$$\mu(F \geq r) \leq e^{Br - r^2/2}$$

for every $r \geq A + 1$. Together with the same inequality for $-F$, the conclusion follows from the proof of Proposition 1.2. Proposition 2.8 is therefore established. □

Inequality (2.25) actually fits into the more general framework of inequalities between entropy and energy introduced in [Ba1]. Given a non-negative function Ψ on $\mathbb{R}_+$, let us say that we have an entropy-energy inequality whenever for all f in $\mathcal{A}$ with $E_\mu(f^2) = 1$,

$$\text{Ent}_\mu(f^2) \leq \Psi\big(\mathcal{E}(f)\big). \tag{2.28}$$

By homogeneity, logarithmic Sobolev inequalities correspond to linear functions Ψ whereas defective logarithmic Sobolev inequalities correspond to affine Ψ's. Assume Ψ to be concave. Then (2.28) is equivalent to a family of defective logarithmic Sobolev inequalities

$$\text{Ent}_\mu(f^2) \leq \varepsilon E_\mu(f^2) + C(\varepsilon)\mathcal{E}(f), \quad \varepsilon \geq 0. \tag{2.29}$$

It is plain that, in the various settings studied above, the Laplace transform approach may be adapted to such an entropy-energy function. Depending upon to the rate at which Ψ increases to infinity, or, equivalently upon the behavior of $C(\varepsilon)$ as $\varepsilon \to 0$, various integrability results on Lipschitz functions may be obtained. It may even happen that Lipschitz functions are bounded if Ψ does not increase too quickly.

On the pattern of Proposition 2.3, we describe a general result that yields a variety of Laplace transform and tail inequalities for Lipschitz functions under some entropy-energy inequality. An alternate description of the next statement is presented in the paper [G-R] on the basis of (2.29). As will be studied in Parts 4 and 5, the form of the entropy-energy inequalities of Proposition 2.9 below is adapted to the concept of modified logarithmic Sobolev inequalities which often arise when the chain rule formula for the energy fails.

Let $\mathcal{A}$ be a class of functions in $L^1(\mu)$. For every f in $\mathcal{A}$, let $N(f) \geq 0$. According to the argument developed for Proposition 2.3, the proof of the following statement is straighforward.

Proposition 2.9. *Let $\mathcal{A}$ be such that, for every $f \in \mathcal{A}$ and $\lambda \in \mathbb{R}$, $\lambda f \in \mathcal{A}$, $E(e^{\lambda f}) < \infty$ and $N(\lambda f) = |\lambda| N(f)$. Assume there is a function $B(\lambda) \geq 0$ on $\mathbb{R}_+$ such that for every $f \in \mathcal{A}$ with $N(f) \leq \lambda$,*

$$\text{Ent}(e^f) \leq B(\lambda) E(e^f).$$

Then, for every F in $\mathcal{A}$ such that $N(F) \leq 1$,

$$E(e^{\lambda F}) \leq \exp\left(\lambda E(F) + \lambda \int_0^\lambda \frac{B(s)}{s^2} \, ds\right)$$

for every $\lambda \in \mathbb{R}$.

By homogeneity of N, Proposition 2.3 corresponds to the choice of $B(\lambda) = \lambda^2/2$, $\lambda \geq 0$.

The various examples discussed on the basis of Proposition 2.3 may also be reconsidered in this context. Suppose, for example, that, for some generalized modulus of gradient $|\nabla f|$, the entropy-energy inequality (2.28) holds. Then, by the change of variable formula, for every f with $\|\nabla f\|_\infty \leq \lambda$,

$$\mathrm{Ent}(e^f) \leq \Psi\left(\frac{\lambda^2}{4}\right) \mathrm{E}(e^f).$$

Now, depending upon how $B(\lambda)$ grows as λ goes to infinity, Proposition 2.9 will describe various tail estimates of Lipschitz functions. Rather than to discuss this in detail, let us briefly examine three specific behaviors of $B(\lambda)$.

Corollary 2.10. *In the setting of Proposition 2.9, if*

$$\int^\infty \frac{B(\lambda)}{\lambda^2}\, d\lambda < \infty, \tag{2.30}$$

then there exists $C > 0$ such that $\|F\|_\infty \leq C$ for every F such that $N(F) \leq 1$.

Proof. It is an easy matter to see from (2.30) and Proposition 2.9, that

$$\mathrm{E}(e^{\lambda|F|}) \leq e^{C\lambda}$$

for some $C > 0$ and all $\lambda \geq 0$ large enough. By Chebyshev's inequality, this implies that

$$\mu(|F| \geq 2C) \leq e^{-C\lambda} \to 0$$

as $\lambda \to \infty$. Corollary 2.10 is proved. Actually, if N is the Lipschitz norm on a metric space (X, d), the diameter of X will be finite (less than or equal to $2C$), see [Le2]. $\square$

In the second example, we consider a Gaussian behavior only for the small values of λ. The statement describes the typical tail of the exponential distribution (cf. Section 4.1).

Corollary 2.11. *In the setting of Proposition 2.9, assume that for some $c > 0$ and $\lambda_0 > 0$,*

$$B(\lambda) \leq c\lambda^2 \tag{2.31}$$

for every $0 \leq \lambda \leq \lambda_0$. Then, if F is such that $N(F) \leq 1$, we have $\mathrm{E}(|F|) < \infty$ and, for every $r \geq 0$,

$$\mu(F \geq \mathrm{E}(F) + r) \leq \exp\left(-\min\left(\frac{\lambda_0 r}{2}, \frac{r^2}{4c}\right)\right).$$

Proof. Arguing as next to Proposition 2.3, we may assume that $F \in \mathcal{A}$. With the notation of the proof of Proposition 2.3, for every $0 \leq \lambda \leq \lambda_0$,

$$K'(\lambda) \leq c\lambda^2.$$

Therefore $K(\lambda) \le K(0) + c\lambda$ so that

$$E(e^{\lambda F}) \le e^{\lambda E(F) + c\lambda^2}$$

for every $0 \le \lambda \le \lambda_0$ thus. By Chebyshev's inequality,

$$\mu(F \ge E(F) + r) \le e^{-\lambda r + c\lambda^2}.$$

If $r \le 2c\lambda_0$, choose $\lambda = \frac{r}{2c}$ while if $r \le 2c\lambda_0$, we simply take $\lambda = \lambda_0$. The conclusion easily follows. $\quad\square$

A third example of interest concerns Poisson tails on which we will come back in Part 5.

Corollary 2.12. *In the setting of Proposition 2.9, assume that for some $c, d > 0$,*

$$B(\lambda) \le c\lambda^2 e^{d\lambda} \tag{2.32}$$

for every $\lambda \ge 0$. Then, if F is such that $N(F) \le 1$, we have $E(|F|) < \infty$ and, for every $r \ge 0$,

$$\mu(F \ge E(F) + r) \le \exp\left(-\frac{r}{4d} \log\left(1 + \frac{dr}{2c}\right)\right).$$

In particular, $E(e^{\alpha|F|\log_+|F|}) < \infty$ for sufficiently small $\alpha > 0$.

Proof. It is similar to the preceding ones. We have

$$K'(\lambda) \le c e^{d\lambda}, \quad \lambda \ge 0.$$

Hence, $K(\lambda) \le K(0) + \frac{c}{d}(e^{d\lambda} - 1)$, that is

$$E(e^{\lambda F}) \le e^{\lambda E(F) + \frac{c\lambda}{d}(e^{d\lambda} - 1)}, \quad \lambda \ge 0.$$

By Chebyshev's inequality, for every $r \ge 0$ and $\lambda \ge 0$,

$$\mu(F \ge E(F) + r) \le e^{-\lambda r + \frac{c\lambda}{d}(e^{d\lambda} - 1)}.$$

When $r \le \frac{4c}{d}$ (the constants are not sharp), choose $\lambda = \frac{r}{4c}$ so that

$$e^{-\lambda r + \frac{c\lambda}{d}(e^{d\lambda} - 1)} \le e^{-\lambda r + 2c\lambda^2} = e^{-\frac{r^2}{8c}}.$$

while, when $r \ge \frac{4c}{d}$, choose $\lambda = \frac{1}{d}\log(\frac{dr}{2c})$ for which

$$e^{-\lambda r + \frac{c\lambda}{d}(e^{d\lambda} - 1)} \le e^{-\frac{r}{2d}\log(\frac{dr}{2c})}.$$

These two estimates together yield the inequality of Corollary 2.11. The proof is complete. $\quad\square$

The inequality of Corollary 2.11 describes the classic Gaussian tail behavior for the small values of r and the Poisson behavior for the large values of r (with respect to the ratio $\frac{c}{d}$). The constants have no reason to be sharp.

We refer to the recent work [G-R] for further examples in this line of investigation.

2.5 Poincaré inequalities and concentration

In the last section, we apply the preceding functional approach in case of a spectral gap inequality. As we have seen (Proposition 2.1), spectral gap inequalities are usually weaker than logarithmic Sobolev inequalities, and, as a result, they only imply exponential integrability of Lipschitz functions. The result goes back to M. Gromov and V. Milman [G-M] (on a compact Riemannian manifold but with an argument that works similarly in a more general setting; see also [Br]). It has been investigated recently in [A-M-S] and [A-S] using moment bounds, and in [Sc] using a differential inequality on Laplace transforms similar to Herbst's argument. We follow here the approach of S. Aida and D. Stroock [A-S].

Assume that for some energy function $\mathcal{E}$ on a class $\mathcal{A}$.

$$\mathrm{Var}(f) \leq C\mathcal{E}(f).$$

Apply this inequality to $e^{f/2}$. If $\mathcal{E}$ is the Dirichlet form associated to a symmetric Markov semigroup (2.7), we can apply (2.8) to get

$$\mathrm{E}(e^f) - \mathrm{E}(e^{f/2})^2 \leq \frac{C}{2}\,|||f|||_\infty^2\,\mathrm{E}(e^f).$$

In case $\mathcal{E}$ is the energy of a local gradient satisfying the chain rule formula, the constant $\frac{1}{2}$ is improved to $\frac{1}{4}$. The following statement thus summarizes the various examples of operators $\mathcal{E}$.

Let again $\mathcal{A}$ be a subset of $L^1(\mu)$. For every $f \in \mathcal{A}$, let $N(f) \geq 0$. We agree that $N(f) \leq 1$ for some function f on X if f is the limit of a sequence of functions $(f_n)_{n \in \mathbb{N}}$ in $\mathcal{A}$ with $N(f_n) \leq 1$ for every n.

Proposition 2.13. *Let $\mathcal{A}$ be such that, for every $f \in \mathcal{A}$ and every $\lambda \in \mathbb{R}$, $\mathrm{E}(e^{\lambda f}) < \infty$ and $N(\lambda f) = |\lambda| N(f)$. Assume that for some $C > 0$ and every $f \in \mathcal{A}$,*

$$\mathrm{E}(e^f) - \mathrm{E}(e^{f/2})^2 \leq C N(f)^2 \mathrm{E}(e^f). \tag{2.33}$$

Then, for every F such that $N(F) \leq 1$, $\mathrm{E}(|F|) < \infty$ and

$$\mathrm{E}(e^{\lambda(F - \mathrm{E}(F))}) \leq \prod_{k=0}^{\infty} \left(\frac{1}{1 - \frac{C\lambda^2}{4^k}}\right)^{2^k} \tag{2.34}$$

for all $|\lambda| < 1/\sqrt{c}$. In particular,

$$\mathrm{E}(e^{\alpha|F|}) < \infty$$

for every $\alpha < 1/\sqrt{C}$.

Proof. Assume first $F \in \mathcal{A}$ with $N(F) \leq 1$. Set $H(\lambda) = E(e^{\lambda F})$, $\lambda \geq 0$. Applying (2.33) to λF yields

$$H(\lambda) - H\left(\frac{\lambda}{2}\right)^2 \leq C\lambda^2 H(\lambda).$$

Hence, for every $\lambda < 1/\sqrt{C}$,

$$H(\lambda) \leq \frac{1}{1 - C\lambda^2} H\left(\frac{\lambda}{2}\right)^2.$$

Applying the same inequality for $\lambda/2$ and iterating, yields, after n steps,

$$H(\lambda) \leq \prod_{k=0}^{n-1} \left(\frac{1}{1 - \frac{C\lambda^2}{4^k}}\right)^{2^k} H\left(\frac{\lambda}{2^n}\right)^{2^n}.$$

Now $H(\lambda/\alpha)^{\alpha} \to e^{\lambda E(F)}$ as $\alpha \to \infty$. Hence, (2.34) is satisfied for this F which we assumed in $\mathcal{A}$. In particular, if $0 < \lambda_0 < 1/\sqrt{C}$, and if

$$K_0 = K_0(\lambda_0) = \prod_{k=0}^{\infty} \left(\frac{1}{1 - \frac{C\lambda_0^2}{4^k}}\right)^{2^k} < \infty,$$

then

$$\mu\left(|F - E(F)| \geq r\right) \leq 2K_0 \, e^{-\lambda_0 r} \tag{2.35}$$

for every $r \geq 0$. Applying (2.35) to a sequence $(F_n)_{n \in \mathbb{N}}$ converging to F with $N(F_n) \leq 1$, and arguing as next to Proposition 2.3 immediately yields the full conclusion of the Proposition. The proof is thus complete. □

The infinite product (2.34) has been estimated in [B-L1] by

$$\frac{1 + \sqrt{C}\lambda}{1 - \sqrt{C}\lambda}.$$

The example of the exponential measure investigated in Section 4.1 below shows that the condition $|\lambda| < 1/\sqrt{C}$ in Proposition 2.13 is optimal. Namely, let ν be the measure with density $\frac{1}{2} e^{-|x|}$ with respect to Lebesgue measure on $\mathbb{R}$. Then, by Lemma 4.1,

$$\mathrm{Var}_{\nu}(f) \leq 4 \, E_{\nu}(f'^2)$$

for every smooth f. Therefore, if $N(f) = \|f'\|_{\infty}$, (2.33) holds with $C = 1$, which is optimal as shown by the case $f(x) = x$.

Proposition 2.13 actually strengthens the early observation by R. Brooks [Br]. Namely, if M is a complete Riemannian manifold with finite volume $V(M)$, and if $V(x, r)$ is the volume of the ball $B(x, r)$ with center x and radius $r \geq 0$, then M has spectral gap zero as soon as

$$\liminf_{r \to \infty} \frac{-1}{r} \log[V(M) - V(x, r)] = 0 \tag{2.36}$$

for some (all) x in M.

3. DEVIATION INEQUALITIES FOR PRODUCT MEASURES

In the recent years, M. Talagrand has developed striking new methods for investigating the concentration of measure phenomenon for product measures. These ideas led to significant progress in an number of various areas such as probability in Banach spaces, empirical processes, geometric probability, statistical mechanics... The interested reader will find in the important contribution [Ta6] a complete account of these methods and results (see also [Ta7]). In this chapter, we indicate an alternate approach to some of Talagrand's inequalities based on logarithmic Sobolev inequalities and the methods of Chapter 2. The main point is that while concentration inequalities do not necessarily tensorize, the results follow from stronger logarithmic Sobolev inequalities which, as we know, do tensorize. In particular, we emphasize dimension free results.

The main deviation inequalities for convex functions form the core of Section 3.2, introduced by the discrete concentration property with respect to the Hamming metric in 3.1. Applications to sharp bounds on empirical processes conclude the chapter.

While it is uncertain whether this approach could recover Talagrand's abstract principles, the deviation inequalities themselves follow rather easily from it. On the abstract inequalities themselves, let us mention here the recent alternate approach by K. Marton [Mar1], [Mar2] and A. Dembo [De] (see also [D-Z]) based on information inequalities and coupling in which the concept of entropy also plays a crucial role. Hypercontraction methods were already used in [Kw-S] to study integrability of norms of sums of independent vector valued random variables. The work by K. Marton also involves Markov chains. Her arguments have been brought into relation recently with the logarithmic Sobolev inequality approach, and her results have been extended to larger classes of Markov chains, by P.-M. Samson [Sa]. We review some of these ideas in Section 3.3.

3.1 Concentration with respect to the Hamming metric

A first result on concentration in product spaces is the following. Let $(X_i, \mathcal{B}_i, \mu_i)$, $i = 1, \ldots, n$, be are arbitrary probability space, and let $P = \mu_1 \otimes \cdots \otimes \mu_n$ be a product measure on the product space $X = X_1 \times \cdots \times X_n$. A generic point in X is denoted by $x = (x_1, \ldots, x_n)$. Then, for every F on X such that $|F(x) - F(y)| \leq 1$

whenever $x = (x_1, \ldots, x_n)$ and $y = (y_1, \ldots, y_n)$ only differ by one coordinate,

$$P(F \geq \mathrm{E}_P(F) + r) \leq e^{-r^2/2n}. \tag{3.1}$$

This inequality can be established by rather elementary martingale arguments [Mau1], [M-S], and was important in the early developments of concentration in product spaces (cf. [Ta6]). Our first aim will be to realize that it is also an elementary consequence of the logarithmic Sobolev approach developed in Section 2.3. We owe this observation to S. Kwapień.

Let f on the product space X. Recall we define f_i on X_i, $i = 1, \ldots, n$, by $f_i(x_i) = f(x_1, \ldots, x_{i-1}, x_i, x_{i+1}, \ldots, x_n)$ with $x_1, \ldots, x_{i-1}, x_{i+1}, \ldots, x_n$ fixed.

Proposition 3.1. For every f on the product space X,

$$\mathrm{Ent}_P(e^f) \leq \frac{1}{2} \sum_{i=1}^n \mathrm{E}_P \left(\int\int \left(f_i(x_i) - f_i(y_i) \right)^2 e^{f_i(x_i)} d\mu_i(x_i) d\mu_i(y_i) \right).$$

Proof. The proof is elementary. We may assume f bounded. By the product property of entropy, it is enough to deal with the case $n = 1$. By Jensen's inequality,

$$\mathrm{Ent}_P(e^f) \leq \mathrm{E}_P(fe^f) - \mathrm{E}_P(e^f)\mathrm{E}_P(f).$$

The right-hand-side of the latter may then be rewritten as

$$\frac{1}{2} \int\int \left(f(x) - f(y) \right) \left(e^{f(x)} - e^{f(y)} \right) dP(x) dP(y).$$

Since

$$(u - v)(e^u - e^v) \leq \tfrac{1}{2}(u - v)^2(e^u + e^v), \quad u, v \in \mathbb{R},$$

the conclusion easily follows. □

As a consequence of Proposition 3.1, if

$$N(f) = \sup_{x \in X} \left(\int \sum_{i=1}^n (f(x) - f(y))^2 dP(y) \right)^{1/2},$$

then

$$\mathrm{Ent}_P(e^f) \leq \frac{1}{2} N(f)^2 \mathrm{E}_P(e^f).$$

Therefore, applying Proposition 2.3, if F is a Lipschitz function on X such that

$$|F(x) - F(y)| \leq \mathrm{Card}\{1 \leq i \leq n; x_i \neq y_i\},$$

then $N(F) \leq n$ from which (3.1) follows.

This basic example actually indicates the route we will follow next, in particular with convex functions. Before turning to this case, let us mention that Proposition 3.1 has a clear analogue for variance that states that

$$\mathrm{Var}_P(f) \leq \frac{1}{2} \sum_{i=1}^n \mathrm{E}_P \left(\int\int \left(f_i(x_i) - f_i(y_i) \right)^2 d\mu_i(x_i) d\mu_i(y_i) \right). \tag{3.2}$$

3.2 Deviation inequalities for convex functions

One of the first important results underlying M. Talagrand's developments is the following inequality for arbitrary product measures [Ta1], [J-S] (see also [Mau2]). Let F be a convex Lipschitz function on $\mathbb{R}^n$ with $\|F\|_{\mathrm{Lip}} \leq 1$. Let μ_i, $i = 1, \ldots, n$, be probability measures on $[0, 1]$ and denote by P the product probability measure $\mu_1 \otimes \cdots \otimes \mu_n$. Then, for every $r \geq 0$,

$$P(F \geq m + t) \leq 2 e^{-t^2/4} \qquad (3.3)$$

where m is a median of F for P. As in the Gaussian case (1.20), this bound is dimension free, a feature of fundamental importance in this investigation. However, contrary to the Gaussian case, it is known that the convexity assumption on F is essential (cf. [L-T], p. 25). The proof of (3.3) [in the preceding references] is based on the inequality

$$\int e^{\frac{1}{4} d(\cdot, \mathrm{Conv}(A))^2} dP \leq \frac{1}{P(A)}$$

which is established by geometric arguments and a simple induction on the number of coordinates. It has since been embedded in an abstract framework which M. Talagrand calls convex hull approximation (cf. [Ta6], [Ta7]). M. Talagrand also introduced the concept of approximation by a finite number of points [Ta2], [Ta6], [Ta7]. These powerful abstract tools have been used in particular to study sharp deviations inequalities for large classes of functions (cf. Section 3.4).

The aim of this section is to provide a simple proof of inequality (3.3) based on the functional inequalities presented in Part 2. The point is that while the deviation inequality (3.3) has no reason to be tensorizable, it is actually a consequence of a logarithmic Sobolev inequality, which only needs to be proved in dimension one. The main result in this direction is the following statement. Let thus $\mu_1, \ldots, \mu_n$ be arbitrary probability measures on $[0, 1]$ and let P be the product probability $P = \mu_1 \otimes \cdots \otimes \mu_n$. We say that a function f on $\mathbb{R}^n$ is separately convex if it is convex in each coordinate. Recall that a convex function on $\mathbb{R}$ is continuous and almost everywhere differentiable. We denote by ∇f the usual gradient of f on $\mathbb{R}^n$ and by $|\nabla f|$ its Euclidean length.

Theorem 3.2. *Let f be a function on $\mathbb{R}^n$ such that $\log f^2$ is separately convex $(f^2 > 0)$. Then, for any product probability P on $[0, 1]^n$,*

$$\mathrm{Ent}_P(f^2) \leq 4 \, \mathbb{E}_P(|\nabla f|^2)$$

Notice that Theorem 3.2 amounts to saying that for every separately convex function f on $\mathbb{R}^n$,

$$\mathrm{Ent}_P(e^f) \leq \mathbb{E}_P(|\nabla f|^2 e^f). \qquad (3.4)$$

Proof. By a simple approximation, it is enough to deal with sufficiently smooth functions. We establish a somewhat stronger result, namely that for any product probability P on $\mathbb{R}^n$, and any smooth separately convex function f,

$$\mathrm{Ent}_P(e^f) \leq \int\int \sum_{i=1}^{n} (x_i - y_i)^2 (\partial_i f)^2(x) \, e^{f(x)} \, dP(x) dP(y). \qquad (3.5)$$

By Proposition 3.1, it is enough to show that for every $i = 1, \ldots, n$,

$$\frac{1}{2} \int\int (f_i(x_i) - f_i(y_i))^2 e^{f_i(x_i)} d\mu_i(x_i) d\mu_i(y_i)$$

$$\leq \int\int (x_i - y_i)^2 f_i'(x_i)^2 e^{f_i(x_i)} d\mu_i(x_i) d\mu_i(y_i).$$

We may thus assume that $n = 1$. Now,

$$\frac{1}{2} \int\int (f(x) - f(y))^2 e^{f(x)} d\mu(x) d\mu(y)$$

$$\leq \int\int_{\{f(x) \geq f(y)\}} (f(x) - f(y))^2 e^{f(x)} d\mu(x) d\mu(y).$$

Since f is convex, for all $x, y \in \mathbb{R}$,

$$f(x) - f(y) \leq (x - y) f'(x).$$

The proof is easily completed. Theorem 3.2 is established. $\qquad\square$

It should be emphasized that inequality (3.5), established in the preceding proof for arbitrary product measures on $\mathbb{R}^n$ is actually a stronger version of Theorem 3.2 which is particulary used for norms of sums of independent random vectors (Section 3.4). This inequality puts forward the generalized gradient (in dimension one)

$$|\nabla f(x)| = \left(\int (x - y)^2 f'(y)^2 d\mu(y) \right)^{1/2}$$

of statistical interest.

With a little more effort, the constant of the logarithmic Sobolev inequality of Theorem 3.2 may be improved to 2 (which is probably the optimal constant). We need simply improve the estimate of the entropy in dimension one. To this end, recall the variational caracterization of entropy ([H-S]) as

$$\text{Ent}(e^f) = \inf_{c>0} E_P(fe^f - (\log c + 1) e^f + c). \tag{3.6}$$

Let P be a probability measure concentrated on $[0, 1]$. Let f be (smooth and) convex on $\mathbb{R}$. Let then $y \in [0, 1]$ be a point at which f is minimum and take $c = e^{f(y)}$ (in (3.6)). For every $x \in [0, 1]$,

$$f(x) e^{f(x)} - (\log c + 1) e^{f(x)} + c = [f(x) - f(y)] e^{f(x)} - [e^{f(x)} - e^{f(y)}]$$

$$= [(f(x) - f(y)) - 1 + e^{-(f(x) - f(y))}] e^{f(x)}$$

$$\leq \frac{1}{2} [f(x) - f(y)]^2 e^{f(x)}$$

since $u - 1 + e^{-u} \leq \frac{u^2}{2}$ for every $u \geq 0$. Hence, by convexity, and since $x, y \in [0, 1]$,

$$f(x) e^{f(x)} - (\log c + 1) e^{f(x)} + c \leq \frac{1}{2} f'(x)^2 e^{f(x)}$$

from which we deduce, together with (3.6), that

$$\mathrm{Ent}_P(e^f) \le \frac{1}{2}\,\mathrm{E}_P(f'^2 e^f).$$

We may now apply the results of Section 2.3 to get Gaussian deviation inequalities for convex Lipschitz functions with respect to product measures. (On the discrete cube, see also [Bob1]).

Corollary 3.3. *Let F be a separately convex Lipschitz function on $\mathbb{R}^n$ with Lipschitz constant $\|F\|_{\mathrm{Lip}} \le 1$. Then, for every product probability P on $[0,1]^n$, and every $r \ge 0$,*

$$P\big(F \ge \mathrm{E}_P(F) + r\big) \le e^{-r^2/2}.$$

This inequality is the analogue of (3.3) with the mean instead of the (a) median m and the improved bound $e^{-t^2/2}$.

The proof of Corollary 3.3 is a direct application of Corollary 2.5. Only some regularization procedure has to be made precise. Replacing F by a convolution with a Gaussian kernel, we may actually suppose that $|\nabla F| \le 1$ everywhere. Then, the argument is entirely similar to the one detailed, for example in the Gaussian case (after Corollary 2.6). The result follows by approximation.

M. Talagrand [Ta1] (see also [J-S], [Mau2], [Ta6], [Ta7]) actually showed deviation inequalities under the level m, that is an inequality for $-F$ (F convex). It yields a concentration result of the type

$$P\big(|F - m| \ge r\big) \le 4e^{-r^2/4}, \quad r \ge 0. \tag{3.7}$$

It does not seem that such a deviation inequality for $-F$, F convex, follows from the preceding approach (since e^{-F} need not be convex). At a weak level though, we may use Poincaré inequalities. Indeed, we may first state the analogue of Theorem 3.2 for variance, whose proof is similar. This result was first mentioned in [Bob2].

Proposition 3.4. *Let f be a separately convex function on $\mathbb{R}^n$. Then, for any product probability P on $[0,1]^n$,*

$$\mathrm{Var}_P(f) \le \mathrm{E}_P(|\nabla f|^2).$$

Therefore, for any separately convex function F with $\|F\|_{\mathrm{Lip}} \le 1$,

$$P\big(|F - \mathrm{E}_P(F)| \ge r\big) \le \frac{1}{r^2}$$

for every $r \ge 0$. As seems to be indicated by the results in the next section, the convexity in each coordinate might not be enough to ensure deviation under the mean or the median. Using alternate methods, we will see indeed that sharp concentration inequalities do hold for concave functions, even under less stringent assumptions than Lipschitz. Although deviation inequalities above the mean or the median are the useful inequalities in probability and its applications, concentration inequalities are sometimes important issues (e.g. in geometry of Banach spaces [M-S], percolation, spin glasses... [Ta6]).

Corollary 3.3 of course extends to probability measures μ_i supported on $[a_i, b_i]$, $i = 1, \ldots, n$, (following for example (3.5) of the proof of Theorem 3.2, or by scaling). In particular, if P is a product measure on $[a, b]^n$ and if F is separately convex on $\mathbb{R}^n$ with Lipschitz constant less than or equal to 1, for every $r \geq 0$,

$$P\big(F \geq \mathrm{E}_P(F) + r\big) \leq e^{-r^2/2(b-a)^2}.$$

Let us also recall one typical application of these deviation inequalities to norms of random series. Let η_i, $i = 1, \ldots, n$, be independent random variables on some probability space $(\Omega, \mathcal{A}, \mathbb{P})$ with $|\eta_i| \leq 1$ almost surely. Let v_i, $i = 1, \ldots, n$, be vectors in some arbitrary Banach space E with norm $\| \cdot \|$. Then, for every $r \geq 0$,

$$\mathbb{P}\left(\left\| \sum_{i=1}^{n} \eta_i v_i \right\| \geq \mathbb{E} \left\| \sum_{i=1}^{n} \eta_i v_i \right\| + r \right) \leq e^{-r^2/8\sigma^2}$$

where

$$\sigma^2 = \sup_{\|\xi\| \leq 1} \sum_{i=1}^{n} \langle \xi, v_i \rangle^2.$$

This inequality is the analogue of the Gaussian deviation inequalities (1.8) and (1.24). For the proof, simply consider F on $\mathbb{R}^n$ defined by

$$F(x) = \left\| \sum_{i=1}^{n} x_i v_i \right\|, \quad x = (x_1, \ldots, x_n) \in \mathbb{R}^n.$$

Then, by duality, for $x, y \in \mathbb{R}^n$,

$$|F(x) - F(y)| \leq \left\| \sum_{i=1}^{n} (x_i - y_i) v_i \right\| = \sup_{\|\xi\| \leq 1} \sum_{i=1}^{n} (x_i - y_i) \langle \xi, v_i \rangle \leq \sigma |x - y|,$$

where the last step is obtained from the Cauchy-Schwarz inequality.

3.3 Information inequalities and concentration

Recently, K. Marton [Mar1], [Mar2] (see also [Mar3]) studied the preceding concentration inequalities in the context of contracting Markov chains. Her approach is based on information inequalities and coupling ideas. Specifically, she is using convexity of entropy together with Pinsker's inequality [Pi]

$$\|\mu - \nu\|_{\mathrm{T.V.}} \leq \sqrt{\frac{1}{2} \mathrm{Ent}_\mu \left(\frac{d\nu}{d\mu} \right)} \tag{3.8}$$

where the probability measure ν is assumed to be absolutely continuous with respect to μ with density $\frac{d\nu}{d\mu}$. That such an inequality entails concentration properties may be shown in the following way. Given a separable metric space (X, d) and two Borel probability measures μ and ν on X, set

$$W_1(\mu, \nu) = \inf \int\int d(x, y) d\pi(x, y)$$

where the infimum runs over all probability measures π on the product space $X \times X$ with marginals μ and ν. Consider now the inequality

$$W_1(\mu, \nu) \leq \sqrt{2C \, \mathrm{Ent}_\mu\left(\frac{d\nu}{d\mu}\right)} \tag{3.9}$$

for some $C > 0$. By the coupling characterization of the total variation distance, Pinsker's inequality corresponds to the trivial distance on X (and to $C = \frac{1}{4}$). Let then A and B with $\mu(A), \mu(B) > 0$, and consider the conditional probabilities $\mu_A = \mu(\cdot|A)$ and $\mu_B = \mu(\cdot|B)$. By the triangle inequality and (3.9),

$$\begin{aligned} W_1(\mu_A, \mu_B) &\leq W_1(\mu, \mu_A) + W_1(\mu, \mu_B) \\ &\leq \sqrt{2C \, \mathrm{Ent}_\mu\left(\frac{d\mu_A}{d\mu}\right)} + \sqrt{2C \, \mathrm{Ent}_\mu\left(\frac{d\mu_B}{d\mu}\right)} \\ &= \sqrt{2C \log \frac{1}{\mu(A)}} + \sqrt{2C \log \frac{1}{\mu(B)}} \, . \end{aligned} \tag{3.10}$$

Now, all measures with marginals μ_A and μ_B must be supported on $A \times B$, so that, by the definition of W_1,

$$W_1(\mu_A, \mu_B) \geq d(A, B) = \inf\{d(x, y); x \in A, y \in B\}.$$

Then (3.10) implies a concentration inequality. Fix A with, say, $\mu(A) \geq \frac{1}{2}$ and take B the complement of A_r for $r \geq 0$. Then $d(A, B) \geq r$ so that

$$r \leq \sqrt{2C \log \frac{1}{\mu(A)}} + \sqrt{2C \log \frac{1}{1 - \mu(A_r)}} \leq \sqrt{2C \log 2} + \sqrt{2C \log \frac{1}{1 - \mu(A_r)}} \, .$$

Hence, whenever $r \geq 2\sqrt{2C \log 2}$ for example,

$$1 - \mu(A_r) \leq e^{-r^2/8C}.$$

Now, the product property of entropy allows us to tensorize Pinsker-type inequalities to produce concentration in product spaces. For example, this simple scheme may be used to recover, even with sharp constants, the concentration (3.1) with respect to the Hamming metric. Indeed, if we let d be the Hamming metric on the product space $X = X_1 \times \cdots \times X_n$, starting with (3.8), convexity of entropy shows that for any probability measure Q on X absolutely continuous with respect to the product measure $P = \mu_1 \otimes \cdots \otimes \mu_n$,

$$W_1(P, Q) \leq \sqrt{\frac{n}{2} \mathrm{Ent}_P\left(\frac{dQ}{dP}\right)}$$

from which (3.1) follows according to the preceding argument.

It might be worthwhile noting that S. Bobkov and F. Götze [B-G] recently proved that an inequality such as (3.9) holding for all measures ν absolutely continuous with respect to μ is actually equivalent to the Gaussian bound

$$E_\mu(e^{\lambda F}) \leq e^{\lambda E_\mu(F) + C\lambda^2/2}$$

on the Lipschitz functions F on (X, d) with $\|F\|_{\mathrm{Lip}} \leq 1$. This observation connects the information theory approach to the logarithmic Sobolev approach emphasized in this work. It also shows that a logarithmic Sobolev inequality in this case is a stronger statement than a Pinsker-type inequality.

For the Gaussian measure γ on $\mathbb{R}^n$ equipped with the Euclidean distance d_2, M. Talagrand [Ta9] proved that, not only (3.9) holds but

$$W_2(\gamma, \nu) \leq \sqrt{2 \mathrm{Ent}_\gamma \left(\frac{d\nu}{d\gamma}\right)} \tag{3.11}$$

where now

$$W_2(\gamma, \nu) = \inf \left(\iint d_2(x, y)^2 \, d\pi(x, y) \right)^{1/2}.$$

He further investigated in this paper the case of the exponential distribution to recover its concentration properties (cf. Section 4.1). Recently, it was proved in [O-V] that (3.11) may be shown to follow from the Gaussian logarithmic Sobolev inequality (2.15).

In order to cover with these methods the inequalities for convex functions of Section 3.2, K. Marton [Mar2] introduced another metric on measures in the form of

$$d_2(\mu, \nu) = E_\mu \left(\left(1 - \frac{d\nu}{d\mu}\right)_+^2 \right)^{1/2}.$$

This distance is analogous to the variational distance and one can actually show that

$$d_2(\mu, \nu) = \inf \left(\int \mathbb{P}(\xi \neq y \mid \zeta = y)^2 d\nu(y) \right)^{1/2}$$

where the infimum is over all couples of random variables (ξ, ζ) such that ξ has distribution μ and ζ distribution ν. Note that $d_2(\mu, \nu)$ is not symmetric in μ, ν. Together with the appropriate information inequality on d_2 and convexity of relative entropy, she proved in this way the concentration inequalities for convex functions of the preceding section. Her arguments has been further developed by A. Dembo [De] to recover in this way most of M. Talagrand's abstract inequalities.

But K. Marton's approach was initially devoted to some non-product Markov chains for which it appears to be a powerful tool. More precisely, let P be a Markov chain on $[0, 1]^n$ with transition kernels K_i, $i = 1, \ldots, n$, that is

$$dP(x_1, \ldots, x_n) = K_n(x_n, dx_{n-1}) \cdots K_2(x_2, dx_1) K_1(dx_1).$$

Assume that, for some $0 \le a < 1$, for every $i = 1, \ldots, n$, and every $x, y \in [0, 1]$,

$$\left\| K_i(x, \cdot) - K_i(y, \cdot) \right\|_{\mathrm{T.V.}} \le a. \tag{3.12}$$

The case $a = 0$ of course corresponds to independent kernels K_i. The main result of [Mar2] (expressed on functions) is the following.

Theorem 3.5. *Let P be a Markov chain on $[0, 1]^n$ satisfying (3.12) for some $0 \le a < 1$. For every convex Lipschitz map F on $\mathbb{R}^n$ with $\|F\|_{\mathrm{Lip}} \le 1$,*

$$P(F \ge \mathrm{E}_P(F) + r) \le e^{-(1 - \sqrt{a})^2 r^2 / 4}$$

for every $r \ge 0$ and similarly for $-F$.

This result has been extended in [Mar4] and, independently in [Sa], to larger classes od dependent processes. Moreover, in [Sa], P.-M. Samson brings into relation the information approach with the logarithmic Sobolev approach. Let P and Q be probability measures on $\mathbb{R}^n$. Following the one-dimensional definition of d_2, set

$$d_2(P, Q) = \inf \sup_\alpha \iint \sum_{i=1}^n \alpha_i(y) \mathrm{I}_{x_i \ne y_i} d\pi(x, y)$$

where the infimum is over all probability measures π on $\mathbb{R}^n \times \mathbb{R}^n$ with marginals P and Q and the supremum runs over all $\alpha = (\alpha_1, \ldots, \alpha_n)$ where the α_i's are non-negative functions on $\mathbb{R}^n$ such that

$$\int \sum_{i=1}^n \alpha_i^2(y) dQ(y) \le 1.$$

As shown by K. Marton, we have similarly a coupling description as

$$d_2(P, Q) = \inf \left(\int \sum_{i=1}^n \mathbb{P}\left(\xi_i \ne y_i \mid \zeta_i = y_i \right)^2 dQ(y) \right)^{1/2}$$

where the infimum runs over all random variables $\xi = (\xi_1, \ldots, \xi_n)$ and $\zeta = (\zeta_1, \ldots, \zeta_n)$ such that ξ has distribution P and ζ distribution Q.

Let now P denote the distribution of a sample $X_1, \ldots, X_n$ of real random variables. Following Marton's techniques, for any Q absolutely continuous with respect to P,

$$\max(d_2(P, Q), d_2(Q, P)) \le \sqrt{2\|\Gamma\| \operatorname{Ent}_P \left(\frac{dQ}{dP} \right)} \tag{3.13}$$

where $\|\Gamma\|$ is the operator norm of a certain mixing matrix Γ that measures the L^2-dependence of the variables $X_1, \ldots, X_n$. Now, $\|\Gamma\|$ may be shown to be bounded independently of the dimension in a number of interesting cases, including Doeblin recurrent Markov chains and Φ-mixing processes (cf. [Mar4], [Sa]). (3.13) then yields concentration inequalities for new classes of measures and processes.

Moreover, it is shown in [Sa] how (3.13) may be considered as a kind of dual version of the logarithmic Sobolev inequalities for convex (and concave) functions of Section 3.2 above. Let f be (smooth and) convex on $[0,1]^n$. By Jensen's inequality,

$$\frac{\text{Ent}_P(e^f)}{\text{E}_P(e^f)} \leq \int f(x)\frac{e^{f(x)}}{\text{E}_P(e^f)}\,dP(x) - \int f(y)dP(y).$$

Let P^f be the probability measure on $[0,1]^n$ whose density with respect to P is $e^f/\text{E}_P(e^f)$. Let π be the probability measure on $\mathbb{R}^n \times \mathbb{R}^n$ with marginals P and P^f. Then,

$$\frac{\text{Ent}_P(e^f)}{\text{E}_P(e^f)} \leq \iint [f(y) - f(x)]\,d\pi(x,y).$$

Since f is convex, for every $x = (x_1,\ldots,x_n)$ and $y = (y_1,\ldots,y_n) \in [0,1]^n$,

$$f(x) - f(y) \leq \sum_{i=1}^n |x_i - y_i||\partial_i f(x)| \leq \sum_{i=1}^n |\partial_i f(x)|\text{I}_{x_i \neq y_i}.$$

As a consequence, for all probability measures π on $\mathbb{R}^n \times \mathbb{R}^n$ with marginals P and P^f,

$$\frac{\text{Ent}_P(e^f)}{\text{E}_P(e^f)} \leq \iint \sum_{i=1}^n |\partial_i f(x)|\text{I}_{x_i \neq y_i}\,d\pi(x,y).$$

According to the definition of $d_2(P, P^f)$, and by the Cauchy-Schwarz inequality,

$$\frac{\text{Ent}_P(e^f)}{\text{E}_P(e^f)} \leq d_2(P, P^f)\left(\int \sum_{i=1}^n \iint |\partial_i f(x)|^2 dP^f(x)\right)^{1/2}. \tag{3.14}$$

Since

$$\frac{dP^f}{dP} = \frac{e^f}{\text{E}_P(e^f)}$$

we get from (3.13) and (3.14) that

$$\frac{\text{Ent}_P(e^f)}{\text{E}_P(e^f)} \leq \|\Gamma\|\left(2\frac{\text{Ent}_P(e^f)}{\text{E}_P(e^f)}\right)^{1/2}\left(\int |\nabla f|^2 \frac{e^f}{\text{E}_P(e^f)}dP\right)^{1/2}.$$

It follows that for every (smooth) convex function on $[0,1]^n$,

$$\text{Ent}_P(e^f) \leq 2\|\Gamma\|^2\text{E}_P(|\nabla f|^2 e^f) \tag{3.15}$$

which amounts to the inequality of Theorem 3.2.

It is worthwhile noting that the same proof for a concave function f yields instead of (3.15)

$$\text{Ent}_P(e^f) \leq 2\|\Gamma\|^2\text{E}_P(|\nabla f|^2)\text{E}_P(e^f). \tag{3.16}$$

These observations clarify the discussion on separately convex or concave functions in Section 3.2. In contrast to Theorem 3.2, the proof of these results fully uses the convexity or concavity assumptions on f rather than only convexity in each coordinate. Together with Herbst's argument, these inequalities imply the conclusions of Theorem 3.5. On the other hand, deviation inequalities under the mean for concave functions F only require $\text{E}_P(|\nabla F|^2) \leq 1$.

3.4 Applications to bounds on empirical processes

Sums of independent random variables are a natural application of the preceding deviation inequalities for product measures. In this section, we survey some of these applications, with a particular emphasis on bounds for empirical processes.

Tail probabilities for sums of independent random variables have been extensively studied in classical Probability theory. One finished result is the so-called Bennett inequality (after contributions by Bernstein, Kolmogorov, Prohorov, Hoeffding etc). Let $X_1, \ldots, X_n$ be independent mean-zero real-valued random variables on some probability space $(\Omega, \mathcal{A}, \mathbb{P})$ such that $|X_i| \leq C$, $i = 1, \ldots, n$, and $\sum_{i=1}^n \mathbb{E}(X_i^2) \leq \sigma^2$. Set $S_n = X_1 + \cdots + X_n$. Then, for every $r \geq 0$,

$$\mathbb{P}(S_n \geq r) \leq \exp\left(-\frac{r}{2C} \log\left(1 + \frac{Cr}{\sigma^2}\right)\right). \tag{3.17}$$

Such an inequality describes the Gaussian tail for the values of r which are small with respect to σ^2, and the Poisson behavior for the large values (think, for example, of a sample of independent Bernoulli variables, with probability of success either $\frac{1}{2}$ or on the order of $\frac{1}{n}$.)

Now, in statistical applications, one is interested in such a bound uniformly over classes of functions, and importance of such inequalities has been emphasized recently in the statistical treatment of selection of models by L. Birgé and P. Massart [B-M1], [B-M2], [B-B-M]. More precisely, let $X_1, X_2, \ldots, X_n, \ldots$ be independent random variables with values in some measurable space $(S, \mathcal{S})$ with identical distribution $\mathcal{P}$, and let, for $n \geq 1$,

$$\mathcal{P}_n = \frac{1}{n} \sum_{i=1}^n \delta_{X_i}$$

be the empirical measures (on $\mathcal{P}$). A class $\mathcal{F}$ of real measurable functions on S is said to be a Glivenko-Cantelli class if $\sup_{f \in \mathcal{F}} |\mathcal{P}_n(f) - \mathcal{P}(f)|$ converges almost surely to 0. It is a Donsker class if, in a sense to be made precise, $\sqrt{n}(\mathcal{P}_n(f) - \mathcal{P}(f))$, $f \in \mathcal{F}$, converges in distribution toward a centered Gaussian process with covariance function $\mathcal{P}(fg) - \mathcal{P}(f)\mathcal{P}(g)$, $f, g \in \mathcal{F}$. These definitions naturally extend the classic example of the class of all indicator functions of intervals $(-\infty, t]$, $t \in \mathbb{R}$ (studied precisely by Glivenko-Cantelli and Donsker). These asymptotic properties however often need to be turned into tail inequalities at fixed n on classes $\mathcal{F}$ which are as rich as possible (to determine accurate approximation by empirical models). In particular, these bounds aim to be as close as possible to the one-dimensional inequality (3.17) (corresponding to a class $\mathcal{F}$ reduced to only one function).

Sharp bounds for empirical processes have been obtained by M. Talagrand [Ta5], [Ta8] as a consequence of his abstract inequalities for product measures. We observe here that the functional approach based on logarithmic Sobolev inequalities developed in the preceding sections may be use to produce similar bounds. The key idea is to exploit the logarithmic Sobolev inequality (3.5) emphasized in the proof of Theorem 3.2 and to apply it to norm of sums of independent vector valued random variables. The convexity properties of the norm of a sum allow us to easily estimate the gradient on the right-hand side of (3.5). It yields to the following result for which we refer to [Le4] for further details. Let as before X_i, $i = 1, \ldots, n$, be independent random variables with values in some space S, and let $\mathcal{F}$ be a countable class of measurable functions on S. Set

$$Z = \sup_{f \in \mathcal{F}} \left| \sum_{i=1}^{n} f(X_i) \right|.$$

Theorem 3.6. If $|f| \leq C$ for every f in $\mathcal{F}$, and if $\mathbb{E}f(X_i) = 0$ for every $f \in \mathcal{F}$ and $i = 1, \ldots, n$, then, for all $r \geq 0$,

$$\mathbb{P}\big(Z \geq \mathbb{E}(Z) + r\big) \leq 3 \exp\left(-\frac{r}{KC} \log\left(1 + \frac{Cr}{\sigma^2 + C\mathbb{E}(Z)}\right)\right)$$

where $\sigma^2 = \sup_{f \in \mathcal{F}} \sum_{i=1}^{n} \mathbb{E}f^2(X_i)$ and $K > 0$ is a numerical constant.

This statement is as close as possible to (3.17). With respect to this inequality, the main feature is the deviation property with respect to the mean $\mathbb{E}(Z)$. Such an inequality of course belongs to the concentration phenomenon, with the two parameters $\mathbb{E}(Z)$ and σ^2 which are similar to the Gaussian case (1.24). Bounds on $\mathbb{E}(Z)$ require different tools (chaining, entropy, majorizing measures cf. [L-T]). The proof of Theorem 3.6 is a rather easy consequence of (3.5) for the Gaussian tail. It is a little bit more difficult for the Poissonian part. It is based on the integration of the following differential inequality, consequence of a logarithmic Sobolev inequality for convex functionals,

$$\lambda H'(\lambda) - H(\lambda) \log H(\lambda) \leq \lambda^2 \mathbb{E}\left(\sup_{f \in \mathcal{F}} \sum_{i=1}^{n} \big(f(X_i) - f(Y_i)\big)^2 e^{\lambda Z}\right) \tag{3.18}$$

for $\lambda \geq 0$, where, as usual, $H(\lambda) = \mathbb{E}(e^{\lambda Z})$, and where $(Y_i)_{1 \leq i \leq n}$ is an independent copy of the sequence $(Y_i)_{1 \leq i \leq n}$ (cf. [Le4]). Integration of this inequality is performed in an improved way in [Mas] yielding sharper numerical constants, that are even optimal in the case of a class consisting of non-negative functions.

Deviations under the mean (i.e. bounds for $\mathbb{P}\{Z \leq \mathbb{E}(Z) - r\}$) may be deduced similarly from the logarithmic Sobolev approach. This was overlooked in [Le4] and we are grateful to P.-M. Samson for pointing out that the argument in [Le4] actually also yields such a conclusion. Namely, since the functions in $\mathcal{F}$ are assumed to be (uniformly) bounded (by C), an elementary inspection of the arguments of [Le4] shows that (3.18) for $\lambda \leq 0$ holds with λ^2 replaced by $\lambda^2 e^{-2C\lambda}$ in front of the right-hand term. Since the Gaussian bounds (where deviation above or under the mean is really sensible) only require (3.18) for the small values of λ, the same argument is actually enough to conclude to a deviation under the mean. In particular, the bound of Theorem 3.6 also controls $\mathbb{P}\{|Z - \mathbb{E}(Z)| \geq r\}$ (up to numerical constants).

4. MODIFIED LOGARITHMIC SOBOLEV INEQUALITIES FOR LOCAL GRADIENTS

M. Talagrand discovered a few years ago [Ta3] that products of the usual exponential distribution somewhat surprisingly satisfy a concentration property which, in some respect, is stronger than Gaussian concentration. Our first aim here will be to show, following [B-L1], that this result can be seen as a consequence of some appropriate logarithmic Sobolev inequality which we call modified. Modified logarithmic Sobolev inequalities actually appear in various contexts and further examples will be presented, for discrete gradients, in the next chapter. Their main interest is that they tensorize with two parameters on the gradient, one on its supremum norm, and one on the usual quadratic norm. This feature is the appropriate explanation for the concentration property of the exponential measure.

The first paragraph is devoted to the modified logarithmic Sobolev inequality for the exponential measure. We then describe the product properties of modified logarithmic Sobolev inequalities. In the last section, we show, in a general setting, that all measures with a spectral gap (with respect to a local gradient) do satisfy the same modified inequality as the exponential distribution. Most of the results presented here are taken from the joint paper [B-L1] with S. Bobkov.

4.1 The exponential measure

In the paper [Ta3], M. Talagrand proved an isoperimetric inequality for the product measure of the exponential distribution which implies the following concentration property. Let ν^n be the product measure on $\mathbb{R}^n$ when each factor is endowed with the measure ν of density $\frac{1}{2}e^{-|x|}$ with respect to Lebesgue measure. Then, for every Borel set A with $\nu^n(A) \geq \frac{1}{2}$ and every $r \geq 0$,

$$\nu^n\left(A + \sqrt{r}B_2 + rB_1\right) \geq 1 - e^{-r/K} \tag{4.1}$$

for some numerical constant $K > 0$ where B_2 is the Euclidean unit ball and B_1 is the ℓ^1 unit ball in $\mathbb{R}^n$, i.e.

$$B_1 = \left\{ x = (x_1, \ldots, x_n) \in \mathbb{R}^n; \sum_{i=1}^{n} |x_i| < 1 \right\}.$$

A striking feature of (4.1) is that it may be used to improve some aspects of the Gaussian concentration (1.10) especially for cubes [Ta3], [Ta4]. Consider indeed the increasing map $\psi : \mathbb{R} \to \mathbb{R}$ that transform ν into the one-dimensional canonical Gaussian measure γ. It is a simple matter to check that

$$|\psi(x) - \psi(y)| \leq C \min\left(|x - y|, |x - y|^{1/2}\right), \quad x, y \in \mathbb{R}, \tag{4.2}$$

for some numerical constant $C > 0$. The map $\Psi : \mathbb{R}^n \to \mathbb{R}^n$ defined by $\Psi(x) = (\psi(x_i))_{1 \leq i \leq n}$ transforms ν^n into γ^n. Consider now Borel a set A of $\mathbb{R}^n$ such that $\gamma^n(A) \geq \frac{1}{2}$. Then

$$\gamma^n\left(\Psi\left(\Psi^{-1}(A) + \sqrt{r}B_2 + rB_1\right)\right) = \nu^n\left(\left(\Psi^{-1}(A)\right) + \sqrt{r}B_2 + rB_1\right) \geq 1 - e^{-r/K}.$$

However, it follows from (4.2) that

$$\Psi\left(\Psi^{-1}(A) + \sqrt{r}B_2 + rB_1\right) \subset A + C'rB_2.$$

Thus (4.1) improves upon (1.6). To illustrate the improvement. let

$$A = \left\{x \in \mathbb{R}^n; \max_{1 \leq i \leq n} |x_i| \leq m\right\}$$

where $m = m(n)$ is chosen so that $\gamma^n(A) \geq \frac{1}{2}$ (and hence $m(n)$ is of order $\sqrt{\log n}$). Then, when $r \geq 1$ is very small compared to $\log n$. it is easily seen that actually

$$\Psi\left(\Psi^{-1}(A) + \sqrt{r}B_2 + rB_1\right) \subset A + C_1\left(\frac{\sqrt{r}}{\sqrt{\log n}}B_2 + \frac{r}{\sqrt{\log n}}B_1\right)$$

$$\subset A + C_2\sqrt{\frac{r}{\log n}}\,rB_2.$$

As for Gaussian concentration, inequality (4.1) may be translated equivalently on functions in the following way (see the end of the section for details). For every real-valued function F on $\mathbb{R}^n$ such that $\|F\|_{\mathrm{Lip}} \leq \alpha$ and

$$|F(x) - F(y)| \leq \beta \sum_{i=1}^n |x_i - y_i|, \quad x, y \in \mathbb{R}^n,$$

for every $r \geq 0$,

$$\nu^n(F \geq m + r) \leq \exp\left(-\frac{1}{K}\min\left(\frac{r}{\beta}, \frac{r^2}{\alpha^2}\right)\right) \tag{4.3}$$

for some numerical constant $K > 0$ where m is either the mean or a median of F for ν_n. Again, this inequality extends in the appropriate sense the case of linear functions F. By Rademacher's theorem, the hypotheses on F are equivalent to saying that F is almost everywhere differentiable with

$$\sum_{i=1}^n |\partial_i F|^2 \leq \alpha^2 \quad \text{and} \quad \max_{1 \leq i \leq n} |\partial_i F| \leq \beta \quad \text{a.e..}$$

Our first aim here will be to present an elementary proof of (4.3) (and thus (4.1)) based on logarithmic Sobolev inequalities. An alternate proof, however close to Talagrand's ideas, has already been given by B. Maurey using inf-convolution [Mau2] (see also [Ta6]). M. Talagrand himself obtained recently another proof as a consequence of a stronger transportation cost inequality [Ta9] (cf. Section 3.3). Our approach is simpler even than the transportation method and is based on the results of Section 2.3. Following the procedure there in case of the exponential distribution would require to determine the appropriate logarithmic Sobolev inequality satisfied by ν^n. We cannot hope for an inequality such as the Gaussian logarithmic Sobolev inequality (2.15) to hold simply because it would imply that linear functions have a Gaussian tail for ν^n. To investigate logarithmic Sobolev inequalities for ν^n, it is enough, by the fundamental product property of entropy, to deal with the dimension one. One first inequality may be deduced from the Gaussian logarithmic Sobolev inequality. Given a smooth function f on $\mathbb{R}$, apply (2.15) in dimension 2 to $g(x, y) = f(\frac{x^2+y^2}{2})$. Let $\tilde{\nu}$ denote the one-sided exponential distribution with density e^{-x} with respect to Lebesgue measure on $\mathbb{R}_+$, and let $\tilde{\nu}^n$ denote the product measure on $\mathbb{R}_+^n$. Then

$$E_{\tilde{\nu}}(f^2) \leq 4 \int x f'(x)^2 d\tilde{\nu}(x).$$

Hence, for every smooth f on $\mathbb{R}_+^n$,

$$E_{\tilde{\nu}^n}(f^2) \leq 4 \int \sum_{i=1}^{n} x_i |\partial_i f(x)|^2 d\tilde{\nu}^n(x). \tag{4.4}$$

It does not seem however that this logarithmic Sobolev inequality (4.4) can yield the concentration property (4.3) via the Laplace transform approach of Section 2.3. In a sense, this negative observation is compatible with the fact that (4.3) improves upon some aspects of the Gaussian concentration. We thus have to look for some other version of the logarithmic Sobolev inequality for the exponential distribution. To this aim, let us observe that, at the level of Poincaré inequalities, there are two distinct inequalities. For simplicity, let us deal again only with $n = 1$. The first one, in the spirit of (4.4), is

$$\mathrm{Var}_{\tilde{\nu}}(f) \leq \int x f'(x)^2 d\tilde{\nu}(x). \tag{4.5}$$

This may be shown, either from the Gaussian Poincaré inequality as before, with however a worse constant, or by noting that the first eigenvalue of the Laguerre generator with invariant measure $\tilde{\nu}$ is 1 (cf. [K-S]. By the way, that 4 is the best constant in (4.4) is an easy consequence of our arguments. Namely, if (4.4) holds with a constant $C < 4$, a function f, on $\mathbb{R}_+$ for simplicity, such that $x f'(x)^2 \leq 1$ almost everywhere would be such that $\int e^{f^2/4} d\tilde{\nu}_1 < \infty$ by Corollary 2.6. But the example of $f(x) = 2\sqrt{x}$ contradicts this consequence. We thus recover in this simple way the main result of [K-S].) The second Poincaré inequality appeared in the work by M. Talagrand [Ta3], actually going back to [Kl], and states that

$$\mathrm{Var}_{\tilde{\nu}}(f) \leq 4 E_{\tilde{\nu}}(f'^2). \tag{4.6}$$

These two inequalities are not comparable and, in a sense, we are looking for an analogue of (4.6) for entropy.

To introduce this result, let us first recall the proof of (4.6). We will work with the double exponential distribution ν. It is plain that all the results hold, with the obvious modifications, for the one-sided exponential distribution $\tilde{\nu}$. Denote by $\mathcal{L}^n$ the space of all continuous almost everywhere differentiable functions $f : \mathbb{R}^n to \mathbb{R}$ such that $\int |f| d\nu^n < \infty$, $\int |\nabla f| d\nu^n < \infty$ and $\lim_{x_i \to \pm\infty} e^{-|x_i|} f(x_1, \ldots, x_i, \ldots x_n) = 0$ for every $i = 1, \ldots, n$ and $x_1, \ldots, x_{i-1}, x_{i+1}, \ldots, x_n \in \mathbb{R}$. The main argument of the proof is the following simple observation. If $\varphi \in \mathcal{L}^1$, by the integration by parts formula,

$$\int \varphi \, d\nu = \varphi(0) + \int \mathrm{sgn}(x) \varphi'(x) d\nu(x). \tag{4.7}$$

Lemma 4.1. *For every $f \in \mathcal{L}^1$,*

$$\mathrm{Var}_\nu(f) \leq 4\,\mathrm{E}_\nu\left(f'^2\right).$$

Proof. Set $g(x) = f(x) - f(0)$. Then, by (4.7) and the Cauchy-Schwarz inequality,

$$\mathrm{E}_\nu(g^2) = 2\int \mathrm{sgn}(x) g'(x) g(x) d\nu(x) \leq 2\left(\mathrm{E}_\nu(g'^2)\right)^{1/2} \left(\mathrm{E}_\nu(g^2)\right)^{1/2}.$$

Since $\mathrm{Var}_\nu(f) = \mathrm{Var}_\nu(g) \leq \mathrm{E}_\nu(g^2)$, and $g' = f'$, the lemma follows. □

We turn to the corresponding inequality for entropy and the main result of this section.

Theorem 4.2. *For every $0 < c < 1$ and every Lipschitz function f on $\mathbb{R}$ such that $|f'| \leq c < 1$ almost everywhere,*

$$\mathrm{Ent}_\nu\left(e^f\right) \leq \frac{2}{1-c} \mathrm{E}_\nu\left(f'^2 e^f\right).$$

Note that Theorem 4.2, when applied to functions εf as $\varepsilon \to 0$, implies Lemma 4.1. Theorem 4.2 is the first example of what we will call a modified logarithmic Sobolev inequality. We only use Theorem 4.2 for some fixed valued of c, for example $c = \frac{1}{2}$.

Proof. Changing f into $f +$ const we may assume that $f(0) = 0$. Since

$$u \log u \geq u - 1, \quad u \geq 0,$$

we have

$$\mathrm{Ent}_\nu\left(e^f\right) \leq \mathrm{E}_\nu(f e^f - e^f + 1).$$

Since $|f'| \leq \lambda < 1$ almost everywhere, the functions e^f, $f e^f$ and $f^2 e^f$ all belong to $\mathcal{L}^1$. Therefore, by repeated use of (4.7),

$$\mathrm{E}_\nu(f e^f - e^f + 1) = \int \mathrm{sgn}(x) f'(x) f(x) e^{f(x)} d\nu(x)$$

and

$$E_\nu(f^2 e^f) = 2\int \mathrm{sgn}(x)f'(x)f(x)e^{f(x)}\,d\nu(x) + \int \mathrm{sgn}(x)f'(x)f(x)^2\,e^{f(x)}d\nu(x).$$

By the Cauchy-Schwarz inequality and the assumption on f',

$$E_\nu(f^2 e^f) \le 2\big(E_\nu(f'^2 e^f)\big)^{1/2}\big(E_\nu(f^2 e^f)\big)^{1/2} + cE_\nu(f^2 e^f)$$

so that

$$E_\nu(f^2 e^f) \le \left(\frac{2}{1-c}\right)^2 E_\nu(f'^2 e^f).$$

Now, by the Cauchy-Schwarz inequality again,

$$\mathrm{Ent}_{\nu_1}(e^f) \le \int \mathrm{sgn}(x)f'(x)f(x)e^{f(x)}\,d\nu_1(x)$$

$$\le \big(E_\nu(f'^2 e^f)\big)^{1/2}\big(E_\nu(f^2 e^f)\big)^{1/2} \le \frac{2}{1-c}E_\nu(f'^2 e^f)$$

which is the result. Theorem 4.2 is established. □

We are now ready to describe the application to Talagrand's concentration inequality (4.3). As a consequence of Theorem 4.2 and of the product property of entropy (Proposition 2.2), for every smooth enough function F on $\mathbb{R}^n$ such that $\max_{1\le i\le n}|\partial_i F| \le 1$ almost everywhere and every λ, $|\lambda| \le c < 1$,

$$\mathrm{Ent}_{\nu^n}(e^{\lambda F}) \le \frac{2\lambda^2}{1-c}E_{\nu^n}\left(\sum_{i=1}^n (\partial_i F)^2 e^{\lambda F}\right). \tag{4.8}$$

Let us take for simplicity $c = \frac{1}{2}$ (although $c < 1$ might improve some numerical constants below). Assume now moreover that $\sum_{i=1}^n (\partial_i F)^2 \le \alpha^2$ almost everywhere. Then, by (4.8),

$$\mathrm{Ent}_{\nu^n}(e^{\lambda F}) \le 4\alpha^2\lambda^2 E_{\nu^n}(e^{\lambda F})$$

for every $|\lambda| \le \frac{1}{2}$. As a consequence of Corollary 2.11, we get that

$$\nu^n\big(F \ge E_{\nu^n}(F) + r\big) \le \exp\left(-\frac{1}{4}\min\left(r, \frac{r^2}{4\alpha^2}\right)\right) \tag{4.9}$$

for every $r \ge 0$. By homogeneity, this inequality amounts to (4.3) (with $K = 16$) and our claim is proved. As already mentioned, we have a similar result for the one-sided exponential measure.

To complete this section, let us sketch the equivalence between (4.1) and (4.3). (Although we present the argument for ν^n only, it extends to more general situations, as will be used in the next section.) To see that (4.1) implies (4.3), simply apply (4.1) to $A = \{F \le m\}$ where m is a median of F for ν^n and note that

$$A + \sqrt{r}B_2 + rB_1 \subset \{F \le m + \alpha\sqrt{r} + \beta r\}.$$

Using a routine argument (cf. the end of Section 1.3), the deviation inequality (4.3) from either the median or the mean are equivalent up to numerical constants (with possibly a further constant in front of the exponential function). Now starting from (4.3) with m the mean for example, consider, for $A \subset \mathbb{R}^n$ and $x = (x_1, \ldots, x_n) \in \mathbb{R}^n$,

$$F_A(x) = \inf_{a \in A} \sum_{i=1}^{n} \min(|x_i - a_i|, |x_i - a_i|^2).$$

For $r > 0$, set then $F = \min(F_A, r)$. We have $\sum_{i=1}^{n} |\partial_i F|^2 \leq 4r$ and $\max_{1 \leq i \leq n} |\partial_i F| \leq 2$ almost everywhere. Indeed, it is enough to prove this result for $G = \min(G_a, r)$ for every fixed a where

$$G_a(x) = \sum_{i=1}^{n} \min(|x_i - a_i|, |x_i - a_i|^2).$$

Now, almost everywhere, and for every $i = 1, \ldots, n$, $|\partial_i G_a(x)| \leq 2|x_i - a_i|$ if $|x_i - a_i| \leq 1$ whereas $|\partial_i G_a(x)| \leq 1$ if $|x_i - a_i| > 1$. Therefore, $\max_{1 \leq i \leq n} |\partial_i G_a(x)| \leq 2$ and

$$\sum_{i=1}^{n} |\partial_i G_a(x)|^2 \leq 4 \sum_{i=1}^{n} \min(|x_i - a_i|, |x_i - a_i|^2) = 4 G_a(x)$$

which yields the announced claim. Now, if $\nu^n(A) \geq \frac{1}{2}$,

$$E_{\nu^n}(F) \leq r(1 - \nu^n(A)) \leq \frac{r}{2}.$$

It then follows from (4.3) that

$$\nu^n(F_A \geq r) = \nu^n(F \geq r) \leq \nu^n\left(F \geq E_{\nu^n}(F) + \frac{r}{2}\right) \leq e^{-r/16K}.$$

Since $\{F_A \leq r\} \subset A + \sqrt{r} B_2 + r B_1$, the result follows.

4.2 Modified logarithmic Sobolev inequalities

The inequality put forward in Theorem 4.2 for the exponential measure is a first example of what we call modified logarithmic Sobolev inequalities. In order to describe this notion in some generality, we take again the general setting of Part 2. Let thus $(X, \mathcal{B}, \mu)$ be a probability space, and let $\mathcal{A}$ be a subset of $L^1(\mu)$. Consider a "gradient" operator Γ on $\mathcal{A}$ such that $\Gamma(f) \geq 0$ and $\Gamma(\lambda f) = \lambda^2 \Gamma(f)$ for every $f \in \mathcal{A}$ and $\lambda \in \mathbb{R}$. Examples are $\Gamma(f) = |\nabla f|^2$ for a generalized modulus of gradient (2.3), or $\Gamma(f) = |Df|^2$ for a discrete gradient (such as (2.9)).

Definition 4.3. We say that μ satisfies a modified logarithmic Sobolev inequality with respect to Γ (on $\mathcal{A}$) if there is a function $B(\lambda) \geq 0$ on $\mathbb{R}_+$ such that, whenever $\|\Gamma(f)\|_\infty^{1/2} \leq \lambda$,

$$\text{Ent}_\mu(e^f) \leq B(\lambda) E_\mu(\Gamma(f) e^f)$$

for all f in $\mathcal{A}$ such that $E_\mu(e^f) < \infty$.

According to Theorem 4.2, the exponential measure ν on the line satisfies a modified logarithmic Sobolev inequality with respect to the usual gradient with $B(\lambda)$ bounded for the small values of λ. On the other hand, the Gaussian measure γ satisfies a modified logarithmic Sobolev inequality with $B(\lambda) = \frac{1}{2}$, $\lambda \geq 0$.

Definition 4.3 might appear very similar to the inequalities investigated via Proposition 2.9. Actually, Definition 4.3 implies that

$$\mathrm{Ent}_\mu(e^f) \leq \lambda^2 B(\lambda) \, \mathrm{E}_\mu(e^f)$$

for every f with $\|\Gamma(f)\|_\infty \leq \lambda$. In particular, if $B(\lambda)$ is bounded for the small values of λ, Lipschitz functions will have an exponential tail according to Corollary 2.12.

The main new feature here is that the modified logarithmic Sobolev inequality of Definition 4.3 tensorizes in terms of two parameters rather than only the Lipschitz bound. This property is summarized in the next proposition which is an elementary consequence of the product property of entropy (Proposition 2.2).

Let $(X_i, \mathcal{B}_i, \mu_i)$, $i = 1, \ldots, n$, be probability spaces and denote by $P = \mu_1 \otimes \cdots \otimes \mu_n$ on the product space $X = X_1 \times \cdots \times X_n$. Consider operators Γ_i on classes $\mathcal{A}_i$, $i = 1, \ldots, n$. If f is a function on the product space, for each i, f_i is the function f depending on the i-th variable with the other coordinates fixed.

Proposition 4.4. *Assume that for every f on $(X_i, \mathcal{B}_i, \mu_i)$ such that $\|\Gamma_i(f)\|_\infty^{1/2} \leq \lambda$,*

$$\mathrm{Ent}_{\mu_i}(e^f) \leq B(\lambda) \, \mathrm{E}_{\mu_i} \left(\Gamma_i(f) \, e^f \right),$$

$i = 1, \ldots, n$. Then, for every f on the product space such that $\max_{1 \leq i \leq n} \|\Gamma_i(f_i)\|_\infty^{1/2} \leq \lambda$,

$$\mathrm{Ent}_P(e^f) \leq B(\lambda) \, \mathrm{E}_P \left(\sum_{i=1}^n \Gamma_i(f_i) \, e^f \right).$$

According to the behavior of $B(\lambda)$, this proposition yields concentration properties in terms of two parameters,

$$\max_{1 \leq i \leq n} \|\Gamma_i(f_i)\|_\infty^{1/2} \quad \text{and} \quad \left\| \sum_{i=1}^n \Gamma_i(f_i) \right\|_\infty.$$

For example, if $B(\lambda) \leq c$ for $0 \leq \lambda \leq \lambda_0$, following the proof of (4.9), the product measure P will satisfy the same concentration inequality as the one for the exponential measure (4.3). In the next chapter, we investigate cases such as $B(\lambda) \leq ce^{d\lambda}$, $\lambda \geq 0$, related to the Poisson measure. Rather than to discuss some further abstract result according to the behavior of $B(\lambda)$ (in the spirit of Corollaries 2.11 and 2.12), we refer to Corollary 4.6 and Theorem 5.5 for examples of applications.

4.3 Poincaré inequalities and modified logarithmic Sobolev inequalities

In this section, we show that the concentration properties of the exponential measure described in Section 4.1 is actually shared by all measures satisfying a Poincaré inequality (with respect to a local gradient). More precisely, we show, following [B-L1], that every such measure satisfies the modified logarithmic Sobolev inequality of Theorem 4.2.

Let thus $|\nabla f|$ be a generalized modulus of gradient on a metric space (X, d), satisfying thus the chain rule formula (2.5). Throughout this paragraph, we assume that μ is a probability measure on X equipped with the Borel σ-field $\mathcal{B}$ such that for some $C > 0$ and all f in $L^2(\mu)$,

$$\mathrm{Var}_\mu(f) \leq C\, \mathrm{E}_\mu\big(|\nabla f|^2\big). \tag{4.10}$$

We already know from Proposition 2.13 that such a spectral gap inequality implies exponential integrability of Lipschitz functions. We actually show that it also implies a modified logarithmic Sobolev inequality which yields concentration properties for the product measures μ^n.

Theorem 4.5. *For any function f on X such that $\|\nabla f\|_\infty \leq \lambda < 2/\sqrt{C}$,*

$$\mathrm{Ent}_\mu(e^f) \leq B(\lambda)\, \mathrm{E}_\mu\big(|\nabla f|^2 e^f\big)$$

where

$$B(\lambda) = \frac{C}{2}\left(\frac{2 + \lambda\sqrt{C}}{2 - \lambda\sqrt{C}}\right)^2 e^{\sqrt{5C}\lambda}.$$

We refer to the paper [B-L1] for the proof of Theorem 4.5.

Now, $B(\lambda)$ is uniformly bounded for the small values of λ, for example $B(\lambda) \leq 3e^5 C/2$ when $\lambda \leq 1/\sqrt{C}$. As a corollary, we obtain, following the proof of (4.9) and the discussion on Proposition 4.4, a concentration inequality of Talagrand's type for the product measure μ^n of μ on X^n. If f is a function on the product space X^n, denote by $|\nabla_i f|$ the length of the gradient with respect to the i-th coordinate.

Corollary 4.6. *Denote by μ^n the product of μ on X^n. Then, for every function F on X^n such that*

$$\sum_{i=1}^n |\nabla_i F|^2 \leq \alpha^2 \quad \text{and} \quad \max_{1 \leq i \leq n} |\nabla_i F| \leq 3$$

μ-almost everywhere, $\mathrm{E}_{\mu^n}(|F|) < \infty$ and

$$\mu^n\big(F \geq \mathrm{E}_{\mu^n}(F) + r\big) \leq \exp\left(-\frac{1}{K}\min\left(\frac{r}{3}, \frac{r^2}{\alpha^2}\right)\right)$$

where $K > 0$ only depends on the constant C in the Poincaré inequality (4.10).

One may obtain a similar statement for products of possibly different measures μ with a uniform lower bound on the constants in the Poincaré inequalities (4.10).

Following the argument at the end of Section 4.1, Corollary 4.6 may be turned into an inequality on sets such as (4.1). More precisely, if $\mu^n(A) \geq \frac{1}{2}$, for every $r \geq 0$ and some numerical constant $K > 0$,

$$\mu^n\big(F_A^h \geq r\big) \leq e^{-r/K}.$$

where $h(x,y) = \min(d(x,y), d(x,y)^2)$, $x, y \in X$, and, for $x = (x_1, \ldots, x_n) \in X^n$ and $A \subset X^n$,

$$F_A^h(x) = \inf_{a \in A} \sum_{i=1}^n h(x_i, a_i).$$

Using analogues of the norm $||| \cdot |||_\infty$, Theorem 4.5 and Corollary 4.6 have been recently extended in [H-T] to the example of the invariant measure of a reversible Markov chain on a finite state space. The main idea consists in showing that the various uses of the chain rule formula in the proof of Theorem 4.6 may be properly extended to this case (see also [A-S] for extensions of the chain rule formula).

Let us observe that for the case of the exponential measure ν, $C = 4$ by Lemma 4.1 so that, for $\lambda < 1$,

$$B(\lambda) = 2\left(\frac{1+\lambda}{1-\lambda}\right)^2 e^{2\sqrt{5}\lambda}$$

which is somewhat worse than the constant given by Theorem 4.2.

In any case, an important feature of the constant $B(\lambda)$ of Theorem 4.5 is that $B(\lambda) \to C/2$ as $\lambda \to 0$. In particular (and as in Theorem 4.2). the modified logarithmic Sobolev inequality of Theorem 4.5 implies the Poincaré inequality (4.10) by applying it to functions εf with $\varepsilon \to 0$. Poincaré inequality and the modified logarithmic Sobolev inequality of Theorem 4.5 are thus equivalent.

On the other hand, let us consider the case of the canonical Gaussian measure γ on the real line for which, by (2.16),

$$\mathrm{Var}_\gamma(f) \leq \mathrm{E}_\gamma(f'^2).$$

Let φ be a smooth function on $\mathbb{R}$, for example C^2 with bounded derivatives. Apply the multidimensional analogue (Proposition 4.4) of Theorem 4.5 to the functions

$$f(x) = \varphi\left(\frac{x_1 + \cdots + x_n}{\sqrt{n}}\right), \quad x = (x_1, \ldots, x_n) \in \mathbb{R}^n,$$

for which $\max_{1 \leq i \leq n} |\partial_i f| \leq \|\varphi\|_{\mathrm{Lip}}/\sqrt{n} = \beta_n < 2$ for n large enough. By the rotational invariance of Gaussian measures, and since $\beta_n \to 0$, we get in the limit

$$\mathrm{Ent}_\gamma(e^\varphi) \leq \frac{1}{2} \mathrm{E}_\gamma(\varphi'^2 e^\varphi)$$

that is (after the change of functions $e^\varphi = g^2$) Gross's logarithmic Sobolev inequality (2.15) with optimal constant. Therefore, for the Gaussian measure, Poincaré and logarithmic Sobolev inequalities are in a sense equivalent.

5. MODIFIED LOGARITHMIC SOBOLEV INEQUALITIES IN DISCRETE SETTINGS

We investigate here concentration and logarithmic Sobolev inequalities for discrete gradients which typically do not satisfy a chain rule formula. One such example considered here is $Df(x) = f(x+1) - f(x)$, $x \in \mathbb{N}$. With respect to such gradient, natural measures such as Poisson measures do not satisfy a logarithmic Sobolev inequality in its classical formulation, but rather some modified inequality. Following the recent works [B-L2] and [G-R] on the subject, we study mainly here Poisson type logarithmic Sobolev inequalities and their related concentration properties. The results of this part are taken from the paper [B-L2] with S. Bobkov.

5.1 Logarithmic Sobolev inequality for Bernoulli and Poisson measures

As was presented in Section 2.2, in his seminal 1975 paper, L. Gross [Gr1] proved a logarithmic Sobolev inequality on the two-point space. Namely, let μ be the uniform measure on $\{0, 1\}$. Then, for any f on $\{0, 1\}$,

$$\text{Ent}_\mu(f^2) \leq \frac{1}{2} \, \text{E}_\mu(|Df|^2) \tag{5.1}$$

where

$$Df(x) = f(1) - f(0) = f(x+1) - f(x)$$

(x modulo 2). It is easily seen that the constant $\frac{1}{2}$ is optimal.

The question of the best constant in the previous logarithmic Sobolev inequality for non-symmetric Bernoulli measure was settled seemingly only quite recently. Let μ_p be the Bernoulli measure on $\{0, 1\}$ with $\mu_p(\{1\}) = p$ and $\mu_p(\{0\}) = q = 1 - p$. Then, for any f on $\{0, 1\}$,

$$\text{Ent}_{\mu_p}(f^2) \leq pq \, \frac{\log p - \log q}{p - q} \, \text{E}_{\mu_p}(|Df|^2). \tag{5.2}$$

The constant is optimal, and is equal to $\frac{1}{2}$ when $p = q = \frac{1}{2}$. This result is mentioned in [H-Y] without proof, and worked out in [D-SC]. A simple proof, due S. Bobkov, is presented in the notes [SC2]. O. Rothaus mentioned to the authors of [D-SC] that he computed this constant several years back from now. The main feature of this constant is that, when $p \neq q$, it significantly differs from the spectral gap given by the inequality

$$\text{Var}_{\mu_p}(f) \leq pq \, \text{E}_{\mu_p}(|Df|^2). \tag{5.3}$$

Although inequality (5.2) is optimal, it presents a number of weak points. First of all, the product property of entropy which allows us, together with the central limit theorem, to deduce the logarithmic Sobolev inequality for Gaussian measures from the one for Bernoulli is optimal in the symmetric case. As soon as $p \neq q$, the central limit theorem on the basis of (5.2) only yields the Gaussian logarithmic Sobolev inequality (2.15) with a worse constant. A second limit theorem of interest is of course the Poisson limit. However, after tensorization, (5.2) cannot yield a logarithmic Sobolev inequality for Poisson measures. (Although the constant in (5.2) is bounded as $p \to 0$, we would need it to be of the order of p for $p \to 0$.) There is of course a good reason at that, namely that Poisson measures do not satisfy logarithmic Sobolev inequalities! This is well known to a number of people but let us briefly convince ourselves of this claim. Denote thus by π_θ the Poisson measure on $\mathbb{N}$ with parameter $\theta > 0$ and let us assume that, for some constant $C > 0$, and all f, say bounded, on $\mathbb{N}$,

$$\text{Ent}_{\pi_\theta}(f^2) \leq C \, \text{E}_{\pi_\theta}(|Df|^2) \tag{5.4}$$

where here $Df(x) = f(x+1) - f(x)$, $x \in \mathbb{N}$. Apply (5.4) to the indicator function of the interval $[k+1, \infty)$ for each $k \in \mathbb{N}$. We get

$$-\pi_\theta([k+1, \infty)) \log \pi_\theta([k+1, \infty)) \leq C\pi_\theta(\{k\})$$

which is clearly impossible as k goes to infinity. Similarly, (5.4) cannot hold with the addition of an extra $C\text{E}_\theta(f^2)$ on the right-hand side. It is important for the further developments to notice, according to [G-R], that the exponential integrability results of Section 2.3 with the norm $|||\cdot|||_\infty$ cannot be used at this point to rule out (5.4). Indeed, (5.4) implies via (2.8) that

$$\text{Ent}_{\pi_\theta}(e^f) \leq \frac{C}{2} |||f|||_\infty^2 \text{E}_{\pi_\theta}(e^f).$$

By (2.11),

$$|||f|||_\infty^2 = \sup_{x \in \mathbb{N}} \left(Df(x)^2 + \frac{x}{\theta} Df(x-1)^2 \right).$$

As an application of Corollary 2.4, if F on $\mathbb{N}$ is such that $|||F|||_\infty \leq 1$, we would conclude from the logarithmic Sobolev inequality (5.4) that $\text{E}_{\pi_\theta}(e^{\alpha F^2}) < \infty$ for some $\alpha > 0$. But now, if $|||F|||_\infty \leq 1$, then $DF(x) \leq \sqrt{\frac{\theta}{x+1}}$ for every x. This directly implies that $\text{E}_{\pi_\theta}(e^{\alpha F^2}) < \infty$ for every α which thus would not contradict Corollary 2.4. The norm $|||F|||_\infty$ is therefore not well adapted to our purposes here, and we will rather consider $\sup_{x \in \mathbb{N}} |DF(x)|$ under which we will describe exponential integrability of Poisson type.

One may therefore be led to consider some variations of inequality (5.2) that could behave better under the preceding limits, in particular one could think of modified logarithmic Sobolev inequalities. However, we follow a somewhat different route and turn to an alternate variation of possible own interest.

An equivalent formulation of the Gaussian logarithmic Sobolev inequality (2.15), on the line for simplicity, is that, for any smooth f on $\mathbb{R}$ with strictly positive values,

$$\mathrm{Ent}_\gamma(f) \leq \frac{1}{2} \mathrm{E}_\gamma\left(\frac{1}{f} f'^2\right). \tag{5.5}$$

That (5.5) is equivalent to (2.15) simply follows from a change of functions together with the chain rule formula for the usual gradient on $\mathbb{R}$. Of course, such a change may not be performed equivalently on discrete gradients, so that there is some interest to study an inequality such as

$$\mathrm{Ent}_{\mu_p}(f) \leq C\, \mathrm{E}_{\mu_p}\left(\frac{1}{f}|Df|^2\right) \tag{5.6}$$

on $\{0,1\}$ for the Bernoulli measure μ_p and to ask for the best constant C as a function of p. Our first result will be to show that the best constant C in (5.6) is pq. The behavior in p is thus much better than in (5.2) as $p \to 0$ or 1. and will allow us to derive a logarithmic Sobolev inequality for Poisson measure in the limit. The following is taken from the recent work [B-L2]. An alternate proof of Theorem 5.1 and Corollary 5.3 below using the Γ_2 calculus of [Ba1], [Ba2] may be found in [A-L].

For any $n \geq 1$, we denote by μ_p^n the product measure of μ_p on $\{0,1\}^n$. If f is a function on $\{0,1\}^n$, and $x = (x_1, \ldots, x_n) \in \{0,1\}^n$, set

$$|Df|^2(x) = \sum_{i=1}^{n} |f(x + e_i) - f(x)|^2$$

where $(e_1, \ldots, e_n)$ is the canonical basis of $\mathbb{R}^n$ and the addition is modulo 2. p is arbitrary in $[0,1]$, and $q = 1 - p$.

Theorem 5.1. *For any positive function f on $\{0,1\}^n$,*

$$\mathrm{Ent}_{\mu_p^n}(f) \leq pq\, \mathrm{E}_{\mu_p^n}\left(\frac{1}{f}|Df|^2\right).$$

Proof. By the product property of entropy, it is enough to deal with the case $n = 1$. The proof is based on the following calculus lemma.

Lemma 5.2 *Consider a function*

$$U(p) = \mathrm{Ent}_{\mu_p}(f) - pq\, \mathrm{E}_{\mu_p}(g), \quad 0 \leq p \leq 1,$$

where f and g are arbitrary non-negative functions on $\{0,1\}$. Then $U(p) \leq 0$ for every p if and only if

$$(5.7) \qquad\qquad U'(0) \leq 0 \leq U'(1).$$

If, additionally, $f(0) \geq f(1)$ and $g(0) \geq g(1)$ (respectively $f(0) \leq f(1)$ and $g(0) \leq g(1)$), then the condition (5.7) may be weakened into $U'(0) \leq 0$ (respectively $U'(1) \geq 0$).

Proof. Set $a = f(1)$, $b = f(0)$, $\alpha = g(1)$, $\beta = g(0)$, so that

$$U(p) = (pa \log a + qb \log b) - (pa + qb) \log(pa + qb) - pq(p\alpha + q\beta).$$

Since $U(0) = U(1) = 0$, the condition (5.7) is necessary for U to be non-positive. Now, assume (5.7) is fulfilled. Differentiating in p, we have

$$U'(p) = (a \log a - b \log b) - (a - b)(\log(pa + qb) + 1)$$
$$+ (p - q)(p\alpha + q\beta) - pq(\alpha - \beta),$$
$$U''(p) = -(a - b)^2 (pa + qb)^{-1} + 2(p\alpha + q\beta) + 2(p - q)(\alpha - \beta),$$
$$U'''(p) = (a - b)^3 (pa + qb)^{-2} + 6(\alpha - \beta),$$
$$U''''(p) = -2(a - b)^4 (pa + qb)^{-3}.$$

Since $U'''' \leq 0$, U'' is concave. Hence, formally three situations are possible.

1) $U'' \geq 0$ on [0,1]. In this case, U is convex and thus $U \leq 0$ on [0,1] in view of $U(0) = U(1) = 0$.

2) $U'' \leq 0$ on [0,1]. By (5.7), this case is not possible unless U is identically 0.

3) For some $0 \leq p_0 < p_1 \leq 1$, $U'' \leq 0$ on $[0, p_0]$, $U'' \geq 0$ on $[p_0, p_1]$, and $U'' \leq 0$ on $[p_1, 1]$. In this case, U is concave on $[0, p_0]$, and, due to the assumption $U'(0) \leq 0$, one may conclude that U is non-increasing on $[0, p_0]$. In particular, $U \leq 0$ on $[0, p_0]$. It is then necessary that $U(p_1) \leq 0$. Indeed, U is concave on $[p_1, 1]$, hence the assumption $U(p_1) > 0$ together with $U(1) = 0$ would imply $U'(1) < 0$ which contradicts (5.7). As a result, by convexity of U on $[p_0, p_1]$, we get $U \leq 0$ on $[p_0, p_1]$. At last, $U \leq 0$ on $[p_1, 1]$, since U is concave on $[p_1, 1]$, $U(p_1) \leq 0$ and $U'(1) \geq 0$ (in particular, U is non-decreasing on this interval). The first part of Lemma 2 is thus proved.

We turn to the second part. Again, since $U(0) = U(1) = 0$, any of the conditions $U'(0) \leq 0$ or $U'(1) \geq 0$ is necessary for U to be non-positive on [0, 1]. Now, assume that $a \geq b$, $\alpha \geq \beta$, and $U'(0) \leq 0$ (the other case is similar). Then $U''' \geq 0$, and hence U'' is non-decreasing on [0,1]. Again three cases are formally possible.

1) $U'' \geq 0$ on [0,1]. In this case, U is convex, and thus $U \leq 0$ on [0,1] in view of $U(0) = U(1) = 0$.

2) $U'' \leq 0$ on [0,1]. This can only occur if $U \equiv 0$.

3) For some $0 \leq p_0 \leq 1$, $U'' \leq 0$ on $[0, p_0]$ and $U'' \geq 0$ on $[p_0, 1]$. In this case, U is concave on $[0, p_0]$, and, due to the fact that $U'(0) \leq 0$, one may conclude that U is non-increasing on $[0, p_0]$. In particular $U \leq 0$ on $[0, p_0]$. At last, $U \leq 0$ on $[p_0, 1]$ since U is convex on this interval and $U(p_0) \leq 0$ and $U(1) = 0$. Lemma 2 is established. □

We turn to the proof of Theorem 5.1. Note first the following. In the notation of the proof of Lemma 5.2, set

$$R(a, b) = a \log \tfrac{a}{b} - (a - b).$$

Clearly, $R(a,b) \geq 0$ for all $a, b > 0$. Then,

$$U'(0) \leq 0 \quad \text{if and only if} \quad 3 \geq R(a,b) \tag{5.8}$$

while

$$U'(1) \geq 0 \quad \text{if and only if} \quad \alpha \geq R(b,a) \tag{5.9}$$

Fix f with stricly positive values on $\{0, 1\}$. Apply then Lemma 5.2 to $g = \delta/f$. $\delta > 0$. According to (5.8) and (5.9), the optimal value of $\delta > 0$ in the inequality

$$\text{Ent}_{\mu_p}(f) \leq \delta pq \, \text{E}_{\mu_p}\left(\frac{1}{f}\right) \tag{5.10}$$

provided $p \in [0, 1]$ is arbitrary is given by

$$\delta = \max\{bR(a,b), aR(b,a)\},$$

where $a = f(1)$, $b = f(0)$. By symmetry, one may assume that $a \geq b \geq 0$. Then, $bR(a,b) \leq aR(b,a)$. Indeed, for fixed $b > 0$, the function $\rho(a) = aR(b,a) - bR(a,b)$ has derivative $\rho'(a) = 2R(b,a) \geq 0$. Hence, $\rho(a) \geq \rho(b) = 0$. Thus, $\delta = aR(b,a)$, $a > b > 0$. Now, fixing $b > 0$, consider

$$u(a) = aR(b,a) = a\left(b \log \frac{b}{a} - (b - a)\right), \quad a > b.$$

We have $u'(a) = b \log \frac{b}{a} - 2(b - a)$, thus $u(b) = u'(b) = 0$ and, for every $a > 0$,

$$u''(a) = 2 - \frac{b}{a} \leq 2.$$

Hence, by a Taylor expansion, denoting by a_0 some middle point between a and b. we get

$$\delta = u(a) = u(b) + u'(b)(a - b) + \frac{1}{2}u''(a_0)(a - b)^2 \leq \left(1 - \frac{b}{2a}\right)(a - b)^2.$$

Therefore, $\delta \leq (a - b)^2 = |f(1) - f(0)|^2$ in (5.10) which is the result. Theorem 5.1 is established. □

Observe that in the process of the proof of Theorem 5.1, we actually proved a somewhat better inequality. Namely, for any positive function f on $\{0, 1\}$.

$$\text{Ent}_{\mu_p}(f) \leq pq\left(1 - \frac{1}{2M(f)}\right)\text{E}_{\mu_p}\left(\frac{1}{f}|Df|^2\right)$$

where

$$M(f) = \max\left\{\frac{f(1)}{f(0)}, \frac{f(0)}{f(1)}\right\}.$$

By the product property of entropy, for any f with strictly positive values on $\{0, 1\}^n$.

$$\text{Ent}_{\mu_p^n}(f) \leq pq\left(1 - \frac{1}{2M(f)}\right)\text{E}_{\mu_p^n}\left(\frac{1}{f}|Df|^2\right) \tag{5.11}$$

where

$$M(f) = \max_{x \in \{0,1\}^n} \max_{1 \le i \le n} \frac{f(x + e_i)}{f(x)}.$$

As announced, the logarithmic Sobolev inequality of Theorem 5.1 may be used in the limit to yield a logarithmic Sobolev inequality for Poisson measure. Take namely φ on $\mathbb{N}$ such that $0 < c \le \varphi \le C < \infty$ and apply Theorem 1 to

$$f(x) = f(x_1, \ldots, x_n) = \varphi(x_1 + \cdots + x_n), \quad x = (x_1, \ldots, x_n) \in \{0,1\}^n,$$

with this time $p = \frac{\theta}{n}$, $\theta > 0$ (for every n large enough). Then, setting $S_n = x_1 + \cdots + x_n$,

$$|Df|^2(x) = (n - S_n)\big[\varphi(S_n + 1) - \varphi(S_n)\big]^2 + S_n\big[\varphi(S_n) - \varphi(S_n - 1)\big]^2.$$

Therefore,

$$\mathrm{Ent}_{\mu_p^n}(\varphi(S_n)) \le \frac{\theta}{n}\left(1 - \frac{\theta}{n}\right) \mathrm{E}_{\mu_p^n}\left(\frac{1}{\varphi(S_n)}\Big((n - S_n)\big[\varphi(S_n + 1) - \varphi(S_n)\big]^2\right.$$
$$\left. + S_n\big[\varphi(S_n) - \varphi(S_n - 1)\big]^2\Big)\right).$$

The distribution of S_n under $\mu_{\theta/n}^n$ converges to π_θ. Using that $0 < c \le \varphi \le C < \infty$ and that $\frac{1}{n}\mathrm{E}_{\mu_p^n}(S_n) \to 0$, we immediately obtains the following corollary.

Corollary 5.3. *For any f on $\mathbb{N}$ with strictly positive values,*

$$\mathrm{Ent}_{\pi_\theta}(f) \le \theta \, \mathrm{E}_{\pi_\theta}\left(\frac{1}{f}|Df|^2\right)$$

where we recall that here $Df(x) = f(x + 1) - f(x)$, $x \in \mathbb{N}$.

The example of $f(x) = e^{-cx}$, $x \in \mathbb{N}$, as $c \to \infty$ shows that one cannot expect a better factor of θ in the preceding corollary.

Theorem 5.1 may also be used to imply the Gaussian logarithmic Sobolev inequality up to a constant 2. Actually, using the refined inequality (5.11), we can reach the optimal constant. Let indeed $\varphi > 0$ be smooth enough on $\mathbb{R}$, for example C^2 with bounded derivatives, and apply (5.11) to

$$f(x_1, \ldots, x_n) = \varphi\left(\frac{x_1 + \cdots + x_n - np}{\sqrt{npq}}\right)$$

for fixed p, $0 < p < 1$. Under the smoothness properties on φ, it is easily seen that $M(f) \to 1$ as $n \to \infty$. Therefore, by the Gaussian central limit theorem, we deduce in the classical way inequality (5.5) for φ. Changing φ into φ^2, and using a standard approximation procedure, we get Gross's logarithmic Sobolev inequality (2.15) with its best constant. Another consequences of this sharp form are the spectral gap inequalities for μ_p^n and π_θ. Applying (5.11) to $1 + \varepsilon f$ and letting ε go to 0, we get, since $M(1 + \varepsilon f) \to 1$,

$$\mathrm{Var}_{\mu_p^n}(f) \le pq \, \mathrm{E}_{\mu_p^n}(|Df|^2) \tag{5.12}$$

and

$$\text{Var}_{\pi_\theta}(f) \le \theta \, \text{E}_{\pi_\theta}(|Df|^2). \tag{5.13}$$

5.2 Modified logarithmic Sobolev inequalities and Poisson tails

In analogy with the Gaussian concentration properties of Section 2.3, the logarithmic Sobolev inequalities of the type of those of Theorem 5.1 and Corollary 5.3 entail some information on the Poisson behavior of Lipschitz functions. For simplicity, we only deal with the case of measures on $\mathbb{N}$. According to the preceding section, the results below apply in particular to the Poisson measure π_θ.

Let μ be a probability measure on $\mathbb{N}$ such that, for some constant $C > 0$,

$$\text{Ent}_\mu(f) \le C \, \text{E}_\mu\left(\frac{1}{f}|Df|^2\right) \tag{5.14}$$

for all functions f on $\mathbb{N}$ with positive values, where $Df(x) = f(x+1) - f(x)$, $x \in \mathbb{N}$. As usual, we would like to apply (5.14) to e^f. In this discrete setting, $|D(e^f)| \le |Df|e^f$ is obviously false in general. However,

$$|D(e^f)| \le |Df|e^{|Df|}e^f. \tag{5.15}$$

Indeed, for every $x \in \mathbb{N}$,

$$|D(e^f)(x)| = |e^{f(x+1)} - e^{f(x)}| = |Df(x)|e^\tau$$

for some $\tau \in]f(x), f(x+1)[$ or $]f(x+1), f(x)[$. Since $\tau \le f(x) + |Df(x)|$, the claims follows. Let now f on $\mathbb{N}$ be such that $\sup_{x \in \mathbb{N}} |Df(x)| \le \lambda$. It follows from (5.14) and (5.15) that

$$\text{Ent}_\mu(e^f) \le Ce^{2\lambda}\text{E}_\mu(|Df|^2 e^f). \tag{5.16}$$

In particular,

$$\text{Ent}_\mu(e^f) \le C\lambda^2 e^{2\lambda}\text{E}_\mu(e^f). \tag{5.17}$$

As a consequence of Corollary 2.12, we obtain a first result on Poisson tails of Lipschitz functions.

Proposition 5.4. *Let μ be a probability measure on $\mathbb{N}$ such that, for some constant $C > 0$,*

$$\text{Ent}_\mu(f) \le C \, \text{E}_\mu\left(\frac{1}{f}|Df|^2\right)$$

for all functions f on $\mathbb{N}$ with positive values, where $Df(x) = f(x+1) - f(x)$, $x \in \mathbb{N}$. Then, for any F such that $\sup_{x \in \mathbb{N}} |DF(x)| \le 1$, we have $\text{E}_\mu(|F|) < \infty$ and, for all $r \ge 0$,

$$\mu(F \ge \text{E}_\mu(F) + r) \le \exp\left(-\frac{r}{4}\log\left(1 + \frac{r}{2C}\right)\right).$$

In particular, $\text{E}_\mu(e^{\alpha|F|\log_+|F|}) < \infty$ for sufficiently small $\alpha > 0$.

The inequality of Proposition 5.4 describes the classical Gaussian tail behavior for the small values of r and the Poisson behavior for the large values of r (with respect to C). The constants have no reason to be sharp.

Of course, inequality (5.16) is part of the family of modified logarithmic Sobolev inequalities investigated in Section 4.2, with a function $B(\lambda)$ of the order of $e^{2\lambda}$, $\lambda \geq 0$. According to Proposition 4.4, it may be tensorized in terms of two distinct norms on the gradients. The following statement is then an easy consequence of this observation.

Theorem 5.5. *Let μ be some measure on $\mathbb{N}$. Assume that for every f on $\mathbb{N}$ with $\sup_{x \in \mathbb{N}} |Df(x)| \leq \lambda$,*

$$\text{Ent}_\mu(e^f) \leq B(\lambda)\, \mathbb{E}_\mu(|Df|^2 e^f) \tag{5.18}$$

where, as function of $\lambda \geq 0$,

$$B(\lambda) \leq c e^{d\lambda}$$

for some $c, d > 0$. Denote by μ^n the product measure on $\mathbb{N}^n$. Let F be a function on $\mathbb{N}^n$ such that, for every $x \in \mathbb{N}^n$,

$$\sum_{i=1}^{n} |F(x + e_i) - F(x)|^2 \leq \alpha^2 \quad \text{and} \quad \max_{1 \leq i \leq n} |F(x + e_i) - F(x)| \leq \beta.$$

Then $\mathbb{E}_{\mu^n}(|F|) < \infty$ and, for every $r \geq 0$,

$$\mu^n(F \geq \mathbb{E}_{\mu^n}(F) + r) \leq \exp\left(-\frac{r}{2d\beta}\log\left(1 + \frac{\beta d r}{4c\alpha^2}\right)\right).$$

Proof. We tensorize (5.18) according to Proposition 4.4 to get that for every f on $\mathbb{N}^n$ such that $\max_{1 \leq i \leq n} |f(x + e_i) - f(x)| \leq \lambda$ for every $x \in \mathbb{N}^n$,

$$\text{Ent}_{\mu^n}(e^f) \leq B(\lambda)\, \mathbb{E}_{\mu^n}\left(\sum_{i=1}^{n} |D_i f|^2 e^f\right) \tag{5.19}$$

where $D_i f(x) = f(x + e_i) - f(x)$, $i = 1, \ldots, n$. We then proceed exactly as in Corollary 2.12. Fix F on $\mathbb{N}^n$ satisfying the hypotheses of the statement. We may assume, by homogeneity, that $\beta = 1$. Furthermore, arguing as in Section 2.3, we may assume throughout the argument that F is bounded. Apply (5.19) to λF for every $\lambda \in \mathbb{R}$. Setting $H(\lambda) = \mathbb{E}_{\mu^n}(e^{\lambda F})$, we get

$$\lambda H'(\lambda) - H(\lambda) \log H(\lambda) \leq \alpha^2 \lambda^2 B(\lambda) H(\lambda).$$

Therefore, with, as usual, $K(\lambda) = \frac{1}{\lambda} \log H(\lambda)$,

$$K'(\lambda) \leq \alpha^2 B(\lambda) \leq \alpha^2 c e^{d\lambda}.$$

It follows that, for every $\lambda \geq 0$,

$$K(\lambda) \leq K(0) + \alpha^2 \frac{c}{d}(e^{d\lambda} - 1).$$

In other words,

$$E_{\mu^n}(e^{\lambda F}) \leq e^{\lambda E_{\mu^n}(F) + c\alpha^2 \lambda (e^{d\tau} - 1)/d} \tag{5.20}$$

which holds for every $\lambda \in \mathbb{R}$ (changing F into $-F$.) We conclude with Chebyshev's exponential inequality. For every λ,

$$\mu^n(F \geq E_{\mu^n}(F) + r) \leq e^{-\lambda r + c\alpha^2 \lambda (e^{d\lambda} - 1)/d}.$$

If $dr \leq 4c\alpha^2$ (for example), choose $\lambda = r/4c\alpha^2$ whereas when $dr \geq 4c\alpha^2$, take

$$\lambda = \frac{1}{d} \log\left(\frac{dr}{2c\alpha^2}\right).$$

The proof is easily completed. □

5.3 Sharp bounds

To conclude this work, we study the sharp form of the modified logarithmic Sobolev inequalities for Bernoulli and Poisson measures. As in Section 5.1. we start with the Bernoulli measure. The following statement will be our basic result.

Theorem 5.6. *For any function f on $\{0,1\}^n$,*

$$\text{Ent}_{\mu_p^n}(e^f) \leq pq \, E_{\mu_p^n}\left(\left(|Df|e^{|Df|} - e^{|Df|} + 1\right)e^f\right).$$

Proof. It is similar to the proof of Theorem 5.1 and relies on the next lemma.

Lemma 5.7. *The optimal constant $\delta > 0$ in the inequality*

$$\text{Ent}_{\mu_p}(e^f) \leq \delta pq \, E_{\mu_p}(e^f)$$

provided p is arbitrary in $[0,1]$ and $f : \{0,1\} \to \mathbb{R}$ is fixed is given by

$$\delta = a e^a - e^a + 1$$

where $a = |f(1) - f(0)|$.

Proof One may assume that $f(0) = 0$ and $f(1) = a$. The inequality we want to optimize becomes

$$p(1+x)\log(1+x) - (1+px)\log(1+px) \leq \delta pq(1+px) \tag{5.21}$$

where $x = e^a - 1 \geq 0$. Consider the function $U = U(p)$ which is the difference between the left-hand-side and the right-hand side of (5.21). Then $U(0) = U(1) = 0$ and $U''' \geq 0$. As in the proof of Lemma 5.2, to find the best constant δ amounts to show the inequality $U'(0) \leq 0$. But

$$U'(0) = (1+x)\log(1+x) - x - \delta = a e^a - e^a + 1 - \delta$$

which is the result. $\qquad\square$

According to Lemma 5.7, the theorem is proved in dimension one. We now simply observe that the inequality may be tensorized. By the product property of entropy, we get namely, for every f on $\{0,1\}^n$,

$$\mathrm{Ent}_{\mu_p^n}(e^f)$$

$$\leq pq \int \sum_{i=1}^n \Big(|f(x+e_i)-f(x)|e^{|f(x+e_i)-f(x)|} - e^{|f(x+e_i)-f(x)|}+1\Big)e^{f(x)}d\mu_p^n(x)$$

$$(5.22)$$

where we recall that $(e_1,\dots,e_n)$ is the canonical basis of $\mathbb{R}^n$ and that $x+e_i$ is understood here modulo 2. The function $Q(v)=\sqrt{v}e^{\sqrt{v}}-e^{\sqrt{v}}+1$, $v\geq 0$, is increasing and convex on $[0,\infty)$ with $Q(0)=0$. Hence, setting $a_i=|f(x+e_i)-f(x)|$, $i=1,\dots,n$,

$$\sum_{i=1}^n Q(a_i^2) \leq Q\Big(\sum_{i=1}^n a_i^2\Big) = Q(|Df(x)|^2) = |Df(x)|e^{|Df(x)|} - e^{|Df(x)|}+1.$$

Theorem 5.6 is therefore established. $\qquad\square$

As for Corollary 5.3, the Poisson limit theorem on (5.22) yields the following consequence for π_θ.

Corollary 5.8. *For any function f on $\mathbb{N}$,*

$$\mathrm{Ent}_{\pi_\theta}(e^f) \leq \theta\, \mathrm{E}_{\pi_\theta}\Big(\big(|Df|e^{|Df|}-e^{|Df|}+1\big)e^f\Big).$$

Corollary 5.8 is sharp in many respect. It becomes an equality for linear functions of the type $f(x)=cx+d$, $c\geq 0$. Furthermore, applying Theorem 5.6 and Corollary 5.8 to εf with $\varepsilon\to 0$ yields the Poincaré inequalities (5.12) and (5.13) for μ_p^n and π_θ respectively. This is easily verified using the fact that $ae^a - e^a+1$ behaves like $\frac{1}{2}a^2$ for small a.

As announced, the preceding statements actually describe sharp forms of modified logarithmic Sobolev inequalities in this context. As a consequence of Theorem 5.6 and Corollary 5.8, we namely get

Corollary 5.9. *For any function F on $\{0,1\}^n$ with $\max_{1\leq i\leq n}|f(x+e_i)-f(x)|\leq \lambda$ for every x in $\{0,1\}^n$,*

$$\mathrm{Ent}_{\mu_p^n}(e^f) \leq pq\,\frac{\lambda e^\lambda - e^\lambda +1}{\lambda^2}\,\mathrm{E}_{\mu_p^n}(|Df|^2 e^f).$$

The case $n=1$ is just Lemma 5.7 together with the fact that $\lambda^{-2}[\lambda e^\lambda - e^\lambda +1]$ is non-decreasing in $\lambda\geq 0$. The corollary follows by tensorization. Similarly,

Corollary 5.10. *For any function f on $\mathbb{N}$ with $\sup_{x\in\mathbb{N}}|Df(x)|\leq \lambda$,*

$$\mathrm{Ent}_{\pi_\theta}(e^f) \leq \theta\,\frac{\lambda e^\lambda - e^\lambda +1}{\lambda^2}\,\mathrm{E}_{\pi_\theta}(|Df|^2 e^f).$$

Again, via the central limit theorem, both Corollary 5.9 and Corollary 5.10 contain the Gaussian logarithmic Sobolev inequality. Let indeed φ be smooth enough on $\mathbb{R}$ and apply Corollary 5.9 to

$$f(x_1, \ldots, x_n) = \varphi\left(\frac{x_1 + \cdots + x_n - np}{\sqrt{npq}}\right).$$

Then,

$$\max_{1 \leq i \leq n} |f(x + e_i) - f(x)| \leq \frac{1}{\sqrt{npq}} \|\varphi\|_{\mathrm{Lip}} \to 0$$

as $n \to \infty$ and the result follows since $ae^a - e^a + 1 \sim \frac{1}{2}a^2$ as $a \to 0$. The same argument may be developed on the product form of Corollary 5.10 together with the central limit theorem for sums of independent Poisson random variables.

Due to the sharp constant in Corollary 5.9, the tail estimate of Theorem 5.5 may be improved. We namely get instead of (5.20) in the proof of Theorem 5.5

$$\mathrm{E}_{\mu_p^n}(e^{\lambda F}) \leq e^{\lambda \mathrm{E}_{\mu_p^n}(F) + \lambda \alpha^2 (e^\lambda - 1 - \lambda)}.$$

The same holds for π_θ^n and this bound is sharp since, when $n = 1$ for example, it becomes an equality for $F(x) = x$, $x \in \mathbb{N}$. Together with Chebyshev's inequality and a straightforward minimization procedure, we get, for F on $\{0, 1\}^n$ say, such that, for every $x = (x_1, \ldots, x_n) \in \{0, 1\}^n$,

$$\sum_{i=1}^n |F(x + e_i) - F(x)|^2 \leq \alpha^2 \quad \text{and} \quad \max_{1 \leq i \leq n} |F(x + e_i) - F(x)| \leq \beta$$

where $(e_1, \ldots, e_n)$ is the canonical basis of $\mathbb{R}^n$, then, for every $r \geq 0$,

$$\mu_p^n(F \geq \mathrm{E}_{\mu_p^n}(F) + r) \leq \exp\left(-\left(\frac{r}{\beta} + \frac{pq\alpha^2}{\beta^2}\right) \log\left(1 + \frac{\beta r}{pq\alpha^2}\right) + \frac{r}{3}\right). \qquad (5.23)$$

A similar inequality thus holds for π_θ^n changing pq into θ. Such an inequality may be considered as an extension of the classical exponential inequalities for sums of independent random variables with parameters the size and the variance of the variables, and describing a Gaussian tail for the small values of r and a Poisson tail for its large values (cf. (3.13). It belongs to the family of concentration inequalities for product measures deeply investigated by M. Talagrand [Ta6]. With respect to [Ta6], the study presented here develops some new aspects related to concentration for Bernoulli measures and penalties [Ta6, §2].

6. SOME APPLICATIONS TO LARGE DEVIATIONS AND TO BROWNIAN MOTION ON A MANIFOLD

In this part, we present some applications of the ideas developed around concentration and logarithmic Sobolev inequalities to large deviations and to Brownian motion on a manifold. We show indeed how logarithmic Sobolev inequalities and exponential integrability can be reformulated as a large deviation upper bound. We then discuss some recent logarithmic Sobolev inequality for Wiener measure on the paths of a Riemannian manifold. We apply it to give a large deviation bound for the uniform distance of Brownian motion from its starting point on a manifold with non-negative Ricci curvature.

6.1 Logarithmic Sobolev inequalities and large deviation upper bounds

On some measurable space $(X, \mathcal{B})$, let $(\mu_n)_{n \in \mathbb{N}}$ be a family of probability measures. Consider some generalized gradient Γ on a class $\mathcal{A}$ of functions on X such that, for every $f \in \mathcal{A}$ and $\lambda \in \mathbb{R}$, $\Gamma(\lambda f) = \lambda^2 \Gamma(f) \geq 0$. Γ could be either the square of a generalized modulus of gradient (2.3), or of some discrete one (2.9). Assume now that, for each $n \in \mathbb{N}$, there exists $c_n > 0$ such that, for every f in $\mathcal{A}$,

$$\mathrm{Ent}_{\mu_n}(e^f) \leq \frac{c_n}{2} \mathrm{E}_{\mu_n}(\Gamma(f) e^f). \tag{6.1}$$

Given two measurable sets A and B in X, set

$$d(A, B) = \inf_{x \in A, y \in B} \sup_{\|\Gamma(f)\|_\infty \leq 1} |f(x) - f(y)|. \tag{6.2}$$

Define $\mathcal{V}$ the class of all those $V \in \mathcal{B}$ such that $\lim_{n \to \infty} \mu_n(V) = 1$, and for every $A \in \mathcal{B}$, set

$$r(A) = \sup\{r \geq 0; \text{ there exists } V \in \mathcal{V} \text{ such that } d(A, V) \geq r\}.$$

Theorem 6.1. *Under (6.1), for every $A \in \mathcal{B}$,*

$$\limsup_{n \to \infty} \frac{c_n}{2} \log \mu_n(A) \leq -r(A)^2.$$

Proof. It is straighforward. Let $0 < r < r(A)$. Then, for some V in $\mathcal{V}$, and every n,

$$\mu_n(A) \leq \mu_n(d_V \geq r).$$

Denote by d_V the distance to the set V, and let $F_V = \min(d_V, r)$. As a consequence of Corollary 2.5,

$$\mu_n\big(F_V \geq \mathbb{E}_{\mu_n}(F_V) + r\big) \leq e^{-r^2/2c_n}.$$

Repeating the argument leading to (1.28),

$$\mu_n(A) \leq e^{-\mu_n(V)^2 r^2/2c_n}$$

for every n. Since $\mu_n(V) \to 1$ as $n \to \infty$, the conclusion follows. Theorem 6.1 is established. $\square$

Theorem 6.1 was used in [BA-L] to describe large deviations without topology for Gaussian measures. The operator Γ in this case relates to the Gross-Malliavin derivative and distance has to be understood with respect to the reproducing kernel Hilbert space (cf. [Le3]). In this case, the set functional $r(\cdot)$ is easily seen to connect with the classical rate functional in abstract Wiener spaces and, provided a topology has been fixed, coincide with this functional on the closure of A. Let more precisely μ be a Gaussian measure on the Borel sets $\mathcal{B}$ of a Banach space X with reproducing kernel Hilbert space $\mathcal{H}$. Denote by $\mathcal{K}$ the unit ball of $\mathcal{H}$. For every $\varepsilon > 0$, set $\mu_\varepsilon(\cdot) = \mu(\varepsilon^{-1}\cdot)$ and define the class $\mathcal{V}$ as those elements $V \in \mathcal{B}$ such that $\lim_{\varepsilon \to 0} \mu_\varepsilon(V) = 1$. In this case,

$$r(A) = \sup\big\{r \geq 0; \text{there exists } V \in \mathcal{V} \text{ such that } (V + r\mathcal{K}) \cap A = \emptyset\big\}.$$

Then, for any Borel set A,

$$\limsup_{\varepsilon \to 0} \varepsilon^2 \log \mu_\varepsilon(A) \leq -\frac{1}{2} r(A)^2.$$

Similar lower bounds can be described, however as simple consequences of the Cameron-Martin formula.

6.2 Some tail estimate for Brownian motion on a manifold

We have seen in (1.10) that if $(X_t)_{t \in T}$ is a (centered) Gaussian process such that $\sup_{t \in T} |X_t| < \infty$, then

$$\lim_{r \to \infty} \frac{1}{r^2} \log \mathbb{P}\big\{\sup_{t \in T} |X_t| \geq r\big\} = -\frac{1}{2\sigma^2}$$

where $\sigma = \sup_{t \in T}(\mathbb{E}(X_t^2))^{1/2}$. In particular, if $(B_t)_{t \geq 0}$ is Brownian motion in $\mathbb{R}^n$ starting from the origin, for every $T > 0$,

$$\lim_{r \to \infty} \frac{1}{r^2} \log \mathbb{P}\big\{\sup_{0 \leq t \leq T} |B_t| \geq r\big\} = -\frac{1}{2T}. \tag{6.3}$$

As illustrated by these notes, this result (6.3) may be seen as a consequence of the logarithmic Sobolev inequality for Gaussian measures (cf. Corollary 2.6 and Section 1.3). Our aim here will be show that the same method may be followed for Brownian motion on a manifold. Theorem 6.2 below is known and follows from the Lyons-Takeda forward and backward martingale method [Ly], [Tak]. We only aim to show here unity of the method, deriving this large deviation estimate from logarithmic Sobolev inequalities for heat kernel measures and Wiener measures on path spaces developed recently by several authors. We actually do not discuss here analysis on path spaces and only use what will be necessary to this bound. We refer for example to [Hs2] for details and references. Once the proper logarithmic Sobolev inequality is released, the proof of the upper bound is straightforward, and inequality (6.6) might be of independent interest. The lower bound requires classical volume estimates in Riemannian geometry.

Let thus M be a complete non-compact Riemannian manifold with dimension n and distance d. Denote by $(B_t)_{t \geq 0}$ Brownian motion on M starting from $x_0 \in M$.

Theorem 6.2. *If M has non-negative Ricci curvature, for every $T > 0$,*

$$\lim_{r \to \infty} \frac{1}{r^2} \log \mathbb{P}\Big\{ \sup_{0 \leq t \leq T} d(B_t, x_0) \geq r \Big\} = -\frac{1}{2T}.$$

As a consequence,

$$\mathbb{E}\Big(\exp(\alpha(\sup_{0 \leq t \leq T} d(B_t, x_0))^2) \Big) < \infty$$

for every $\alpha < \frac{1}{2T}$.

Proof. . We first establish the upper bound in the preceding limit. Let $p_t(x, y)$ be the heat kernel on M, fundamental solution of the heat equation $\frac{\partial}{\partial t} = \frac{1}{2}\Delta$ where Δ is the Laplace-Beltrami operator on M. For fixed $t \geq 0$ and $x \in M$, let $\nu_t = \nu_t(x)$ be the heat kernel measure $p_t(x, y)dy$. The following is the logarithmic Sobolev inequality for the heat kernel measure on a Riemannian manifold with Ricci curvature bounded below [Ba2].

Lemma 6.3. *Assume that* $\mathrm{Ric} \geq -K$, $K \in \mathbb{R}$. *For every $t \geq 0$ and $x \in M$, and every smooth function f on M,*

$$\mathrm{Ent}_{\nu_t}(f^2) \leq 2C(t)\,\mathbb{E}_{\nu_t}\left(|\nabla f|^2 \right)$$

where

$$C(t) = C_K(t) = \frac{e^{Kt} - 1}{K} \quad (= t \ \text{si} \ K = 0).$$

We now perform a Markov tensorization on entropy to describe, according to [Hs1], the logarithmic Sobolev inequality for cylindrical functions on the path space over M. Denote by $W_{x_0}(M)$ the space of continuous functions $x : \mathbb{R}_+ \mapsto M$ with $x(0) = x_0$, and by ν the Wiener measure on $W_{x_0}(M)$. A function f is called cylindrical on $W_{x_0}(M)$, if, for some φ on M^n and fixed times $0 \leq t_1 < \cdots < t_n$,

$f(x) = \varphi(x_{t_1}, \ldots, x_{t_n})$. (It will be called smooth if φ is smooth.) If f is a smooth cylindrical function, we denote with some abuse by $\nabla_i f$ the gradient of φ with respect to the i-th coordinate, $i = 1, \ldots, n$. According to [Hs2], let $U = (U_t)_{t \geq 0}$ the horizontal lift of Brownian motion to the tangent bundle $\mathcal{O}(M)$ and let $(\phi_{s,t})_{t \geq s}$ be the Ricci flow (matrix-valued process)

$$\frac{d}{dt}\phi_{s,t} = -\frac{1}{2}\operatorname{Ric}_{U_t}\phi_{s,t}, \quad \phi_{s,s} = I.$$

The following is Lemma 4.1 in [Hs1] to which we refer for the proof.

Lemma 6.4. If $\operatorname{Ric} \geq -K$, $K \in \mathbb{R}$, for any smooth f on $W_{x_0}(M)$,

$$\operatorname{Ent}_\nu(f^2) \leq 2\sum_{i=1}^{n} C(t_i - t_{i-1})\mathrm{E}_\nu\left(\left|\sum_{j=i}^{n}\phi_{t_i,t_j}^* U_{t_j}^{-1}\nabla^j f\right|^2\right)$$

where ϕ^* is the transpose of ϕ.

We can now establish the upper bound in the limit of Theorem 6.2. Let $F(x) = \max_{1 \leq i \leq n} d(x_{t_i}, x_0)$, $0 \leq t_1 < \cdots < t_n \leq T$. Then

$$\left|\sum_{j=i}^{n}\phi_{t_i,t_j}^* U_{t_j}^{-1}\nabla^j F\right|^2 \leq C(T). \tag{6.4}$$

Indeed, for some appropriate partition $(A_j)_{1 \leq j \leq n}$ of $W_{x_0}(M)$, $|\nabla^j F| \leq I_{A_j}$ for every j. On the other hand, since $\operatorname{Ric} \geq -K$, $|\phi_{t_i,t_j}^*| \leq e^{K(t_j - t_i)/2}$, $t_i < t_j$. Therefore, the left-hand side in (6.4) is bounded above by

$$\sum_{i=1}^{n} c(t_i - t_{i-1})\sum_{j=i}^{n} e^{K(t_j - t_i)}I_{A_j} = \frac{1}{K}\sum_{j=1}^{n} I_{A_j}\sum_{i=1}^{j}(e^{K(t_i - t_{i-1})} - 1)e^{K(t_j - t_i)}$$

$$= \frac{1}{K}\sum_{j=1}^{n} I_{A_j}e^{Kt_j}\sum_{i=1}^{j}(e^{-Kt_{i-1}} - e^{-Kt_i})$$

$$= \sum_{j=1}^{n} I_{A_j}C(t_j)$$

$$\leq \max_{1 \leq j \leq n} C(t_j) \leq C(T)$$

which is the result.

Now, apply the logarithmic Sobolev inequality of Lemma 6.4 to λF for every $\lambda \in \mathbb{R}$. We get

$$\operatorname{Ent}_\nu(e^{\lambda F}) \leq 2C(T)\lambda^2\mathrm{E}_\nu(e^{\lambda F}).$$

We then conclude, as in Section 2.3, that for every $r \geq 0$,

$$\nu\big(F \geq \mathrm{E}_\nu(F) + r\big) \leq e^{-r^2/2C(T)}.$$

In other words,

$$\mathbb{P}\Big\{\max_{1\leq i\leq n} d(B_{t_i}, x_0) \geq \mathbb{E}\big(\max_{1\leq i\leq n} d(B_{t_i}, x_0)\big) + r\Big\} \leq e^{-r^2/2c(T)}. \tag{6.5}$$

Since $\sup_{0\leq t\leq T} d(B_t, x_0) < \infty$ almost surely, it follows from (6.5), exactly as for (1.24), that $\mathbb{E}(\sup_{0\leq t\leq T} d(B_t, x_0)) < \infty$ and, for every $r \geq 0$,

$$\mathbb{P}\Big\{\sup_{0\leq t\leq T} d(B_t, x_0) \geq \mathbb{E}\big(\sup_{0\leq t\leq T} d(B_t, x_0)\big) + r\Big\} \leq e^{-r^2/2C(T)}. \tag{6.6}$$

When $K = 0$, it immediately yields that

$$\limsup_{r\to\infty} \frac{1}{r^2} \log \mathbb{P}\Big\{\sup_{0\leq t\leq T} d(B_t, x_0) \geq r\Big\} \leq -\frac{1}{2T}.$$

We are left with the lower bound that will follow from known heat kernel minorations. We assume from now on that Ric ≥ 0. For every $r \geq 0$,

$$\mathbb{P}\Big\{\sup_{0\leq t\leq T} d(B_t, x_0) \geq r\Big\} \geq \mathbb{P}\{d(B_T, x_0) \geq r\} = \int_{\{x; d(x, x_0)\geq r\}} p_T(x, x_0) dx \tag{6.7}$$

Since Ric ≥ 0,

$$p_t(x, y) \geq \frac{1}{(2\pi t)^{n/2}} e^{-d(x,y)^2/2t}$$

for every $x, y \in M$ and $t > 0$ [Da, p. 173]. Therefore, for every $\varepsilon > 0$,

$$\int_{\{x; d(x, x_0)\geq r\}} p_T(x, x_0) dx \geq \int_{\{x: r+\varepsilon \geq d(x, x_0)\geq r\}} \frac{1}{(2\pi T)^{n/2}} e^{-d(x, x_0)^2/2T}$$

$$\geq \frac{1}{(2\pi T)^{n/2}} e^{-(1+\varepsilon)^2 r^2/2T} [V(x_0, (1+\varepsilon)r)) - V(x_0, r)]$$

where $V(x, s)$, $s \geq 0$ is the Riemannian volume of the (open) geodesic ball $B(x, s)$ with center x and radius s in M. By the Riemannian volume comparison theorem (cf. e.g. [Cha2]), for every x in M and $0 < s \leq t$,

$$\frac{V(x, t)}{V(x, s)} \leq \Big(\frac{t}{s}\Big)^n. \tag{6.8}$$

Let now z on the boundary of $B(x_0, (1 + \frac{\varepsilon}{2})r)$. Since

$$B\big(z, \tfrac{\varepsilon}{2} r\big) \subset B\big(x_0, (1+\varepsilon)r\big) \setminus B(x_0, r) \quad \text{and} \quad B(x_0, r) \subset B\big(z, (2 + \tfrac{\varepsilon}{2})r\big),$$

we get by (6.8),

$$V(x_0, r) \leq V\big(z, (2 + \tfrac{\varepsilon}{2})r\big)$$
$$\leq \Big(\frac{4+\varepsilon}{\varepsilon}\Big)^n B\big(z, \tfrac{\varepsilon}{2} r\big)$$
$$\leq \Big(\frac{4+\varepsilon}{\varepsilon}\Big)^n [V(x_0, (1+\varepsilon)r) - V(x_0, r)].$$

Therefore,

$$\left[V(x_0, (1+\varepsilon)r) - V(x_0, r)\right] \geq \left(\frac{\varepsilon}{4+\varepsilon}\right)^n V(x_0, r).$$

Summarizing, for every $r \geq 0$,

$$\int_{\{x; d(x, x_0) \geq r\}} p_T(x, x_0) dx \geq \frac{1}{(2\pi T)^{n/2}} e^{-(1+\varepsilon)^2 r^2 / 2T} \left(\frac{\varepsilon}{4+\varepsilon}\right)^n V(x_0, r).$$

It is now a simple matter to conclude from this lower bound and (6.7) that

$$\liminf_{r \to \infty} \frac{1}{r^2} \log \mathbb{P}\left\{d(B_T, x_0) \geq r\right\} \geq -\frac{1}{2T}.$$

Theorem 6.2 is therefore established. □

7. ON REVERSED HERBST'S INEQUALITIES
AND BOUNDS ON THE LOGARITHMIC SOBOLEV CONSTANT

In this chapter, we investigate one instance in which a concentration property, or rather exponential integrability, implies a logarithmic Sobolev inequality. We present the result in the context of the Boltzmann measures already considered in Section 1.2. The argument is based on a recent observation by F.-Y. Wang [Wan] (see also [Ai]). In a more geometric setting, Wang's result also leads to dimension free lower bounds on the logarithmic Sobolev constant in compact manifolds with non-negative Ricci curvature that we review in the second paragraph. In the last section, we present a new upper bound on the diameter of a compact Riemannian manifold by the logarithmic Sobolev constant, the dimension and the lower bound on the Ricci curvature. We deduce a sharp upper bound on the logarithmic Sobolev constant in spaces with non-negative Ricci curvature. The last section is due to L. Saloff-Coste. It is shown how the preceding ideas may be developed similarly for discrete models, leading to estimates between the diameter and the logarithmic Sobolev constant.

7.1. Reversed Herbst's inequality

As in Section 1.2, let us consider a C^2 function W on $\mathbb{R}^n$ such that e^{-W} is integrable with respect to Lebesgue measure and let

$$d\mu(x) = Z^{-1}e^{-W(x)}dx$$

where Z is the normalization factor. μ is the invariant measure of the generator $L = \frac{1}{2}\Delta - \frac{1}{2}\nabla W \cdot \nabla$. We denote by $W''(x)$ the Hessian of W at the point x.

As we have seen in Theorem 1.1 and (2.17), when, for some $c > 0$, $W''(x) \geq c\,\mathrm{Id}$ for every x, μ satisfies a Gaussian-type isoperimetric inequality as well as a logarithmic Sobolev inequality (with respect to $E_\mu(|\nabla f|^2)$), and therefore a concentration property. In particular,

$$\int e^{\alpha|x|^2}d\mu(x) < \infty$$

for every $\alpha < c/2$. The following theorem, due to F.-Y. Wang [Wan] (in a more general setting) is a sort of conserve to this result.

Theorem 7.1. *Assume that for some* $c \in \mathbb{R}$, $W''(x) \geq c\,\mathrm{Id}$ *for every* x *and that for* some $\varepsilon > 0$,

$$\iint e^{(c^- + \varepsilon)|x-y|^2} d\mu(x) d\mu(y) < \infty$$

where $c^- = -\min(c, 0)$. Then μ satisfies the logarithmic Sobolev inequality

$$\mathrm{Ent}_\mu(f^2) \leq C \, \mathrm{E}_\mu(|\nabla f|^2)$$

for some $C > 0$.

According to (2.17), the theorem is only of interest when $c \leq 0$ (which we assume below). The integrability assumption of the theorem is in particular satisfied when

$$\int e^{2(c^- + \varepsilon)|x|^2} d\mu(x) < \infty.$$

As a consequence of Section 4 of [Ba-L], we may also conclude under the assumptions of Theorem 7.1 to a Gaussian isoperimetric inequality

$$\mu_s(\partial A) \geq \sqrt{c'} \, \mathcal{U}(\mu(A))$$

for some $c' > 0$, in the sense of Section 1. In the recent work [Bob5], the Poincaré inequality for μ is established when $W'' \geq 0$ without any further conditions.

Theorem 7.1 allows us to consider cases when the potential W is not convex. Another instance of this type is provided by the perturbation argument of [H-S]. Assume namely that a Boltzmann measure μ as before satisfies a logarithmic Sobolev inequality with constant C and let $d\nu = T^{-1} e^{-V} dx$ be such that $\|W - V\|_\infty \leq K$. Then ν satisfies a logarithmic Sobolev inequality with constant Ce^{4K}. To prove it, note first that $e^{-K} T \leq Z \leq e^K T$. As put forward in [H-S], for every $a, b > 0$, $b \log b - b \log a - b + a \geq 0$ and

$$\mathrm{Ent}(f^2) = \inf_{a>0} \mathrm{E}(f^2 \log f^2 - f^2 \log a - f^2 + a).$$

Therefore,

$$\begin{aligned}
\mathrm{Ent}_\nu(f^2) &= \inf_{a>0} \mathrm{E}_\mu([f^2 \log f^2 - f^2 \log a - f^2 + a]\, e^{W-V} Z T^{-1}) \\
&\leq e^{2K} \mathrm{Ent}_\mu(f^2) \\
&\leq Ce^{2K} \mathrm{E}_\mu(|\nabla f|^2) \\
&\leq Ce^{2K} \mathrm{E}_\nu(|\nabla f|^2 e^{V-W} T Z^{-1}) \\
&\leq Ce^{4K} \mathrm{E}_\nu(|\nabla f|^2).
\end{aligned}$$

(The same argument applies for the variance and Poincaré inequalities.) One odd feature of both Theorem 7.1 and this perturbation argument is that they yield rather poor constants as functions of the dimension (even for simple product measures) and seem therefore of little use in statistical mechanic applications.

Proof of Theorem 7.1. The main ingredient of the proof is the following result of [Wan] which describes a Harnack-type inequality for the Markov semigroup $(P_t)_{t \geq 0}$ with generator $L = \frac{1}{2}\Delta - \frac{1}{2}\nabla W \cdot \nabla$.

Lemma 7.2. *Under the hypothesis of the theorem, for every bounded measurable function f on $\mathbb{R}^n$, every $x, y \in \mathbb{R}^n$ and every $t > 0$,*

$$P_t f(x)^2 \leq P_t(f^2)(y) \, e^{c(e^{ct}-1)^{-1}|x-y|^2}$$

(where we agree that $c(e^{ct} - 1)^{-1} = t^{-1}$ when $c = 0$).

Proof. We may assume $f > 0$ and smooth. Fix $x, y \in \mathbb{R}^n$ and $t > 0$. Let, for every $0 \leq s \leq t$, $x_s = (s/t)x + (1 - (s/t))y$. Take also a C^1 function h on $[0, t]$ with non-negative values such that $h(0) = 0$ and $h(t) = t$. Set, for $0 \leq s \leq t$,

$$\varphi(s) = P_s\big((P_{t-s}f)^2\big)(x_{h(s)}).$$

Then,

$$\frac{d\varphi}{ds} = P_s(|\nabla P_{t-s}f|^2)(x_{h(s)}) + t^{-1}h'(s)\langle x - y, \nabla P_s((P_{t-s}f)^2)(x_{h(s)})\rangle$$

$$\geq P_s(|\nabla P_{t-s}f|^2)(x_{h(s)}) - t^{-1}|h'(s)||x - y||\nabla P_s((P_{t-s}f)^2)(x_{h(s)})|.$$

Now, under the assumption $W'' \geq c$, it is well-known that, for every smooth g and every $u \geq 0$,

$$|\nabla P_u g| \leq e^{-cu/2} P_u(|\nabla g|). \tag{7.1}$$

For example, the condition $W'' \geq c$ may be interpreted as a curvature condition and (7.1) then follows e.g. from [Ba2], Proposition 2.3. Therefore,

$$\frac{d\varphi}{ds} \geq P_s(|\nabla P_{t-s}f|^2)(x_{h(s)}) - t^{-1}|h'(s)||x - y|e^{-cs/2}P_s(|\nabla(P_{t-s}f)^2|)(x_{h(s)})$$

$$\geq P_s\big(|\nabla P_{t-s}f|^2 - 2t^{-1}|h'(s)||x - y|e^{-cs/2}P_{t-s}f|\nabla P_{t-s}f|\big)(x_{h(s)}).$$

Using that $X^2 - aX \geq -\frac{a^2}{4}$, it follows that

$$\frac{d\varphi}{ds} \geq -t^{-2}|x - y|^2 e^{-cs}h'(s)^2\varphi(s).$$

Integrating this differential inequality yields

$$P_t f(x)^2 \leq P_t(f^2)(y) \exp\left(t^{-2}|x - y|^2 \int_0^t e^{-cs}h'(s)^2 ds\right).$$

We then simply optimize the choice of h by taking

$$h(s) = t(e^{ct} - 1)^{-1}(e^{cs} - 1), \quad 0 \leq s \leq t.$$

The proof of Lemma 7.1 is complete. $\qquad\qquad\square$

In order to prove Theorem 7.1, we will first show that there is a spectral gap inequality for μ. To this goal, we follow the exposition in [B-L-Q]. Let f be a smooth function on $\mathbb{R}^n$, with $E_\mu(f) = 0$. By spectral theory, it is easily seen that for every $t \geq 0$,

$$E_\mu(f^2) \leq E_\mu((P_t f)^2) + 2t\, E_\mu(f(-Lf)) = E_\mu((P_t f)^2) + t\, E_\mu(|\nabla f|^2). \tag{7.2}$$

Since $E_\mu(f) = 0$, for every x,

$$|P_t f(x)| \leq \int |P_t f(x) - P_t f(y)|\, d\mu(y). \tag{7.3}$$

Now,

$$|P_t f(x) - P_t f(y))| \leq |x - y||\nabla P_t f(z)|$$

for some z on the line joining x to y. By (7.1) and by Lemma 7.2 applied to $|\nabla f|$ and to the couple (z, y),

$$|\nabla P_t f(z)| \leq e^{-ct/2} P_t(|\nabla f|)(z) \leq e^{-ct/2} P_t(|\nabla f|^2)(y)^{1/2} e^{c(e^{ct}-1)^{-1}|z-y|^2/2}.$$

Therefore by the Cauchy-Schwarz inequality (with respect to the variable y),

$$\left(\int |P_t f(x) - P_t f(y)|\, d\mu(y) \right)^2 \leq e^{-ct} E_\mu(|\nabla f|^2) \int |x - y|^2 e^{c(e^{ct}-1)^{-1}|x-y|^2}\, d\mu(y).$$

Integrating in $d\mu(x)$, together with (7.2) and (7.3), we get

$$E_\mu(f^2) \leq e^{-ct} E_\mu(|\nabla f|^2) \iint |x - y|^2 e^{c(e^{ct}-1)^{-1}|x-y|^2}\, d\mu(x)d\mu(y) + t\, E_\mu(|\nabla f|^2).$$

Letting t be sufficiently large, it easily follows from the hypothesis that

$$E_\mu(f^2) \leq C\, E_\mu(|\nabla f|^2) \tag{7.4}$$

for some finite constant C.

It would certainly be possible to prove the logarithmic Sobolev inequality in the same spirit. There is however a simpler route via hypercontractivity which, together with the spectral gap, immediately yields the conclusion. Let us consider again Wang's inequality of Lemma 7.2. Let $1 < \theta < 2$ and write, for every f (bounded to start with) and every $t > 0$,

$$E_\mu(|P_t f|^{2\theta}) = E_\mu\left(|P_t f|^\theta (|P_t f|^2)^{\theta/2} \right)$$
$$\leq \iint |P_t f(x)|^\theta (P_t(f^2)(y))^{\theta/2} e^{\theta c(e^{ct}-1)^{-1}|x-y|^2/2}\, d\mu(x)d\mu(y).$$

By Hölder's inequality, we get that

$$\|P_t f\|_{2\theta} \leq N^{\frac{2-\theta}{4\theta}} \|f\|_2 \tag{7.5}$$

where

$$N = \int\int e^{\theta(2-\theta)^{-1}c(e^{ct}-1)^{-1}|x-y|^2} d\mu(x)d\mu(y).$$

Provided θ is sufficiently close to 1 and t large enough. N is finite by the hypothesis. Therefore, P_t satisfies a weak form of hypercontractivity which, as is well-known, is equivalent to a defective logarithmic Sobolev inequality of the type (2.26). We get namely from (7.5) (see [Gr1] or [DeS]),

$$\mathrm{Ent}_\mu(f^2) \leq \frac{\theta t}{\theta-1} \mathrm{E}_\mu(|\nabla f|^2) + \frac{2-\theta}{2(\theta-1)} \log N \, \mathrm{E}_\mu(f^2) \tag{7.6}$$

for every smooth f. We are left to show that such a defective logarithmic Sobolev inequality may be turned into a true logarithmic Sobolev inequality with the help of the spectral gap (7.4). Again, this is a classical fact that relies on the inequality (cf. [Ro2], [De-S]),

$$\mathrm{Ent}_\mu(f^2) \leq \mathrm{Ent}_\mu\big((f - \mathrm{E}_\mu(f))^2\big) + 2\,\mathrm{Var}_\mu(f). \tag{7.7}$$

Inequality (7.6) applied to $f - \mathrm{E}_\mu(f)$ together with (7.7) and (7.4) complete the proof of the theorem. □

Note that N appears in (7.6) in the defective term as $\log N$ whereas in the Poincaré inequality (7.4), it appears as N (or some power of N). This is very sensible for product measures for which usually N is exponential in the dimension.

7.2 Dimension free lower bounds

In this section, we adopt a more geometric point of view and concentrate on lower bounds of the logarithmic Sobolev constant of a (compact) Riemannian manifold M with non-negative Ricci curvature in term of the diameter D of M.

Given some probability measure μ on $(X, \mathcal{B})$, and some energy functional on a class $\mathcal{A}$ of functions, we introduced in Section 2.3 the definitions of spectral gap (or Poincaré) and logarithmic Sobolev inequalities. Let us now agree to denote by λ_1 the largest constant $\lambda > 0$ such that for every f in $\mathcal{A}$,

$$\lambda \, \mathrm{Var}_\mu(f) \leq \mathcal{E}(f),$$

and by ρ_0 the largest $\rho > 0$ such that for every f in $\mathcal{A}$,

$$\rho \, \mathrm{Ent}_\mu(f^2) \leq 2\,\mathcal{E}(f).$$

Although it is usually the case, we cannot always ensure, at this level of generality, that $\lambda_1 \mathrm{Var}(f) \leq \mathcal{E}(f)$ and $\rho_0 \mathrm{Ent}(f^2) \leq 2\mathcal{E}(f)$ for every f. The estimates we present below are proved using arbitrary $\lambda < \lambda_1$ and $\rho < \rho_0$. This will be mostly understood. By Proposition 2.1, one always has that $\rho_0 \leq \lambda_1$. Emphasis has been put in the last years on identifying the logarithmic Sobolev constant and comparing it to the spectral gap.

Let M be a complete connected Riemannian manifold with dimension n and finite volume $V(M)$, and let $d\mu = \frac{dv}{V(M)}$ be the normalized Riemannian measure on M. Compact manifolds are prime examples. Let λ_1 and ρ_0 be respectively the spectral gap and the logarithmic Sobolev constant of μ with respect to Dirichlet form of the Laplace-Beltrami operator Δ (rather than $\frac{1}{2}\Delta$ here) on M, that is

$$\mathcal{E}(f) = \mathrm{E}_\mu\big(f(-\Delta f)\big) = \mathrm{E}_\mu\big(|\nabla f|^2\big)$$

for every smooth enough function f on M. If M is compact, it is known that $0 < \rho_0 \leq \lambda_1$ [Ro1]. Let D be the diameter of M if M is compact.

It is known since [Li], [Z-Y] that when M has non-negative Ricci curvature,

$$\lambda_1 \geq \frac{\pi^2}{D^2}. \tag{7.8}$$

Since $\lambda_1 \geq \rho_0$, it has been an open question for some time to prove that a similarly lower bound holds for the logarithmic Sobolev constant ρ_0. This has been proved recently by F.-Y. Wang [Wan] on the basis of his Lemma 7.2. Following [B-L-Q], we present here a simple proof of a somewhat stronger result.

Theorem 7.3. *Let M be a compact Riemannian manifold with diameter D and non-negative Ricci curvature, and denote by λ_1 and ρ_0 the spectral gap and the logarithmic Sobolev constant. Then*

$$\rho_0 \geq \frac{\lambda_1}{1 + 2D\sqrt{\lambda_1}}.$$

In particular,

$$\rho_0 \geq \frac{\pi^2}{(1 + 2\pi)D^2}.$$

Proof. We use Lemma 7.2 in this geometric context. Under the curvature assumption Ric ≥ 0, it yields similarly that if $(P_t)_{t>0}$ is the heat semigroup on M (with generator Δ), for every f on M, every $x, y \in M$ and $t > 0$,

$$P_t f(x)^2 \leq P_t f^2(y)\, \mathrm{e}^{d(x,y)^2/2t}$$

where $d(x, y)$ is the geodesic distance from x to y. In particular.

$$\|P_t\|_{2\to\infty} \leq \mathrm{e}^{D^2/4t}.$$

By symmetry,

$$\|P_t\|_{1\to\infty} \leq \|P_{t/2}\|_{1\to 2}\|P_{t/2}\|_{2\to\infty} \leq \mathrm{e}^{D^2/t}. \tag{7.9}$$

To prove the theorem, we then simply follow the usual route based on the heat semigroup as developed in [Ba1], and already described in our proof of the Gaussian logarithmic Sobolev inequality (2.15). Fix $f > 0$ smooth and $t > 0$. We write

$$\mathrm{E}_\mu(f \log f) - \mathrm{E}_\mu(P_t f \log P_t f) = -\int_0^t \mathrm{E}_\mu(\Delta P_s f \log P_s f)\,ds$$

$$= \int_0^t \mathrm{E}_\mu\Big(\frac{|\nabla P_s f|^2}{P_s f}\Big)\,ds.$$

Now since $\text{Ric} \geq 0$, $|\nabla P_s f| \leq P_s(|\nabla f|)$ (cf. e.g. [Ba2]). Moreover, by the Cauchy-Schwarz inequality,

$$P_s(|\nabla f|)^2 \leq P_s\left(\frac{|\nabla f|^2}{f}\right) P_s f$$

so that

$$E_\mu(f \log f) - E_\mu(P_t f \log P_t f) \leq \int_0^t E_\mu\left(P_s\left(\frac{|\nabla f|^2}{f}\right)\right) ds = t\, E_\mu\left(\frac{|\nabla f|^2}{f}\right).$$

Now, by (7.9),

$$E_\mu(P_t f \log P_t f) \leq E_\mu(f) \log E_\mu(f) + \frac{D^2}{t} E_\mu(f)$$

(since μ is invariant for P_t). Therefore, for every $t > 0$,

$$\text{Ent}_\mu(f) \leq \frac{D^2}{t} E_\mu(f) + t\, E_\mu\left(\frac{|\nabla f|^2}{f}\right).$$

Changing f into f^2,

$$\text{Ent}_\mu(f^2) \leq \frac{D^2}{t} E_\mu(f^2) + 4t\, E_\mu(|\nabla f|^2). \tag{7.10}$$

As we know, this defective logarithmic Sobolev inequality may then be turned into a true logarithmic Sobolev inequality with the help of λ_1 using (7.7). That is, (7.10) applied to $f - E_\mu(f)$ yields together with (7.7)

$$\text{Ent}_\mu(f^2) \leq \left(\frac{D^2}{t} + 2\right) \text{Var}_\mu(f) + 4t\, E_\mu(|\nabla f|^2)$$

$$\leq \left(\frac{D^2}{\lambda_1 t} + \frac{2}{\lambda_1} + 4t\right) E_\mu(|\nabla f|^2).$$

Optimizing in $t > 0$, the first claim of Theorem 7.3 follows. The second claim is then a consequence of (7.8). The proof is complete. $\qquad\square$

Similar results may be obtained in manifolds with Ricci curvature bounded below. Formulae are however somewhat more complicated (see [Wan], [B-L-Q]).

7.3 Upper bounds on the logarithmic Sobolev constant

We pursue our brief investigation on the spectral gap and logarithmic Sobolev constant by means of upper bounds. This question has mainly be raised in the framework of a Markov generator with associated Dirichlet form $\mathcal{E}$. It covers in particular Laplace-Beltrami and second-order elliptic operators on manifolds. Let us briefly review a few examples, some of them already alluded to in the previous chapters.

Spectral gaps and logarithmic Sobolev constants coincide for Gaussian measures by (2.15) and (2.16). A first example for which $\rho_0 < \lambda_1$ was brought in light in the paper [K-S] with the Laguerre generator with invariant measure the one-sided exponential distribution. As we have seen indeed in (4.4) and (4.5), $\rho_0 = \frac{1}{2} < 1 = \lambda_1$. On the two-point space $\{0,1\}$ with measure $\mu_p(\{1\}) = p$ and $\mu_p(\{0\}) = q = 1 - p$ and energy

$$\mathcal{E}(f) = \mathrm{E}_{\mu_p}\left(|Df|^2\right) = |f(1) - f(0)|^2.$$

we have seen ((5.2), (5.3)) that $\lambda_1 = pq$ whereas

$$\rho_0 = pq \, \frac{\log p - \log q}{p - q}.$$

In particular, $\rho_0 = \lambda_1$ only in the symmetric case $p = q = \frac{1}{2}$. Although rather recent, this example clearly indicates that, in general $\rho_0 < \lambda_1$. As discussed in Part 5, Poisson measures may be considered as an extreme case for which λ_1 is strictly positive (and may be shown to be equal to 1) while $\rho_0 = 0$. On the other hand, by (2.15) and (2.16), $\rho_0 = \lambda_1 = 1$ for the canonical Gaussian measure on $\mathbb{R}^n$.

We turn to another family of examples. Let M be a smooth complete connected Riemannian manifold with dimension n and finite volume $V(M)$, and let $d\mu = \frac{dv}{V(M)}$ be the normalized Riemannian measure on M. Compact manifolds are prime examples. Let λ_1 and ρ_0 be respectively the spectral gap and the logarithmic Sobolev constant of μ with respect to Dirichlet form of the Laplace-Beltrami Δ operator on M. We have seen that when M is compact, $0 < \rho_0 \leq \lambda_1$. When $\mathrm{Ric} \geq R > 0$, it goes back to A. Lichnerowicz (cf. [Cha1] that $\lambda_1 \geq R_n$ where $R_n = \frac{R}{1-\frac{1}{n}}$ with equality if and only if M is a sphere (Obata's theorem). This lower bound has been shown to hold similarly for the logarithmic Sobolev constant by D. Bakry and M. Emery [Ba-E] so that $\lambda_1 \geq \rho_0 \geq R_n$. The case of equality for ρ_0 is a consequence of Obata's theorem due to an improvement of the preceding by O. Rothaus [Ro2] who showed that when M is compact and $\mathrm{Ric} \geq R$ ($R \in \mathbb{R}$),

$$\rho_0 \geq \alpha_n \lambda_1 + (1 - \alpha_n) R_n \tag{7.11}$$

where $\alpha_n = 4n/(n+1)^2$. As examples, $\rho_0 = \lambda_1 = n$ on the n-sphere [M-W]. On the n-dimensional torus, $\lambda_1 = \rho_0 = 1$. The question whether $\rho_0 < \lambda_1$ in this setting has been open for some time until the geometric investigation by L. Saloff-Coste [SC1]. He showed that actually the existence of a logarithmic Sobolev inequality in a Riemannian manifold with finite volume and Ricci curvature bounded below forces the manifold to be compact whereas it is known that there exists non-compact manifolds of finite volume with $\lambda_1 > 0$. In particular, there exist compact manifolds of constant negative sectional curvature with spectral gaps uniformly bounded away from zero, and arbitrarily large diameters (cf. [SC1]. This yield examples for which the ratio ρ_0/λ_1 can be made arbitrarily small.

Our first result here is a significant improvement of the quantitative bound of of [SC1].

Theorem 7.4. *Assume that* $\mathrm{Ric} \geq -K$, $K \geq 0$. *If* $\rho_0 > 0$, *then* M *is compact. Furthermore, if* D *is the diameter of* M, *there exists a numerical constant* $C > 0$ *such that*

$$D \leq C\sqrt{n} \, \max\!\left(\frac{1}{\sqrt{\rho_0}}, \frac{\sqrt{K}}{\rho_0}\right).$$

It is known from the theory of hypercontractive semigroups (cf. [De-S]) that conversely there exists $C(n, K, \varepsilon)$ such that

$$\rho_0 \geq \frac{C(n, K, \varepsilon)}{D}$$

when $\lambda_1 \geq \varepsilon$.

The proof of [SC1] uses refined bounds on heat kernel and volume estimates. A somewhat shorter proof is provided in [Le2], still based on heat kernel. We present here a completely elementary argument based on the Riemannian volume comparison theorems and the concentration properties behind logarithmic Sobolev inequalities described in Part 2.

Proof. As a consequence of Corollary 2.6 and (2.25), for every measurable set A in M and every $r \geq 0$,

$$1 - \mu(A_r) \leq e^{-\rho_0 \mu(A)^2 r^2/2} \tag{7.12}$$

where $A_r = \{x \in M, d(x, A) < r\}$. This is actually the only property that will be used throughout the proof.

We show first that M is compact. We proceed by contradiction and assume that M is not compact. Denote by $B(x, u)$ the geodesic ball in M with center x and radius $u \geq 0$. Choose $A = B(x_0, r_0)$ a geodesic ball such that $\mu(A) \geq \frac{1}{2}$. By non-compactness (and completeness), for every $r \geq 0$, we can take z at distance $r_0 + 2r$ from x_0. In particular, $A \subset B(z, 2(r_0 + r))$. By the Riemannian volume comparison theorem [Cha2], for every $x \in M$ and $0 < s < t$,

$$\frac{V(x, t)}{V(x, s)} \leq \left(\frac{t}{s}\right)^n e^{\sqrt{(n-1)K} t} \tag{7.13}$$

where we recall that $V(x, u)$ is the volume of the ball $B(x, u)$ with center x and radius $u \geq 0$. Therefore,

$$V(z, r) \geq \left(\frac{r}{2(r_0 + r)}\right)^n e^{-2(r+r_0)\sqrt{(n-1)K}} V(z, 2(r_0 + r))$$

$$\geq \frac{1}{2}\left(\frac{r}{2(r_0 + r)}\right)^n e^{-2(r_0+r)\sqrt{(n-1)K}} V(M).$$

Since $B(z, r)$ is included in the complement of $A_r = B(x_0, r_0 + r)$, we get from (7.12)

$$\frac{1}{2}\left(\frac{r}{2(r_0 + r)}\right)^n e^{-2(r_0+r)\sqrt{(n-1)K}} \leq e^{-\rho_0 r^2/8} \tag{7.14}$$

which is impossible as $r \to \infty$.

Thus M is compact. Denote by D be its diameter. Let $x_0 \in M$ and let $B(x_0, \frac{D}{8})$. We distinguish between two cases. If $\mu(B(x_0, \frac{D}{8})) \geq \frac{1}{2}$, take $A = B(x_0, \frac{D}{8})$. By definition of D, we may choose $r = r_0 = \frac{D}{8}$ in (7.14) to get

$$\frac{1}{2} \cdot \frac{1}{4^n} e^{-\sqrt{(n-1)K} D/2} \leq e^{-\rho_0 D^2/512}.$$

If $\mu(B(x_0, \frac{D}{8})) < \frac{1}{2}$, apply (7.11) to A the complement of $B(x_0, \frac{D}{8})$. The ball $B(x_0, \frac{D}{16})$ is included in the complement of $A_{D/16}$. Moreover, by (7.13),

$$V\left(x_0, \frac{D}{16}\right) \geq \frac{1}{16^n} e^{-\sqrt{(n-1)KD}} V(M).$$

Therefore, by (7.12) with $r = \frac{D}{16}$,

$$\frac{1}{16^n} e^{-\sqrt{(n-1)KD}} \leq e^{-\rho_0 D^2/2048}.$$

In both cases,

$$\rho_0 D^2 - C\sqrt{(n-1)K}\, D - Cn \leq 0$$

for some numerical constant $C > 0$. Hence

$$D \leq \frac{C\sqrt{(n-1)K} + \sqrt{C^2(n-1)K + 4C\rho_0 n}}{2\rho_0}$$

and thus

$$D \leq \frac{C\sqrt{(n-1)K} + \sqrt{C\rho_0 n}}{\rho}$$

which yields the conclusion. The theorem is established. □

Note that the proof shows, under the assumption of Theorem 7.4, that M is compact as soon as

$$\limsup_{r \to \infty} \frac{-1}{r} \left[1 - \log \mu(B(x,r))\right] = \infty$$

for some (or all) $x \in M$. In particular $\lambda_1 > 0$ under this condition. This observation is a kind of converse to (2.36).

Corollary 7.5. *Let M be a compact Riemannian manifold with dimension n and non-negative Ricci curvature. Then*

$$\rho_0 \leq \frac{Cn}{D^2}$$

for some numerical constant $C > 0$.

Corollary 7.5 has to be compared to Cheng's upper bound on the spectral gap [Che] of compact manifolds with non-negative Ricci curvature

$$\lambda_1 \leq \frac{2n(n+4)}{D^2} \tag{7.15}$$

so that, generically, the difference between the upper bound on λ_1 and ρ_0 seems to be of the order of n. Moreover, it is mentioned in [Che] that there exists examples with $\lambda_1 \approx n^2/D^2$. Although we are not aware of such examples, they indicate perhaps that both Rothaus' lower bound (7.11) and Corollary 7.5 could be sharp. Note also that (7.11) together with Corollary 7.5 allows us to recover Cheng's upper bound on λ_1 of the same order in n. Actually, the proof of Theorem 7.4 together with the concentration property under the spectral gap (Proposition 2.13) would also yield Cheng's inequality (7.15) up to a numerical constant.

Corollary 7.5 is stated for (compact) manifolds without boundary but it also holds for compact manifolds of non-negative Ricci curvature with convex boundary (and Neuman's conditions). In particular, this result applies to convex bounded domains in $\mathbb{R}^n$ equipped with normalized Lebesgue measure. If we indeed closely inspect the proof of Theorem 7.4 in the latter case for example, we see that what is only required is (7.12), that holds similarly, and the volume comparisons. These are however well-known and easy to establish for bounded convex domains in $\mathbb{R}^n$. In this direction, it might be worthwhile mentioning moreover that the first non-zero Neumann eigenvalue λ_1 of the Laplacian on radial functions on the Euclidean ball B in $\mathbb{R}^n$ behaves as n^2. It may be identified indeed as the square of the first positive zero κ_n of the Bessel function $J_{n/2}$ of order $n/2$ (cf. [Cha1] e.g.). (On a sphere of radius r, there will be a factor r^{-2} by homogeneity.) In particular, standard methods or references [Wat] show that $\kappa_n \approx n$ as n is large. Denoting by ρ_0 the logarithmic Sobolev constant on radial functions on B, a simple adaption of the proof of Theorem 7.4 shows that $\rho_0 \leq Cn$ for some numerical constant $C > 0$. Actually, ρ_0 is of the order of n and this may be shown directly in dimension one by a simple analysis of the measure with density nx^{n-1} on the interval $[0, 1]$. We are indebted to S. Bobkov for this observation. One can further measure on this example the difference between the spectral gap and the logarithmic Sobolev constant as the dimension n is large. (On general functions, λ_1 and ρ_0 are both of the order of n, see [Bob5].)

As another application, assume $\mathrm{Ric} \geq R > 0$. As we have seen, by the Bakry-Emery inequality [Ba-E], $\rho_0 \geq R_n$ where $R_n = \frac{R}{1-\frac{1}{n}}$. Therefore, by Corollary 7.5,

$$D \leq C\sqrt{\frac{n-1}{R}}.$$

Up to the numerical constant, this is just Myers' theorem on the diameter of a compact manifold $D \leq \pi\sqrt{\frac{n-1}{R}}$ (cf. [Cha2]). This could suggest that the best numerical constant in Corollary 7.5 is π^2.

7.4 Diameter and logarithmic Sobolev constant for Markov chains

As in Section 2.1, let $K(x, y)$ be a Markov chain on a finite state space X with symmetric invariant probability measure μ. As before, let ρ_0 be the logarithmic Sobolev constant of (K, μ) defined as the largest $\rho > 0$ such that

$$\rho \, \mathrm{Ent}_\mu(f^2) \leq 2\mathcal{E}(f, f)$$

for every f on X. Recall that here

$$\mathcal{E}(f, f) = \frac{1}{2} \sum_{x,y \in X} \left(f(x) - f(y)\right)^2 K(x, y)\mu(\{x\}).$$

Recall also we set

$$\|\|f\|\|_\infty^2 = \sup\{\mathcal{E}(gf, f) - \tfrac{1}{2}\mathcal{E}(g, f^2); \|g\|_1 \leq 1\}$$

which, as we have seen, takes here the form

$$\|\|f\|\|_\infty^2 = \frac{1}{2}\sup_{x \in X}\sum_{y \in X}(f(x) - f(y))^2 K(x, y).$$

As a consequence of Corollary 2.4, for every F such that $\|\|F\|\|_\infty \leq 1$,

$$\mu(F \geq \mathrm{E}_\mu(F) + r) \leq e^{-\rho_0 r^2/4} \tag{7.16}$$

for every $r \geq 0$. If we then define the distance function associated with $\|\| \cdot \|\|_\infty$ as

$$d(x, y) = \sup_{\|\|f\|\|_\infty \leq 1}[f(x) - f(y)],$$

we get immediately from (7.16) and (1.28) that for every set A with $\mu(A) > 0$,

$$\mu(A_r) \geq 1 - e^{-\rho_0\mu(A)^2 r^2/4} \tag{7.17}$$

where $A_r = \{x; d(x, A) < r\}$. We are thus exactly in the same conditions as in the proof of Theorem 7.4.

Denote by D the diameter of X for the distance d defined above. We can thus state.

Proposition 7.6. *If μ is nearly constant, that is if there exists C such that, for every x, $\mu(\{x\}) \leq C \min_{y \in X}\mu(\{y\})$, then*

$$\rho_0 \leq \frac{64\log(C|X|)}{D^2}$$

where $|X|$ is the cardinal of X.

Proof. Consider two points $x, y \in X$ such that $d(x, y) = D$. Let B the ball with center x and radius $D/2$. Let A be the set with the largest measure amongst B and B^c. Then $\mu(A) \geq 1/2$. Observe that either x or y is in the complement $(A_r)^c$ of A_r with $r = D/2$. Indeed, if $A = B$, then $(A_r)^c = \{z; d(x, z) \geq D\}$ and $y \in (A_r)^c$ because $d(x, y) = D$; if $A = B^c$, $x \in (A_r)^c$ because $d(x, A) > D/2$. Hence (7.17) yields

$$\min_{z \in X}\mu(\{z\}) \leq e^{-\rho_0 D^2/64}.$$

Since, by the hypothesis on μ, $\min_{z \in X}\mu(\{z\}) \geq (C|X|)^{-1}$, the conclusion follows. $\square$

The distance most often used in the present setting is not d but the combinatoric distance d_c associated with the graph with vertex-set X and edge-set

$\{(x, y) : K(x, y) > 0\}$. This distance can be defined as the minimal number of edges one has to cross to go from x to y. Equivalently,

$$d_c(x, y) = \sup_{\|\nabla f\|_\infty \leq 1} [f(x) - f(y)]$$

where

$$\|\nabla f\|_\infty = \sup\{|f(x) - f(y)|; K(x, y) > 0\}.$$

Recall, from Section 2.1, that since $\sum_y K(x, y) = 1$,

$$\||f|\|_\infty^2 \leq \frac{1}{2} \|\nabla f\|_\infty^2.$$

In particular, the combinatoric diameter D_c satisfies $D_c^2 \leq D^2/2$.

Let us now survey a number of examples at the light of Proposition 7.6.

Consider the first the hypercube $\{0, 1\}^n$ with $K(x, y) = 1/n$ if x, y differ by exactly one coordinate and $K(x, y) = 0$ otherwise. The reversible measure is the uniform distribution and $\rho_0 = 2/n$. Proposition 7.6 tells us that $\rho_0 \leq 23/n$.

Consider the Bernoulli-Laplace model of diffusion. This is a Markov chain on the n-sets of an N-set with $n \leq N/2$. If the current state is an n-set A, we pick an element x at random in A, an element y at random in A^c and change A to $B = (A \backslash \{x\}) \cup \{y\}$. The kernel K is given by $K(A, B) = 1/[n(N-n)]$ if $|A \cap B| = n-2$ and $K(A, B) = 0$ otherwise. The uniform distribution $\pi(A) = \binom{N}{n}^{-1}$ is the reversible measure. Clearly, $D_c = n$. Hence

$$\rho_0 \leq \frac{32 \log \binom{N}{n}}{n^2}.$$

In the limit case, $n = N/2$, this yields $\rho_0 \leq C/n$ which is the right order of magnitude [L-Y].

Let now the chain random transpositions on the symmetric group S_n. Here, $K(\sigma, \theta) = 2/[n(n-1)]$ if $\theta = \sigma\tau$ for some transposition τ and $K(\sigma, \theta) = 0$ otherwise and $\pi \equiv (n!)^{-1}$, The diameter is $D_c = n - 1$ and one knows that ρ_0 is of order $1/n \log n$ [D-SC], [L-Y]. Proposition 7.6 gives

$$\rho_0(n - 1)^2 \leq 32 \log(n!).$$

Since we know that $\rho_0 \geq (4n \log n)^{-1}$ [D-SC], we can also conclude from Proposition 7.6 that

$$D \leq \sqrt{64n \log n / \rho_0} \leq 16n \log n.$$

At present writing it is not clear whether or not this bound can be obtained more easily. Note that the upper bound

$$d(x, y) \leq \left(\frac{1}{2} \min_{K(x,y) > 0} K(x, y) \right)^{-1/2} d_c(x, y)$$

only yields $D \leq n^2$, up to a multiplicative constant. It might be worthile observing that in this example, ρ_0 is of order $1/n \log n$ while it has been shown by B. Maurey

[Mau1] that concentration (with respect to the combinatoric metric) is satisfied at a rate of the order of $1/n$.

Consider a N-regular graph with N fixed. Let $K(x,y) = 1/N$ if they are neighbors and $K(x,y) = 0$ otherwise. Then $\mu(\{x\}) = 1/|X|$. Assume that the number $N(x,t)$ of elements in the ball $B(x,t)$ with center x and radius t in the combinatoric metric d_c satisfies

$$\forall x \in X, \quad \forall t > 0, \quad N(x,2t) \le CN(x,t). \tag{7.18}$$

Fix $x,y \in X$ such that $d_c(x,y) = D_c$. Set $A = B(x,D_c/2)$, and let $0 < r < D/4$. Then $B(y,D_c/4)$ is contained in the complement of A_r. Now, by our hypothesis, $N(x,D_c/2) \ge C^{-1}|X|$ and $N(y,D_c/4) \ge C^{-2}|X|$ so that

$$1 - \mu(A_r) \ge C^{-2}, \quad \mu(A) \ge C^{-1}.$$

Reporting in (7.17), we obtain

$$\rho_0 \le \frac{128C^2 \log C}{D_c^2}.$$

For N and C fixed, this is the right order of magnitude in the class of Cayley graphs of finite groups satisfying the volume doubling condition (7.18). See [D-SC, Theorem 4.1].

As a last example, consider any N-regular graph on a finite set X. Let $K(x,y) = 1/N$ if they are neighbors and $K(x,y) = 0$ otherwise. Then $\mu(\{x\}) = 1/|X|$ and $|X| \le N^{D_c}$ (at least if $D_c \ge 2$). Thus, we get from Proposition 7.6 that

$$\rho_0 \le \frac{64D_c \log N}{D^2} \le \frac{46 \log N}{D}.$$

Compare with the results of [D-SC] and Section 7.3. This is, in a sense, optimal generically. Indeed, if $|X| \ge 4$, one also have the lower bound [D-SC]

$$\rho_0 \ge \frac{\lambda}{2D_c \log N}$$

where $1 - \lambda$ is the second largest eigenvalue of K. There are many known families of N-regular graphs (N fixed) such that $|X| \to \infty$ whereas $\lambda \ge \epsilon > 0$ stays bounded away from zero (the so-called expanders graphs). Moreover graphs with this property are "generic" amongst N-regular graphs [Al].

REFERENCES

[Ai] S. Aida. Uniform positivity improving property, Sobolev inequalities and spectral gaps. J. Funct. Anal. 158, 152–185 (1998).

[A-M-S] S. Aida, T. Masuda, I. Shigekawa. Logarithmic Sobolev inequalities and exponential integrability. J. Funct. Anal. 126, 83–101 (1994).

[A-S] S. Aida, D. Stroock. Moment estimates derived from Poincaré and logarithmic Sobolev inequalities. Math. Res. Lett. 1, 75–86 (1994).

[Al] N. Alon. Eigenvalues and expanders. J. Combin. Theory, Ser. B, 38, 78–88 (1987).

[A-L] C. Ané, M. Ledoux. On logarithmic Sobolev inequalities for continuous time random walks on graphs. Preprint (1998).

[Ba1] D. Bakry. L'hypercontractivité et son utilisation en théorie des semigroupes. Ecole d'Eté de Probabilités de St-Flour. Lecture Notes in Math. 1581, 1–114 (1994). Springer-Verlag.

[Ba2] D. Bakry. On Sobolev and logarithmic Sobolev inequalities for Markov semigroups. New trends in Stochastic Analysis. 43–75 (1997). World Scientific.

[Ba-E] D. Bakry, M. Emery. Diffusions hypercontractives. Séminaire de Probabilités XIX. Lecture Notes in Math. 1123, 177–206 (1985). Springer-Verlag.

[Ba-L] D. Bakry, M. Ledoux. Lévy-Gromov's isoperimetric inequality for an infinite dimensional diffusion generator. Invent. math. 123, 259–281 (1996).

[B-L-Q] D. Bakry, M. Ledoux, Z. Qian. Preprint (1997).

[Be] W. Beckner. Personal communication (1998).

[BA-L] G. Ben Arous, M. Ledoux. Schilder's large deviation principle without topology. Asymptotic problems in probability theory: Wiener functionals and asymptotics. Pitman Research Notes in Math. Series 284, 107–121 (1993). Longman.

[B-M1] L. Birgé, P. Massart. From model selection to adaptive estimation. Festschrift for Lucien LeCam: Research papers in Probability and Statistics (D. Pollard, E. Torgersen and G. Yang, eds.) 55–87 (1997). Springer-Verlag.

[B-M2] L. Birgé, P. Massart. Minimum contrast estimators on sieves: exponential bounds and rates of convergence (1998). Bernoulli, to appear.

[B-B-M] A. Barron, L. Birgé, P. Massart. Risk bounds for model selection via penalization (1998). Probab. Theory Relat. Fields, to appear.

[Bob1] S. Bobkov. On Gross' and Talagrand's inequalities on the discrete cube. Vestnik of Syktyvkar University, Ser. 1, 1, 12–19 (1995) (in Russian).

[Bob2] S. Bobkov. Some extremal properties of Bernoulli distribution. Probability Theor. Appl. 41, 877–884 (1996).

[Bob3] S. Bobkov. A functional form of the isoperimetric inequality for the Gaussian measure. J. Funct. Anal. 135, 39–49 (1996).

[Bob4] S. Bobkov. An isoperimetric inequality on the discrete cube and an elementary proof of the isoperimetric inequality in Gauss space. Ann. Probability 25, 206–214 (1997).

[Bob5] S. Bobkov. Isoperimetric and analytic inequalities for log-concave probability measures (1998). Ann. Probability, to appear.

[B-G] S. Bobkov, F. Götze. Exponential integrability and transportation cost related to logarithmic Sobolev inequalities (1997). J. Funct. Anal., to appear.

[B-H] S. Bobkov, C. Houdré. Isoperimetric constants for product probability measures. Ann. Probability 25, 184–205 (1997).

[B-L1] S. Bobkov, M. Ledoux. Poincaré's inequalities and Talagrand's concentration phenomenon for the exponential measure. Probab. Theory Relat. Fields 107, 383–400 (1997).

[B-L2] S. Bobkov, M. Ledoux. On modified logarithmic Sobolev inequalities for Bernoulli and Poisson measures. J. Funct. Anal. 156, 347–365 (1998).

[Bon] A. Bonami. Etude des coefficients de Fourier des fonctions de $L^p(G)$. Ann. Inst. Fourier 20, 335–402 (1970).

[Bor] C. Borell. The Brunn-Minkowski inequality in Gauss space. Invent. math. 30, 207–216 (1975).

[Br] R. Brooks. On the spectrum of non-compact manifolds with finite volume. Math. Z. 187, 425–437 (1984).

[Cha1] I. Chavel. Eigenvalues in Riemannian geometry. Academic Press (1984).

[Cha2]   I. Chavel. Riemannian geometry - A modern introduction. Cambridge Univ. Press (1993).

[Che]   S.-Y. Cheng. Eigenvalue comparison theorems and its geometric applications. Math. Z. 143, 289–297 (1975).

[Da]   E. B. Davies. Heat kernel and spectral theory. Cambridge Univ. Press (1989).

[D-S]   E. B. Davies, B. Simon. Ultracontractivity and the heat kernel for Schrödinger operators and Dirichlet Laplacians. J. Funct. Anal. 59, 335–395 (1984).

[De]   A. Dembo. Information inequalities and concentration of measure. Ann. Probability 25. 927–939 (1997).

[D-Z]   A. Dembo, O. Zeitouni. Transportation approach to some concentration inequalities in product spaces. Elect. Comm. in Probab. 1, 83–90 (1996).

[De-S]   J.-D. Deuschel, D. Stroock. Large deviations. Academic Press (1989).

[D-SC]   P. Diaconis, L. Saloff-Coste. Logarithmic Sobolev inequalities for finite Markov chains. Ann. Appl. Prob. 6, 695–750 (1996).

[Eh]   A. Ehrhard. Symétrisation dans l'espace de Gauss. Math. Scand. 53, 281–301 (1983).

[G-M]   M. Gromov, V. D. Milman. A topological application of the isoperimetric inequality. Amer. J. Math. 105, 843–854 (1983).

[Gr1]   L. Gross. Logarithmic Sobolev inequalities. Amer. J. Math. 97, 1061–1083 (1975).

[Gr2]   L. Gross. Logarithmic Sobolev inequalities and contractive properties of semigroups. Dirichlet Forms, Varenna 1992. Lect. Notes in Math. 1563, 54–88 (1993). Springer-Verlag.

[G-R]   L. Gross, O. Rothaus. Herbst inequalities for supercontractive semigroups. Preprint (1997).

[H-Y]   Y. Higuchi, N. Yoshida. Analytic conditions and phase transition for Ising models. Lecture Notes in Japanese (1995).

[H-S]   R. Holley, D. Stroock. Logarithmic Sobolev inequalities and stochastic Ising models. J. Statist. Phys. 46, 1159–1194 (1987).

[H-T]   C. Houdré, P. Tetali. Concentration of measure for products of Markov kernels via functional inequalities. Preprint (1997).

[Hs1]   E. P. Hsu. Logarithmic Sobolev inequalities on path spaces over Riemannian manifolds. Commun. Math. Phys. 189, 9–16 (1997).

[Hs2]   E. P. Hsu. Analysis on Path and Loop Spaces (1996). To appear in IAS/Park City Mathematics Series, Vol. 5, edited by E. P. Hsu and S. R. S. Varadhan, American Mathematical Society and Institute for Advanced Study (1997).

[J-S]   W. B. Johnson, G. Schechtman. Remarks on Talagrand's deviation inequality for Rademacher functions. Longhorn Notes, Texas (1987).

[Kl]   C. A. J. Klaassen. On an inequality of Chernoff. Ann. Probability 13, 966–974 (1985).

[K-S]   A. Korzeniowski, D. Stroock. An example in the theory of hypercontractive semigroups. Proc. Amer. Math. Soc. 94, 87–90 (1985).

[Kw-S]   S. Kwapień, J. Szulga. Hypercontraction methods in moment inequalities for series of independent random variables in normed spaces. Ann. Probability 19, 369–379 (1991).

[K-L-O] S. Kwapień, R. Latala, K. Oleszkiewicz. Comparison of moments of sums of independent random variables and differential inequalities. J. Funct. Anal. 136, 258–268 (1996).

[Le1]   M. Ledoux. Isopérimétrie et inégalités de Sobolev logarithmiques gaussiennes. C. R. Acad. Sci. Paris, 306, 79–92 (1988).

[Le2]   M. Ledoux. Remarks on logarithmic Sobolev constants. exponential integrability and bounds on the diameter. J. Math. Kyoto Univ. 35, 211–220 (1995).

[Le3]   M. Ledoux. Isoperimetry and Gaussian Analysis. Ecole d'Eté de Probabilités de St-Flour 1994. Lecture Notes in Math. 1648, 165–294 (1996). Springer-Verlag.

[Le4]   M. Ledoux. On Talagrand's deviation inequalities for product measures. ESAIM Prob. & Stat. 1, 63–87 (1996).

[L-T]   M. Ledoux, M. Talagrand. Probability in Banach spaces (Isoperimetry and processes). Ergebnisse der Mathematik und ihrer Grenzgebiete. Springer-Verlag (1991).

[L-Y]   T.Y. Lee, H.-T. Yau. Logarithmic Sobolev inequality fo some models of random walks. Preprint (1998).

[Li]   P. Li. A lower bound for the first eigenvalue of the Laplacian on a compact manifold. Indiana Univ. Math. J. 28, 1013–1019 (1979).

[Ly]   T. Lyons. Random thoughts on reversible potential theory. Summer School in Potentiel Theory, Joensuu 1990. Publications in Sciences 26, 71–114 University of Joensuu.

[MD]   C. McDiarmid. On the method of bounded differences. Surveys in Combinatorics. London Math. Soc. Lecture Notes 141, 148–188 (1989). Cambridge Univ. Press.

[Mar1] K. Marton. Bounding $\bar{d}$-distance by information divergence: a method to prove measure concentration. Ann. Probability 24, 857–866 (1996).

[Mar2] K. Marton. A measure concentration inequality for contracting Markov chains. Geometric and Funct. Anal. 6, 556–571 (1997).

[Mar3] K. Marton. Measure concentration for a class of random processes. Probab. Theory Relat. Fields 110, 427–439 (1998).

[Mar4] K. Marton. On a measure concentration of Talagrand for dependent random variables. Preprint (1998).

[Mas]   P. Massart. About the constants in Talagrand's deviation inequalities for empirical processes (1998). Ann. Probability, to appear.

[Mau1] B. Maurey. Constructions de suites symétriques. C. R. Acad. Sci. Paris 288, 679–681 (1979).

[Mau2] B. Maurey. Some deviations inequalities. Geometric and Funct. Anal. 1, 188–197 (1991).

[Mi]   V. D. Milman. Dvoretzky theorem - Thirty years later. Geometric and Funct. Anal. 2, 455–479 (1992) .

[M-S]   V. D. Milman, G. Schechtman. Asymptotic theory of finite dimensional normed spaces. Lecture Notes in Math. 1200 (1986). Springer-Verlag.

[M-W]   C. Muller, F. Weissler. Hypercontractivity of the heat semigroup for ultraspherical polynomials and on the n-sphere. J. Funct. Anal. 48, 252–283 (1982).

[O-V]   F. Otto, C. Villani. Generalization of an inequality by Talagrand, viewed as a consequence of the logarithmic Sobolev inequality. Preprint (1998).

[Pi]   M. S. Pinsker. Information and information stability of random variables and processes. Holden-Day, San Franscico (1964).

[Ro1]   O. Rothaus. Diffusion on compact Riemannian manifolds and logarithmic Sobolev inequalities. J. Funct. Anal. 42, 358–367 (1981).

[Ro2]   O. Rothaus. Hypercontractivity and the Bakry-Emery criterion for compact Lie groups. J. Funct. Anal. 65, 358–367 (1986).

[Ro3]   O. Rothaus. Logarithmic Sobolev inequalities and the growth of L^p norms (1996).

[SC1]   L. Saloff-Coste. Convergence to equilibrium and logarithmic Sobolev constant on manifolds with Ricci curvature bounded below. Colloquium Math. 67, 109–121 (1994).

[SC2]   L. Saloff-Coste. Lectures on finite Markov chains. Ecole d'Eté de Probabilités de St-Flour 1996. Lecture Notes in Math. 1665, 301–413 (1997). Springer-Verlag.

[Sa]   P.-M. Samson. Concentration of measure inequalities for Markov chains and Φ-mixing processes. Preprint (1998).

[Sc]   M. Schmuckenschläger. Martingales, Poincaré type inequalities and deviations inequalities. J. Funct. Anal. 155, 303–323 (1998).

[St]   D. Stroock. Logarithmic Sobolev inequalities for Gibbs states. Dirichlet forms, Varenna 1992. Lecture Notes in Math. 1563, 194–228 (1993).

[S-Z]   D. Stroock, B. Zegarlinski. The logarithmic Sobolev inequality for continuous spin systems on a lattice. J. Funct. Anal. 104, 299–326 (1992).

[S-T]   V. N. Sudakov, B. S. Tsirel'son. Extremal properties of half-spaces for spherically invariant measures. J. Soviet. Math. 9, 9–18 (1978); translated from Zap. Nauch. Sem. L.O.M.I. 41, 14–24 (1974).

[Tak]   M. Takeda. On a martingale method for symmetric diffusion process and its applications. Osaka J. Math. 26, 605–623 (1989).

[Ta1]   M. Talagrand. An isoperimetric theorem on the cube and the Khintchine-Kahane inequalities. Proc. Amer. Math. Soc. 104, 905–909 (1988).

[Ta2]   M. Talagrand. Isoperimetry and integrability of the sum of independent Banach space valued random variables. Ann. Probability 17, 1546–1570 (1989).

[Ta3]   M. Talagrand. A new isoperimetric inequality for product measure. and the concentration of measure phenomenon. Israel Seminar (GAFA), Lecture Notes in Math. 1469, 91–124 (1991). Springer-Verlag.

[Ta4]   M. Talagrand. Some isoperimetric inequalities and their applications. Proc. of the International Congress of Mathematicians, Kyoto 1990, vol. II, 1011–1024 (1992). Springer-Verlag.

[Ta5]   M. Talagrand. Sharper bounds for Gaussian and empirical processes. Ann. Probability 22, 28–76 (1994).

[Ta6]   M. Talagrand. Concentration of measure and isoperimetric inequalities in product spaces. Publications Mathématiques de l'I.H.E.S. 81, 73–205 (1995).

[Ta7]   M. Talagrand. A new look at independence. Ann. Probability, 24, 1–34 (1996).

[Ta8]   M. Talagrand. New concentration inequalities in product spaces. Invent. math. 126, 505–563 (1996).

[Ta9]   M. Talagrand. Transportation cost for Gaussian and other product measures. Geometric and Funct. Anal. 6, 587–600 (1996).

[Wan]   F.-Y. Wang. Logarithmic Sobolev inequalities on noncompact Riemannian manifolds. Proab. Theory Relat. Fields 109, 417–424 (1997).

[Wat]   G. N. Watson. A treatise on the theory of Bessel functions. Cambridge Univ. Press (1944).

[Z-Y]   J. Q. Zhong, H. C. Yang. On the estimate of the first eigenvalue of a compact Riemanian manifold. Sci. Sinica Ser. A 27 (12), 1265–1273 (1984).

UNE SIMPLIFICATION DE L'ARGUMENT DE TSIRELSON SUR LE CARACTERE NON-BROWNIEN DES PROCESSUS DE WALSH

BERNARD DE MEYER

Il est bien connu que le mouvement Brownien a la propriété de représentation prévisible sur sa filtration naturelle. Dans [3], M. Yor posait la question réciproque: les filtrations ayant la propriété de représentation prévisible pour un mouvement Brownien linéaire sont-elles les filtrations naturelles d'un autre mouvement Brownien?

Cette question a été la motivation principale des études récentes du processus de Walsh. Ce processus X est une martingale à valeurs dans le plan complexe $\mathbb{C}$, dont le cube X^3 est un processus réel positif et dont le module $|X|$ est un mouvement Brownien réfléchi.

La filtration naturelle d'un tel processus X a en effet la propriété de représentation prévisible pour un mouvement Brownien réel et l'on a longtemps conjecturé qu'un processus de Walsh ne pouvait être mesurable sur une filtration Brownienne. Cette conjecture fut démontrée par Tsirelson [2] en 1997, en introduisant la notion de "confort" des filtrations Browniennes. Cette démonstration était relativement lourde sur le plan technique et cette complexité dissimulait la nature profonde de l'argument.

En fait, comme veut le souligner cette note, le "caractère non Brownien" du processus de Walsh est une conséquence d'une propriété très "dyadique" de la filtration Brownienne, propriété initialement conjecturée par M.T. Barlow et à présent démontrée dans [1]: si G est un temps honnête (i.e. la fin d'un ensemble optionnel) sur une filtration Brownienne $\{\mathcal{F}_t\}$, et si $\mathcal{F}_G$ (respectivement $\mathcal{F}_G^+$) représente la tribu engendrée par les variables R_G, lorsque R parcourt la classe des processus optionnels (respectivement progressifs), alors $\mathcal{F}_G^+$ diffère de $\mathcal{F}_G$ au plus par un bit d'information. En d'autres termes, il existe un événement A tel que $\mathcal{F}_G^+ = \sigma(\mathcal{F}_G, A)$.

Supposer qu'un processus de Walsh X soit adapté à la filtration $\{\mathcal{F}_t\}$ conduit alors à une contradiction: en prenant pour temps honnête G le dernier zéro avant 1 de X, il est clair que la variable $e_1 := X_1/|X_1|$ est dans $\mathcal{F}_G^+$ et peut dès lors s'écrire

$$e_1 = f \mathbb{1}_A + g \mathbb{1}_{A^c}, \tag{1}$$

où f et g sont deux variables $\mathcal{F}_G$-mesurables. Mais ceci est impossible, car un argument de symétrie indique que la variable e_1 est indépendante de $\mathcal{F}_G$ et prend les trois valeurs $\exp(ik2\pi/3)$, $k = 0, 1, 2$ avec une probabilité de $1/3$ chacune.

La propriété "dyadique" de la filtration Brownienne est en fait démontrée dans [1] comme une conséquence de l'argument de Tsirelson généralisé au problème des martingales araignées (voir le théorème ci-après) et ne peut donc servir de point de départ de notre démonstration. Cependant, nous montrerons que le caractère "confortable" de la filtration Brownienne permet d'établir une relation très semblable à (1):

$$e_1 = f \mathbb{1}_A + e_1 \mathbb{1}_{A^c},$$

où l'événement A se réalise avec une probabilité très proche de $1/2$ et f est $\mathcal{H}_G$-mesurable ($\{\mathcal{H}_t\}$ dénote ici une filtration par rapport à laquelle X est également un processus de Walsh).

Nous obtenons à nouveau une contradiction, car sur l'événement $\{P[A|\mathcal{H}_G] \geq P[A]\}$ qui a une probabilité non nulle, e_1 prend, conditionnellement à $\mathcal{H}_G$, la valeur f avec une probabilité plus grande que $P[A] \approx 1/2$, alors que

$$P[e_1 = \exp(ik2\pi/3)|\mathcal{H}_G] = 1/3, \text{ pour } k = 0, 1, 2.$$

Nous adapterons directement l'argument ébauché ici au cadre plus général des martingales araignées telles qu'introduites dans [1]:

Une toile T de dimension l dans un espace vectoriel réel est par définition $T := \{\lambda.e^j | \lambda \geq 0; j = 1, \ldots, l\}$, où $\{e^j\}_{j=1,\ldots,l}$ est une famille de vecteurs de rang $l-1$ satisfaisant $\sum_{j=1}^{l} e^j = 0$. Nous noterons Δ l'ensemble convexe engendré par $\{e^j; j = 1, \ldots, l\}$, et h, l'application $T \to \{0, e^1, \cdots, e^l\}$ qui à $\lambda.e^j \in T$, $\lambda \neq 0$, associe $h(\lambda.e^j) := e^j$, et à 0 associe $h(0) := 0$.

Une martingale à valeurs dans une toile T est dite martingale araignée.

Le résultat que nous démontrerons ici est alors:

Théorème : *Soit $\{\mathcal{F}_t\}$ la filtration naturelle d'un mouvement Brownien W. Sur une toile T de dimension $l \geq 3$, la martingale identiquement nulle est la seule martingale araignée sur T issue de 0 et adaptée à $\{\mathcal{F}_t\}$.*

Démonstration : Soit Y une martingale araignée sur T, issue de 0 et adaptée à $\{\mathcal{F}_t\}$. Nous allons montrer que, pour tout temps déterministe S, p.s. $Y_S = 0$.

Si ce n'était pas le cas, nous pourrions construire par la technique de "concaténation" suivante une martingale araignée Z bornée, issue de 0, adaptée à $\{\mathcal{F}_t\}$ et telle que p.s., $Z_1 \neq 0$. Par arrêt, nous pouvons considérer Y_S borné. En outre cette variable peut s'exprimer comme une fonction H de la trajectoire $\{W_s, s \leq S\}$. Définissons alors

$$Y^n := H(\{\sqrt{2^{n+1}.S}(W_{1-1/2^n+s/(S.2^{n+1})} - W_{1-1/2^n}), s \leq S\}),$$

et notons

$$Z_1 := \sum_{n=0}^{\infty} \mathbb{1}_{\{\forall k \in \{0,\ldots,n-1\}: Y^k = 0\}} Y^n.$$

La martingale $Z_t := E[Z_1|\mathcal{F}_t]$ a alors les propriétés requises.

Pour $\epsilon > 0$ fixé, notons X_t la martingale $Z_{t \wedge \tau_\epsilon}/\epsilon$, où $\tau_\epsilon := \inf\{t \leq 1 : Z_t/\epsilon \notin \Delta\}$. Dénotons e le processus $h(X)$. Remarquons que sur l'événement $\{\tau_\epsilon < 1\}$, la relation $X_1 = h(X_1)$ est vérifiée. L'événement $\{X_1 = e_1\}$ a alors une probabilité aussi proche de 1 que l'on veut, pour peu que ϵ soit suffisamment petit, puisque $\{Z_1/\epsilon \notin \Delta\} \subset \{\tau_\epsilon < 1\}$ et $1 = P(Z_1 \neq 0) = P(\cup_{\epsilon > 0}\{Z_1/\epsilon \notin \Delta\})$.

Puisque X est mesurable par rapport à la filtration Brownienne, il existe une application F qui, à chaque trajectoire W du Brownien, associe la trajectoire $X_\cdot := F(W_\cdot)$ de la martingale X.

Tout comme Tsirelson, introduisons à présent un mouvement Brownien W' indépendant du premier et de même dimension. Soit alors $\{\mathcal{H}_t\}$ la filtration engendrée par (W, W').

Pour $r \in [0, 1[$, dénotons par W^r le mouvement Brownien $W^r := \sqrt{r}W + \sqrt{1-r}W'$ et soit X^r le processus $F(W^r)$. Clairement les deux processus X et X^r sont des martingales araignées de même loi sur la filtration $\{\mathcal{H}_t\}$. Définissons également le processus $e^r := h(X^r)$.

Soit ensuite $g_t := \sup\{s \leq t : X_s = 0\}$ et posons $G := g_1$ et $G^r := \sup\{t \leq 1 : X_t^r = 0\}$.

Nous établirons par la suite le lemme suivant qui indique que les martingales X et X^r sont à la fois très proches et suffisamment différentes:

Lemme :

1) $\lim_{r \to 1} P[\{e_1 = e_1^r\}] = 1$.

2) Pour tout $r < 1$: $P[\{G = G^r\}] = 0$ et $P[\{G^r < G\}] = 1/2$.

Nous noterons $\mathcal{H}_G$ la σ-algèbre engendrée par les variables aléatoires R_G, où R est un processus prévisible sur la filtration $\{\mathcal{H}_t\}$.

La formule de balayage (voir théorème 1 dans [4]) indique alors que si R est un processus prévisible borné à valeurs réelles, le processus $R_{g_t}X_t$ est une martingale. En particulier, $E[R_G X_1] = R_{g_0}X_0 = 0$. Vu la définition de $\mathcal{H}_G$, cela signifie que $E[X_1 | \mathcal{H}_G] = 0$.

Soit à présent A l'événement: $\{G^r < G\} \cap \{e_1 = e_1^r\} \cap \{X_1 = e_1\}$. Il suit des définitions de G^r et A que, sur A, $X_1 = e_1 = e_1^r = e_G^r$ et partant: $X_1 = e_G^r \mathbb{1}_A + X_1 \mathbb{1}_{A^c}$. Puisque e^r est un processus prévisible, e_G^r est mesurable par rapport à $\mathcal{H}_G$ et il suit alors que $0 = E[X_1 | \mathcal{H}_G] = e_G^r P[A | \mathcal{H}_G] + (1 - P[A | \mathcal{H}_G])U$, où $U := E[X_1 | \mathcal{H}_G, A^c]$.

Puisque X_1 est à valeurs dans Δ, par convexité, il en est de même de la variable U et l'on peut donc écrire U de manière unique comme $U = \sum_{j=1}^l \lambda_j e^j$, où $\lambda_j \geq 0$ et $\sum_{j=1}^l \lambda_j = 1$. Sur l'événement $\{e_G^r = e^k\}$, la relation précédente peut donc s'écrire: $0 = \sum_{j=1}^l \xi_j e^j$, avec $\xi_j := (1 - P[A | \mathcal{H}_G])\lambda_j$ si $j \neq k$ et $\xi_k := P[A | \mathcal{H}_G] + (1 - P[A | \mathcal{H}_G])\lambda_k$.

Puisque $\{e^j; j = 1, \ldots, l\}$ est une famille de rang $l - 1$ dont la somme est nulle, nous concluons que les égalités $\xi_j = \xi_k$ ont lieu pour tout $j \neq k$. Il suit alors:

$$
\begin{aligned}
1 &= \sum_{j=1}^l \lambda_j \\
&= (l-1)P[A | \mathcal{H}_G]/(1 - P[A | \mathcal{H}_G]) + l\lambda_k \\
&\geq (l-1)P[A | \mathcal{H}_G]/(1 - P[A | \mathcal{H}_G]).
\end{aligned}
$$

En particulier, sur $\{e_G^r = e^k\}$: $P[A | \mathcal{H}_G] \leq 1/l$.

En outre, sur $\{e_G^r = 0\}$, $X_G^r = 0$, et donc $G^r \geq G$. Ceci indique que l'événement A est incompatible avec $\{e_G^r = 0\}$ ou, en d'autres termes, $P[A|\mathcal{H}_G] = 0$ sur $\{e_G^r = 0\}$. Dès lors, la relation $P[A|\mathcal{H}_G] \leq 1/l$ est toujours vérifiée et partant $P[A] \leq 1/l$. Ceci est impossible pour $l \geq 3$ car, en vertu du lemme, lorsque $r \to 1$ et $\epsilon \to 0$, $P[A]$ tend vers $1/2$. ∎

Démonstration du lemme: Remarquons que les processus W et W^r ont même loi et que W^r converge en probabilité vers W lorsque r tend vers 1. Puisque e_1 peut s'écrire comme une fonction K de la trajectoire Brownienne $\{W_s, s \leq 1\}$ et que $e_1^r = K(\{W_s^r, s \leq 1\})$, l'assertion 1) est une conséquence immédiate du lemme de Slutsky (voir [1], théorème 1).

Il est aisé d'établir que les processus (W, W^r) et (W^r, W) ont même loi. Il en découle que (G, G^r) a même loi que (G^r, G) et partant la relation $P[G^r < G] = 1/2$ est une conséquence immédiate de la relation $P[G = G^r] = 0$, qui se démontre exactement comme le lemme 4 de [1]. ∎

REFERENCES

[1] Barlow, M.T., M. Emery, F. B. Knight, S. Song and M. Yor (1998), Autour d'un théorème de Tsirelson sur des filtrations browniennes et non-browniennes, *Sém. Prob. XXXII, Lecture Notes in Mathematics 1686*, 264-305, Springer Verlag, Berlin.

[2] Tsirelson, B. (1997), Triple points: from non-Brownian Filtrations to harmonic measures, *Geometric and Functional Analysis 7*, 1096-1142, Birkhäuser Verlag, Basel.

[3] Yor, M. (1979), Sur l'étude des martingales continues extrémales. *Stochastics 2*, 191-196.

[4] Yor, M. (1979), Sur le balayage des semimartingales continues. *Sém. Prob. XIII, Lecture Notes in Mathematics 721*, 453-471, Springer Verlag, Berlin.

ON CERTAIN PROBABILITIES
EQUIVALENT TO WIENER MEASURE,
D'APRÈS DUBINS, FELDMAN,
SMORODINSKY AND TSIRELSON

W. SCHACHERMAYER

ABSTRACT. L. Dubins, J. Feldman, M. Smorodinsky and B. Tsirelson gave an example of an equivalent measure Q on standard Wiener space such that each adapted Q-Brownian motion generates a strictly smaller filtration then the original one. The construction of this important example is complicated and technical.

We give a variant of their construction which differs in some of the technicalities but essentially follows their ideas, hoping that some readers may find our presentation easier to digest than the original papers.

1. INTRODUCTION

This paper grew out of the author's attempt to understand the construction of the admirable paper [DFST 96] as well as its extensions given in [FT 96] and [F 96].

Here is their main result:

1.1 Theorem. *(Dubins, Feldman, Smorodinsky, Tsirelson):*
Let $B = (B_t)_{t \geq 0}$ be a standard real-valued Brownian motion starting at $B_0 = 0$ defined on a stochastic base $(\Omega, \mathcal{F}, \mathbb{P})$ and its natural filtration $(\mathcal{F}_t)_{t \geq 0}$.

For $\varepsilon > 0$, there is a probability measure Q on $\mathcal{F}$, equivalent to $\mathbb{P}$, with $1 - \varepsilon \leq \frac{dQ}{d\mathbb{P}} \leq 1 + \varepsilon$ and such that for every $(\mathcal{F}_t)_{t \geq 0}$-adapted process $B' = (B'_t)_{t \geq 0}$ which is a standard Brownian motion under Q (relative to the filtration $(\mathcal{F}_t)_{t \geq 0}$), the process B' generates a strictly smaller filtration than $(\mathcal{F}_t)_{t \geq 0}$.

We refer to [SY 81], [RY 91], p. 336, [RY 94] p. 210 and [DFST 96] for an account on the significance of this theorem, which settled a 15-year-old question related to the Girsanov-transformation.

Let us also mention that recently B. Tsirelson [T 97] (see also [EY 98] and [BEKSY 98]) gave another example of a filtered probability space $(\Omega, \mathcal{F}, (\mathcal{F}_t)_{t \geq 0}, \mathbb{P})$, namely the space generated by a Walsh-martingale, which displays similar features as the present example $(\Omega, \mathcal{F}, (\mathcal{F}_t)_{t \geq 0}, Q)$: both examples are filtered probability spaces of "instant dimension 1" and not generated by a Brownian motion; the example in [T 97] is even robust under an equivalent change of measure (while the present one, of course, is not). These two examples are, nevertheless, different in spirit: roughly speaking in the present example the argument is based on the independence of the increments of Brownian motion while the example from [T 97] is based on the difference of Walsh-martingales and Brownian motion, when these processes hit zero.

The author frankly admits that he found it quite hard to understand the construction in [DFST 96]. After having paved his own way through the construction he thought that it might be helpful to the probability community to write up his understanding of the construction in order to give a somewhat different presentation of the example. However, no claims of originality are made (we just translate the ideas from [DFST 96] into a slightly alternative language) and we are not even sure whether our construction is "simpler" (of course, it seems simpler to the author; as usual in Mathematics, everything that you know how to do, seems simple to you). There are some technical differences in the construction of the present paper, as compared to [DFST 96]: firstly, we include the strengthening of the construction obtained in [FT 96], i.e., the control on the L^∞-norm rather than on the L^2-norm of the Radon-Nikodym-derivative $\frac{dQ}{d\mathbb{P}}$, from the very beginning into our construction (at little extra cost). This is natural, as the splitting into two steps (as in [DFST 96] and [FT 96]) apparently is only due to the way these authors gradually improved their example. Secondly we isolate a crucial step of the construction of [DFST 96] into the elementary combinatorial lemma 2.7 below, which — at least to the author — also allows for some intuitive understanding.

As regards the final strengthening by J. Feldman [F 96] we don't have any contribution: let us just note that this strengthening can be put on top of our example exactly in the same way as it was originally put on top of the example from [DFST 96] and [FT 96].

We have made an effort to keep our presentation entirely selfcontained; but, of course, we strongly recommend the reader to have a copy of [DFST 96] at hand.

My sincere thanks go to J. Feldman and M. Smorodinsky for a pleasant conversation on this topic and in particular to M. Smorodinsky for an inspiring talk in June 1997 at the Schrödinger Institute, Vienna, as well as to M. Emery and M. Yor for making me familiar with the content of the papers [T 97], [EY 98] and [BEKSY 98] and in particular to M. Emery for a lot of help and advice in the final redaction of the paper. After the completion of a first version of the present paper, S. Beghdadi-Sakrani and M. Emery also have given a further variant of the construction of [DFST 96] as well as some more general results [BE 99].

2. The Example

Let $X = (X_0, X_1, \cdots)$ be a real valued stochastic process defined on a stochastic base $(\Omega, \mathcal{F}, \mathbb{P})$. We shall look at the process "in reverse order", i.e., we define the filtration $(\mathcal{F}_n)_{n=0}^\infty$ to be

$$\mathcal{F}_n = \sigma(X_n, X_{n+1}, \cdots).$$

In the present paper we always shall assume that X is tail-trivial, i.e., that the sigma-algebra $\mathcal{F}_\infty = \bigcap_{n=0}^\infty \mathcal{F}_n$ only consists of sets of probability zero or one.

The subsequent definition describes the way in which the independence of Brownian increments will come into play. As we have learned from M. Smorodinsky the idea behind this definition goes back to P. Lévy (in [V 95], p. 756 it is referred to as the Lévy-Bernstein-Rosenblatt problem):

2.1 Definition (compare [S 98]). A *parametrisation* of the process X is given by a two-dimensional process $(\tilde{X}, Y) = (\tilde{X}_n, Y_n)_{n=0}^\infty$ defined on a stochastic base $(\tilde{\Omega}, \tilde{\mathcal{F}}, \tilde{\mathbb{P}})$ and a sequence $(f_n)_{n=1}^\infty$ of deterministic Borel-measurable functions defined on $[0,1] \times \mathbb{R}^{\mathbb{N}}$ such that

   (i) the processes X and $\tilde{X}$ are identical in law,
   (ii) the sequence $(Y_n)_{n=0}^\infty$ is a sequence of i.i.d. random variables uniformly distributed on $[0,1]$ and such that Y_n is independent of $(\tilde{X}_i)_{i=n+1}^\infty$,
   (iii) the equation
   $$\tilde{X}_{n-1}(\omega) = f_{n-1}(Y_{n-1}(\omega), \tilde{X}_n(\omega), \tilde{X}_{n+1}(\omega), \cdots)$$
   holds true, for each $n \geq 1$ and almost each ω.

We call the parametrisation *generating* if, in addition, for each n, the random variable X_n is $\sigma(Y_n, Y_{n+1}, \cdots)$-measurable.

We have been somewhat pedantic in the above definition, as regards the joining of the processes X and Y, by distinguishing between the processes X and $\tilde{X}$ to have a safe ground for the subsequent, rather subtle, considerations about the sigma-algebras which are generated by Y and $\tilde{X}$ rather than by Y and X (the latter being, strictly speaking, defined on different stochastic bases). But, if no confusion can arise, we shall follow the common habit in probability theory and write X instead of $\tilde{X}$.

Assertion (iii) requires that, for each n, there is a deterministic rule, prescribed by the functions f_{n-1}, such that, for almost each ω, we may determine the value $X_{n-1}(\omega)$ from the history $(X_n(\omega), X_{n+1}(\omega), \cdots)$ and the "innovation" $Y_{n-1}(\omega)$, the latter coming from a sequence of *independent* random variables. It is easy to see that any real-valued process X (in fact, any process taking its values in a polish space) admits a parametrisation. The notion of a *generating parametrisation* captures the intuitive idea of a parametrisation which is chosen in such a way that we can determine (a.s.) the value of $X_n(\omega)$ by only looking at the history $(Y_n(\omega), Y_{n+1}(\omega), \cdots)$ of the "innovations".

It is rather obvious that a process X admitting a generating parametrisation has to be tail-trivial: indeed, suppose to the contrary that there is a set $A \in \mathcal{F}_\infty = \bigcap_{n=0}^\infty \mathcal{F}_n$ with $0 < \mathbb{P}[A] < 1$ and suppose that X admits a generating parametrisation: then the set A is in $\sigma(X_0, X_1, \cdots)$ and is independent of $(Y_n)_{n=0}^\infty$ and therefore not in $\sigma(Y_0, Y_1, \cdots)$, a contradiction to the requirements of definition 2.1.

But the converse does not hold true, i.e., a tail-trivial process X does not, in general, allow a generating parametrisation. This highly non-trivial and remarkable fact was first proved by A. Vershik [V 70], [V 73]. We refer to ([DFST 96], p. 885) and [S 98] for a presentation of this example.

In fact, the construction given in [DFST 96] and its presentation in the present paper is just an example displaying the phenomenon of a tail trivial $\{-1, +1\}$-valued process X not admitting a generating parametrisation and such that, in addition, the process $(X_n)_{n=0}^\infty$ is obtained from an i.i.d. sequence $(\varepsilon_n)_{n=0}^\infty$ of Bernoulli-variables defined on $(\Omega, \mathcal{F}, \mathbb{P})$ by putting a slightly altered equivalent measure Q on $(\Omega, \mathcal{F})$.

We start by giving an easy motivating example which should help to develop some intuition for the concept of a *generating parametrisation* (we have learned it from M. Smorodinsky and found it illuminating despite its simplicity). As the example is not needed for the sequel the reader may just as well skip it.

Example. (compare [V 95], p. 756) Let $0 \le \eta < \frac{1}{2}$ and define the $\{-1, +1\}$-valued Markov process $(X_n)_{n=0}^\infty$ via the transition probabilities

$$\mathbb{P}[X_{n-1} = +1 | X_n = +1] = \frac{1}{2} + \eta,$$

$$\mathbb{P}[X_{n-1} = -1 | X_n = +1] = \frac{1}{2} - \eta,$$

$$\mathbb{P}[X_{n-1} = +1 | X_n = -1] = \frac{1}{2} - \eta,$$

$$\mathbb{P}[X_{n-1} = -1 | X_n = -1] = \frac{1}{2} + \eta,$$

for each n. Clearly this well-defines a stationary tail-trivial Markov process $(X_n)_{n=0}^\infty$.

A possible way to define a parametrisation of this process is given by the following coupling technique: let $(Y_n)_{n=0}^\infty$ be an i.i.d. sequence of random variables uniformly distributed in $[0,1]$. Define, for $m \in \mathbb{N}$, the process $\left(X_n^{(m)} \right)_{n=0}^m$ by letting $X_m^{(m)} = 1$ and, for $n = 1, \ldots, m$,

$$X_{n-1}^{(m)} = f_{n-1}(Y_{n-1}, X_n^{(m)}) = \begin{cases} \mathbb{I}_{(0, \frac{1}{2}+\eta)}(Y_{n-1}) - \mathbb{I}_{(\frac{1}{2}+\eta, 1)}(Y_{n-1}) \\ \qquad \text{if } X_n^{(m)} = 1 \\ \mathbb{I}_{(0, \frac{1}{2}-\eta)}(Y_{n-1}) - \mathbb{I}_{(\frac{1}{2}-\eta, 1)}(Y_{n-1}) \\ \qquad \text{if } X_n^{(m)} = -1 \end{cases}$$

One easily checks that, for $n \ge 0$ fixed, the sequence of random variables $\left(X_n^{(m)} \right)_{m=n}^\infty$ converges almost surely to a random variable X_n and the sequence $(X_n)_{n=0}^\infty$ satisfies the above Markov transition probabilities as well as the relations

$$X_{n-1} = f_{n-1}(Y_{n-1}, X_n) = \begin{cases} \mathbb{I}_{(0, \frac{1}{2}+\eta)}(Y_{n-1}) - \mathbb{I}_{(\frac{1}{2}+\eta, 1)}(Y_{n-1}) \\ \qquad \text{if } X_n = 1 \\ \mathbb{I}_{(0, \frac{1}{2}-\eta)}(Y_{n-1}) - \mathbb{I}_{(\frac{1}{2}-\eta, 1)}(Y_{n-1}) \\ \qquad \text{if } X_n = -1 \end{cases}$$

Let us verify explicitly that this parametrisation is generating: let $n \in \mathbb{N}$ and $\omega \in \Omega$ be such that $Y_n(\omega) \notin (\frac{1}{2} - \eta, \frac{1}{2} + \eta)$. In this case $Y_n(\omega)$ determines already $X_n(\omega)$, regardless of the history $(X_{n+1}(\omega), X_{n+2}(\omega), \cdots)$. But from now on we know everything about the trajectory $(X_n(\omega), X_{n-1}(\omega), \cdots, X_0(\omega))$ by only looking at $(Y_n(\omega), Y_{n-1}(\omega), \cdots, Y_0(\omega))$: the number $X_{n-1}(\omega)$ then is a deterministic function of the numbers $Y_n(\omega)$ and $Y_{n-1}(\omega)$, and so on.

More formally, for each n, the sigma-algebra $\sigma(Y_n, Y_{n+1}, \cdots)$ contains the sigma-algebra $\sigma(X_n)$ on the set

$$A_n = \bigcup_{i \geq n} \left\{ Y_i \notin \left(\frac{1}{2} - \eta, \frac{1}{2} + \eta \right) \right\}.$$

Noting that $\mathbb{P}[A_n] = 1$, for each $n \in \mathbb{N}$, we deduce that X_n is $\sigma(Y_n, Y_{n+1}, \cdots)$-measurable, which readily shows that the above parametrisation is generating. $\square$

We now give the basic example which relates the assertion of theorem 1.1 with the notion of a generating parametrisation.

2.2 Lemma. *Let $(B_t)_{t \geq 0}$ be a Brownian motion defined on $(\Omega, \mathcal{F}, \mathbb{P})$ equipped with its natural filtration $(\mathcal{F}_t)_{t \geq 0}$, and let Q be a probability measure on $\mathcal{F}$ equivalent to $\mathbb{P}$.*

Fix a sequence $(t_n)_{n=0}^{\infty}$ strictly decreasing to zero and define the process $(X_n)_{n=0}^{\infty}$ by letting

$$X_n = \begin{cases} +1 & \text{if } B_{t_n} - B_{t_{n+1}} \geq 0 \\ -1 & \text{if } B_{t_n} - B_{t_{n+1}} < 0 \end{cases}$$

Suppose that there is an $(\mathcal{F}_t)_{t \geq 0}$-adapted process $(B'_t)_{t \geq 0}$ defined on $(\Omega, \mathcal{F})$ which is a Brownian motion under Q, and such that $(B'_t)_{t \geq 0}$ generates the filtration $(\mathcal{F}_t)_{t \geq 0}$.

Then the process $X = (X_n)_{n=0}^{\infty}$ under the measure Q admits a generating parametrisation.

Proof. Let $(\tilde{Y}_n)_{n=0}^{\infty}$ be the sequence of random variables, defined on $(\Omega, \mathcal{F}, Q)$,

$$\tilde{Y}_n = \left(B'_t - B'_{t_{n+1}} \right)_{t_{n+1} \leq t \leq t_n}, \qquad n = 0, 1, \cdots$$

where $\tilde{Y}_n$ takes its values in the polish space $C[t_{n+1}, t_n]$. As the law of $\tilde{Y}_n$ is diffuse we may find Borel-isomorphisms $(i_n)_{n=0}^{\infty}$ from $C[t_{n+1}, t_n]$ to $[0, 1]$ such that $Y_n = i_n \circ \tilde{Y}_n$ is uniformly distributed on $[0, 1]$, which furnishes an i.i.d. sequence $(Y_n)_{n=0}^{\infty}$ of uniformly distributed $[0, 1]$-valued random variables under the measure Q.

Note that, for each $n \in \mathbb{N}$, the sigma-algebras $\sigma((Y_k)_{k \geq n})$ and $\sigma((B'_t)_{0 \leq t \leq t_n})$ coincide, and by assumption are equal to $\mathcal{F}_{t_n}$.

It follows that, defining the random variables φ_n and ψ_n by

$$\varphi_n = ((B_t - B_{t_{n+1}})_{t_{n+1} \leq t \leq t_n}, X_{n+1}, X_{n+2}, \dots)$$
$$\psi_n = (Y_n, X_{n+1}, X_{n+2}, \dots)$$

taking their values in $C[t_{n+1}, t_n] \times \{-1, +1\}^{\mathbb{N}}$ and $[0, 1] \times \{-1, +1\}^{\mathbb{N}}$ respectively, φ_n and ψ_n generate the same sigma-algebras (up to null-sets) on Ω, if we equip the respective target spaces with their Borel sigma-algebras. In particular, we may find a nullset N in Ω such that, for $\omega, \omega' \in \Omega \backslash N$, we have $\varphi_n(\omega) = \varphi_n(\omega')$ iff we have $\psi_n(\omega) = \psi_n(\omega')$. We infer that we may define a Borel map $F_n = F_n(y_n, x_{n+1}, x_{n+2}, \dots)$ from $[0, 1] \times \{-1, +1\}^{\mathbb{N}}$ to $C[t_{n+1}, t_n] \times \{-1, +1\}^{\mathbb{N}}$ inducing $\varphi_n \circ \psi_n^{-1}$ (to be precise: we define F_n by letting $F_n \circ \psi_n(\omega) = \varphi_n(\omega)$, for $\omega \in \Omega \backslash N$, and extend F_n in a Borel-measurable (but otherwise arbitrary) way from the range $\psi_n(\Omega)$ to the entire space $[0, 1] \times \{-1, +1\}^{\mathbb{N}}$). Defining f_n to be the sign of the first coordinate (i.e., the $C[t_{n+1}, t_n]$-coordinate) of the function F_n, evaluated at t_n, we have found the parametrisation

$$X_n = f_n(Y_n, X_{n+1}, X_{n+2}, \dots).$$

The sequence $(f_n)_{n=0}^{\infty}$ therefore defines a parametrisation of the process $(X_n)_{n=0}^{\infty}$ and it is clear that the parametrisation is generating, as by hypothesis $\sigma(Y_n, Y_{n+1}, \dots) = \mathcal{F}_{t_n}$, for each $n \in \mathbb{N}$. $\quad\square$

Remark. We have used the concept of *generating parametrisation*, as in [S 98], instead of the concept of *substandard processes*, i.e., processes admitting a *standard extension*, as in [DFST 96], because we find the former notion more intuitive. Both concepts are equivalent and may be mutually translated one into the other (compare also [S 98]).

The message of lemma 2.2 is that the proof of theorem 1.1 may be reduced to a coin-tossing game, indexed by the negative numbers.

2.2a Corollary. *In order to prove theorem 1.1 it suffices to give a proof for the subsequent assertion:*

Denote by λ the Haar probability measure on the Borel sigma-algebra $\mathcal{B}$ of $\mathfrak{X} = \{-1, +1\}^{\mathbb{N}}$ and by $\varepsilon_n : \mathfrak{X} \to \{-1, +1\}$ the n'th coordinate projection. For $\varepsilon > 0$, there is a probability measure μ on $\mathfrak{X}$ with $1 - \varepsilon \leq \frac{d\mu}{d\lambda} \leq 1 + \varepsilon$ and such that the process $(\varepsilon_n)_{n=1}^{\infty}$, as defined on $(\mathfrak{X}, \mathcal{B}, \mu)$, does not admit a generating parametrisation.

Proof. Using the notation of lemma 2.2 consider $X = (X_n)_{n=1}^{\infty}$ defined there as a measurable map from $(\Omega, \mathcal{F})$ to $(\mathfrak{X}, \mathcal{B})$. Assuming that there is a measure μ on $\mathfrak{X}$ satisfying the above assertion define the measure Q on $\mathcal{F}$ by letting

$$\frac{dQ}{d\mathbb{P}}(\omega) = \frac{d\mu}{d\lambda}(X(\omega)), \qquad \text{for } \omega \in \Omega.$$

This definition is done in such a way that the process $(\varepsilon_n)_{n=1}^{\infty}$, defined on $(\mathfrak{X}, \mathcal{B}, \mu)$ and the process $(X_n)_{n=1}^{\infty}$, defined on $(\Omega, \mathcal{F}, Q)$ are identical in law.

By assumption, $(\varepsilon_n)_{n=1}^{\infty}$ does not admit a generating parametrisation under μ, hence $(X_n)_{n=1}^{\infty}$ does not admit one either under Q. It follows from lemma 2.2 that $(\mathcal{F}_t)_{t\geq 0}$ cannot be generated by a Q-Brownian motion $(B_t')_{t\geq 0}$. $\square$

The remainder of the paper will be dedicated to construct a measure μ on $\mathfrak{X}$ satisfying the assertion of the above corollary.

The principal component of the construction is given in the subsequent lemma.

Let us fix some notation: for $n \in \mathbb{N}$, we denote by $\mathfrak{X}_n$ the space $\{-1, +1\}^n$ and by λ_n, or just λ, if there is no danger of confusion, the uniform probability distribution on $\mathfrak{X}_n$. By $(\varepsilon_i)_{i=1}^n$ we denote the coordinate functions on $\mathfrak{X}_n$. Note that $(\varepsilon_i)_{i=1}^n$ is an i.i.d. sequence of Bernoulli-variables if we equip $\mathfrak{X}_n$ with the probability measure λ_n.

A little notational warning: in the subsequent lemma the time $i = 1, \cdots, p$ will "run into the future" as opposed to the setting above, and — when speaking about a stopping time — we shall refer to the filtration $(\mathcal{F}_i)_{i=0}^p$, where $\mathcal{F}_i = \sigma(\varepsilon_1, \cdots, \varepsilon_i)$.

2.3. Lemma. *Let* $p \in \mathbb{N}$, $\frac{1}{8} > \kappa > \kappa/4 > \eta > 0$, *and define the density process* $Z = (Z_i)_{i=0}^p$ *on* $(\mathfrak{X}_p, \lambda_p)$ *by* $Z_0 = 1$,

$$Z_i/Z_{i-1} = 1 + \eta\varepsilon_i, \qquad\qquad i = 1, \cdots, p$$

and the stopping time T *by*

$$T = \inf\{1 \leq i \leq p : Z_i \notin [1 - (\kappa - \eta(1+\kappa)), 1 + (\kappa - \eta(1+\kappa))]\} \wedge p.$$

Denote by $\hat{\mu}$ *and* μ *respectively the probability measures on* $\mathfrak{X}_p$ *with Radon-Nikodym derivatives*

$$\frac{d\hat{\mu}}{d\lambda} = Z_p \quad and \quad \frac{d\mu}{d\lambda} = Z_T.$$

We then have:

(i) $\lambda[T < p] < \frac{4p\eta^2}{\kappa^2}$,

(ii) $1 - \kappa \leq \frac{d\mu}{d\lambda} \leq 1 + \kappa$

(iii) *for every pair* $(f_i^\mu)_{i=1}^p$ *and* $(f_i^\lambda)_{i=1}^p$ *of parametrisations of the coordinate process* $(\varepsilon_1, \cdots, \varepsilon_p)$ *under the measures* μ *and* λ *respectively we have that*

$$\mathbb{P}[(f_i^\mu)_{i=1}^p = (f_i^\lambda)_{i=1}^p] \leq \left(1 - \frac{\eta}{2}\right)^p + \frac{4p\eta^2}{\kappa^2}$$

Before aboarding the proof we want to clarify — again somewhat pedantically — the precise meaning of assertion (iii): we equip $\Omega = \mathfrak{X}_p \times [0,1]^p = \{-1, +1\}^p \times [0,1]^p$ with measure $\mathbb{P} = \lambda \otimes m$, where m denotes the p-fold product of Lebesgue-measure on $[0,1]$. We denote by $X_1, \cdots, X_p, Y_1, \cdots, Y_p$ the projections to the coordinates of Ω and by $(x_1, \cdots, x_p, y_1, \cdots, y_p)$ the elements of Ω. By the parametrisations $(f_i^\mu)_{i=1}^p$ and $(f_i^\lambda)_{i=1}^p$ we mean deterministic functions $f_i^\mu(y_i, x_{i-1}, \cdots, x_1)$ and $f_i^\lambda(y_i, x_{i-1}, \cdots, x_1)$ such that the processes

$(f_i^\mu(Y_i, X_{i-1}, \cdots, X_1))_{i=1}^p$ and $(f_i^\lambda(Y_i, X_{i-1}, \cdots, X_1))_{i=1}^p$ are versions of the co-ordinate processes $(\varepsilon_i)_{i=1}^p$ defined on $(\mathfrak{X}_p, \mu)$ and $(\mathfrak{X}_p, \lambda)$ respectively.

Proof of Lemma 2.3. (i) Writing the defining equation of $(Z_i)_{i=0}^p$ as

$$Z_{i+1} - Z_i = Z_i \eta \varepsilon_i$$

we see that Z is a martingale with respect to the measure λ and that

$$\|Z_T - 1\|_{L^2(\lambda)}^2 = \|Z_p^T - Z_0\|_{L^2(\lambda)}^2$$
$$= \sum_{i=1}^p \|Z_i^T - Z_{i-1}^T\|_{L^2(\lambda)}^2$$
$$\leq p[\eta(1 + (\kappa - \eta(1 + \kappa)))]^2 \leq 2p\eta^2.$$

Here we denoted by $Z^T = (Z_i^T)_{i=0}^p$ the stopped process $Z^T = (Z_{i \wedge T})_{i=0}^p$. Noting that on $\{T < p\}$ we have that $|Z_T - Z_0| > \kappa - \eta(1 + \kappa)$ we get

$$\lambda[T < p] \leq \frac{2p\eta^2}{(\kappa - \eta(1 + \kappa))^2} \leq \frac{4p\eta^2}{\kappa^2}.$$

(ii) is rather obvious as we have defined T in such a way that Z_T is certain to stay within $[1 - \kappa, 1 + \kappa]$.

(iii) We first reason with the measure $\hat{\mu}$ instead of μ and we shall write $f_i^{\hat{\mu}}$ for a parametrisation of $(\varepsilon_i)_{i=1}^p$ under $\hat{\mu}$. Note that, for every $1 \leq i \leq p$ and every $x_1, \cdots, x_{i-1}$ we have that

$$\mathbb{P}[f_i^{\hat{\mu}} = 1 | X_1 = x_1, \cdots, X_{i-1} = x_{i-1}] = \frac{1 + \eta}{2}$$
$$\text{and} \quad \mathbb{P}[f_i^\lambda = 1 | X_1 = x_1, \cdots, X_{i-1} = x_{i-1}] = \frac{1}{2}$$

Hence, conditionally on each set $\{X_1 = x_1, \ldots, X_{i-1} = x_{i-1}\}$ the event $\{f_i^{\hat{\mu}} = f_i^\lambda\}$ depends only on Y_i and has probability at most $1 - \frac{\eta}{2}$. Using the independence of the random variables $Y_1, \ldots, Y_i$ we therefore get

$$\mathbb{P}[f_i^{\hat{\mu}} = f_i^\lambda | X_1, \ldots, X_{i-1}, Y_1, \ldots, Y_{i-1}] \leq 1 - \frac{\eta}{2}$$

which gives

$$\mathbb{P}[(f_j^{\hat{\mu}})_{j=1}^i = (f_j^\lambda)_{j=1}^i] \leq (1 - \frac{\eta}{2})\mathbb{P}[(f_j^{\hat{\mu}})_{j=1}^{i-1} = (f_j^\lambda)_{j=1}^{i-1}]$$

and therefore

$$\mathbb{P}[(f_i^{\hat{\mu}})_{i=1}^p = (f_i^\lambda)_{i=1}^p] \leq (1 - \frac{\eta}{2})^p.$$

To pass from $\hat{\mu}$ to μ note that on the set $\{T = p\}$ the measures μ and $\hat{\mu}$ coincide which, using (i), readily implies the inequality

$$P\left[(f_i^\mu)_{i=1}^n = (f_i^\lambda)_{i=1}^n\right] \leq \left(1 - \frac{\eta}{2}\right)^p + \frac{4p\eta^2}{\kappa^2}. \qquad \Box$$

The message of the above lemma is quite counter-intuitive and surprising, at least to the author (when choosing the parameters such that the bounds in (i) and (iii) are close to zero and in (ii) close to one): on one hand side (ii) asserts that the random variables $(X_1, \cdots, X_p) = (\varepsilon_1, \cdots, \varepsilon_p)$ have a very similar joint distribution under μ and under λ; on the other hand (iii) implies that if we try to parameterise the process $(X_1, \cdots, X_p)$ under μ and λ respectively then, for each parametrisation $(f_i^\mu)_{i=1}^p, (f_i^\lambda)_{i=1}^p$, there are only few ω's such that $f_i^\mu(Y_i(\omega), X_{i-1}(\omega), \cdots X_1(\omega)) = f_i^\lambda(Y_i(\omega), X_{i-1}(\omega), \cdots X_1(\omega))$, for $i = 1, \cdots, p$. Loosely speaking: although the result of the random variable $(X_1, \cdots, X_n)$ is likely to be the same under μ as well as under λ we cannot materialise this probable coincidence by a sequential pathwise procedure parametrised by independent increments on the coordinates $i = 1, \cdots, p$.

A similar interpretation of the above lemma goes as follows: There is a Borel-measurable transformation $T : (\mathfrak{X}_p \times [0,1], \lambda \otimes m) \longrightarrow \mathfrak{X}_p$ such that $T(\lambda \otimes m) = \mu$ and such that

$$\mathbb{P}[T(X_1, \cdots, X_p, Y) = (x_1, \cdots, x_p)|X_1 = x_1, \cdots, X_p = x_p] \geq 1 - \kappa,$$

for each $(x_1, \cdots, x_p) \in \mathfrak{X}_p$, where $\mathbb{P}$ denotes $\lambda \otimes m$. This is just a straightforward reinterpretation of the assertion $\frac{d\mu}{d\lambda} \geq 1 - \kappa$. In particular we have

$$\mathbb{P}[T(X_1, \cdots, X_p, Y) = (X_1, \cdots, X_p)] \geq 1 - \kappa.$$

On the other hand, (iii) can be interpreted as the fact that for every Borel-measurable transformation $T : \mathfrak{X}_p \times [0,1]^p \longrightarrow \mathfrak{X}_p$ which maps $\lambda \otimes m^p$ to μ and in addition, is $\sigma(X_1, \cdots, X_{i-1}, Y_i) \longrightarrow \sigma(X_1, \cdots, X_i)$ measurable, for each $i = 1, \cdots, p$, we have

$$\mathbb{P}[T(X_1, \cdots, X_p, Y_1, \cdots, Y_p) = (X_1, \cdots, X_p)] \leq \left(1 - \frac{\eta}{2}\right)^2 + \frac{4p\eta^2}{\kappa^2}.$$

At the danger of being repetitive, let us rephrase this once more in terms of a mind experiment: suppose you are told the laws λ and μ as above and you are given a machine which produces an i.i.d. sequence $(Y_1, \cdots, Y_p)$ of $[0,1]$-valued uniformly distributed random variables. Define (w.l.g.) the functions

$$f_i^\lambda(Y_i) = \begin{cases} +1 \text{ if } Y_i \in [0, \frac{1}{2}] \\ -1 \text{ if } Y_i \in]\frac{1}{2}, 1] \end{cases}$$

so that $(X_i)_{i=1}^p = (f_i^\lambda(Y_i))_{i=1}^p$ is a fair sequence of p coin tosses. Now you are asked to define a (deterministic) mechanism which associates to every outcome $(x_1, \cdots, x_p) = (X_1(\omega), \cdots, X_p(\omega))$, possibly using the information of the underlying random numbers $(y_1, \cdots, y_p) = (Y_1(\omega), \cdots, Y_p(\omega))$, a "manipulated" outcome $(\check{x}_1, \cdots, \check{x}_p) = T(x_1, \cdots, x_p, y_1, \cdots, y_p)$ such that the process $(\check{X}_1, \cdots, \check{X}_p)$ has law μ and, in addition, this application of "corriger la fortune" should only be applied rather seldomly, i.e. $\mathbb{P}[(\check{X}_1, \cdots \check{X}_p) \neq (X_1, \cdots, X_p)]$ should be small. The question is: can you do this? The answer depends on the interpretation of what we mean by "deterministic mechanism". If we are allowed to first wait until we know the entire realisation $(x_1, \cdots, x_p)$, the answer is yes, as the map T constructed above, $(\check{x}_1, \cdots, \check{x}_p) = T(x_1, \cdots, x_p, y)$ satisfies $\mathbb{P}[(\check{X}_1, \cdots, \check{X}_p) \neq (X_1, \cdots, X_p)] < \kappa$ (as random source Y we may, e.g., take the fractional part of the random variable $2Y_1$). But if we are confined to make our choice "in real time" (compare ([T 97], def. 1.1 and the subsequent discussion) for a precise definition of this notion), i.e., we have to decide whether we let $\check{x}_i = x_i$ or $\check{x}_i \neq x_i$ after having only seen the outcomes $x_1, \cdots, x_{i-1}$ and using the information y_i, then the answer is no: assertion (iii) above implies that for each such rule $(f_i(y_i, x_{i-1}, \cdots, x_i))_{i=1}^p$ producing a process $(\check{X}_i)_{i=1}^p$ under the law μ, the probability that we have to change x_i into $\check{x}_i \neq x_i$, for at least one i, is close to one.

For the proof of theorem 1.1 we shall apply the above lemma in a slightly more technical form which we describe in the next lemma.

2.4 Lemma. *Let p, κ, η be as in lemma 2.3 above and suppose we are given in addition $0 < \alpha < 1$.*

Let $(\tau_i)_{i=1}^p, (\tau_i')_{i=1}^p$ be two elements in $\{-1, +1\}^p$ such that

$$\#\{i : \tau_i \neq \tau_i'\} \geq \alpha p.$$

Define two density processes Z, Z' by letting $Z_0 = Z_0' = 1$ and

$$Z_i/Z_{i-1} = 1 + \tau_i \eta \varepsilon_i, \quad Z_i'/Z_{i-1}' = 1 + \tau_i' \eta \varepsilon_i$$

and two stopping times T and T' by

$$T = \inf\{1 \leq i \leq p : \{Z_i \notin [1 - (\kappa - \eta(1+\kappa)), 1 + (\kappa - \eta(1+\kappa))]\} \wedge p$$
$$T' = \inf\{1 \leq i \leq p : \{Z_i' \notin [1 - (\kappa - \eta(1+\kappa)), 1 + (\kappa - \eta(1+\kappa))]\} \wedge p$$

and by $\mu, \hat{\mu}, \mu', \hat{\mu}'$ the measures with densities

$$\frac{d\hat{\mu}}{d\lambda} = Z_p, \frac{d\mu}{d\lambda} = Z_T, \quad \frac{d\hat{\mu}'}{d\lambda} = Z_p', \frac{d\mu'}{d\lambda} = Z_{T'}'.$$

We then have

(i) $\lambda[T < p, T' < p] < \frac{8p\eta^2}{\kappa^2}$,

(ii) $1 - \kappa \leq \frac{d\mu}{d\lambda} \leq 1 + \kappa$ *and* $1 - \kappa \leq \frac{d\mu'}{d\lambda} \leq 1 + \kappa$,

(iii) *for every pair $(f_i^\mu)_{i=1}^p$ and $(f_i^{\mu'})_{i=1}^p$ of parametrisations of the coordinate process $(\varepsilon_1, \cdots, \varepsilon_p)$ under the measures μ and μ' respectively we have that*

$$\mathbb{P}[(f_i^\mu)_{i=1}^p = (f_i^{\mu'})_{i=1}^p] \leq (1 - \eta)^{\alpha p} + \frac{8p\eta^2}{\kappa^2}$$

The proof of lemma 2.4 is analogous to that of 2.3 and therefore skipped.

We now indicate for which values of the parameters p, κ, η, α we shall apply lemma 2.4 in our subsequent inductive construction indexed by $k = k_0, k_0+1, \cdots$; in the sequel we shall (almost) always remain the following relations between the integers k, n and p:

$$n = 2^k$$
$$p = 2^{k-1} = n/2.$$

This rather peculiar notation comes from the fact that we want to stick as close as possible to the notation in [DFST 96], who used the symbols k and n in a similar way as we do and, on the other hand, we want to avoid constant use of the notation $k - 1$ and $n/2$ for quantities which will constantly be used.

2.5 Corollary. *For $k \in \mathbb{N}$ we shall choose*

$$p = p_k = 2^{k-1},$$
$$\alpha = \alpha_k = p_k^{-\frac{1}{4}},$$
$$\kappa = \kappa_k = k^{-2},$$
$$\eta = \eta_k = k^{-3} 2^{-k/2}.$$

Using these parameters in lemma 2.4 we obtain, for k sufficiently large, the estimates

(ii) $1 - k^{-2} \le \frac{d\mu}{d\lambda} \le 1 + k^{-2}$ and $1 - k^{-2} \le \frac{d\mu'}{d\lambda} \le 1 + k^{-2}$

(iii) $\mathbb{P}[(f_i^{\mu})_{i=1}^p = (f_i^{\mu'})_{i=1}^p] \le 5k^{-2}$.

Proof. We have to estimate the quantities $(1 - \eta)^{\alpha p}$ and $\frac{8p\eta^2}{\kappa^2}$ in assertion (iii) of lemma 2.4:

$$(1-\eta_\kappa)^{\alpha_\kappa p_\kappa} = (1-k^{-3}2^{-k/2})^{2^{k-1} \cdot 2^{-\frac{k-1}{4}}} \approx (1-2^{-k/2})^{2^{\frac{3k}{4}}} = ((1-2^{-k/2})^{2^{k/2}})^{2^{k/4}} \approx e^{-2^{k/4}}$$

and

$$\frac{8p\eta^2}{\kappa^2} = \frac{8k^{-6}2^{k-1} \cdot 2^{-k}}{k^{-4}} = 4k^{-2}. \quad \square$$

We have used in the above proof the symbol $\approx$ to describe an approximate equality and we shall freely continue to do so when it is clear that the asymptotic approximations work good enough to prove the desired estimates.

We now can formulate the result which parallels the "Fundamental Lemma" of ([DFST 96], p. 894):

2.6 Fundamental Lemma. *For k large enough, there is a family $(\mu_j)_{j=1}^{2^{2n}} = (\mu_j)_{j=1}^{2^{2^{k+1}}}$ of probability measures on $\mathfrak{X}_n = \mathfrak{X}_{2^k}$ such that*

(ii) $$1 - k^{-2} \le \frac{d\mu_j}{d\lambda} \le 1 + k^{-2}, \qquad j = 1, \cdots, 2^{2n},$$

(iii) for every pair $j \neq j'$ and parametrisations $(f_i^{\mu_j})_{i=1}^n$ and $(f_i^{\mu_{j'}})_{i=1}^n$ of the co-ordinate process $(X_1, \cdots, X_n)$ on $\mathfrak{X}_n$ under the measures μ_j and $\mu_{j'}$ respectively, we have

$$\mathbb{P}\left[(f_i^{\mu_j})_{i=1}^n = (f_i^{\mu_{j'}})_{i=1}^n\right] \leq 6k^{-2}.$$

· Of course, the idea to prove the fundamental lemma is to apply lemma 2.4 and corollary 2.5, where we let $n = p$ (in contrast to our above agreement on notation $p = n/2$; the reason why we finally have to take $p = n/2$ will soon become clear): we would like to find 2^{2n} many different sequences $\tau_i(j)_{i=1}^n$ taking their values in $\{-1, +1\}$, where $1 \leq j \leq 2^{2n}$, such that, for every fixed pair $j \neq j'$, we have, for at least $n^{3/4}$ many i's, that $\tau_i(j) \neq \tau_i(j')$. We advise the reader to convince her- or himself that — if such a choice $(\tau_i(j))_{i=1}^n)_{j=1}^{2^{2n}}$ were indeed possible — it were straightforward to deduce the fundamental lemma from lemma 2.4.

However, life is not always as nice and easy as we would like it to be: there is no sequence $(\tau_i(\cdot))_{i=1}^n$ of $\{-1, +1\}$-valued functions on a set of cardinality 2^{2n} such that, for $j \neq j'$, we have $\tau_i(j) \neq \tau_i(j')$ for at least $n^{3/4}$ many i's. In fact, such a sequence $(\tau_i(\cdot))_{i=1}^n$ cannot even separate the points of a set of cardinality 2^{2n} (it needs $2n$ functions to do this job); hence there always will be some $j \neq j'$ such that $\tau_i(j) = \tau_i(j')$, for all $i = 1, \cdots, n$.

So we have to proceed in a more sophisticated way: note that — in spite of the above sad news — for a *typical* choice of $j \neq j'$ there will be approximately $n/2$ many (i.e., much more than the required $n^{3/4}$ many) i's such that $\tau_i(j) \neq \tau_i(j')$ if we take $(\tau_i)_{i=1}^n$ to be an independent sequence of functions defined on $(\mathfrak{X}_{2n}, \lambda_{2n})$ assuming the values $+1$ and -1 with probability $\frac{1}{2}$ (we now identify the set $\{j : 1 \leq j \leq 2^{2n}\}$ with $\mathfrak{X}_{2n}$ equipped with measure λ_{2n}). The basic idea is to consider not only one sequence $(\tau_i)_{i=1}^p$ (from now on we are generous and use only $p = \frac{n}{2}$ many functions) but a large collection $((\tau_i^r)_{i=1}^p)_{r=1}^{2^p}$ of such sequences, which we may think of as applying an i.i.d. sequence $(\tau_i)_{i=1}^p$ as above to 2^p many random permutations of the set $\mathfrak{X}_{2n}$. If we do this it seems quite intuitive that for *the overwhelming majority of pairs* $j \neq j'$ we have that *for most of the* $1 \leq r \leq 2^p$ we have $\tau_i^r(j) \neq \tau_i^r(j')$ for at least $n^{3/4}$ many i's.

The subsequent combinatorial lemma, whose proof is based on the above ideas, shows that we even can replace the term *for the overwhelming majority of pairs* $j \neq j'$ by the term *for each pair* $j \neq j'$.

2.7 Combinatorial Lemma. *Letting $p = 2^{k-1}$ and $n = 2^k$, for k sufficiently large, there is a family $((\tau_i^r(\cdot))_{i=1}^p)_{r=1}^{2^p}$ of $\{-1, +1\}$-valued functions defined on the set $\mathfrak{X}_{2n} = \{-1, +1\}^{2n}$ such that, for each pair $j \neq j'$ in $\mathfrak{X}_{2n}$, we have*

$$\frac{\#\{r : \#\{i : \tau_i^r(j) \neq \tau_i^r(j')\} \geq n^{3/4}\}}{2^p} \geq 1 - p^{-1/2}.$$

The proof of the lemma relies on elementary combinatorics and is somewhat lengthy. Also we suspect that there are much stronger results known in the combinatorial literature (but not known to the author). For these reasons we banned the proof of the combinatorial lemma 2.7 to the appendix.

Proof of the Fundamental Lemma 2.6. We shall define the measures $(\mu_j)_{j=1}^{2^{2n}}$ on $\mathfrak{X}_n$ by defining the density processes $(Z_i^j)_{i=1}^n$ with respect to the measure λ on $\mathfrak{X}_n$.

For the first $p = n/2$ coordinates, we *don't do anything!* We simply let

$$Z_i^j = 1, \qquad\qquad \text{for } i = 1, \cdots, p, \quad j = 1, \cdots, 2^{2n}.$$

The first $p = n/2$ coordinates are only used to create 2^p many atoms in $\sigma(X_1, \cdots, X_p)$ defined by $\{X_1 = \pm 1, \cdots, X_p = \pm 1\}$, for all choices of ± 1, where $X_1, \cdots, X_p$ denote the first p coordinate functions on $\mathfrak{X}_n$. We enumerate these atoms by $I_1, \cdots, I_r, \cdots, I_{2^p}$.

Identifying the set $\{j : 1 \le j \le 2^{2n}\}$ with $\mathfrak{X}_{2n}$ apply lemma 2.7 to choose the functions $(\tau_i^r(j))_{i=1}^p$ satisfying

$$\frac{\#\{r : \#\{i : \tau_i^r(j) \ne \tau_i^r(j')\} \ge n^{3/4}\}}{2^p} \ge 1 - p^{-1/2}.$$

Now define, for $1 \le j \le 2^{2n}$, and $p \le i < n$,

$$\left(Z_{i+1}^j / Z_i^j\right) \chi_{I_r} = 1 + \tau_i^r(j) \eta \varepsilon_i, \qquad r = 1, \cdots, 2^p$$

where from now on the parameters p, α, κ, η are understood to denote the parameters $p_k, \alpha_k, \kappa_k, \eta_k$ defined in corollary 2.5. We again stop the density processes at time

$$T_j = \inf\{1 \le i \le n : Z_i^j \notin [1 - (\kappa - \eta(1 + \kappa)), 1 + (\kappa - \eta(1 + \kappa))]\} \wedge n$$

and define

$$\frac{d\mu_j}{d\lambda} = Z_{T_j}^j.$$

Assertion (ii) of the fundamental lemma now follows from assertion (ii) of corollary 2.5.

To prove (iii) fix $j \ne j'$: on at least $(1 - p^{-1/2})2^p$ many of the atoms I_r the (renormalized) restrictions of μ_j and $\mu_{j'}$ to the atom I_r satisfies the hypotheses of lemma 2.4 and corollary 2.5. Hence, for any pair of parametrisations $(f_i^{\mu_j})_{i=1}^n$ and $(f_i^{\mu_{j'}})_{i=1}^n$ of the coordinate process $(X_1, \cdots, X_n)$ under the measures μ_j and $\mu_{j'}$ respectively we have

$$\mathbb{P}[(f_i^{\mu_j})_{i=1}^n = (f_i^{\mu_{j'}})_{i=1}^n | (X_1, \cdots, X_p) \in I_r] \le 5k^{-2}$$

for at least $(1 - p^{-1/2})2^p = (1 - 2^{-\frac{k-1}{2}})2^p$ many $r's$. Hence

$$\mathbb{P}[(f_i^{\mu_j})_{i=1}^n = (f_i^{\mu_{j'}})_{i=1}^n] \le 5k^{-2} + 2^{-\frac{k-1}{2}}$$

which shows assertion (iii) and finishes the proof of the Fundamental Lemma. $\qquad\square$

Proof of Theorem 1.1. Similarly as in [DFST 96] we only have to paste the ingredients together which are provided by the Fundamental lemma, in order to construct a probability measure μ on $\mathfrak{X} = \{-1, +1\}^{\mathbb{N}}$ satisfying the requirements of corollary 2.2a: choose k_0 to be large enough such that, for $k \geq k_0$, the assertions of the Fundamental Lemma hold true and such that

$$\prod_{k=k_0}^{\infty} (1 + k^{-2}) < 1 + \varepsilon \text{ and } \prod_{k=k_0}^{\infty} (1 - 6k^{-2}) > \frac{3}{4},$$

where $\frac{1}{2} > \varepsilon > 0$ is taken from the statement of Corollary 2.2a.

Let $\tilde{\mathfrak{X}}$ be the compact space

$$\mathfrak{X} = \prod_{k=k_0}^{\infty} \mathfrak{X}_{2^k} = \prod_{k=k_0}^{\infty} \{-1, +1\}^{2^k},$$

and define, for $k \geq k_0$, the Markov transition probabilities $(\mu_{x_{k+1}})_{x_{k+1} \in \mathfrak{X}_{2^{k+1}}}$ to be the family of probability measures on $\mathfrak{X}_{2^k}$ given by the Fundamental Lemma 2.6, where we identify the set $\{j : 1 \leq j \leq 2^{2^{k+1}}\}$ with the set $\{x_{k+1} : x_{k+1} \in \mathfrak{X}_{2^{k+1}}\}$ by an arbitrary bijection.

Denote by $V_{x_{k+1}}(x_k)$ the Radon-Nikodym derivative of $\mu_{x_{k+1}}$ with respect to Haar measure λ_{2^k} on $\mathfrak{X}_{2^k}$, i.e.

$$V_{x_{k+1}} = \frac{d\mu_{x_{k+1}}}{d\lambda_{2^k}},$$

and by Z the density function on $\mathfrak{X}$,

$$Z(x) = \prod_{k=k_0}^{\infty} V_{x_{k+1}}(x_k)$$

where $x = (x_k)_{k=k_0}^{\infty} \in \mathfrak{X}$. By assertion (ii) of the fundamental lemma 2.6 and the above choice of k_0 we have $\|Z - 1\|_{\infty} < \varepsilon$ and the measure μ on $\mathfrak{X}$ defined by

$$\frac{d\mu}{d\lambda} = Z$$

is the unique probability measure on the Borel sets of $\mathfrak{X}$ inducing the transition probabilities $(\mu_{x_{k+1}})_{x_{k+1} \in \mathfrak{X}_{2^{k+1}}}$.

We still have to show that the coordinate process on $\mathfrak{X}$, which we now denote by X, under the measure μ does not admit a generating parametrisation, which will finish the proof of theorem 1.1 by corollary 2.2a. So, fix a parametrisation

$$\left(\left(\tilde{X}_{k,i}\right)_{i=1}^{2^k}\right)_{k=k_0}^{\infty} = \left(\left(f_{k,i}(Y_{k,i}, \tilde{X}_{k,i+1}, \ldots, \tilde{X}_{k,2^k}, \tilde{X}_{k+1,1}, \ldots)\right)_{i=1}^{2^k}\right)_{k=k_0}^{\infty}.$$

of the process X, where we now are careful to write $\tilde{X}$ instead of X (compare definition 2.1 and the subsequent discussion). Assuming that the parametrisation is generating let us work towards a contradiction.

To alleviate notation, we write y_k and x_k (resp. Y_k and X_k or $\tilde{X}_k$ if we refer to random variables) for the elements $y_k = (y_{k,i})_{i=1}^{2^k}$ and $x_k = (x_{k,i})_{i=1}^{2^k}$ in $\mathfrak{X}_{2^k}$, and f_k for the $\mathfrak{X}_{2^k}$-valued function $f_k = (f_{k,i})_{i=1}^{2^k}$. We then may write the parametrisation as

$$(\tilde{X}_k)_{k=k_0}^\infty = \left(f_k(Y_k, \tilde{X}_{k+1}, \tilde{X}_{k+2}, \dots) \right)_{k=k_0}^\infty ,$$

with the interpretation, that the components $(f_{k,i})_{i=1}^{2^k}$ are defined inductively (for $i = 2^k, 2^k - 1, \dots, 1$) by the above more explicit formula, letting $x_{k,i} = f_{k,i}(y_{k,i}, x_{k,i+1}, \dots,$
$x_{k,2^k}, x_{k+1}, \dots)$.

To further alleviate notation, note that by the construction of the measure μ on $\mathfrak{X}$ the random variable $\tilde{X}_k = f_k(Y_k, \tilde{X}_{k+1}, \tilde{X}_{k+2}, \dots)$ is independent of $\tilde{X}_{k+2}, \tilde{X}_{k+3}, \dots$, conditionally on $\tilde{X}_{k+1}$. We therefore may assume w.l.g. that the parametrisation is of the form

$$\tilde{X}_k = f_k(Y_k, \tilde{X}_{k+1}), \qquad\qquad k = k_0, k_0 + 1, \dots.$$

We now define, similarly as in [S 98], inductively the Borel functions $(g_k)_{k=k_0}^\infty$ by

$$g_{k_0}(y_{k_0}, x_{k_0+1}) \qquad = f_{k_0}(y_{k_0}, x_{k_0+1})$$
$$g_{k_0+1}(y_{k_0}, y_{k_0+1}, x_{k_0+2}) = g_{k_0}(y_{k_0}, f_{k_0+1}(y_{k_0+1}, x_{k_0+2}))$$

$$\vdots$$

$$g_k(y_{k_0}, \dots, y_k, x_{k+1}) \qquad = g_{k-1}(y_{k_0}, \dots, y_{k-1}, f_k(y_k, x_{k+1}))$$

so that, for each k, the random variable $g_k(Y_{k_0}, \dots, Y_k, \tilde{X}_{k+1})$ equals the random variable $\tilde{X}_{k_0}$ a.s.; the function g_k describes how we may determine the random variable $\tilde{X}_{k_0}$ from the "past" $\tilde{X}_{k+1}$ and the "innovations" $Y_k, Y_{k-1}, \dots, Y_{k_0}$.

Claim. For $k \geq k_0$ and $x_{k+1} \neq x'_{k+1}$, where x_{k+1} and x'_{k+1} are fixed elements of $\mathfrak{X}_{2^{k+1}}$, we have

$$\tilde{\mathbb{P}}\left[g_k(Y_{k_0}, \dots, Y_k, x_{k+1}) \neq g_k(Y_{k_0}, \dots, Y_k, x'_{k+1}) \right] > \prod_{j=k_0}^k (1 - 6j^{-2}) > \frac{3}{4},$$

where $\tilde{\mathbb{P}}$ denotes, as in definition 2.1, the probability under which $(Y_k)_{k=k_0}^\infty$ is an i.i.d. sequence uniformly distributed on $[0, 1]$.

To verify the claim we proceed inductively on $k = k_0, k_0 + 1, \dots$: for $k = k_0$ the claim follows from the construction and assertion (iii) of the Fundamental Lemma 2.6. Now suppose that the claim holds true for $k - 1$; applying assertion (iii) of the Fundamental Lemma again we obtain that, for $x_{k+1} \neq x'_{k+1}$,

$$\tilde{\mathbb{P}}\left[f_k(Y_k, x_{k+1}) \neq f_k(Y_k, x'_{k+1}) \right] > 1 - 6k^{-2}.$$

Applying the inductive hypothesis on all pairs $(x_k, x_k'), x_k \neq x_k'$ in $\mathfrak{X}_k$ that are assumed by $(f_k(Y_k, x_{k+1}), f_k(Y_k, x_{k+1}'))$ we have proved the above claim.

Now we shall use the assumption that the parametrisation $(f_k)_{k=k_0}^\infty$ is generating to obtain the desired contradiction: if $\tilde{X}_{k_0}$ is $\sigma(Y_{k_0}, Y_{k_0+1}, \dots)$-measurable we may find $k \geq k_0$ and a Borel function $G(y_{k_0}, \dots, y_k)$ such that

$$\tilde{\mathbb{P}}\left[\tilde{X}_{k_0} = G(Y_{k_0}, \dots, Y_k)\right] > \frac{7}{8},$$

or, written differently,

$$\tilde{\mathbb{P}}\left[g_k(Y_{k_0}, \dots, Y_k, \tilde{X}_{k+1}) = G(Y_{k_0}, \dots, Y_k)\right] > \frac{7}{8}.$$

As, for each $x_{k+1} \in \mathfrak{X}_{2^{k+1}}$, we have $(1-\varepsilon) \cdot 2^{-2^{k+1}} \leq \tilde{\mathbb{P}}[\tilde{X}_{k+1} = x_{k+1}] \leq (1+\varepsilon)2^{-2^{k+1}}$, and $(Y_{k_0}, \dots, Y_k)$ is independent of $\tilde{X}_{k+1}$ under $\mathbb{P}$, it follows that there are at least two elements $x_{k+1} \neq x_{k+1}'$ in $\mathfrak{X}_{2^{k+1}}$ such that

$$\tilde{\mathbb{P}}\left[g_k(Y_{k_0}, \dots, Y_k, x_{k+1}) = G(Y_{k_0}, \dots, Y_k)\right] > \frac{3}{4},$$

$$and \quad \tilde{\mathbb{P}}\left[g_k(Y_{k_0}, \dots, Y_k, x_{k+1}') = G(Y_{k_0}, \dots, Y_k)\right] > \frac{3}{4},$$

which implies

$$\tilde{\mathbb{P}}\left[g_k(Y_{k_0}, \dots, Y_k, x_{k+1}) = g_k(Y_{k_0}, \dots, Y_k, x_{k+1}')\right] > \frac{1}{2}.$$

This contradiction to the above claim finishes the proof of theorem 1.1. $\square$

APPENDIX

We now prove the combinatorial lemma 2.7. We consider the space $\mathfrak{X} = \mathfrak{X}_{p2^p} = \{-1, +1\}^{p2^p} = \{-1, +1\}^{2^{k-1}2^{2^{k-1}}}$ equipped with uniform distribution $\mathbb{P} = \lambda$. We denote by $x = ((x_i^r)_{i=1}^p)_{r=1}^{2^p}$ the elements of $\mathfrak{X}$ and by $((\tau_i^r)_{i=1}^p)_{r=1}^{2^p}$ the coordinate functions.

A.1 Lemma. *For k large enough, $p = 2^{k-1}, n = 2^k$, and fixed $x_0 \in \mathfrak{X}$, the set*

$$A = \left\{ \begin{array}{l} x \in \mathfrak{X} : \text{there are more than } p^{-1/2}2^p \text{ many } r\text{'s for which} \\ \text{there are less than } n^{3/4} \text{ many } i\text{'s with } \tau_i^r(x_0) \neq \tau_i^r(x) \end{array} \right\}$$

satisfies $\lambda[A] < 2^{-2^p}$.

Proof of lemma A.1. We may assume w.l.g. that $x_0 = (1, 1, \cdots, 1)$ so that $\tau_i^r(x_0) \neq \tau_i^r(x)$ iff $\tau_i^r(x) = -1$.

Claim. *For fixed $1 \leq r \leq 2^p$ and*

$$A_r = \{x : \text{for less than } n^{3/4} \text{ many } i\text{'s we have } \tau_i^r(x) = -1\}$$

we have

$$\mathbb{P}[A_r] \leq 2^{-p/2}.$$

To show the claim we first estimate the probability of the set

$$B_r = \{x : \text{for exactly } n^{3/4} \text{many } i\text{'s we have } \tau_i^r(x) = -1\}$$

(assuming that $n^{3/4}$ is an integer). Using the estimate $\binom{n}{k} \leq n^k$ we get

$$\mathbb{P}[B_r] = \binom{p}{n^{3/4}} 2^{-p} \leq 2^{-p}(p)^{(2p)^{3/4}}$$

$$= (2^{-1} p^{2^{3/4} p^{-1/4}})^p.$$

Noting that the term in the bracket tends to $\frac{1}{2}$, as p increases, we obtain

$$\mathbb{P}[B_r] \leq 2^{-\frac{2p}{3}}, \qquad \text{for } k \geq k_0.$$

Finally we can estimate

$$\mathbb{P}[A_r] = (2p)^{3/4} \mathbb{P}[B_r] \leq 2^{-p/2}, \qquad \text{for } k \geq k_0,$$

which proves the claim.

Using the assertion of the claim we can estimate the probability of the event

$$B = \left\{ \begin{array}{l} x \in X : \text{ there are precisely } p^{-1/2} 2^p \text{ many } r\text{'s for which} \\ \text{there are less than } p^{3/4} \text{ many } i\text{'s with } \tau_i^r(x) = -1 \end{array} \right\}.$$

Applying the inequality $\binom{n}{k} \leq (\frac{en}{k})^k$ we obtain

$$\mathbb{P}[B] \leq \binom{2^p}{p^{-1/2} 2^p} \cdot \mathbb{P}[A_r]^{p^{-1/2} 2^p}$$

$$\leq \left(\frac{e}{p^{-1/2}}\right)^{p^{-1/2} 2^p} \cdot \left(2^{-p/2}\right)^{p^{-1/2} 2^p}$$

$$= \left(ep^{1/2} 2^{-p/2}\right)^{p^{-1/2} 2^p} \approx 2^{-p^{1/2} 2^p}.$$

This allows us to estimate

$$\mathbb{P}[A] \le 2^p \cdot \mathbb{P}[B]$$
$$\le 2^p \cdot 2^{-p^{1/2} 2^p}$$
$$= \left((2^p)^{2^{-p}} 2^{-p^{1/2}} \right)^{2^p}.$$

Noting that, for k tending to infinity, the term in the outer bracket tends to zero, and therefore is eventually less than $\frac{1}{2}$, we finished the proof. $\square$

Proof of lemma 2.7. Let $\mathfrak{X} = \mathfrak{X}_{p2^p}$ and $((\tau_i^r)_{i=1}^p)_{r=1}^{2^p}$ be as above and carry out the following inductive procedure: choose an arbitrary element $x_1 \in \mathfrak{X}$ and remove from $\mathfrak{X}$ the set

$$A(x_1) = \left\{ \begin{array}{l} x \in \mathfrak{X}, x \ne x_1 : \text{there are more than } p^{-1/2} 2^p \text{ many } r's \text{ for which} \\ \text{there are less than } p^{3/4} \text{ many } i's \text{ with } \tau_i^r(x_1) \ne \tau_i^r(x) \end{array} \right\}.$$

The remaining set $\mathfrak{X} \backslash A(x_1)$ has probability bigger than $1 - 2^{-2^p}$ and therefore is non-empty, so that we can choose $x_2 \in \mathfrak{X} \backslash A(x_1)$.

Now remove the set $A(x_2)$, which is defined similarly, and choose $x_3 \in \mathfrak{X} \backslash \{A(x_1) \cup A(x_2)\}$. Continuing in an obvious way we may continue the procedure to obtain $2^{2n} = 2^{4p}$ many elements $(x_j)_{j=1}^{2^{2n}}$ before this procedure stops. (In fact we could even obtain in this way 2^{2^p} many elements x_j, which shows in particular how far the assertion of lemma 2.7 is from being sharp). Identifying the points $(x_j)_{j=1}^{2^{2n}}$ with $\mathfrak{X}_{2n} = \{-1, 1\}^{2n}$ and restricting the functions $((\tau_i^r)_{i=1}^p)_{r=1}^{2^p}$, to this set, the proof of lemma 2.7 now is complete. $\square$

References

[BEKSY 98]   M.T. Barlow, M. Émery, F.B. Knight, S. Song, M. Yor, *Autour d'un théorème de Tsirelson sur des filtrations Browniennes et non Browniennes*, Séminaire de Probabilités XXXII, LNM, Springer **1686** (1998), 264 – 305.

[BE 99]   S. Beghdadi-Sakrani, M. Émery, *On certain probabilities equivalent to cointossing, d'après Schachermayer*, in this volume (1999).

[DFST 96]   L. Dubins, J. Feldman, M. Smorodinsky, B. Tsirelson, *Decreasing Sequences of σ-fields and a Measure Change for Brownian Motion*, Annals of Probability **24** (1996), 882 – 904.

[EY 98]   M. Émery, M. Yor, *Sur un théorème de Tsirelson relatif à des mouvements Browniens corrélés et à la nullité de certain temps locaux*, Séminaire de Probabilités XXXII, LNM, Springer **1686** (1998), 306 – 312.

[F 96]   J. Feldman, *ε-close Measures Producing Nonisomorphic Filtrations*, Annals of Probability **24** (1996), 912 – 914.

[FT 96]   J. Feldman, B. Tsirelson, *Decreasing Sequences of σ-fields and a Measure Change for Brownian Motion II*, Annals of Probability **24** (1996), 905 – 911.

[RY 91]   D. Revuz, M. Yor, *Continuous Martingales and Brownian Motion*, Springer (1991).

[RY 94]   D. Revuz, M. Yor, *Continuous Martingales and Brownian Motion*, second edition, Springer (1994).

[S 98]   M. Smorodinsky, *An Example of a non Sub-Standard Process*, to appear in Israel Journal of Math. (1998).

[SY 80]   D.W. Stroock, M. Yor, *On extremal solutions of martingale problems*, Ann. Sci. Ecole Norm. Sup. **13** (1980), 95 – 164.

[SY 81]   D.W. Stroock, M. Yor, *Some remarkable martingales*, Sém. Prob. XV, LNM **850** (1981), 590 – 603.

[T 97]   B. Tsirelson, *Triple points: from non-Brownian filtrations to harmonic measures*, GAFA Geometric And Functional Analysis, Birkhäuser **7** (1997), 1096 – 1142.

[V 70]   A.M. Vershik, *Decreasing sequences of measurable partitions and their applications*, Soviet. Math. Dokl. **11** (1970), 1007 – 1011.

[V 73]   A.M. Vershik, *Approximation in measure theory. Dissertation, Leningrad Univ.*, In Russian (1973).

[V 95]   A.M. Vershik, *The Theory of Decreasing Sequences of Measurable Partitions*, St. Petersburg Math. Journal **6/4** (1995), 705 – 761.

ON CERTAIN PROBABILITIES EQUIVALENT TO COIN-TOSSING, D'APRÈS SCHACHERMAYER

S. Beghdadi-Sakrani and M. Émery

L. Dubins, J. Feldman. M. Smorodinsky and B. Tsirelson have constructed in [4] and [5] a probability $\mathbf{Q}$ on Wiener space, equivalent to Wiener measure, but such that the canonical filtration on Wiener space is not generated by any $\mathbf{Q}$-Brownian motion whatsoever! Dreadfully complicated, their construction is almost as incredible as the existence result itself, and these articles are far from simple. In this volume. W. Schachermayer [14] proposes a shorter, more accessible redaction of their construction. We have rewritten it once more, for two reasons: First, the language used by these five authors is closer to that of dynamical systems than to a probabilist's mother tongue: where they use a dependence of the form $X = f(Y)$, we work with X measurable with respect to a σ-field; having both versions enables the reader to choose her favorite setup. Second, a straightforward adaptation of Tsirelson's ideas in [16] gives a stronger result (non-cosiness instead of non-standardness).

We warmly thank Walter Schachermayer for many fruitful conversations and helpful remarks, and for spotting an error (of M. É.) in an earlier version. We also thank Boris Tsirelson for his comments (though we disagree with him on one point: he modestly insists that we have given too much credit to him and to his co-authors) and Marc Yor for his many observations and questions.

1. — Notations

Probability spaces will always be complete: if $(\Omega, \mathcal{A}, \mathbf{P})$ is a probability space, $\mathcal{A}$ contains all negligible events. Similarly, we consider only sub-σ-fields of $\mathcal{A}$ containing all negligible events of $\mathcal{A}$. For instance, the product $(\Omega^1, \mathcal{A}^1, \mathbf{P}^1) \otimes (\Omega^2, \mathcal{A}^2, \mathbf{P}^2)$ of two probability spaces is endowed with its completed product σ-field $\mathcal{A}^1 \otimes \mathcal{A}^2$, containing all null events for $\mathbf{P}^1 \times \mathbf{P}^2$. $L^0(\Omega, \mathcal{A}, \mathbf{P})$ (or shortly $L^0(\Omega, \mathbf{P})$, or $L^0(\mathcal{A})$ etc. if there is no ambiguity) denotes the space of equivalence classes of all a.s. finite r.v.'s; so $L^0(\Omega, \mathcal{A}, \mathbf{P}) = L^0(\Omega, \mathcal{A}, \mathbf{Q})$ if $\mathbf{P}$ and $\mathbf{Q}$ are equivalent probabilities.

When there is an ambiguity on the probability $\mathbf{P}$, expectations and conditional expectations will be written $\mathbf{P}[X]$ and $\mathbf{P}[X|\mathcal{B}]$ instead of the customary $\mathbf{E}[X]$ and $\mathbf{E}[X|\mathcal{B}]$.

An *embedding* of a probability space $(\Omega, \mathcal{A}, \mathbf{P})$ into another one $(\bar{\Omega}, \bar{\mathcal{A}}, \bar{\mathbf{P}})$ is a mapping Ψ from $L^0(\Omega, \mathcal{A}, \mathbf{P})$ to $L^0(\bar{\Omega}, \bar{\mathcal{A}}, \bar{\mathbf{P}})$ that commutes with Borel operations on finitely many r.v.'s:

$$\Psi\big(f(X_1, \ldots, X_n)\big) = f\big(\Psi(X_1), \ldots, \Psi(X_n)\big) \quad \text{for every Borel } f$$

and preserves the probability laws:

$$\bar{\mathbf{P}}\big[\Psi(X) \in E\big] = \mathbf{P}[X \in E] \quad \text{for every Borel } E.$$

If Ψ embeds $(\Omega, \mathcal{A}, \mathbf{P})$ into $(\bar{\Omega}, \bar{\mathcal{A}}, \bar{\mathbf{P}})$ and if $\mathbf{Q}$ is a probability absolutely continuous with respect to $\mathbf{P}$, then Ψ also embeds $(\Omega, \mathcal{A}, \mathbf{Q})$ into $(\bar{\Omega}, \bar{\mathcal{A}}, \bar{\mathbf{Q}})$, where $\bar{\mathbf{Q}}$ is defined by $d\bar{\mathbf{Q}}/d\bar{\mathbf{P}} = \Psi(d\mathbf{Q}/d\mathbf{P})$. If $\mathbf{P}$ and $\mathbf{Q}$ are equivalent, so are also $\bar{\mathbf{P}}$ and $\bar{\mathbf{Q}}$.

An embedding is always injective and transfers not only random variables, but also sub-σ-fields, filtrations, processes, etc. It is called an *isomorphism* if it is surjective: it then has an inverse. An embedding Ψ of $(\Omega, \mathcal{A}, \mathbf{P})$ into $(\bar{\Omega}, \bar{\mathcal{A}}, \bar{\mathbf{P}})$ is always an isomorphism between $(\Omega, \mathcal{A}, \mathbf{P})$ and $(\bar{\Omega}, \Psi(\mathcal{A}), \bar{\mathbf{P}})$.

Processes and filtrations will be parametrized by time, represented by a subset T of $\mathbf{R}$. We shall use three special cases only: $\mathsf{T} = \mathbf{R}_+$, the usual time-axis for continuous-time processes, $\mathsf{T} = -\mathbf{N} = \{\ldots, -2, -1, 0\}$ and T finite (and non-empty). In the first case ($\mathsf{T} = [0, \infty)$), filtrations are always right-continuous; for instance, the product of two filtrations $\mathcal{F}^1$ and $\mathcal{F}^2$ is the smallest right-continuous filtration $\mathcal{G}$ on $(\Omega^1, \mathcal{A}^1, \mathbf{P}^1) \otimes (\Omega^2, \mathcal{A}^2, \mathbf{P}^2)$ such that $\mathcal{G}_t \supset \mathcal{F}_t^1 \otimes \mathcal{F}_t^2$. If $\mathcal{F} = (\mathcal{F}_t)_{t \in \mathsf{T}}$ is a filtration, we set

$$\mathcal{F}_{-\infty} = \bigcap_{t \in \mathsf{T}} \mathcal{F}_t \quad \text{and} \quad \mathcal{F}_{\infty} = \bigvee_{t \in \mathsf{T}} \mathcal{F}_t ;$$

so $\mathcal{F}_{-\infty} = \mathcal{F}_0$ when $\mathsf{T} = \mathbf{R}_+$ and $\mathcal{F}_{\infty} = \mathcal{F}_0$ when $\mathsf{T} = -\mathbf{N}$. If $(\Omega, \mathcal{A}, \mathbf{P}, \mathcal{F})$ and $(\bar{\Omega}, \bar{\mathcal{A}}, \bar{\mathbf{P}}, \bar{\mathcal{F}})$ are two filtered probability spaces, the filtrations $\mathcal{F}$ and $\bar{\mathcal{F}}$ are called isomorphic if there exists an isomorphism Ψ between $(\Omega, \mathcal{F}_{\infty}, \mathbf{P})$ and $(\bar{\Omega}, \bar{\mathcal{F}}_{\infty}, \bar{\mathbf{P}})$ such that $\Psi(\mathcal{F}_t) = \bar{\mathcal{F}}_t$ for each t. For instance, if two processes have the same law, their natural filtrations are isomorphic.

A filtration indexed by $\mathbf{R}_+$ is *Brownian* if it is generated by one, or finitely many, or countably many independent real Brownian motions, started at the origin. It is well known that two Brownian filtrations are isomorphic if and only if they are generated by the same number ($\leqslant \infty$) of independent Brownian motions. (For if X is an m-dimensional Brownian motion in the filtration generated by an n-dimensional Brownian motion Y, there exist predictable processes H^{ij} such that $dX^i = \sum_j H^{ij} dY^j$; they form an $m \times n$ matrix H verifying $H\,{}^tH = \mathrm{Id}_m$ almost everywhere on $(\Omega \times \mathbf{R}_+, d\mathbf{P} \otimes dt)$, whence $m \leqslant n$.)

A filtration indexed by $-\mathbf{N}$ is *standard* if it is generated by a process $(Y_n)_{n \leqslant 0}$ where the Y_n's are independent r.v.'s with diffuse laws. It is always possible to choose each Y_n with law $\mathcal{N}(0, 1)$, so all standard filtrations are isomorphic.

2. — Immersions

DEFINITION. — *Let $\mathcal{F}$ and $\mathcal{G}$ be two filtrations on a probability space $(\Omega, \mathcal{A}, \mathbf{P})$. The filtration $\mathcal{F}$ is* immersed *in $\mathcal{G}$ if every $\mathcal{F}$-martingale is a $\mathcal{G}$-martingale.*

This is inspired by the definition of an *extension* of a filtration, by Dubins, Feldman, Smorodinsky and Tsirelson in [4], and by that of a *morphism* from a filtered probability space to another, by Tsirelson in [16]; see also the *liftings* in Section 7 of Getoor and Sharpe [6]. (In these definitions, both filtrations are not necessarily on the same Ω.) Our definition is but a rephrasing of Brémaud, Jeulin and Yor's *Hypothèse (H)* in [2] and [9].

Immersion implies in particular that $\mathcal{F}_t$ is included in $\mathcal{G}_t$ for each t, but it is a much stronger property. As shown in [2], it amounts to requiring that the $\mathcal{G}$-optional projection of any $\mathcal{F}_{\infty}$-measurable process is $\mathcal{F}$-optional (and hence equal to the $\mathcal{F}$-optional projection), and it is also equivalent to demanding, for each t, that $\mathcal{F}_\infty \cap \mathcal{G}_t = \mathcal{F}_t$ and that $\mathcal{F}_\infty$ and $\mathcal{G}_t$ are conditionally independent given $\mathcal{F}_t$. We shall only need the weaker sufficient condition given by Lemma 1 below, concerning independent enlargements.

LEMMA 1. — *Let $\mathcal{F}$ and $\mathcal{G}$ be two independent filtrations (that is, $\mathcal{F}_t$ and $\mathcal{G}_t$ are independent for each t). If $\mathcal{H}$ is the smallest filtration containing both $\mathcal{F}$ and $\mathcal{G}$, $\mathcal{F}$ is immersed in $\mathcal{H}$.*

PROOF. — By a density argument, it suffices to show that every square-integrable $\mathcal{F}$-martingale M is an $\mathcal{H}$-martingale. For $s<t$, $F_s \in L^\infty(\mathcal{F}_s)$ and $G_t \in L^\infty(\mathcal{G}_t)$, one can write $\mathbf{E}[M_t F_s G_t] = \mathbf{E}[G_t]\mathbf{E}[M_t F_s] = \mathbf{E}[G_t]\mathbf{E}[M_s F_s] = \mathbf{E}[M_s F_s G_t]$. As products of the form $F_s G_t$ are total in $L^2(\mathcal{F}_s \otimes \mathcal{G}_t)$, $M_t - M_s$ is orthogonal to $L^2(\mathcal{F}_s \otimes \mathcal{G}_t)$. The lemma follows since $\mathcal{H}_s$ is included in $\mathcal{F}_s \otimes \mathcal{G}_t$ by Lemma 2 of Lindvall and Rogers [10]. ∎

Another, very simple, example of immersion is obtained by stopping: if T is an $\mathcal{F}$-stopping time, the stopped filtration $\mathcal{F}^T$ is immersed in $\mathcal{F}$.

The immersion property is in general not preserved when $\mathbf{P}$ is replaced with an equivalent probability; we shall sometimes write "$\mathbf{P}$-immersed" to specify the probability. But it is preserved if the density is $\mathcal{F}_\infty$- or $\mathcal{G}_{-\infty}$-measurable:

LEMMA 2. — *Let $\mathcal{F}$ and $\mathcal{G}$ be two filtrations on $(\Omega, \mathcal{A}, \mathbf{P})$, $\mathcal{F}$ being $\mathbf{P}$-immersed in $\mathcal{G}$; let $\mathbf{Q}$ be a probability absolutely continuous with respect to $\mathbf{P}$. If the density $D = d\mathbf{Q}/d\mathbf{P}$ has the form $D = D'D''$, where D' is $\mathcal{F}_\infty$-measurable and D'' is $\mathcal{G}_{-\infty}$-measurable, then $\mathcal{F}$ is also $\mathbf{Q}$-immersed in $\mathcal{G}$.*

PROOF. — By taking absolute values, we may suppose $D' \geqslant 0$ and $D'' \geqslant 0$. It suffices to show that every bounded $(\mathbf{Q}, \mathcal{F})$ martingale M is also a $(\mathbf{Q}, \mathcal{G})$-martingale; by adding a constant, we may also suppose M positive.

For $t \in \mathsf{T}$, the following equalities hold $\mathbf{Q}$-a.s. (with $[0, \infty]$-valued conditional expectations; for the last equality, approximate D' by $D' \wedge n$ and use the $\mathbf{P}$-immersion hypothesis):

$$\mathbf{Q}[M_\infty | \mathcal{G}_t] = \frac{\mathbf{P}[DM_\infty | \mathcal{G}_t]}{\mathbf{P}[D | \mathcal{G}_t]} = \frac{D''\mathbf{P}[D'M_\infty | \mathcal{G}_t]}{D''\mathbf{P}[D' | \mathcal{G}_t]} = \frac{\mathbf{P}[D'M_\infty | \mathcal{G}_t]}{\mathbf{P}[D' | \mathcal{G}_t]} = \frac{\mathbf{P}[D'M_\infty | \mathcal{F}_t]}{\mathbf{P}[D' | \mathcal{F}_t]} \ .$$

So $\mathbf{Q}[M_\infty | \mathcal{G}_t]$ is $\mathcal{F}_t$-measurable, hence $\mathbf{Q}[M_\infty | \mathcal{G}_t] = \mathbf{Q}[M_\infty | \mathcal{F}_t] = M_t$. ∎

DEFINITION. — *Let $X = (X_t)_{t \in \mathsf{T}}$ be a process and $\mathcal{F} = (\mathcal{F}_t)_{t \in \mathsf{T}}$ a filtration, both on the same probability space. The process X is immersed in the filtration $\mathcal{F}$ if the natural filtration of X is immersed in $\mathcal{F}$.*

A process $X = (X_i)_{1 \leqslant i \leqslant n}$ is also said to be immersed in a filtration $(\mathcal{F}_i)_{0 \leqslant i \leqslant n}$ if the process $(0, X_1, \ldots, X_n)$ is immersed in $\mathcal{F}$.

Recall that X is adapted to $\mathcal{F}$ if and only if the natural filtration of X is included in $\mathcal{F}$. Saying that X is immersed in $\mathcal{F}$ is much stronger: it further means that for an $\mathcal{F}$-observer, the predictions about the future behaviour of X depend on the past and present of X only.

As for an example, remark that the same name, Brownian motion, is used for two different objects: first, any continuous, centred Gaussian process with covariance $\mathbf{E}[B_s B_t] = s$ for $s \leqslant t$ (this definition characterizes the law of B); second, a process such that $B_0 = 0$ and $B_t - B_s$ is independent of $\mathcal{F}_s$ with law $\mathcal{N}(0, t-s)$ (one often says that B is an $\mathcal{F}$-Brownian motion). The latter definition, involving the filtration, amounts to requiring that B has a Brownian law (former definition) *and is immersed in* $\mathcal{F}$.

Similarly, a Markov process is $\mathcal{F}$-Markov if and only it is immersed in $\mathcal{F}$: as a consequence. if $\mathcal{F}$ is immersed in $\mathcal{G}$, every $\mathcal{F}$-Markov process is also a $\mathcal{G}$-Markov process. with the same transition probabilities.

Lemma 2 can be rephrased in terms of immersed processes:

COROLLARY 1. — *If a process X is* $\mathbf{P}$-*immersed in a filtration* $\mathcal{F}$, *and if a probability* $\mathbf{Q}$ *is absolutely continuous with respect to* $\mathbf{P}$. *with a* $\sigma(X)$-*measurable. or* $\mathcal{F}_{-\infty}$-*measurable density* $d\mathbf{Q}/d\mathbf{P}$. *then X is also* $\mathbf{Q}$-*immersed in* $\mathcal{F}$.

PROOF. — Apply Lemma 2 and the above definition. ∎

Tsirelson has introduced in [16] the notion of a joining of two filtrations. The next definition is a particular case of joining (we demand that both filtrations are defined on the same Ω).

DEFINITION. — *Two filtrations $\mathcal{F}$ and $\mathcal{G}$ (or two processes X and Y) on the same probability space $(\Omega, \mathcal{A}, \mathbf{P})$ are* jointly immersed *if there exists a filtration $\mathcal{H}$ on $(\Omega, \mathcal{A}, \mathbf{P})$ such that both $\mathcal{F}$ and $\mathcal{G}$ (or X and Y) are immersed in $\mathcal{H}$.*

To illustrate the immersion property. and also to fix some notations to be used later, here is an easy statement.

LEMMA 3. — *Denote by Ω_n the set $\{-1, 1\}^n$, by $\mathbf{P}_n$ the uniform probability on Ω_n (fair coin-tossing), and by $\varepsilon_1, \ldots, \varepsilon_n$ the coordinates on Ω_n: endow Ω_n with the filtration generated by the process $(0, \varepsilon_1, \ldots, \varepsilon_n)$.*

If $\alpha = (\alpha_i)_{1 \leqslant i \leqslant n}$ is a predictable process on Ω_n such that $|\alpha| \leqslant 1$ and if $Z = (Z_i)_{0 \leqslant i \leqslant n}$ is the $\mathbf{P}_n$-martingale defined by $Z_0 = 1$ and

$$Z_i = Z_{i-1}(1 + \alpha_i \varepsilon_i),$$

the formula $\mathbf{Q} = Z_n \cdot \mathbf{P}_n$ defines a probability law on Ω_n.

Let $X = (X_i)_{1 \leqslant i \leqslant n}$ be a process with values ± 1, defined on some $(\bar{\Omega}, \bar{\mathcal{A}}, \bar{\mathbf{P}})$ and adapted to some filtration $\mathcal{H} = (\mathcal{H}_i)_{0 \leqslant i \leqslant n}$. The process X has law $\mathbf{Q}$ and is immersed in $\mathcal{H}$ if and only if

$$\forall i \in \{1, \ldots, n\} \quad \forall e \in \{-1, 1\} \quad \bar{\mathbf{P}}[X_i = e | \mathcal{H}_{i-1}] = \tfrac{1}{2}\big(1 + \alpha_i(X_1, \ldots, X_{i-1})e\big).$$

REMARK. — Every probability $\mathbf{Q}$ on Ω_n is obtained from such an α, but this correspondence between α and $\mathbf{Q}$ is not a bijection: if $\mathbf{Q}$ neglects some ω's. the martingale Z vanishes from some time on, and modifying α after that time does not change Z nor $\mathbf{Q}$. When restricted to the α's such that $|\alpha| < 1$ and to the $\mathbf{Q}$'s that are equivalent to $\mathbf{P}_n$. this correspondence is a bijection; in this case, α will be called the α-process associated to $\mathbf{Q}$.

PROOF OF LEMMA 3. — Clearly, Z is a positive martingale and $\mathbf{Q}$ is a probability. A process X with values ± 1 has law $\mathbf{Q}$ if and only if

$$\forall i \quad \bar{\mathbf{P}}[X_1 = e_1, \ldots, X_i = e_i] = 2^{-i} Z_i(e_1, \ldots, e_i),$$

or $\quad \forall i \quad \bar{\mathbf{P}}[X_i = e_i | X_1 = e_1, \ldots, X_{i-1} = e_{i-1}] = \tfrac{1}{2}(Z_i/Z_{i-1})(e_1, \ldots, e_i)$

$$= \tfrac{1}{2}\big(1 + \alpha_i(e_1, \ldots, e_{i-1})e_i\big),$$

or $\quad \forall i \quad \bar{\mathbf{P}}[X_i = e | X_1, \ldots, X_{i-1}] = \tfrac{1}{2}\big(1 + \alpha_i(X_1, \ldots, X_{i-1})e\big).$

Now if X has law $\mathbf{Q}$ and is immersed in $\mathcal{H}$.

$$\bar{\mathbf{P}}[X_i = e\,|\mathcal{H}_{i-1}] = \bar{\mathbf{P}}[X_i = e\,|X_1,\ldots,X_{i-1}] = \tfrac{1}{2}\left(1 + \alpha_i(X_1,\ldots,X_{i-1})\,e\right).$$

Conversely. for a ± 1-valued process X adapted to $\mathcal{H}$ such that. for each i in $\{1,\ldots,n\}$ and each e in $\{-1,1\}$. $\bar{\mathbf{P}}[X_i = e\,|\mathcal{H}_{i-1}] = \left(1 + \alpha_i(X_1,\ldots,X_{i-1})\,e\right)$. one has on the one hand $\bar{\mathbf{P}}[X_i = e\,|\mathcal{H}_{i-1}] = \bar{\mathbf{P}}[X_i = e\,|X_1,\ldots,X_{i-1}]$ and on the other hand $\bar{\mathbf{P}}[X_i = e\,|X_1,\ldots,X_{i-1}] = \left(1 + \alpha_i(X_1,\ldots,X_{i-1})\,e\right)$; so X is immersed in $\mathcal{H}$ and has law $\mathbf{Q}$. ∎

3. — Separate σ-fields

DEFINITION. — *Given a probability space* $(\Omega, \mathcal{A}, \mathbf{P})$, *two sub-$\sigma$-fields* $\mathcal{B}$ *and* $\mathcal{C}$ *of* $\mathcal{A}$ *are* separate *if* $\mathbf{P}[B = C] = 0$ *for all random variables* $B \in \mathrm{L}^0(\mathcal{B})$ *and* $C \in \mathrm{L}^0(\mathcal{C})$ *with diffuse laws. Two filtrations* $\mathcal{F}$ *and* $\mathcal{G}$ *on* $(\Omega, \mathcal{A}, \mathbf{P})$ *are* separate *if the σ-fields* $\mathcal{F}_\infty$ *and* $\mathcal{G}_\infty$ *are.*

Two independent sub-σ-fields are always separate; this can be seen by taking $a = p = 1$ in Proposition 1 below. But observe that separation depends only on the null sets of $\mathbf{P}$, whereas independence, and more generally hypercontractivity, is not preserved by equivalent changes of probability.

PROPOSITION 1. — *Let* $\mathcal{B}$ *and* $\mathcal{C}$ *be two sub-σ-fields in a probability space* $(\Omega, \mathcal{A}, \mathbf{P})$. *Suppose that, for some* $p \in [1,2)$ *and some* $a < \infty$, *the following inequality holds:*

$$\forall B \in \mathrm{L}^p(\mathcal{B}) \quad \forall C \in \mathrm{L}^p(\mathcal{C}) \qquad \mathbb{E}[BC] \leqslant a\,\|B\|_{\mathrm{L}^p}\,\|C\|_{\mathrm{L}^p}.$$

The σ-fields $\mathcal{B}$ *and* $\mathcal{C}$ *are separate.*

PROOF. — Let $B \in \mathrm{L}^0(\Omega, \mathcal{B}, \mathbf{P})$ and $C \in \mathrm{L}^0(\Omega, \mathcal{C}, \mathbf{P})$ have diffuse laws μ and ν; the measure $\lambda = \mu + \nu$ is positive, diffuse, with mass 2. For each $n \geqslant 1$ it is possible to partition $\mathbb{R}$ into $2n$ Borel sets $E_1, \ldots, E_{2n}$, each with measure $\lambda(E_i) = 1/n$. When applied to the r.v.'s $\mathbb{1}_{E_i} \circ B$ and $\mathbb{1}_{E_i} \circ C$. the hypothesis entails

$$\mathbf{P}\big[B \in E_i \text{ and } C \in E_i\big] \leqslant a\,\mathbf{P}[B \in E_i]^{\frac{1}{p}}\,\mathbf{P}[C \in E_i]^{\frac{1}{p}}$$
$$= a\,\mu(E_i)^{\frac{1}{p}}\,\nu(E_i)^{\frac{1}{p}} \leqslant a\,\lambda(E_i)^{\frac{2}{p}} = a\,n^{-\frac{2}{p}}.$$

Summing over i gives

$$\mathbf{P}[B = C] \leqslant \mathbf{P}\big[B \text{ and } C \text{ are in the same } E_i\big]$$
$$= \sum_i \mathbf{P}\big[B \in E_i \text{ and } C \in E_i\big] \leqslant 2n\,a\,n^{-\frac{2}{p}} = 2a\,n^{\frac{p-2}{p}}.$$

Since $\dfrac{p-2}{p} < 0$, letting n tend to infinity now yields $\mathbf{P}[B = C] = 0$. ∎

REMARK. — When $a = 1$. the inequality featuring in Proposition 1 is called hypercontractivity (because it means that the conditional expectation operator $\mathbb{E}[\ \cdot\ |\mathcal{B}]$ is not only a contraction from $\mathrm{L}^p(\mathcal{C})$ to $\mathrm{L}^p(\mathcal{B})$, but also a contraction from $\mathrm{L}^p(\mathcal{C})$ to the smaller space $\mathrm{L}^q(\mathcal{B})$, where q is the conjugate exponent of p).

The next proposition is one of Tsirelson's tools in [16]. It uses Proposition 1 to show separation for some Gaussianly generated σ-fields; Thouvenot [15] has another proof. via ergodic theory, that does not use hypercontractivity.

PROPOSITION 2. — *Let X' and X'' be two independent, centred Gaussian processes with the same law. For each $\theta \in \mathbf{R}$ the process $X^\theta = X' \cos \theta + X'' \sin \theta$ has the same law as X' and X''; for $\theta \neq 0$ mod π the σ-fields $\sigma(X^\theta)$ and $\sigma(X')$ satisfy the hypercontractivity property of Proposition 1 with $p = 1 + |\cos \theta| < 2$ and $a = 1$, and are therefore separate.*

PROOF. — The first property is well known—it is the very definition of 2-stability!— and can be readily verified by computing the covariance of X^θ.

Hypercontractivity is a celebrated theorem of Nelson [11]. When X' is just a normal r.v., a proof by stochastic calculus is given by Neveu [12]; a straightforward extension gives the case when X' is a normal random vector in $\mathbf{R}^n$ (see Dellacherie, Maisonneuve and Meyer [3]); and the general case follows by approximating $B \in \mathrm{L}^p(\sigma(X^\theta))$ and $C \in \mathrm{L}^p(\sigma(X'))$ in L^p with r.v.'s of the form $B' = f(X_{t_1}^\theta, \dots, X_{t_n}^\theta)$ and $C' = g(X'_{t_1}, \dots, X'_{t_n})$, where f and g are Borel in n variables.

Separation stems from hypercontractivity, as shown by Proposition 1. ∎

Non probabilistic proofs of Gaussian hypercontractivity, as well as references to the literature, can be found in Gross [7] and Janson [8].

4. — Cosiness

Cosiness was invented by Tsirelson in [16], as a necessary condition for a filtration to be Brownian. There is a whole range of possible variations on his original definition; the one we choose below is taylor-made for Proposition 4.

DEFINITION. — *A filtered probability space $(\Omega, \mathcal{A}, \mathbf{P}, \mathcal{F})$ is cosy if for each $\varepsilon > 0$ and each $U \in \mathrm{L}^0(\Omega, \mathcal{F}_\infty, \mathbf{P})$, there exists a probability space $(\bar{\Omega}, \bar{\mathcal{A}}, \bar{\mathbf{P}})$ with two filtrations $\mathcal{F}'$ and $\mathcal{F}''$ such that*

(i) *$(\bar{\Omega}, \mathcal{F}'_\infty, \bar{\mathbf{P}}, \mathcal{F}')$ and $(\bar{\Omega}, \mathcal{F}''_\infty, \bar{\mathbf{P}}, \mathcal{F}'')$ are isomorphic to $(\Omega, \mathcal{F}_\infty, \mathbf{P}, \mathcal{F})$;*

(ii) *$\mathcal{F}'$ and $\mathcal{F}''$ are jointly immersed;*

(iii) *$\mathcal{F}'$ and $\mathcal{F}''$ are separate;*

(iv) *the copies $U' \in \mathrm{L}^0(\mathcal{F}'_\infty)$ and $U'' \in \mathrm{L}^0(\mathcal{F}''_\infty)$ of U by the isomorphisms in condition (i) are ε-close in probability: $\bar{\mathbf{P}}[|U'-U''| > \varepsilon] \leq \varepsilon$.*

When there is no ambiguity on the underlying space $(\Omega, \mathcal{A}, \mathbf{P})$, we shall often simply say that the filtration $\mathcal{F}$ is cosy.

This definition is not equivalent to Tsirelson's original one. In his definition, the notion of separation featuring in condition (iii) is a bound on the joint bracket of any two martingales in the two filtrations, in terms of their quadratic variations. Other possible choices would be for instance a hypercontractivity inequality, as in Proposition 1, or a bound on the correlation coefficient for two r.v.'s in $\mathrm{L}^2(\mathcal{F}'_\infty)$ and $\mathrm{L}^2(\mathcal{F}''_\infty)$, or, when time is discrete, a bound on the predictable covariation $\bar{\mathbf{E}}[(X'_{n+1}-X'_n)(X''_{n+1}-X''_n)|\bar{\mathcal{F}}_n]$ of any two martingales in the two filtrations. In any case, the basic idea is always the same: we have two jointly immersed copies of the given filtration, that are close to each other as expressed by (iv), but nonetheless separate in some sense. As for an example, with the weak notion of separation we are using, a filtration $\mathcal{F}$ such that $\mathcal{F}_\infty$ has an atom is always cosy. Indeed, it suffices to take $(\bar{\Omega}, \bar{\mathcal{A}}, \bar{\mathbf{P}}, \mathcal{F}') = (\bar{\Omega}, \bar{\mathcal{A}}, \bar{\mathbf{P}}, \mathcal{F}'') = (\Omega, \mathcal{A}, \mathbf{P}, \mathcal{F})$; (i), (ii) and (iv) are trivial, and (iii) is vacuously satisfied since no $\mathcal{F}_\infty$-measurable r.v. has a diffuse law! When dealing with such filtrations, other definitions of separation are more appropriate.

Another, more superficial, difference with Tsirelson's definition is that, instead of fixing U and ε, he deals with a whole sequence of joint immersions, and condition (iv) becomes the convergence to 0 for every U of the distance in probability $d(U',U'')$.

The use of a real random variable U could also be extended to random elements in an arbitrary separable metric space, for instance some space of functions; by Slutsky's lemma, it would not be more general, but it may be notationally more convenient.

Notice that conditions (i) and (ii) involve $\mathbf{P}$ and the whole filtration $\mathcal{F}$ in an essential way, whereas conditions (iii) and (iv) act only on the end-σ-field $\mathcal{F}_\infty$ and on the equivalence class of $\mathbf{P}$. Cosiness is not always preserved when $\mathbf{P}$ is replaced by an equivalent probability (see Theorems 1 and 2 below); but it is an instructive exercise to try proving this preservation by means of Lemma 2—and to see why it does not work.

LEMMA 4. — *Let* $(\mathcal{F}_t)_{t\in\mathsf{T}}$ *be a cosy filtration. If* $(t_n)_{n\leqslant 0}$ *is a sequence in* T *such that* $t_{n-1} < t_n$ *for all* $n \leqslant 0$, *the filtration* $\mathcal{H} = (\mathcal{H}_n)_{n\leqslant 0}$ *defined by* $\mathcal{H}_n = \mathcal{F}_{t_n}$ *is cosy too.*

PROOF. — It suffices to remark that if a filtration $(\mathcal{F}_t)_{t\in\mathsf{T}}$ is immersed in a filtration $(\mathcal{G}_t)_{t\in\mathsf{T}}$, then $(\mathcal{F}_{t_n})_{n\leqslant 0}$ is immersed in $(\mathcal{G}_{t_n})_{n\leqslant 0}$. The lemma follows then immediately from the definition of cosiness and the transitivity of immersions. ∎

LEMMA 5. — *A filtration immersed in a cosy filtration is itself cosy.*

PROOF. — If $(\Omega,\mathcal{A},\mathbf{P})$ is endowed with two filtrations $\mathcal{F}$ and $\mathcal{G}$, if $\mathcal{F}$ is immersed in $\mathcal{G}$ and if Ψ is an embedding of $(\Omega,\mathcal{G}_\infty,\mathbf{P})$ into some probability space, then the filtration $\Psi(\mathcal{F})$ is immersed in $\Psi(\mathcal{G})$. The lemma follows immediately from this remark, the definition of cosiness and the transitivity of immersions. ∎

COROLLARY 2. — *Let* $(\Omega,\mathcal{A},\mathbf{P},\mathcal{F})$ *and* $(\Omega',\mathcal{A}',\mathbf{P}',\mathcal{F}')$ *be filtered probability spaces and* Ψ *be an embedding of* $(\Omega,\mathcal{A},\mathbf{P})$ *into* $(\Omega',\mathcal{A}',\mathbf{P}')$, *such that the filtration* $\Psi(\mathcal{F})$ *is immersed in* $\mathcal{F}'$. *If* $(\Omega',\mathcal{F}'_\infty,\mathbf{P}',\mathcal{F}')$ *is cosy, so is also* $(\Omega,\mathcal{F}_\infty,\mathbf{P},\mathcal{F})$.

PROOF. — This filtered probability space is isomorphic to $\big(\Omega',\Psi(\mathcal{F}_\infty),\mathbf{P}',\Psi(\mathcal{F})\big)$, which is cosy by Lemma 5. ∎

We now turn to a very important sufficient condition for cosiness, or conversely a necessary condition for a filtration to be Gaussianly generated. It is due to Tsirelson [16], who devised cosiness to have this necessary condition at his disposal.

PROPOSITION 3. — *The natural filtration of a Gaussian process is cosy. More generally, let* $X = (X_i)_{i\in I}$ *be a Gaussian process and* $\mathcal{F} = (\mathcal{F}_t)_{t\in\mathsf{T}}$ *a filtration. If there are subsets* $I_t \subset I$ *such that* $\mathcal{F}_t = \bigcap_{\varepsilon>0} \sigma(X_i, i \in I_{t+\varepsilon})$ *for each* $t \in \mathsf{T}$, *the filtration* $\mathcal{F}$ *is cosy.*

(In this statement, $I_{t+\varepsilon}$ should be taken equal to I_t when the time T is discrete.)

PROOF. — We may suppose X centred. Call $(\Omega,\mathcal{A},\mathbf{P})$ the sample space where X is defined. On the filtered product space $(\overline{\Omega},\overline{\mathcal{A}},\overline{\mathbf{P}},\overline{\mathcal{F}}) = (\Omega,\mathcal{A},\mathbf{P},\mathcal{F}) \otimes (\Omega,\mathcal{A},\mathbf{P},\mathcal{F})$, the processes $X^0(\overline{\omega}) = X^0(\omega_1,\omega_2) = X(\omega_1)$ and $X^{\pi/2}(\overline{\omega}) = X^{\pi/2}(\omega_1,\omega_2) = X(\omega_2)$ are independent copies of X. For $\theta \in \mathbf{R}$, the process $X^\theta = X^0\cos\theta + X^{\pi/2}\sin\theta$ is yet another copy of X; notice that X^θ and $X^{\theta+\pi/2}$ are independent.

Every r.v. $U \in \mathrm{L}^0(\mathcal{F}_\infty)$ has the form $u(X_{i_1},\ldots,X_{i_k},\ldots)$, where u is Borel and $(i_1,\ldots,i_k,\ldots)$ is a sequence in I. This makes it possible to define an embedding Φ^θ

of $(\Omega, \mathcal{F}_\infty, \mathbf{P})$ into $(\bar{\Omega}, \bar{\mathcal{F}}_\infty, \bar{\mathbf{P}})$ by $\Phi^\theta(U) = u(X_{i_1}^\theta, \ldots, X_{i_k}^\theta, \ldots)$. When $\theta \to 0$, X^θ tends to X^0 almost surely, hence also in probability, and $\Phi^\theta(U) \to \Phi^0(U)$ by Slutsky's lemma (see Théorème 1 of [1]); so, given U and ε, $\Phi^\theta(U)$ is ε-close to $\Phi^0(U)$ in probability if θ is close enough to 0.

To establish cosiness, we shall take $\mathcal{F}' = \Phi^0(\mathcal{F})$ and $\mathcal{F}'' = \Phi^\theta(\mathcal{F})$ with θ close (but not equal) to 0. By the preceding remark, condition (iv) is fulfilled for some such θ; as $\theta \neq 0$, condition (iii) stems from Proposition 2, with $X' = X^0$ and $X'' = X^{\pi/2}$.

To prove (ii), we shall establish that each filtration $\mathcal{F}^\theta = \Phi^\theta(\mathcal{F})$ is immersed in the product filtration $\bar{\mathcal{F}}$. The latter is the smallest right-continuous filtration such that $\bar{\mathcal{F}}_t \supset \mathcal{F}_t^0 \vee \mathcal{F}_t^{\pi/2}$. Now $\mathcal{F}_t^\theta = \bigcap_{\varepsilon > 0} \sigma(X_i^\theta, i \in I_{t+\varepsilon}) \subset \mathcal{F}_{t+\varepsilon}^0 \vee \mathcal{F}_{t+\varepsilon}^{\pi/2}$, whence $\mathcal{F}_t^\theta \subset \mathcal{F}_t^0 \vee \mathcal{F}_t^{\pi/2}$ by applying twice Lemma 2 of Lindvall and Roger [10]; this yields the inclusion $\mathcal{F}_t^\theta \vee \mathcal{F}_t^{\theta+\pi/2} \subset \mathcal{F}_t^0 \vee \mathcal{F}_t^{\pi/2}$. The inversion formulae

$$X^0 = X^\theta \cos\theta - X^{\theta+\pi/2} \sin\theta \qquad \text{and} \qquad X^{\pi/2} = X^\theta \sin\theta + X^{\theta+\pi/2} \cos\theta$$

give the reverse inclusion, so $\mathcal{F}_t^\theta \vee \mathcal{F}_t^{\theta+\pi/2} = \mathcal{F}_t^0 \vee \mathcal{F}_t^{\pi/2}$, and $\bar{\mathcal{F}}$ is also the smallest right-continuous filtration such that $\bar{\mathcal{F}}_t$ contains $\mathcal{F}_t^\theta \vee \mathcal{F}_t^{\theta+\pi/2}$. As $\mathcal{F}^\theta$ and $\mathcal{F}^{\theta+\pi/2}$ are independent filtrations, $\mathcal{F}^\theta$ is immersed in $\bar{\mathcal{F}}$ by Lemma 1. ∎

COROLLARY 3. — *A Brownian filtration* $(\mathcal{F}_t)_{t \geq 0}$, *a standard filtration* $(\mathcal{F}_n)_{n \in -\mathbb{N}}$, *are always cosy.*

REMARKS. — Proposition 3 becomes false if it is only supposed that $\mathcal{F}_\infty$ is Gaussianly generated, without assuming the same for each $\mathcal{F}_t$. In that case, the same construction as in the above proof can be performed, yielding filtrations enjoying (ii), (iii) and (iv). But the immersion property (i) may fail, because $\mathcal{F}_t^0 \vee \mathcal{F}_t^{\pi/2}$ and $\mathcal{F}_t^\theta \vee \mathcal{F}_t^{\theta+\pi/2}$ are no longer equal. For instance, the σ-field of any Lebesgue space is always generated by a normal random variable, and yet there exist non-cosy filtrations $\mathcal{F}$ such that $(\Omega, \mathcal{F}_\infty, \mathbf{P})$ is a Lebesgue space. Some examples are the filtered probability spaces constructed by Dubins, Feldman, Smorodinsky and Tsirelson (Theorems 1 and 2 below); other examples are the natural filtration of a Walsh process, shown by Tsirelson [16] to be non-cosy, and of a standard Poisson process (by the same argument as in the next paragraph).

Proposition 3 also becomes false if "Gaussian" is replaced with "α-stable for some $\alpha < 2$", because the separation property (iii) is not suited to those processes. For instance, let X be a Lévy α-stable process; call $\Psi(\lambda)$ its characteristic exponent (so $\exp[i\lambda X_t + t\Psi(\lambda)]$ is a martingale for each λ), call T the first time when $|\Delta X_T| \geq 1$ (so $0 < T < \infty$), and let $h > 0$ be such that $\mathbf{P}[T < h] \leq 1/3$. Suppose we have two copies X' and X'' of X, jointly immersed in some filtration $\mathcal{H}$, with separate filtrations.

The $\mathcal{H}$-stopping times $T' = \inf\{t : |\Delta X'| \geq 1\}$ and $T'' = \inf\{t : |\Delta X''| \geq 1\}$ verify $T' \neq T''$ a.s. by separation. Let $Y_t = X'_{T''+t} - X'_{T''}$. As T'' is an $\mathcal{H}$-stopping time and X' is immersed in $\mathcal{H}$, the processes $\exp[i\lambda Y_t + t\Psi(\lambda)]$ are $\mathcal{H}$-martingales, and Y has the same law as X. But on the event $\{T'' < T' < T'' + h\}$, a jump larger than 1 occurs for Y at time $T' - T''$, that is, between the times 0 and h. So, by definition of h, one has $\mathbf{P}[T'' < T' < T'' + h] \leq 1/3$; and similarly, by exchanging X' and X'' in the definition of Y, $\mathbf{P}[T' < T'' < T' + h] \leq 1/3$. Taking the union of these two events and using $T' \neq T''$ gives $\mathbf{P}[|T' - T''| < h] \leq 2/3$. This bounds below the distance in probability between T' and T'', and condition (iv) in the definition of cosiness cannot be satisfied.

The next proposition, a sufficient condition for non-cosiness. summarizes the strategy of Dubins. Feldman. Smorodinsky and Tsirelson in [4].

PROPOSITION 4. — *Let* $(\Omega, \mathcal{A}, \mathbf{P}, \mathcal{F})$ *be a filtered probability space and* $(\varepsilon_n)_{n<0}$ *a sequence in* $[0, 1)$ *such that* $\sum_{n<0} \varepsilon_n < \infty$. *Suppose given a strictly increasing sequence* $(t_n)_{n \leqslant 0}$ *in* T *(that is,* $t_{n-1} < t_n$*). and an* $\mathbf{R}^{-\mathbb{N}}$*-valued random vector* $(U_n)_{n \leqslant 0}$, *with diffuse law, such that* U_n *is* $\mathcal{F}_{t_n}$*-measurable for each* n *and* U_0 *takes only finitely many values.*

Assume that for any filtered probability space $(\bar{\Omega}, \bar{\mathcal{A}}, \bar{\mathbf{P}}, \bar{\mathcal{F}})$ *and for any two filtrations* $\mathcal{F}'$ *and* $\mathcal{F}''$ *isomorphic to* $\mathcal{F}$ *and jointly immersed in* $\bar{\mathcal{F}}$, *one has for each* $n < 0$

$$\bar{\mathbf{P}}[U'_{n+1} = U''_{n+1} | \bar{\mathcal{F}}_{t_n}] \leqslant \varepsilon_n \quad \text{on the event } \{U'_n \neq U''_n\}$$

(where U'_n *and* U''_n *denote the copies of* U_n *in the* σ*-fields* $\mathcal{F}'_\infty$ *and* $\mathcal{F}''_\infty$*).*
Then $\mathcal{F}$ *is not cosy.*

The hypothesis that the whole process $(U_n)_{n \leqslant 0}$ has a diffuse law is of course linked to the separation condition we are using. But notice that each U_n taken separately is not necessarily diffuse; in the situation considered later (in the proof of Theorem 1), each U_n can take only finitely many values.

PROOF. — By Lemma 4. we may suppose $\mathsf{T} = -\mathbb{N}$ and $t_n = n$ without loss of generality.

If $\mathcal{F}'$ and $\mathcal{F}''$ are isomorphic to $\mathcal{F}$ and immersed in $\bar{\mathcal{F}}$, we know that

$$\bar{\mathbb{E}}[\mathbb{1}_{\{U'_{n+1} \neq U''_{n+1}\}} | \bar{\mathcal{F}}_n] \geqslant (1 - \varepsilon_n) \quad \text{on the event } \{U'_n \neq U''_n\} .$$

By induction on n. this implies

$$\mathbb{1}_{\{U'_n \neq U''_n\}} \bar{\mathbb{E}}[\mathbb{1}_{\{U'_{n+1} \neq U''_{n+1}\}} \cdots \mathbb{1}_{\{U'_0 \neq U''_0\}} | \bar{\mathcal{F}}_n] \geqslant \mathbb{1}_{\{U'_n \neq U''_n\}} (1 - \varepsilon_n) \ldots (1 - \varepsilon_{-1})$$

for $n < 0$. and a fortiori

$$(*) \qquad \forall n \leqslant 0 \qquad \bar{\mathbf{P}}[U'_0 \neq U''_0 | \bar{\mathcal{F}}_n] \geqslant \varepsilon \quad \text{on the event } \{U'_n \neq U''_n\} ,$$

where $\varepsilon > 0$ denotes the value of the convergent infinite product $\prod_{n<0} (1 - \varepsilon_n)$.

To establish non-cosiness, consider any two isomorphic copies of $\mathcal{F}$, jointly immersed in $\bar{\mathcal{F}}$ and separate. As the law of $(U_n)_{n \leqslant 0}$ is diffuse, the separation assumption gives $\bar{\mathbf{P}}[\exists n \leqslant 0 \ U'_n \neq U''_n] = 1$, and there exists an $m < 0$ such that

$$\bar{\mathbf{P}}[\exists n \in \{m, m+1, \ldots, 0\} \ U'_n \neq U''_n] \geqslant \tfrac{1}{2} .$$

The $\bar{\mathcal{F}}$-stopping time $T = \inf \{n : m \leqslant n \leqslant 0 \text{ and } U'_n \neq U''_n\}$ verifies $\bar{\mathbf{P}}[T < \infty] \geqslant \tfrac{1}{2}$ and $U'_n \neq U''_n$ on $\{T = n\}$. The minoration $(*)$ gives $\bar{\mathbf{P}}[U'_0 \neq U''_0 | \bar{\mathcal{F}}_T] \geqslant \varepsilon$ on $\{T < \infty\}$. whence $\bar{\mathbf{P}}[U'_0 \neq U''_0] \geqslant \bar{\mathbf{P}}[U'_0 \neq U''_0, T < \infty] \geqslant \tfrac{1}{2} \varepsilon$. As U'_0 and U''_0 assume only finitely many values, their distance in probability is bounded below, and condition (iv) in the definition of cosiness is not satisfied. ∎

5. — The main results

The following two theorems are the rewriting, in the language of cosiness, of the amazing results of Dubins, Feldman, Smorodinsky and Tsirelson [4] and [5]; see also [14].

The canonical space for a coin-tossing game indexed by the time $\mathbf{T} = -\mathbf{N}$ will be denoted by $(\Omega, \mathcal{A}, \mathbf{P}, \mathcal{F})$, where $\Omega = \{-1, 1\}^{-\mathbf{N}}$ is endowed with the coordinates ε_n, $\mathcal{F}_n$ is generated by $\sigma(\varepsilon_m, m \leqslant n)$ and the null events. $\mathcal{A} = \mathcal{F}_0 = \mathcal{F}_\infty$ and $\mathbf{P}$ is the fair coin-tossing probability, making the ε_n's independent and uniformly distributed on $\{-1, 1\}$.

THEOREM 1. — *Given $\delta > 0$, there exists on $(\Omega, \mathcal{A})$ a probability $\mathbf{Q}$ such that*

(i) $\mathbf{Q}$ *is equivalent to* $\mathbf{P}$ *and* $\left| \dfrac{d\mathbf{Q}}{d\mathbf{P}} - 1 \right| < \delta$;

(ii) $\mathcal{F} = (\mathcal{F}_n)_{n \leqslant 0}$ *is not cosy on* $(\Omega, \mathcal{A}, \mathbf{Q})$.

By Corollary 3, (ii) implies that $(\Omega, \mathcal{A}, \mathbf{Q}, \mathcal{F})$ is not standard.

If $X = (X_n)_{n \leqslant 0}$ is a process with law $\mathbf{Q}$ (defined on some probability space), its natural filtration is isomorphic to $\mathcal{F}$ under $\mathbf{Q}$; by Theorem 1 and Corollary 2, X cannot be immersed in any cosy filtration whatsoever, nor a fortiori in any standard filtration (Corollary 3).

Let $(W, \mathcal{B}, \lambda, \mathcal{G})$ denote the one-dimensional Wiener space: On $W = C(\mathbf{R}_+, \mathbf{R})$, w_t are the coordinates, λ makes w a Brownian motion started at the origin, $\mathcal{G} = (\mathcal{G}_t)_{t \geqslant 0}$ is the natural filtration of w, and $\mathcal{B} = \mathcal{G}_\infty$.

THEOREM 2. — *Given $\delta > 0$, there exists on $(W, \mathcal{B})$ a probability μ such that*

(i) μ *is equivalent to* λ *and* $\left| \dfrac{d\mu}{d\lambda} - 1 \right| < \delta$;

(ii) $\mathcal{G} = (\mathcal{G}_t)_{t \geqslant 0}$ *is not cosy on* $(W, \mathcal{B}, \mu)$.

By Corollary 3, (ii) implies that the filtration $\mathcal{G}$ on $(W, \mathcal{B}, \mu)$ is not Brownian.

If $X = (X_t)_{t \geqslant 0}$ is a process with law μ (defined on some probability space), its natural filtration is isomorphic to $\mathcal{G}$ under μ; by Theorem 2 and Corollary 2, X cannot be immersed in any cosy filtration whatsoever, nor a fortiori in any Brownian filtration (Corollary 3).

PROOF OF THEOREM 2, ASSUMING THEOREM 1. — Given $\delta > 0$, Theorem 1 yields a probability $\mathbf{Q}$ on Ω, such that $(\Omega, \mathcal{A}, \mathbf{Q}, \mathcal{F})$ is not cosy, and whose density $D = D(\varepsilon_n, n \leqslant 0) = d\mathbf{Q}/d\mathbf{P}$ verifies $|D - 1| < \delta$. Denote by $\mathcal{H}_n$ the σ-field $\mathcal{G}_{2^n}$ on W and by $\mathcal{H}$ the filtration $(\mathcal{H}_n)_{n \leqslant 0}$.

Define a mapping $S : W \to \Omega$ by $S = (S_n)_{n \leqslant 0}$ with $S_n(w) = \mathrm{sgn}(w_{2^n} - w_{2^{n-1}})$; the law $\lambda \circ S^{-1}$ of S under λ is $\mathbf{P}$. Define μ on W by $d\mu/d\lambda = D' = D \circ S$. On $(W, \mathcal{B}, \lambda)$, the vector $(S_{n+1}, \ldots, S_0)$ and the σ-field $\mathcal{H}_n$ are independent; hence, S is λ-immersed in $\mathcal{H}$. By Corollary 1, S is also μ-immersed in $\mathcal{H}$. Now the law $\mu \circ S^{-1}$ of S under μ is $\mathbf{Q}$ since, for $A \in \mathcal{A}$,

$$\mu[S \in A] = \mu[\mathbb{1}_A \circ S] = \lambda[\mathbb{1}_A \circ S \, D'] = \lambda[(D\mathbb{1}_A) \circ S] = \mathbf{P}[D\mathbb{1}_A] = \mathbf{Q}(A) .$$

So the mapping $\Psi : L^0(\Omega, \mathcal{A}, \mathbf{Q}) \to L^0(W, \mathcal{B}, \mu)$ defined by $\Psi(X) = X \circ S$ is an embedding of $(\Omega, \mathcal{A}, \mathbf{Q})$ into $(W, \mathcal{B}, \mu)$. The filtration $\Psi(\mathcal{F})$ is the natural filtration of S, so it is μ-immersed in $\mathcal{H}$. Since $(\Omega, \mathcal{F}_\infty, \mathbf{Q}, \mathcal{F})$ is not cosy by definition of $\mathbf{Q}$, neither is $(W, \mathcal{H}_0, \mu, \mathcal{H})$ by Corollary 2, nor $(W, \mathcal{G}_\infty, \mu, \mathcal{G})$ by Lemma 4. ∎

6. — Proof of Theorem 1

From now on, we follow closely Schachermayer's simplified exposition [14] of the construction by Dubins, Feldman, Smorodinsky and Tsirelson.

DEFINITION. — Let $p \geqslant 1$ be an integer and $\mathcal{M}$ the set of all rectangular matrices $\tau = (\tau_i^j)_{1 \leqslant i \leqslant p, 1 \leqslant j \leqslant 2^p}$ with entries τ_i^j in $\{-1, 1\}$. Two matrices τ' and τ'' in $\mathcal{M}$ are *close* if for at least $2^p/p$ values of j, there are at most $p^{12/13}$ values of i such that $\tau_i'^j \neq \tau_i''^j$.

LEMMA 6 (Schachermayer's Combinatorial Lemma). — *For each p large enough, there exist 2^{4p} matrices in $\mathcal{M}$ that are pairwise not close.*

Schachermayer's proof below uses very rough combinatorial estimates and can give $2^{2^p/p}$ pairwise not close matrices, instead of the mere 2^{4p} needed in the sequel; this overabundance is already present in the original proof by Dubins, Feldman, Smorodinsky and Tsirelson.

PROOF. — Let π denote the uniform probability on $\mathcal{M}$; π chooses the entries τ_i^j by tossing a fair coin. It suffices to show that for each $\tau \in \mathcal{M}$, the "neighbourhood" $C_\tau = \{\sigma \in \mathcal{M} : \sigma \text{ and } \tau \text{ are close}\}$ has probability $\pi(C_\tau) \leqslant 2^{-4p}$. So fix τ and let $s_i^j = -\tau_i^j$. As

$$C_\tau = \{\sigma \in \mathcal{M} : \text{for at least } 2^p/p \text{ values of } j,$$
$$\text{there are at most } p^{12/13} \text{ values of } i \text{ such that } \sigma_i^j = s_i^j \},$$

one has

$$C_\tau = \bigcup_{\substack{J \subset \{1, \ldots, 2^p\} \\ |J| = \lceil 2^p/p \rceil}} \bigcap_{j \in J} C_\tau^j,$$

where

$$C_\tau^j = \{\sigma \in \mathcal{M} : \text{for at most } p^{12/13} \text{ values of } i, \sigma_i^j = s_i^j\}$$

and

$$\pi(C_\tau^j) = \frac{1}{2^p} \left\{ \binom{p}{0} + \binom{p}{1} + \ldots + \binom{p}{[p^{12/13}]} \right\}.$$

For p large enough, the sum has less than $p/2$ terms and the last one is the largest, giving

$$\pi(C_\tau^j) \leqslant \frac{1}{2^p} \frac{p}{2} \binom{p}{[p^{12/13}]} \leqslant \frac{1}{2^p} \frac{p}{2} p^{p^{12/13}};$$

$$\ln \pi(C_\tau^j) \leqslant -p \ln 2 + \ln \frac{p}{2} + p^{12/13} \ln p.$$

This is equivalent to $-p \ln 2$ when p tends to infinity, so, for p large enough, $\ln \pi(C_\tau^j) \leqslant -\frac{1}{2} p \ln 2$ and $\pi(C_\tau^j) \leqslant 2^{-p/2}$.

The columns of a random matrix in $\mathcal{M}$ are independent for π; so the C_τ^j's are independent events, and, setting $q = \lceil 2^p/p \rceil$,

$$\pi(C_\tau) = \pi \left[\bigcup_{\substack{J \subset \{1, \ldots, 2^p\} \\ |J| = q}} \bigcap_{j \in J} C_\tau^j \right] \leqslant \sum_{\substack{J \subset \{1, \ldots, 2^p\} \\ |J| = q}} \pi \left[\bigcap_{j \in J} C_\tau^j \right] \leqslant \binom{2^p}{q} (2^{-p/2})^q.$$

Using $\binom{a}{q} \leqslant \frac{a^q}{q!} \leqslant \frac{a^q}{(q/3)^q}$, one gets

$$\pi(C_\tau) \leqslant \left(\frac{2^p}{q/3} 2^{-\frac{p}{2}} \right)^q \leqslant (3p \, 2^{-p/2})^q$$

and

$$\frac{1}{4p} \ln \pi(C_\tau) \leqslant \frac{1}{4p} \left\lceil \frac{2^p}{p} \right\rceil \left(\ln(3p) - \frac{p}{2} \ln 2 \right).$$

This tends to $-\infty$ when p tends to infinity, so $\pi(C_\tau) \leqslant 2^{-4p}$ for p large enough. ∎

The next lemma is the Fundamental Lemma of Dubins. Feldman. Smorodinsky and Tsirelson [4]; we borrow it from [14]. Recall the notations of Lemma 3: $\mathbf{P}_n$ is the uniform law (fair coin-tossing) on $\Omega_n = \{-1, 1\}^n$.

LEMMA 7. — *For every p large enough, there exists a set $\mathcal{Q}_{2p}$ of probabilities on Ω_{2p} with the following three properties:*

(i) $\mathcal{Q}_{2p}$ *has 2^{4p} elements;*

(ii) *each $\mathbf{Q} \in \mathcal{Q}_{2p}$ satisfies*

$$\left| \frac{d\mathbf{Q}}{d\mathbf{P}_{2p}} - 1 \right| \leqslant p^{-1/4} \; ;$$

(iii) *for any two different probabilities $\mathbf{Q}'$ and $\mathbf{Q}''$ in $\mathcal{Q}_{2p}$ and any two $\{-1, 1\}$-valued processes X' and X'', indexed by $\{1, \ldots, 2p\}$, with laws $\mathbf{Q}'$ and $\mathbf{Q}''$, defined on the same probability space $(\bar{\Omega}, \bar{\mathcal{A}}, \bar{\mathbf{P}})$ and jointly immersed, one has*

$$\bar{\mathbf{P}}\left[X_i' = X_i'' \text{ for all } i \in \{1, \ldots, 2p\} \right] \leqslant p^{-1/4} \; .$$

Since $\mathbf{Q}'$ and $\mathbf{Q}''$ are not far from $\mathbf{P}_{2p}$ by (ii), hence not far from each other, it is always possible to find two processes X' and X'' with laws $\mathbf{Q}'$ and $\mathbf{Q}''$ and such that $\bar{\mathbf{P}}[X' = X'']$ is close to 1. What the lemma says. is that if X' and X'' have these laws *and are jointly immersed*, then $\bar{\mathbf{P}}[X' = X'']$ must be small.

In [16], Tsirelson shows that a Walsh process cannot be immersed in a cosy filtration. A key step in his method is a lower estimate of the expected distance $\mathbf{E}[d(X', X'')]$, where X' and X'' are two Walsh processes embedded in a common filtration. The majoration of $\bar{\mathbf{P}}[X' = X'']$ in Lemma 7 is a discrete analogue of this minoration.

PROOF. — The 2^{4p} probabilities to be constructed on Ω_{2p} will be defined through their α-processes (notations of Lemma 3). More precisely, to each matrix $\tau \in \mathcal{M}$ (notations of Lemma 6), we shall associate an α^τ and a $\mathbf{Q}^\tau$ such that $|d\mathbf{Q}^\tau/d\mathbf{P}_{2p} - 1| \leqslant p^{-1/4}$. Then we shall prove that if τ' and τ'' are not close, any two processes X' and X'', having laws $\mathbf{Q}^{\tau'}$ and $\mathbf{Q}^{\tau''}$ and jointly immersed, verify $\bar{\mathbf{P}}[X' = X''] \leqslant p^{-1/4}$. As Lemma 6 gives 2^{4p} such matrices τ, the 2^{4p} associated probabilities will pairwise have property (iii), and Lemma 7 will be established.

As was already the case for Lemma 6, the proof works for $p \geqslant p_0$ where p_0 is an unspecified constant. The symbol $\lesssim$ will be used for inequalities valid for p large enough.

Step one: *Definition of a probability $\mathbf{Q}^\tau$ for each matrix τ, and two estimates on $\mathbf{Q}^\tau$.*

We shall slightly change the notations: a matrix $\tau \in \mathcal{M}$ will not be written (τ_i^j) with $1 \leqslant i \leqslant p$ and $1 \leqslant j \leqslant 2^p$ as in Lemma 6, but $(\tau_i^{e_1, \ldots, e_p})$, where i ranges from $p+1$ to $2p$ and $e_1, \ldots, e_p$ are in $\{-1, 1\}$ (use an arbitrary bijection between $\{1, \ldots, 2^p\}$ and $\{-1, 1\}^p$).

The matrix $\tau \in \mathcal{M}$ is fixed. The coordinates on Ω_{2p} are $\varepsilon_1, \ldots, \varepsilon_{2p}$. Define a predictable process on Ω_{2p} by

$$\beta_i = \begin{cases} 0 & \text{for } 1 \leqslant i \leqslant p \\ \eta\, \tau_i^{\varepsilon_1, \ldots, \varepsilon_p} & \text{for } p+1 \leqslant i \leqslant 2p, \end{cases}$$

where $\eta = p^{-11/12}$ is a small positive number. (This β is not the promised α^τ yet; be patient!) A $\mathbf{P}_{2p}$-martingale $Z = (Z_i)_{1 \leqslant i \leqslant 2p}$ is defined on Ω_{2p} by $Z_0 = 1$ and

$$Z_i = Z_{i-1}(1 + \beta_i \varepsilon_i) \qquad \text{for } 1 \leqslant i \leqslant 2p \; ;$$

it verifies $Z_i = 1$ for $1 \leqslant i \leqslant p$. Set $\gamma = p^{-2/7}$, introduce the stopping time

$$T = 2p \wedge \inf \{i : |Z_i - 1| > \gamma\} .$$

and remark that $p < T \leqslant 2p$. The probability $\mathbf{Q}^\tau$ will be $Z_T \cdot \mathbf{P}_{2p}$; in other words. the martingale associated with $\mathbf{Q}^\tau$ is the stopped Z^T. and the corresponding α-process (notations of Lemma 3) is β up to time T and 0 after T.

Now, by definition of T. $1-\gamma \leqslant Z_{T-1} \leqslant 1+\gamma \leqslant 2$. so

$$|Z_T-1| \leqslant |Z_T-Z_{T-1}| + |Z_{T-1}-1| \leqslant \eta\, Z_{T-1} + |Z_{T-1}-1| \leqslant 2\eta + \gamma \overset{\bullet}{\leqslant} p^{-1/4} .$$

yielding property (ii).

Observe that

$$\left\|Z_T-1\right\|^2_{\mathrm{L}^2(\mathbf{P}_{2p})} = \sum_{i=p}^{2p-1} \left\|Z^T_{i+1}-Z^T_i\right\|^2_{\mathrm{L}^2(\mathbf{P}_{2p})} \leqslant \sum_{i=p}^{2p-1} \left\|Z^T_{i+1}-Z^T_i\right\|^2_{\mathrm{L}^\infty}$$

$$\leqslant \sum_{i=p}^{2p-1} \eta^2 \left\|Z^T_i \mathbb{1}_{\{T>i\}}\right\|^2_{\mathrm{L}^\infty} \leqslant \sum_{i=p}^{2p-1} 4\eta^2 = 4p\eta^2 ;$$

this gives the estimate

$$\mathbf{P}_{2p}[T < 2p] \leqslant \mathbf{P}_{2p}\big[|Z_T-1| > \gamma\big] \leqslant \frac{4p\eta^2}{\gamma^2} ,$$

whence, using property (ii),

$$\mathbf{Q}^\tau[T < 2p] \leqslant \big(1 + p^{-1/4}\big)\, \mathbf{P}_{2p}[T < 2p] \leqslant 8p\eta^2\gamma^{-2} \overset{\bullet}{\leqslant} \tfrac{1}{4}\, p^{-1/4} .$$

<u>Step two</u>: *If two matrices* $\tau' \in \mathcal{M}$ *and* $\tau'' \in \mathcal{M}$ *are not close, any two jointly immersed processes* X' *and* X'' *with laws* $\mathbf{Q}^{\tau'}$ *and* $\mathbf{Q}^{\tau''}$ *verify* $\bar{\mathbf{P}}[X' = X''] \leqslant p^{-1/4}$.

We are given two matrices τ' and τ'', not close to each other; the construction of Step one, performed for τ' and τ'', yields on Ω_{2p} two martingales Z' and Z'', two stopping times T' and T'', and two laws $\mathbf{Q}^{\tau'}$ and $\mathbf{Q}^{\tau''}$.

Let X' and X'' be two processes indexed by $\{1, \ldots, 2p\}$, defined on some $(\bar{\Omega}, \bar{\mathcal{A}}, \bar{\mathbf{P}})$, jointly immersed in some filtration $(\mathcal{H}_i)_{1 \leqslant i \leqslant 2p}$, with respective laws $\mathbf{Q}^{\tau'}$ and $\mathbf{Q}^{\tau''}$. The processes X' and X'' can be considered as Ω_{2p}-valued random variables; as X' and X'' are $\mathcal{H}$-adapted, $S' = T' \circ X'$ and $S'' = T'' \circ X''$ are $\mathcal{H}$-stopping times, as well as $S = S' \wedge S''$.

One has $\bar{\mathbf{P}}[S' < 2p] = \mathbf{Q}^{\tau'}[T' < 2p] \leqslant \tfrac{1}{4} p^{-1/4}$ and similarly for S''; hence $\bar{\mathbf{P}}[S < 2p] \leqslant \tfrac{1}{2} p^{-1/4}$, and to establish the claim of Step two, it suffices to show that $\bar{\mathbf{P}}[X' = X'' \text{ and } S = 2p] \leqslant \tfrac{1}{2} p^{-1/4}$.

Fix $e_1, \ldots, e_p \in \{-1, 1\}^p$. For i such that $p+1 \leqslant i \leqslant 2p$. set $t'_i = \tau'^{e_1, \ldots, e_p}_i$ and $t''_i = \tau''^{e_1, \ldots, e_p}_i$. On the event $E_{e_1, \ldots, e_p} = \{X'_1 = X''_1 = e_1, \ldots, X'_p = X''_p = e_p\}$, Lemma 3 and the definitions of $\mathbf{Q}^{\tau'}$ and $\mathbf{Q}^{\tau''}$ yield for $p+1 \leqslant i \leqslant 2p$ and $e = \pm 1$

$$\bar{\mathbf{P}}[X'_i = e \mid \mathcal{H}_{i-1}] = \tfrac{1}{2}(1 + \eta t'_i e \mathbb{1}_{\{S' > i-1\}})$$

$$\bar{\mathbf{P}}[X''_i = e \mid \mathcal{H}_{i-1}] = \tfrac{1}{2}(1 + \eta t''_i e \mathbb{1}_{\{S'' > i-1\}}) ;$$

hence, on the event $E_{e_1, \ldots, e_p} \cap \{S > i-1\}$, one has

$$\bar{\mathbf{P}}[X'_i = e \mid \mathcal{H}_{i-1}] = \tfrac{1}{2}(1 + \eta t'_i e) \qquad \text{and} \qquad \bar{\mathbf{P}}[X''_i = e \mid \mathcal{H}_{i-1}] = \tfrac{1}{2}(1 + \eta t''_i e) .$$

Consequently, on the same event, if $t_i' \neq t_i''$, that is, if $t_i' t_i'' = -1$, one can write

$$\bar{\mathbf{P}}[X_i' = X_i'' \mid \mathcal{H}_{i-1}] \leqslant \bar{\mathbf{P}}[X_i' = t_i'' \text{ or } X_i'' = t_i' \mid \mathcal{H}_{i-1}]$$
$$\leqslant \bar{\mathbf{P}}[X_i' = t_i'' \mid \mathcal{H}_{i-1}] + \bar{\mathbf{P}}[X_i'' = t_i' \mid \mathcal{H}_{i-1}]$$
$$= \tfrac{1}{2}(1 + \eta t_i' t_i'') + \tfrac{1}{2}(1 + \eta t_i'' t_i') = \tfrac{1}{2}(1 - \eta) + \tfrac{1}{2}(1 - \eta) = 1 - \eta :$$

and on $E_{e_1, \ldots, e_p}$, one has $\mathbb{1}_{\{S > i-1\}} \bar{\mathbf{P}}[X_i' = X_i'' \mid \mathcal{H}_{i-1}] \leqslant \mathbb{1}_{\{S > i-1\}} (1 - \eta)^{\mathbb{1}_{t_i' \neq t_i''}}$. Since the events

$$A_i = E_{e_1, \ldots, e_p} \cap \{X_{p+1}' = X_{p+1}'', \ldots, X_i' = X_i''. S > i-1\}$$

verify $A_i \in \mathcal{H}_i$ and $A_i = A_{i-1} \cap \{S > i-1\} \cap \{X_i' = X_i''\}$, one can write for $p+1 \leqslant i \leqslant 2p$

$$\bar{\mathbf{P}}[A_i \mid \mathcal{H}_{i-1}] = \mathbb{1}_{A_{i-1}} \mathbb{1}_{\{S > i-1\}} \bar{\mathbf{P}}[X_i' = X_i'' \mid \mathcal{H}_{i-1}]$$
$$\leqslant \mathbb{1}_{A_{i-1}} \mathbb{1}_{\{S > i-1\}} (1 - \eta)^{\mathbb{1}_{t_i' \neq t_i''}} \leqslant \mathbb{1}_{A_{i-1}} (1 - \eta)^{\mathbb{1}_{t_i' \neq t_i''}}.$$

By induction, this gives for $i \geqslant p$

$$\bar{\mathbf{P}}[A_i \mid \mathcal{H}_p] \leqslant \mathbb{1}_{E_{e_1, \ldots, e_p}} (1 - \eta)^{\mathbb{1}_{t_{p+1}' \neq t_{p+1}''} + \cdots + \mathbb{1}_{t_i' \neq t_i''}};$$

taking $i = 2p$, we finally get

$$\mathbb{1}_{E_{e_1, \ldots, e_p}} \bar{\mathbf{P}}[X' = X'', S = 2p \mid \mathcal{H}_p] \leqslant \mathbb{1}_{E_{e_1, \ldots, e_p}} (1 - \eta)^{\mathrm{Card}\, \{i : t_i' \neq t_i''\}}.$$

Unfix $e_1, \ldots, e_p$ and write

$$\bar{\mathbf{P}}[X' = X'', S = 2p] = \bar{\mathbf{E}}\Big[\sum_{e_1, \ldots, e_p} \mathbb{1}_{E_{e_1, \ldots, e_p}} \bar{\mathbf{P}}[X' = X'', S = 2p \mid \mathcal{H}_p] \Big] \leqslant \bar{\mathbf{E}}\big[(1 - \eta)^N\big],$$

where N denotes the number of i's such that $\tau_i'^{X_1', \ldots, X_p'} \neq \tau_i''^{X_1', \ldots, X_p'}$. Since the α-process associated to $\mathbf{Q}^{\tau'}$ is zero on the interval $\{1, \ldots, p\}$, $X_1', \ldots, X_p'$ are independent and uniformly distributed on $\{-1, 1\}$. Now τ' and τ'' are not close, so for less than $2^p/p$ values of $(e_1, \ldots, e_p)$, there are at most $p^{12/13}$ values of i such that $\tau_i'^{e_1, \ldots, e_p} \neq \tau_i''^{e_1, \ldots, e_p}$, and

$$\bar{\mathbf{P}}[N \leqslant p^{12/13}] \leqslant \frac{1}{p}.$$

This implies

$$\bar{\mathbf{P}}[X' = X'', S = 2p] \leqslant \bar{\mathbf{E}}[(1 - \eta)^N] \leqslant \frac{1}{p} + (1 - \eta)^{p^{12/13}}$$
$$\leqslant p^{-1} + e^{-\eta p^{12/13}} = p^{-1} + e^{-p^{1/156}} \leqslant \tfrac{1}{2} p^{-1/4}.$$

The proof of step two and of Lemma 7 is complete. $\blacksquare$

As we shall use Lemma 7 only in the case when p has the form 2^{k-1}, it is convenient to re-state it in this case:

COROLLARY 4. — *For every k large enough, there exists a set $\mathcal{Q}_{2^k}$ of probabilities on Ω_{2^k} with the following three properties:*

(i) $\mathcal{Q}_{2^k}$ *has $2^{2^{k+1}}$ elements;*

(ii) *each $\mathbf{Q} \in \mathcal{Q}_{2^k}$ satisfies*

$$\left| \frac{d\mathbf{Q}}{d\mathbf{P}_{2^k}} - 1 \right| \leqslant 2^{-(k-1)/4};$$

(iii) *for any two different probabilities $\mathbf{Q}'$ and $\mathbf{Q}''$ in $\mathcal{Q}_{2^k}$ and any two $\{-1, 1\}$-valued processes X' and X'', indexed by $\{1, \ldots, 2^k\}$, with laws $\mathbf{Q}'$ and $\mathbf{Q}''$, defined on the same probability space $(\bar{\Omega}, \bar{\mathcal{A}}, \bar{\mathbf{P}})$ and jointly immersed, one has*

$$\bar{\mathbf{P}}[X_i' = X_i'' \text{ for all } i \in \{1, \ldots, 2^k\}] \leqslant 2^{-(k-1)/4}.$$

PROOF OF THEOREM 1. — We may and shall suppose that $\delta < \frac{1}{2}$. Let k_0 be a number such that Corollary 4 holds for $k \geqslant k_0$ and that

$$\prod_{k \geqslant k_0} \left(1 - 2^{-(k-1)/4}\right) > 1 - \delta \qquad \text{and} \qquad \prod_{k \geqslant k_0} \left(1 + 2^{-(k-1)/4}\right) < 1 + \delta .$$

Instead of working with the sample space $\{-1.1\}^{-\mathbb{N}}$. we shall consider

$$\Omega = \prod_{k=\infty}^{k_0} \Omega_{2^k} = \ldots \times \Omega_{2^k} \times \ldots \times \Omega_{2^{k_0}}$$

endowed with the product probability $\mathbf{P} = \ldots \times \mathbf{P}_{2^k} \times \ldots \times \mathbf{P}_{2^{k_0}}$. The projection of Ω on the factor Ω_{2^k} will be called X_{-k}. The coordinates on Ω_{2^k} will not be denoted by $\varepsilon_1, \ldots, \varepsilon_{2^k}$. but by $\varepsilon_{-2^{k+1}+1}, \varepsilon_{-2^{k+1}+2}, \ldots, \varepsilon_{-2^k}$. This identifies Ω with the canonical space of a coin-tossing game indexed by the integers $\leqslant -2^{k_0}$ (this is $\{-1,1\}^{-\mathbb{N}}$ up to a translation of the time-axis). The factor space Ω_{2^k} corresponds to time ranging from $-2^{k+1}+1$ to -2^k. The filtration $\mathcal{F} = (\mathcal{F}_i)_{i \leqslant -2^{k_0}}$ is the one generated by the coordinates $(\varepsilon_i)_{i \leqslant -2^{k_0}}$.

For each $k \geqslant k_0$, notice that the sets $\Omega_{2^{k+1}}$ and $\mathcal{Q}_{2^k}$ have the same cardinality by condition (i) of Corollary 4, and choose a bijection M_k from $\Omega_{2^{k+1}}$ to $\mathcal{Q}_{2^k}$. For $x_{-k-1} \in \Omega_{2^{k+1}}$. $M_k(x_{-k-1})$ is a probability on Ω_{2^k}; this defines a Markov transition matrix $M_k(x_{-k-1}, x_{-k})$ from $\Omega_{2^{k+1}}$ to Ω_{2^k}. Its density

$$m_k(x_{-k-1}, x_{-k}) = \frac{M_k(x_{-k-1}, x_{-k})}{\mathbf{P}_{2^k}(x_{-k})} = \frac{M_k(x_{-k-1}, x_{-k})}{2^{-2^k}}$$

verifies $\mathbf{E}\big[m_k(X_{-k-1}, X_{-k})\big] = 1$, and also $|m_k(x_{-k-1}, x_{-k}) - 1| \leqslant 2^{-(k-1)/4}$ by condition (ii) of Corollary 4. The infinite product

$$D(\omega) = \prod_{k \geqslant k_0} m_k\big(X_{-k-1}(\omega), X_{-k}(\omega)\big)$$

satisfies

$$\prod_{k \geqslant k_0} (1 - 2^{-(k-1)/4}) \leqslant D(\omega) \leqslant \prod_{k \geqslant k_0} (1 + 2^{-(k-1)/4}) ,$$

whence $|D(\omega) - 1| < \delta$ and $\mathbf{E}[D] = 1$ by dominated convergence. This defines a probability $\mathbf{Q} = D \cdot \mathbf{P}$ on Ω, satisfying condition (i) of Theorem 1. For $\mathbf{Q}$, the process $X = (\ldots, X_{-k}, \ldots, X_{-k_0})$ is Markov, with the M_k's as transition matrices, as can readily be checked by setting $D_k = \mathbf{P}[D|\mathcal{F}_{-2^k}] = \mathbf{P}\big[\prod_{\ell \geqslant k-1} m_\ell(X_{-\ell-1}, X_{-\ell})\big]$ and writing

$$\mathbf{Q}\Big[\bigcap_{\ell=k_0}^{k} \{X_{-\ell} = x_{-\ell}\}\Big] = \prod_{\ell=k_0}^{k-2} M_\ell(x_{-\ell-1}, x_{-\ell}) \, \mathbf{P}\big[D_k \, \mathbb{1}_{\{X_{-k}=x_{-k}\}}\big]$$

$$= \prod_{\ell=k_0}^{k-2} M_\ell(x_{-\ell-1}, x_{-\ell}) \, \mathbf{Q}[X_{-k} = x_{-k}] .$$

To complete the proof, it remains to establish that $(\Omega, \mathcal{F}_\infty, \mathbf{Q}, \mathcal{F})$ is not cosy: this will be done by applying Proposition 4 to $(\Omega, \mathcal{F}_\infty, \mathbf{Q}, \mathcal{F})$, with $t_\ell = -2^{-\ell}$ and $U_\ell = X_\ell$. The random vector $X = (X_\ell)_{\ell \leqslant -k_0}$ can be identified with $(\varepsilon_i)_{i \leqslant -2^{k_0}}$; its law $\mathbf{Q}$ is diffuse, for it is equivalent to $\mathbf{P}$. So, to apply Proposition 4 (with the index $-k_0$ instead of 0; this is irrelevant), it suffices to establish the following lemma.

LEMMA 8. — *On some* $(\bar{\Omega}, \bar{\mathcal{A}}, \bar{\mathbf{P}})$, *let* $\mathcal{F}'$ *and* $\mathcal{F}''$ *be two filtrations isomorphic to* $\mathcal{F}$, *jointly immersed in some filtration* $\bar{\mathcal{F}}$; *call* X' *and* X'' *the copies of* X *in the* σ-*fields* $\mathcal{F}'_{2^{k_0}}$ *and* $\mathcal{F}''_{2^{k_0}}$. *For every* $k > k_0$, *on has*

$$\bar{\mathbf{P}}[X'_{-k+1} = X''_{-k+1} | \bar{\mathcal{F}}_{-2^k}] \leqslant 2^{-(k-2)/4} \quad \text{on the event } \{X'_{-k} \neq X''_{-k}\} .$$

PROOF. — In the filtration $\mathcal{F}$, X is a Markov process with transition probabilities the M_k's. By isomorphic transfer and immersion, so are also X' and X'' in $\bar{\mathcal{F}}$.

Fix $k \geqslant k_0$, fix x' and x'' in Ω_{2^k} such that $x' \neq x''$. Take an arbitrary $\bar{\mathcal{F}}_{-2^k}$-measurable event A included in $\{X'_{-k} = x', X''_{-k} = x''\}$, introduce the new probability $\bar{\mathbf{P}}^A = \bar{\mathbf{P}}[\ |A]$, and observe that the density $\bar{\mathbf{P}}^A/d\bar{\mathbf{P}}$ is $\bar{\mathcal{F}}_{-2^k}$-measurable. By Corollary 1, $X'_{-k+1} = (\varepsilon'_{-2^k+1}, \dots, \varepsilon'_{-2^{k-1}})$ and $X''_{-k+1} = (\varepsilon''_{-2^k+1}, \dots, \varepsilon''_{-2^{k-1}})$ are $\bar{\mathbf{P}}^A$-immersed in the filtration $(\bar{\mathcal{F}}_{-2^k+1}, \bar{\mathcal{F}}_{-2^k+1}, \dots, \bar{\mathcal{F}}_{-2^{k-1}})$. Since their respective laws under $\bar{\mathbf{P}}^A$ are $M_{k-1}(x')$ and $M_{k-1}(x'')$, two different probabilities in $\mathcal{Q}_{2^{k-1}}$. property (iii) of Corollary 4 gives $\bar{\mathbf{P}}^A[X'_{-k+1} = X''_{-k+1}] \leqslant 2^{-(k-2)/4}$. As A is an arbitrary $\bar{\mathcal{F}}_{-2^k}$-measurable event included in $\{X'_{-k} = x', X''_{-k} = x''\}$, this implies

$$\bar{\mathbf{P}}[X'_{-k+1} = X''_{-k+1} | \bar{\mathcal{F}}_{-2^k}] \leqslant 2^{-(k-2)/4} \quad \text{on } \{X'_{-k} = x', X''_{-k} = x''\} . \quad \blacksquare$$

REFERENCES

[1] M.T. Barlow, M. Émery, F.B. Knight, S. Song & M. Yor. Autour d'un théorème de Tsirelson sur des filtrations browniennes et non browniennes. *Séminaire de Probabilités XXXII.* Lecture Notes in Mathematics 1686, Springer 1998.

[2] P. Brémaud & M. Yor. Changes of filtrations and of probability measures. *Z. Wahrscheinlichkeitstheorie verw. Gebiete 45,* 269–295, 1978.

[3] C. Dellacherie, B. Maisonneuve & P. A. Meyer. Probabilités et Potentiel. Chapitres XVII à XXIV : Processus de Markov (fin), Compléments de calcul stochastique. Hermann, 1992.

[4] L. Dubins, J. Feldman, M. Smorodinsky & B. Tsirelson. Decreasing sequences of σ-fields and a measure change for Brownian motion. *Ann. Prob. 24,* 882–904, 1996.

[5] J. Feldman & B. Tsirelson. Decreasing sequences of σ-fields and a measure change for Brownian motion. II *Ann. Prob. 24,* 905–911, 1996.

[6] R.K. Getoor & M.J. Sharpe. Conformal martingales. *Inventiones math. 16,* 271–308, 1972.

[7] L. Gross. Logarithmic Sobolev inequalities and contractive properties of semigroups. *Dirichlet Forms,* Lecture Notes in Mathematics 1563, Springer 1993.

[8] . S. Janson. Gaussian Hilbert Spaces. Cambridge Tracts in Mathematics 129. Cambridge University Press 1997.

[9] T. Jeulin & M. Yor. Nouveaux résultats sur le grossissement des tribus. *Ann. Scient. E.N.S. 11,* 429–443, 1978.

[10] T. Lindvall & L.C.G. Rogers. Coupling of multidimensional diffusions by reflexion. *Ann. Prob. 14,* 860–872, 1986.

[11] E. Nelson. The free Markov field. *J. Funct. Anal. 12,* 211–227, 1973.

[12] J. Neveu. Sur l'espérance conditionnelle par rapport à un mouvement brownien. *Ann. Inst. Henri Poincaré, Section B, 12,* 105–109, 1976.

[13] D. Revuz & M. Yor. Continuous Martingales and Brownian Motion. Springer, 1991.

[14] W. Schachermayer. On certain probabilities equivalent to Wiener measure, d'après Dubins, Feldman, Smorodinsky and Tsirelson. In this volume.

[15] J.-P. Thouvenot. Private communication.

[16] B. Tsirelson. Triple points: From non-Brownian filtrations to harmonic measures. *GAFA, Geom. funct. anal. 7,* 1096–1142, 1997.

On the joining of sticky Brownian motion

J. WARREN[1]

Abstract

We present an example of a one-dimensional diffusion that cannot be inno-
vated by Brownian motion. We do this by studying the ways in which two copies
of sticky Brownian motion may be joined together and applying Tsirel'son's criteria of cosiness.

There has been much recent interest in Tsirel'son's idea [9] of studying the filtration
of Walsh Brownian motion through the behaviour of pairs of such processes. A general
technique has been developed by Tsirel'son [10] and others which involves taking two
copies of a filtration and jointly immersing them in a larger set-up. See also Émery
and Yor [5], Beghdadi-Sakrani and Émery [4] and Barlow et al. [2]. This note is
motivated by applying these ideas to a particular process - sticky Brownian motion.

Let θ be a real constant satisfying $0 < \theta < \infty$. Suppose that $\left(\Omega, (\mathcal{F}_t)_{t \geq 0}, \mathbb{P}\right)$ is
a filtered probability space satisfying the usual conditions, and that $(X_t; t \geq 0)$ is
a continuous, adapted process taking values in $[0, \infty)$ which satisfies the stochastic
differential equation

$$(0.1) \qquad X_t = x + \int_0^t 1_{(X_s > 0)} dW_s + \theta \int_0^t 1_{(X_s = 0)} ds,$$

where $(W_t; t \geq 0)$ is a real-valued $\mathcal{F}_t$-Brownian motion and $x \geq 0$ is some constant.
We say that X is sticky Brownian motion with parameter θ started from x, and refer
to W as its driving Brownian motion. Unless stated otherwise we will assume $x = 0$.
Sticky Brownian motion arose in the work of Feller [6] on strong Markov processes
taking values in $[0, \infty)$ that behave like Brownian motion away from 0. In fact it
can be constructed quite simply as a time change of reflected Brownian motion so
that the resulting process is slowed down at zero, and so spends a real amount of time
there. However here our interest will be focused on it arising as a solution of the above
SDE. This equation does not admit a strong solution, it is not possible to construct
X directly from W, and the filtration $\mathcal{F}$ is not generated by W alone. Warren [12]
obtained a description of the extra randomness (hereafter referred to as the singular
contribution) in terms of a mutation process on trees. Here we will suppose that our
set-up carries two $\mathcal{F}_t$-Brownian motions $W^{(1)}$ and $W^{(2)}$ and two adapted processes
$X^{(1)}$ and $X^{(2)}$ such that each pair $\left(X^{(i)}, W^{(i)}\right)$ satisfies an equation of the same form
as (0.1), the value of θ being the same in both. We refer to this as a joining of sticky
Brownian motion.

In the first section of this note we consider the case $W^{(1)} \equiv W^{(2)}$, and show that
there is a family of different joinings such that this is so, which may be parameterised
by $p \in [0, 1]$. This parameter may be thought of as the correlation between the
singular contributions. If $p = 1$ then the singular contributions are identical and
hence so are $X^{(1)}$ and $X^{(2)}$, whereas for any $p < 1$ the process $\left(X^{(1)}, X^{(2)}\right)$ can and
does spend time away from the 'diagonal'.

---

[1]University of Warwick, United Kingdom

In the second section, following Tsirel'son's method, we consider joinings with the instantaneous correlation between $W^{(1)}$ and $W^{(2)}$ bounded in modulus away from 1. and investigate what happens as this correlation is allowed to approach 1. We will find that the limiting law is that of the joining constructed in the previous section with $p = 0$.

Tsirel'son's concept of cosiness is a necessary condition for Brownian innovation. By this we mean the existence of a probabilistic setup $(\Omega, (\mathcal{F}_t)_{t \geq 0}, \mathbb{P})$ carrying a $\mathcal{F}_t$-adapted sticky Brownian motion X, a $\mathcal{F}_t$-Brownian motion W, with the pair X and W satisfying equation (0.1), and such that the filtration $(\mathcal{F}_t)_{t \geq 0}$ is the natural filtration of a Brownian motion (necessarily different from W). For a discussion of innovation see [11]. The limiting behaviour of the joinings we observe in the second section is exactly the failure of the cosiness criteria, and so we deduce that Brownian innovation of sticky Brownian motion is impossible. In particular the filtration generated by X and W is not Brownian.

1 Correlated mutations and some singular planar diffusions

Theorem 1. Let $p \in [0,1]$. There exists a joining of sticky Brownian motion with $W^{(1)} \equiv W^{(2)}$ such that

$$(1.1) \qquad L_t^0(X^{(1)} \vee X^{(2)}) = 2(2-p)\theta A_t^{00},$$

where $A_t^{00} = \int_0^t 1_{(X_s^{(1)} = X_s^{(2)} = 0)} ds$, and L^0 is the semimartingale local time of $X^{(1)} \vee X^{(2)}$ at 0. The joint law of the processes $X^{(1)}$, $X^{(2)}$, and the common driving Brownian motion is uniquely determined.

Proof. We begin by considering an arbitrary joining with $W^{(1)}$ and $W^{(2)}$ equal to some common process W. We can write $X^{(1)} \vee X^{(2)}$ as the sum of three contributions, of which, at any time, at most one is non-zero.

$$X_t^{(1)} \vee X_t^{(2)} = Z_t^{(=)} + Z_t^{(1)} + Z_t^{(2)},$$

where

$$Z_t^{(=)} = X_t^{(1)} 1_{\left(X_t^{(1)} = X_t^{(2)}\right)} = X_t^{(2)} 1_{\left(X_t^{(1)} = X_t^{(2)}\right)},$$

and for $i = 1, 2$, with j denoting $3 - i$,

$$Z_t^{(i)} = X_t^{(i)} 1_{\left(X_t^{(i)} > X_t^{(j)}\right)}.$$

It follows from the formula for balayage of semimartingales, see [7] and the appendix of this note, that the processes $Z^{(=)}, Z^{(1)}$, and $Z^{(2)}$ are themselves continuous semimartingales, and that,

$$Z_t^{(=)} = \int_0^t 1_{\left(z_s^{(=)} > 0\right)} dW_s + \tfrac{1}{2} L_t^0(Z^{(=)}),$$

$$Z_t^{(i)} = \int_0^t 1_{\left(z_s^{(i)} > 0\right)} dW_s + \tfrac{1}{2} L_t^0(Z^{(i)}),$$

where, as always, $L^a(Z)$ denotes the local time at level a of the semimartingale Z. For each of the three processes $Z^{(=)}$, $Z^{(1)}$ and $Z^{(2)}$, the measure $dL^0(Z)$ is supported on the set of times $\{t : X_t^{(1)} = X_t^{(2)} = 0\}$. We must also observe that

$$Y_t^{(i)} \overset{\text{def}}{=} X_t^{(j)} 1_{\left(X_t^{(i)} > X_t^{(j)}\right)} = \int_0^t 1_{\left(Y_s^{(i)} > 0\right)} dW_s + \theta \int_0^t 1_{\left(Z_s^{(i)} > 0, Y_s^{(i)} = 0\right)} ds.$$

Notice that this time the balayage formula does not introduce an additional term which grows when $X_t^{(1)} = X_t^{(2)} = 0$. That this is so may be deduced from an appropriate application of théorème 2 of [7] (see the appendix again!).

Next we have that

$$L_t^0\left(X^{(1)} \vee X^{(2)}\right) = L_t^0\left(Z^{(=)}\right) + L_t^0\left(Z^{(1)}\right) + L_t^0\left(Z^{(2)}\right),$$

and since

$$X_t^{(i)} = Z_t^{(=)} + Z_t^{(i)} + Y_t^{(j)},$$

we also find that

$$2\theta A_t^{00} = L_t^0\left(Z^{(=)}\right) + L_t^0\left(Z^{(i)}\right).$$

Thus if $L_t^0\left(X^{(1)} \vee X^{(2)}\right) = 2(2-p)\theta A_t^{00}$ for some fixed $p \in [0,1]$ then we infer that

$$L_t^0\left(Z^{(=)}\right) = 2p\theta A_t^{00}, \quad \text{and} \quad L_t^0\left(Z^{(i)}\right) = 2(1-p)\theta A_t^{00}.$$

Let $|Z_t| = X_t^{(1)} \vee X_t^{(2)}$ then the process $\left(|Z_t|; t \geq 0\right)$ is itself a sticky Brownian motion[2] with parameter $(2-p)\theta$. For $i = 1, 2$ let

$$A_t^{(i)} = \int_0^t 1_{\left(Z_s^{(i)} > 0\right)} ds = \int_0^t 1_{\left(X_s^{(i)} > X_s^{(j)}\right)} ds,$$

and $\alpha^{(i)}$ be the right continuous inverse of $A^{(i)}$. Then define $\tilde{Y}_t^{(i)} = Y_{\alpha_t^{(i)}}$ and construct the Brownian motions

$$\tilde{W}_t^{(i)} = \int_0^{\alpha_t^{(i)}} 1_{\left(X_s^{(i)} > X_s^{(j)}\right)} dW_s.$$

Each pair $(\tilde{Y}^{(i)}, \tilde{W}^{(i)})$ satisfies an equation analogous to (0.1).

A (rather laborious) construction of $\left(X^{(1)}, X^{(2)}\right)$ now suggests itself. We will describe it informally- there are no real difficulties here. Start with a Brownian motion W, and choose $|Z|$ according to the conditional law of sticky Brownian motion with parameter $(2-p)\theta$ given W as its driving Brownian motion. Independently assign each excursion of $|Z|$ to be an excursion of $Z^{(=)}$, $Z^{(1)}$ or $Z^{(2)}$ with probability $(1-p)/(1+p)$, $p/(1+p)$ and $p/(1+p)$ respectively. Now for $i = 1, 2$ construct the Brownian motions $\tilde{W}^{(i)}$ as above, and then choose $\tilde{Y}^{(i)}$ according to the conditional law of sticky Brownian motion with parameter θ given $\tilde{W}^{(i)}$ as its driving Brownian

---

[2] $|Z|$ to recall the modulus of the Walsh Brownian motion on three rays.

motion, and independently of anything else. Finally put $Y_t^{(i)} = \tilde{Y}_{A_t^{(i)}}^{(i)}$ and then let $X_t^{(i)} = Z_t^{(=)} + Z_t^{(i)} + Y_t^{(j)}$.

If we consider any joining with the same value of p the joint law of the processes W, $Z^{(=)}$, $Z^{(1)}$, $Z^{(2)}$, $Y^{(1)}$ and $Y^{(2)}$ has the same structure as we have just constructed, and the uniqueness assertion follows from this. Note that we are using here that the joint law of W and X solving (0.1) is unique, and also that there is uniqueness for the martingale problem formulation of the Walsh process on 3 rays, see [3]. $\qquad\square$

Observe that, had we not known that sticky Brownian motion was not generated by its driving Brownian motion, we would be now able to deduce this from the existence of the non-diagonal joinings displayed in the preceding theorem. This is precisely the technique used by Barlow in [1], although he, dealing with a general class of SDEs which have no strong solutions, has to do much work to see non-diagonal joinings exist. Here things are much easier because we understand the nature of the singular contribution very well.

Recall the description of the law of X_t conditional on W, given in [12].

Theorem 2. *Suppose that (X, W) satisfy the SDE equation (0.1). Let $L_t = \sup_{s \le t}(-W_s)$. For fixed t, the conditional law of X_t given W is determined by*

$$(X_t, W) \overset{law}{=} \left(\left(W_t + L_t - T\right)^+, W \right),$$

where T is an independent exponential random variable with mean $1/2\theta$.

We may make repeated application of Theorem 2 to the construction of Theorem 1, and hence obtain the following description of the conditional law of $\left(X_t^{(1)}, X_t^{(2)}\right)$ given the common driving Brownian motion. Those familiar with interpreting Theorem 2 in terms of mutations on trees will easily extend the idea to cover the present case.

Corollary 3. *Let $p \in [0, 1]$ and consider the corresponding joining of sticky Brownian motion constructed in Theorem 1. The conditional law of $\left(X_t^{(1)}, X_t^{(2)}\right)$ given the common driving Brownian motion W is determined by*

$$\left(X_t^{(1)}, X_t^{(2)}, W\right) \overset{law}{=} \left(\left(W_t + L_t - T^{(1)}\right)^+, \left(W_t + L_t - T^{(2)}\right)^+, W \right),$$

where, $L_t = \sup_{s \le t}(-W_s)$, and the law of $\left(T^{(1)}, T^{(2)}\right)$ is described as follows.

Let $(M_y)_{y \ge 0} = \left(M_y^{(1)}, M_y^{(2)}\right)_{y \ge 0}$ be a Markov chain with state space $\{0, 1\}^2$, and let its transition rates be given by the following diagram.

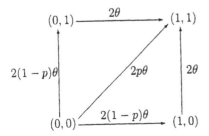

Then we take:

$$\left(T^{(1)}, T^{(2)}\right) \overset{law}{=} \left(\inf\{y : M_y^{(1)} = 1\}, \inf\{y : M_y^{(2)} = 1\}\right).$$

As particular cases, if we take $p = 1$, then $X^{(1)} \equiv X^{(2)}$, while at the other extreme, $p = 0$, and $X^{(1)}$ and $X^{(2)}$ are conditionally independent given the common driving Brownian motion W.

2 Non-cosiness of sticky Brownian motion

We have just seen that when a joining possesses common driving Brownian motions there is a 'hidden' parameter p which may be thought of as describing the correlation of the singular contributions. We want to know whether this possibility exists even if the driving Brownian motions are not identical. The answer to this does not seem, *a priori*, obvious. With any joining the pair $\left(X^{(1)}, X^{(2)}\right)$ spends plenty of time at the origin- which is where they need to be to do something mischievous. However the argument of the next paragraph shows that, at least in a special case, nothing untoward happens.

We consider the case $\langle W^{(1)}, W^{(2)} \rangle \equiv 0$. In this case the four martingales

$$\int_0^t 1_{(X_s^{(1)} = 0)} dW_s^{(1)}, \quad \int_0^t 1_{(X_s^{(1)} > 0)} dW_s^{(1)}, \quad \int_0^t 1_{(X_s^{(2)} = 0)} dW_s^{(2)}, \quad \text{and} \quad \int_0^t 1_{(X_s^{(2)} > 0)} dW_s^{(2)}$$

are mutually orthogonal. Consequently Knight's theorem tells us that if we time change each martingale to obtain a Brownian motion, then these resulting Brownian motions are mutually independent. But, for $i = 1, 2$, the pair $\left(X^{(i)}, W^{(i)}\right)$ is measurable with respect to the two Brownian motions arising from the two stochastic integrals with respect to $W^{(i)}$. Thus $\left(X^{(1)}, W^{(1)}\right)$ and $\left(X^{(2)}, W^{(2)}\right)$ are independent. Hence we see that there is a unique (in law) joining such that the driving processes $W^{(1)}$ and $W^{(2)}$ are orthogonal, and in this case the singular contributions are necessarily independent.

Throughout this section we will consider joinings such that there exists a $\rho_{max} < 1$ such that $|\langle W^{(1)}, W^{(2)} \rangle_t - \langle W^{(1)}, W^{(2)} \rangle_s| \leq \rho_{max}|t - s|$ for all $t, s \geq 0$, we say the maximal correlation of the joining is less than 1.

Lemma 4. *Any random variable belonging to $\mathcal{L}^2(X, W)$ can be expressed as a stochastic integral with respect to W.*

By virtue of this representation property (which is proved in the appendix), the maximal correlation of the joining being less than 1 makes available to us the important *hypercontractivity* inequality, see Tsirelson [10] for an outline of the proof.

Lemma 5. *Suppose that a joining satisfies, for some $\rho_{max} < 1$*

$$|\langle W^{(1)}, W^{(2)} \rangle_t - \langle W^{(1)}, W^{(2)} \rangle_s| \leq \rho_{max}|t - s|, \qquad \text{for all } t, s \geq 0.$$

Then if Φ is a bounded path functional

$$\mathbb{E}\left[\Phi(X^{(1)})\Phi(X^{(2)})\right] \leq \mathbb{E}\left[\Phi(X)^{(1+\rho_{max})}\right]^{2/(1+\rho_{max})},$$

where X possesses the common law of $X^{(1)}$ and $X^{(2)}$.

Theorem 6. *If the maximal correlation of the joining is less than 1, then*

$$L_t^0(X^{(1)} \vee X^{(2)}) = 4\theta A_t^{00},$$

where $A_t^{00} = \int_0^t 1_{(X_s^{(1)}=X_s^{(2)}=0)} ds$, *and* L^0 *is the semimartingale local time of* $X^{(1)} \vee X^{(2)}$ *at* 0.

Proof. Observe that

$$L_t^0(X^{(1)} + X^{(2)}) = 4\theta A_t^{00}.$$

Now as a consequence of the occupation time formula (see Revuz and Yor [8]) we find that

$$|L_t^0(X^{(1)} + X^{(2)}) - L_t^0(X^{(1)} \vee X^{(2)})| \leq \limsup_{\epsilon \downarrow 0} \frac{4}{\epsilon} \int_0^t 1_{(0<X_s^{(1)}<\epsilon)} 1_{(0<X_s^{(2)}<\epsilon)} ds.$$

We use the preceding Lemma to show the expectation of the righthand-side is zero. A simple computation confirms that, if X is a sticky Brownian motion, then,

$$\mathbb{E}\left[1_{(0<X_s<\epsilon)}\right] = O(\epsilon),$$

uniformly for all s. Hence, by virtue of hypercontractivity, for some $\rho_{max} < 1$,

$$\mathbb{E}\left[1_{(0<X_s^{(1)}<\epsilon)} 1_{(0<X_s^{(2)}<\epsilon)}\right] = O(\epsilon^{2/(1+\rho_{max})}),$$

uniformly for all s, and this suffices. $\qquad\square$

This proof displays very clearly how the process $(X^{(1)}, X^{(2)})$, living in the positive quadrant $\mathbb{R}_+^2$, uses motion along axes to visit the origin. But be careful in interpreting this.

Now recall Tsirelson's definition, [10], of cosy. In order for the filtration generated by sticky Brownian motion to be cosy there must exist a sequence of joinings each with maximal correlation less than 1, such that if Φ is any bounded path functional then

$$\Phi(X^{(1)}) - \Phi(X^{(2)}) \xrightarrow{\mathcal{L}^2} 0,$$

as we tend along the sequence. We are actually considering (to use Tsirelson's terminology more properly) self-joinings of the filtration generated by the sticky Brownian motion and the driving motion together. But this distinction, is in fact, unimportant, since any self-joining of the filtration generated by sticky Brownian motion alone is easily enriched to become a joining of the type we are considering.

Corollary 7. *The filtration generated by sticky Brownian motion is non-cosy.*

Proof. Fix $\lambda > 0$ and let $\gamma = \sqrt{2\lambda}$. If X is sticky Brownian motion started from 0 then

$$e^{-\gamma X_t - \lambda t} + (\gamma\theta + \lambda) \int_0^t e^{-\lambda s} 1_{(X_s=0)} ds$$

is a martingale, and

$$\mathbb{E}\left[\int_0^\infty e^{-\lambda s} 1_{(X_s=0)} ds\right] = \frac{1}{\gamma\theta + \lambda}.$$

For any joining of sticky Brownian motion $X_t^{(1)} \vee X_t^{(2)}$ is a submartingale with quadratic variation process $\int_0^t 1_{(X_s^{(1)} \vee X_s^{(2)} > 0)} ds$. If the joining has maximal correlation less than 1, then, by virtue of the preceding Theorem, $L_t^0(X^{(1)} \vee X^{(2)}) = 4\theta A_t^{00}$, and an application of Itô's formula shows that

$$e^{-\gamma X_t^{(1)} \vee X_t^{(2)} - \lambda t} + (2\gamma\theta + \lambda) \int_0^t e^{-\lambda s} 1_{(X_s^{(1)} = X_s^{(2)} = 0)} ds$$

is a supermartingale. In this case we may deduce that

$$\mathbb{E}\left[\int_0^\infty e^{-\lambda s} 1_{(X_s^{(1)} = X^{(2)} = 0)} ds\right] \le \frac{1}{2\gamma\theta + \lambda}.$$

But if the filtration generated by sticky Brownian motion was cosy then as we tend along some sequence of joinings, for each $t \ge 0$,

$$1_{(X_t^{(1)} = 0)} - 1_{(X_t^{(2)} = 0)} \xrightarrow{\mathcal{L}^2} 0,$$

and hence

$$\mathbb{E}\left[\int_0^\infty e^{-\lambda s} 1_{(X_s^{(1)} = X_s^{(2)} = 0)} ds\right] \to \mathbb{E}\left[\int_0^\infty e^{-\lambda s} 1_{(X_s=0)} ds\right].$$

In view of the above computations this is impossible. $\qquad\square$

A little more effort tidies things up. By the law of a joining we mean the joint law of $(X^{(1)}, X^{(2)}, W^{(1)}, W^{(2)})$.

Corollary 8. *Let $\mathbb{P}_n$ for $n \ge 1$, be the laws of a sequence of joinings of sticky Brownian motion, each with maximal correlation less than 1. Suppose that the law of $(W^{(1)}, W^{(2)})$, as we tend along the sequence, converges to the law of the diagonal process (W, W), where W is a Brownian motion. Then, as n tends to infinity, $\mathbb{P}_n$ converges weakly to the law of the joining constructed in Theorem 1 with $p = 0$, that is with independent singular contributions.*

Proof. The sequence $\mathbb{P}_n$ is tight because the marginal laws of $(X^{(i)}, W^{(i)})$ are constant. Suppose $\mathbb{Q}$ is the limit of a convergent subsequence. $\mathbb{Q}$ is evidently the law of a joining of sticky Brownian motion, with identical driving Brownian motions. Because of the uniqueness assertion of Theorem 1 it suffices to identify that, almost surely under $\mathbb{Q}$,

$$L_t^0(X^{(1)} \vee X^{(2)}) = 4\theta A_t^{00}.$$

Consider $f \in C_b(\mathbb{R}_+)$ with $f' \ge 0$, and $f''(0+) = 4\theta f'(0+)$. Note, for such f,

$$M_t^f = f(X_t^{(1)} \vee X_t^{(2)}) - \tfrac{1}{2} \int_0^t f''(X_s^{(1)} \vee X_s^{(2)}) ds$$

is a submartingale under any $\mathbb{P}_n$, whence it is also a submartingale under $\mathbb{Q}$. We may deduce from this that

$$L_t^0\big(X^{(1)} \vee X^{(2)}\big) \geq 4\theta A_t^{00}.$$

and re-examining the proof of Theorem 1 we see that this can only happen with equality. □

Acknowledgments. This work was done while visiting MSRI, where I was very lucky to be able to talk with both Boris Tsirel'son and Ruth Williams.

Appendix. The first part of this appendix contains an explanation as to the use of balayage to deduce that the processes $Z^{(=)}, Z^{(1)}$, and $Z^{(2)}$ are continuous semimartingales.

First we denote by H the random closed set

$$\{t : X_t^{(1)} = X_t^{(2)} = 0\}.$$

and observe that each of the processes

$$\left(1_{\left(X_t^{(1)}=X_t^{(2)}\right)}\right)_{t\geq 0}, \ \left(1_{\left(X_t^{(1)}>X_t^{(2)}\right)}\right)_{t\geq 0}, \ \text{and} \ \left(1_{\left(X_t^{(1)}<X_t^{(2)}\right)}\right)_{t\geq 0}$$

is constant on each component of H^c. For each $t > 0$ we define random times D_t and τ_t by.

$$D_t = \inf\{u > t : u \in H\},$$
$$\tau_t = \sup\{s < t : s \in H\}.$$

Then we consider a process K defined by

$$K_t = \liminf_{u \downarrow\downarrow t} 1_{\left(X_u^{(1)}=X_u^{(2)}\right)}.$$

Because K is bounded and progressive, and $X_{D_t}^{(1)} = 0$ for all t, the balayage formula, see [7], tells us that

$$K_{\tau_t} X_t^{(1)} = \int_0^t \kappa_s dX_s^{(1)} + \mathcal{R}_t,$$

where κ is the previsible projection of K and $\mathcal{R}$ is an adapted, finite-variation process, constant on each component of H^c. In particular $K_{\tau_t} X_t^{(1)}$ is a continuous semimartingale. Now K_t is equal to 1 if t is the left-hand end of an excursion into $\{x^{(1)} = x^{(2)} > 0\}$, but equal to zero at the left-hand end of other components of H^c. Thus we see that

$$K_{\tau_t} X_t^{(1)} = Z_t^{(=)}.$$

Finally we note that $\kappa_t = 1_{\left(X_t^{(1)}=X_t^{(2)}\right)}$ on H^c, whence we see that the semimartingale decomposition of $Z^{(=)}$ must be as claimed in the proof of Theorem 1. By making appropriate changes to the definition of K we may consider $Z^{(1)}$ and $Z^{(2)}$ in the same way.

When we turn to considering the processes $Y^{(1)}$ and $Y^{(2)}$ we need to alter our choice of the closed random set H and the process K. Let us now take H to be set of times at which $X^{(2)}$ is zero, and define

$$K_t = 1_{\left(X_t^{(1)} > X_t^{(2)}\right)}.$$

This choice of K is previsible, and on applying the balayage formula we find that

$$K_{\tau_t} X_t^{(2)} = \int_0^t 1_{\left(X_s^{(1)} > X_s^{(2)}\right)} dX_s^{(2)} = \int_0^t 1_{\left(X_s^{(1)} > X_s^{(2)} > 0\right)} dW_s + \theta \int_0^t 1_{\left(X_s^{(1)} > X_s^{(2)} = 0\right)} ds.$$

with no additional finite-variation term. The argument is completed by observing that $K_{\tau_t} X_t^{(2)} = Y_t^{(1)}$. The process $Y^{(2)}$ may be obtained by making obvious changes to the indices in these formulae.

The final part of this appendix contains a proof of Lemma 4. Introduce the two Brownian motions W^+ and W^0 defined by

$$W_t^+ = \int_0^\infty 1_{\left(X_s > 0, A_s^+ \le t\right)} dW_s,$$

$$W_t^0 = \int_0^\infty 1_{\left(X_s = 0, A_s^0 \le t\right)} dW_s,$$

where $A_t^+ = \int_0^t 1_{\left(X_s > 0\right)} ds$ and $A_t^0 = \int_0^t 1_{\left(X_s = 0\right)} ds$. Notice that the two stochastic integrals above are orthogonal. We find that we are able to write exponential random variables of the form

$$\exp\left\{\sum \lambda_i \left(W_{t_{i+1}}^+ - W_{t_i}^+\right) + \mu_i \left(W_{t_{i+1}}^0 - W_{t_i}^0\right)\right\}$$

as stochastic integrals against W. But these exponential variables are total in $\mathcal{L}^2(W^+, W^0)$. and moreover

$$\mathcal{L}^2(W, X) = \mathcal{L}^2(W^+, W^0),$$

whence the martingale representation property extends to all of $\mathcal{L}^2(W, X)$.

References

[1] M. Barlow, *One dimensional stochastic differential equation with no strong solution*. Journal of the London Mathematical Society, 26:335-347, 1982.

[2] M. Barlow, M. Émery, F. Knight, S. Song, M. Yor, *Autour d'un théorème de Tsirelson sur des filtrations browniennes et non browniennes*. Sém. de Probabilités XXXII, Lecture notes in Mathematics 1686, 264-305. Springer, 1998.

[3] M. Barlow, J. Pitman, M. Yor, *On Walsh's Brownian motions*. Sém. de Probabilités XXIII, Lecture notes in Mathematics 1372, 275-293. Springer, 1989.

[4] S. Beghdadi-Sakrani, M. Émery, *On certain probabilities equivalent to coin-tossing, d'après Schachermayer*. In this volume, 1999.

[5] M. Émery, M. Yor, *Sur un théorème de Tsirelson relatif à des mouvements brown-iens corrélés et à la nullité de certain temps locaux*. Sém. de Probabilités XXXII. Lecture notes in Mathematics 1686, 306-312. Springer. 1998.

[6] W. Feller, *On boundaries and lateral conditions for the Kolmogorov equations*. Annals of Mathematics, Series 2, **65**:527-570, 1957.

[7] P.A. Meyer. C. Stricker, M. Yor, *Sur une formule de la théorie du balayage*. Sém. de Probabilités XIII, Lecture notes in Mathematics 721, 478-487. Springer, 1979.

[8] D. Revuz, M. Yor, *Continuous martingales and Brownian motion*. Springer, Berlin, 1991.

[9] B. Tsirelson. *Walsh process filtration is not Brownian*. Preprint. 1996.

[10] B. Tsirelson. *Triple points: From non-Brownian filtrations to harmonic measures*. Geom. Funct. Anal. **7**:1096-1142, 1997.

[11] B. Tsirelson. *Within and beyond the reach of Brownian innovation*. Documenta Mathematica, extra volume ICM 1998, III:311-320.

[12] J. Warren, *Branching Processes, the Ray-Knight theorem, and sticky Brownian motion.*. Sém. de Probabilités XXXI, Lecture notes in Mathematics 1655, 1-15. Springer. 1997.

BROWNIAN FILTRATIONS ARE NOT STABLE
UNDER EQUIVALENT TIME-CHANGES

M. Émery and W. Schachermayer

1. — Introduction

L. Dubins, J. Feldman, M. Smorodinsky and B. Tsirelson have shown in [DFST 96] that a small perturbation of its probability law can transform Brownian motion into a process whose natural filtration is not generated by any Brownian motion whatsoever. More precisely, they construct on Wiener space $\left(W, \mathcal{F}_\infty, \lambda, (\mathcal{F}_t)_{t\geqslant 0}\right)$ a probability μ equivalent to the Wiener measure λ, with density $d\mu/d\lambda$ arbitrarily close to 1 in L^∞-norm, but such that no process with μ-independent increments generates the canonical filtration $(\mathcal{F}_t)_{t\geqslant 0}$. In fact, the μ constructed in [DFST 96] has the stronger property of being non-cosy [BE 99]. The notion of cosiness was invented by Tsirelson [T 97] as a necessary condition for a filtration to be Brownian; non-cosiness turns out to be a most convenient tool to construct new examples of "paradoxical" filtrations.

Marc Yor raised the following question: Is there something similar to the DFST-phenomenon, with a change of time instead of a change of probability law? More precisely, does there exist on Wiener space an absolutely continuous, strictly increasing time-change such that the time-changed filtration is no longer Brownian?

This question is reasonable only for those time-changes that are absolutely continuous (with respect to dt) and strictly increasing. Indeed, if a time-change is not absolutely continuous, it transforms some non $dt \times d\mathbf{P}$-null subset of $\mathbf{R}_+ \times W$ into a null one A, and the canonical Brownian motion into a martingale M such that $\int \mathbb{1}_A \, d[M, M] \neq 0$; but such a martingale cannot exist in a Brownian filtration. Similarly, if the time-change is not strictly increasing, it transforms a $dt \times d\mathbf{P}$-null set into a non null one A, and all martingales M for the new filtration verify $\int \mathbb{1}_A \, d[M, M] = 0$, so no Brownian motion can be a martingale in this filtration.

The present paper shows that the answer to Yor's question is positive; moreover, as was the case with the perturbation of measure considered in [DFST 96], the perturbation of time can be made arbitrarily small. Our main result, Theorem 4.1 below, is the existence of a family $(T_t)_{t\geqslant 0}$ of stopping times on Wiener space $\left(W, \mathcal{F}_\infty, \lambda, (\mathcal{F}_t)_{t\geqslant 0}\right)$, with the following two properties:

(i) almost surely, the function $t \mapsto T_t(\omega)$ is null at zero and differentiable, with derivative verifying $1 - \varepsilon < dT_t/dt < 1 + \varepsilon$;

(ii) the filtration $(\mathcal{G}_t)_{t\geqslant 0}$ defined by $\mathcal{G}_t = \mathcal{F}_{T_t}$ is not generated by any Brownian motion (more precisely, it is not cosy).

We end this introduction with an outline of the organisation of the paper: in section 2 we present the basic example 2.1 underlying the whole paper. We make an effort to present it as intuitively and non-technically as possible: we only consider sequences of finitely valued random variables which we interpret as "lotteries" and "pointers". Also, we avoid technical concepts such as "cosy filtrations" and "immersions" (although these ideas are behind the construction). We end this section by isolating in Proposition 2.3 a seemingly innocent property of Example 2.1, which will turn out to be crucial.

In section 3 we develop the notion of "cosy filtrations" as introduced in [T 97] (see also [BE 99]). We then show that the property of Example 2.1 isolated in Proposition 2.3 is a sufficient criterion for the non-cosiness of the generated filtration. Next, we show that non-cosiness of Example 2.1 implies in particular non-substandardness in the terminology of ([DFST 96]), i.e., the filtration generated by Example 2.1 cannot be immersed into a filtration generated by a sequence of independent random variables.

Finally in section 4 we use Example 2.1 to construct a time change of Brownian motion that destroys Brownianness of the filtration, as announced in the title. This section is completely elementary and only contains the task of translating Example 2.1 into a time-change.

2. — The discrete example

2.1. EXAMPLE. — We denote by $-\mathbf{N}$ the set $\{\ldots, -2, -1, 0\}$ and we fix a sequence $(p_n)_{n \in -\mathbf{N}}$ of natural numbers, $p_n \geqslant 2$, such that $\sum_{n \in -\mathbf{N}} p_n^{-1} < \infty$; for example $p_n = 2^{-n+1}$ is a good choice.

Now fix a probability space $(\Omega, \mathcal{A}, \mathbf{P})$ on which the following objects are defined: a family $\left((R_{n,q})_{q=1}^{p_n}\right)_{n \leqslant -1}$ of independent random variables such that $R_{n,q}$ is uniformly distributed on $\{1, \ldots, p_{n+1}\}$, and a sequence $(Q_n)_{n \leqslant 0}$ of random variables such that Q_n is uniformly distributed on $\{1, \ldots, p_n\}$, independent of $R_{m,q}$, for $m > n$, and such that

$$(2.1) \qquad Q_{n+1} = R_{n,Q_n}, \qquad \text{a.s., for} \quad n \leqslant -1.$$

It is easy to see that such random variables $\left((R_{n,q})_{q=1}^{p_n}\right)_{n \leqslant -1}$ and $(Q_n)_{n \leqslant 0}$ can indeed be defined on a suitable stochastic basis $(\Omega, \mathcal{A}, \mathbf{P})$ (first consider only $n \geqslant n_0$, then take a projective limit) and that the above properties properties already characterize the joint law of the random variables $\left(\left((R_{n,q})_{q=1}^{p_n}\right)_{n \leqslant -1}, (Q_n)_{n \leqslant 0}\right)$.

Instead of giving a formal proof of these assertions we give an intuitive explanation of the situation: for fixed n, we interpret the random variables $(R_{n,q})_{q=1}^{p_n}$ as p_n successive "lotteries" yielding random results uniformly distributed in $\{1, \ldots, p_{n+1}\}$. The random variable Q_n, taking its value in a uniformly distributed way in $\{1, \ldots, p_n\}$, will be interpreted as the "pointer" which tells us, which of these lotteries (which are drawn independently of Q_n) is relevant for us: if $Q_n(\omega) = q_n$ for some $1 \leqslant q_n \leqslant p_n$, we look at the lottery R_{n,q_n} (and ignore all the other lotteries $(R_{n,q})_{q \neq q_n}$); the outcome $R_{n,q_n}(\omega) = R_{n,Q_n(\omega)}(\omega)$ of this lottery defines by (2.1) the value of the next pointer Q_{n+1}, which tells us which lottery among $(R_{n+1,q})_{q=1}^{p_{n+1}}$ is relevant for us at time $n+1$ and in turn determines the pointer Q_{n+2} via (2.1), and so on.

The basic feature of the example is as follows: if we know the value of the pointer Q_{n_0}, for some $n_0 \in -\mathbb{N}$ which we should think of as lying in the remote past. then we can determine the values of $Q_{n_0+1}, Q_{n_0+2}, \ldots, Q_0$ by only observing the results of lotteries $(R_{n_0,q})_{q=1}^{p_{n_0}}, (R_{n_0+1,q})_{q=1}^{p_{n_0+1}}, \ldots, (R_{-1,q})_{q=1}^{p-1}$. On the other hand. if we only know the results for all the lotteries $\left((R_{n,q})_{q=1}^{p_n}\right)_{n \leqslant -1}$ (without additional knowledge of some Q_{n_0}), then we do not know enough to determine Q_0. This should be rather obvious on an intuitive level (provided $(p_n)_{n \leqslant -1}$ tends fast enough to infinity) and will be proved below. Hence the random variables $(Q_n)_{n \leqslant 0}$ contain some additional information which is not provided by the random variables $\left((R_{n,q})_{q=1}^{p_n}\right)_{n \leqslant -1}$.

But although, for any $n_0 \in -\mathbb{N}$, the information provided by Q_{n_0} in conjunction with $\left((R_{n,q})_{q=1}^{p_n}\right)_{n \leqslant -1}$ determines the value of Q_0, we shall see that the intersection of the sigma-algebras $\mathcal{G}_{n_0} = \sigma(Q_n : n \leqslant n_0)$ is trivial, i.e., $\bigcap_{n_0 \in -\mathbb{N}} \mathcal{G}_{n_0}$ consists only of sets of measure zero or one, and therefore contains no information.

So far. we have only reencountered a well-known pathology of decreasing filtrations (see Exercise 4.12 of [W 91] for a particularly easy example. pointed out by M. Barlow and E. Perkins, also displaying the above described phenomenon; see [vW 83] for a detailed study). The present example has—in contrast to the Barlow-Perkins example—the additional feature that it gives rise to a filtration that is not standard, thus displaying the same (additional) phenomenon as an example due to A. Vershik [V 73].

We now proceed to prove the above assertions.

We consider the two-dimensional process $(X_i)_{i \in I} := ((R_{n,q}, Q_n)_{q=1}^{p_n})_{n \leqslant -1}$, where the index set $I = \{(n,q) : 1 \leqslant q \leqslant p_n, n \leqslant -1\}$ is ordered lexicographically. We denote by $(\mathcal{F}_i)_{i \in I} = ((\mathcal{F}_{n,q})_{q=1}^{p_n})_{n \leqslant -1}$ the filtration generated by the process X; we shall give below an intuitive explanation of the following fact: for $1 \leqslant q \leqslant p_n$, $n \leqslant -1$ and arbitrary $n_0 \leqslant n$

$$
(2.2) \quad
\begin{aligned}
\mathcal{F}_{n,q} &= \sigma(R_{m,r}, Q_\ell : (m,r) \leqslant (n,q) \text{ and } \ell \leqslant n) = \\
&= \sigma(R_{m,r}, Q_\ell : (m,r) \leqslant (n,q) \text{ and } \ell \leqslant n_0).
\end{aligned}
$$

Formula (2.2) implies in particular that, for $1 \leqslant q \leqslant p_n$ and $n \leqslant -1$,

$$
\mathcal{F}_{n,q} = \sigma(R_{n,q}) \vee \sigma(\mathcal{F}_{m,r} : (m,r) < (n,q)),
$$

i.e., the information gained, by passing to (n,q) from its predecessor (which is $(n, q-1)$ for $q > 1$ and $(n-1, p_{n-1})$ for $q = 1$), is given by $R_{n,q}$.

Here is the intuitive explanation of the above formulae (2.2): at time (n,q) the sigma-algebra $\mathcal{F}_{n,q}$ contains, by definition, all the information of the previous lotteries $(R_{m,r})_{(m,r) \leqslant (n,q)}$ as well as the information of all the previous pointers $(Q_\ell)_{\ell \leqslant n}$. If instead we only know the positions of the pointers $(Q_\ell)_{\ell \leqslant n_0}$ for some $n_0 \leqslant n$. then we don't lose any information as the knowledge of Q_{n_0} in conjunction with the knowledge of $(R_{m,r})_{(m,r) \leqslant (n,q)}$ allows us to reconstruct via (2.1) the positions of the pointers $Q_{n_0+1}, \ldots, Q_n$.

LEMMA 2.2. — *The intersection*

$$
\mathcal{F}_{-\infty} := \bigcap_{(n,q) \in I} \mathcal{F}_{n,q}
$$

is trivial, i.e., consists only of sets of measure zero or one.

PROOF. — We start with an observation, which is notable in its own right: $(Q_n)_{n \leqslant 0}$ is an *independent* sequence of random variables. It is instructive to convince oneself on an intuitive level of this property: although, for $n \leqslant -1$ fixed, Q_n determines *which of the lotteries* $(R_{n,q})_{q=1}^{p_n}$ is chosen to define the value of Q_{n+1} via (2.1), we nonetheless have that the *result* Q_{n+1} of the lottery is independent of Q_n as all the random variables $(R_{n,q})_{q=1}^{p_n}$ are independent of Q_n and have the same law. The independence of the whole sequence $(Q_n)_{n \leqslant 0}$ now follows easily.

Next we observe the "skip-independence" of the process $(X_i)_{i \in I}$: for $n_0 \leqslant -1$ the family $((R_{n,q}, Q_n)_{q=1}^{p_n})_{n \geqslant n_0}$ is independent of $((R_{n,q}, Q_n)_{q=1}^{p_n})_{n \leqslant n_0-2}$, which again is rather obvious.

Hence, any event measurable with respect to the sigma-algebra generated by $((R_{n,q}, Q_n)_{q=1}^{p_n})_{n \geqslant n_0}$ for some $n_0 \in -\mathbb{N}$ is independent of $\mathcal{F}_{-\infty}$, which implies the triviality of $\mathcal{F}_{-\infty}$. ∎

PROPOSITION 2.3. — *Let* $(\bar{\Omega}, \bar{\mathcal{A}}, ((\bar{\mathcal{F}}_{n,q})_{q=1}^{p_n})_{n \leqslant -1}, \bar{\mathbb{P}})$ *be a filtered probability space and suppose that two processes* $((R'_{n,q}, Q'_n)_{q=1}^{p_n})_{n \leqslant -1}$ *and* $((R''_{n,q}, Q''_n)_{q=1}^{p_n})_{n \leqslant -1}$ *are defined on* $\bar{\Omega}$, *such that*

(i) $\qquad\qquad ((R'_{n,q}, Q'_n)_{q=1}^{p_n})_{n \leqslant -1}$ *and* $((R''_{n,q}, Q''_n)_{q=1}^{p_n})_{n \leqslant -1}$

are adapted to the filtration $((\bar{\mathcal{F}}_{n,q})_{q=1}^{p_n})_{n \leqslant -1}$ *and*

(ii) $\qquad$ *the processes* $((R'_{n,q}, Q'_n)_{q=1}^{p_n})_{n \leqslant -1}$ *and* $((R''_{n,q}, Q''_n)_{q=1}^{p_n})_{n \leqslant -1}$

both have the law of the process defined in Example 2.1 and, for each (n, q), *the random variables* $(R'_{m,r})_{(m,r)>(n,q)}$ *and* $(R''_{m,r})_{(m,r)>(n,q)}$ *are independent of the sigma-algebra* $\bar{\mathcal{F}}_{n,q}$.

Then, for $n < 0$, *we have*

$$(2.3) \qquad \bar{\mathbb{P}}[Q'_{n+1} \neq Q''_{n+1} | \bar{\mathcal{F}}_{n,1}] = 1 - p_{n+1}^{-1} \quad \text{ on the event } \{Q'_n \neq Q''_n\} \,.$$

PROOF. — It suffices to show (2.3) on the event $\{Q'_n = q'_n, Q''_n = q''_n\}$, where q'_n and q''_n are such that $1 \leqslant q'_n \leqslant p_n$, $1 \leqslant q''_n \leqslant p_n$ and $q'_n \neq q''_n$. So fix such q'_n and q''_n and assume w.l.g. that $q'_n < q''_n$. We shall show, more precisely, that

$$(2.4) \qquad \bar{\mathbb{P}}[Q'_{n+1} \neq Q''_{n+1} | \bar{\mathcal{F}}_{n,q''_n-1}] = 1 - p_{n+1}^{-1} \text{ on the event } \{Q'_n = q'_n, Q''_n = q''_n\} \,.$$

Indeed, for each fixed $1 \leqslant q'_{n+1} \leqslant p_{n+1}$, use assumption (i) to conclude that the event $\{Q'_{n+1} = q'_{n+1}\} = \{R'_{n,Q'_n} = q'_{n+1}\}$ is in $\mathcal{F}_{n,q'_n}$ and therefore in $\mathcal{F}_{n,q''_n-1}$. By assumption (ii) the random variable R''_{n,q''_n} is independent of $\mathcal{F}_{n,q''_n-1}$ and uniformly distributed on $\{1, \ldots, p_{n+1}\}$. Hence the $\bar{\mathcal{F}}_{n,q''_n-1}$-conditional probability for R''_{n,q''_n} to be different from q'_{n+1} identically equals $1 - p_{n+1}^{-1}$. This proves the validity of (2.4) on the event $\{Q'_n = q'_n, Q''_n = q''_n, Q'_{n+1} = q'_{n+1}\}$ and therefore (2.4) and (2.3). ∎

As the next section will show, Proposition 2.3 implies that *the filtration* $(\mathcal{F}_i)_{i \in I}$ *is not generated by any independent sequence of random variables*. This will be proved by Proposition 3.2, which relies on two hypotheses. In the case of the above example, the first hypothesis is just Property (2.3) together with the convergence of the series $\sum p_n^{-1}$; the second hypothesis is the fact that the sequence $Q = (Q_n)_{n \leqslant 0}$ has a diffuse law. And indeed, by independence of $(Q_n)_{n \leqslant 0}$ (seen in the proof of Lemma 2.2), for any deterministic sequence $q = (q_n)_{n \leqslant 0}$ one has

$$\mathbb{P}[Q = q] = \prod_{n \in -\mathbb{N}} \mathbb{P}[Q_n = q_n] = \prod_{n \in -\mathbb{N}} \frac{1}{p_n} \leqslant \prod_{n \in -\mathbb{N}} \frac{1}{2} = 0 \,.$$

3. — Cosiness

This section is borrowed, almost verbatim, from [BE 99], to which we refer for details, comments, and complements.

We shall consider filtrations $(\mathcal{F}_t)_{t\in T}$, where T is totally ordered; this includes the discrete filtrations considered in the previous section, as well as the continuous case $T = \mathbf{R}_+$. We denote by $\mathcal{F}_\infty$ the σ-field $\bigvee_{t\in T} \mathcal{F}_t$ generated by the field $\bigcup_{t\in T} \mathcal{F}_t$ (note that $\mathcal{F}_\infty = \mathcal{F}_0$ when $T = -\mathbf{N}$).

DEFINITION. — *An* embedding *of a probability space* $(\Omega, \mathcal{A}, \mathbf{P})$ *into another one* $(\bar\Omega, \bar{\mathcal{A}}, \bar{\mathbf{P}})$ *is a mapping* Ψ *from* $\mathrm{L}^0(\Omega, \mathcal{A}, \mathbf{P})$ *to* $\mathrm{L}^0(\bar\Omega, \bar{\mathcal{A}}, \bar{\mathbf{P}})$ *that commutes with Borel operations on finitely many r.v.'s:*

$$\Psi\big(f(X_1,\ldots,X_n)\big) = f\big(\Psi(X_1),\ldots,\Psi(X_n)\big) \qquad \textit{for every Borel } f$$

and preserves the probability laws:

$$\bar{\mathbf{P}}\big[\Psi(X) \in E\big] = \mathbf{P}[X \in E] \qquad \textit{for every Borel } E.$$

An embedding is always injective and transfers not only random variables, but also sub-σ-fields, filtrations, processes, etc. It is called an *isomorphism* if it is surjective; it then has an inverse. An embedding Ψ of $(\Omega, \mathcal{A}, \mathbf{P})$ into $(\bar\Omega, \bar{\mathcal{A}}, \bar{\mathbf{P}},)$ is always an isomorphism between $(\Omega, \mathcal{A}, \mathbf{P})$ and $(\bar\Omega, \Psi(\mathcal{A}), \bar{\mathbf{P}})$.

DEFINITIONS. — *Let* $\mathcal{F}$ *and* $\mathcal{G}$ *be two filtrations on a probability space* $(\Omega, \mathcal{A}, \mathbf{P})$.

The filtration $\mathcal{F}$ *is* immersed *in* $\mathcal{G}$ *if every* $\mathcal{F}$-*martingale is a* $\mathcal{G}$-*martingale.* (Note that this implies in particular $\mathcal{F}_t \subset \mathcal{G}_t$ for each $t \in T$.)

The filtrations $\mathcal{F}$ *and* $\mathcal{G}$ *are* separate *if* $\mathbf{P}[F = G] = 0$ *for all random variables* $F \in \mathrm{L}^0(\mathcal{F}_\infty)$ *and* $G \in \mathrm{L}^0(\mathcal{G}_\infty)$ *with diffuse laws.*

DEFINITION. — *A filtered probability space* $(\Omega, \mathcal{A}, \mathbf{P}, \mathcal{F})$ *is* cosy, *if there exist a filtered probability space* $(\bar\Omega, \bar{\mathcal{A}}, \bar{\mathbf{P}}, \bar{\mathcal{F}})$ *and a sequence* $(\Psi^n)_{n\in\mathbf{N}\cup\{\infty\}}$ *of embeddings of* $(\Omega, \mathcal{F}_\infty, \mathbf{P})$ *into* $(\bar\Omega, \bar{\mathcal{A}}, \bar{\mathbf{P}})$ *such that*

(i) *for each* $n \leqslant \infty$, *the filtration* $\Psi^n(\mathcal{F})$ *is immersed in* $\bar{\mathcal{F}}$;

(ii) *for each finite* n, *the filtrations* $\Psi^n(\mathcal{F})$ *and* $\Psi^\infty(\mathcal{F})$ *are separate;*

(iii) *for each* $U \in \mathrm{L}^0(\Omega, \mathcal{F}_\infty, \mathbf{P})$, *the r.v.'s* $\Psi^n(U) \in \mathrm{L}^0(\bar\Omega, \bar{\mathcal{A}}, \bar{\mathbf{P}})$ *converge in probability to* $\Psi^\infty(U)$.

Instead of saying that $(\Omega, \mathcal{A}, \mathbf{P}, \mathcal{F})$, is cosy, we shall often simply say that the filtration $\mathcal{F}$ is cosy. But it should be remembered that cosiness does depend on the probability $\mathbf{P}$.

As mentioned in the introduction, the notion of cosiness is due to Tsirelson [T 97]. We have slightly modified his definition: our separabililty condition is not equivalent to his. (His definition was intended only for filtrations where all martingales are continuous; the sufficient condition for non-cosiness given by Proposition 3.2 works simultaneously for the discrete example of section 2 and for the Brownian time-changes in section 4.)

PROPOSITION 3.1. — Let $(\Omega, \mathcal{A}, \mathsf{P}, (\mathcal{F}_t)_{t \in T})$ be a filtered probability space and $U \in L^0(\mathcal{F}_\infty)$ a random variable assuming only finitely many values. Fix $\gamma > 0$.

Suppose that for any filtered probability space $(\bar{\Omega}, \bar{\mathcal{A}}, \bar{\mathsf{P}}, \bar{\mathcal{F}})$ and for any two filtrations $\mathcal{F}'$ and $\mathcal{F}''$ isomorphic to $\mathcal{F}$, immersed in $\bar{\mathcal{F}}$ and separate, one has $\bar{\mathsf{P}}[U' \neq U''] \geqslant \gamma$ (where U' and U'' are the copies of U in the σ-fields $\mathcal{F}'_\infty$ and $\mathcal{F}''_\infty$). Then $\mathcal{F}$ is not cosy.

PROOF. — Let $(\Omega, \mathcal{A}, \mathsf{P}, \mathcal{F})$, U and γ satisfy the hypothesis of this proposition. Suppose we have some filtered probability space $(\bar{\Omega}, \bar{\mathcal{A}}, \bar{\mathsf{P}}, \bar{\mathcal{F}})$ and some sequence $(\Psi^n)_{n \in \mathbb{N} \cup \{\infty\}}$ of embeddings of $(\Omega, \mathcal{F}_\infty, \mathsf{P})$ into $(\bar{\Omega}, \bar{\mathcal{A}}, \bar{\mathsf{P}})$, fulfilling the first two conditions (i) and (ii) in the definition of cosiness. For every finite n, our hypothesis can be applied to the filtrations $\mathcal{F}' = \Psi^n(\mathcal{F})$ and $\mathcal{F}'' = \Psi^\infty(\mathcal{F})$; this gives $\bar{\mathsf{P}}[\Psi^n(U) \neq \Psi^\infty(U)] \geqslant \gamma$. As U takes finitely many values, the third condition in the definition of cosiness is not satisfied. Consequently, $\mathcal{F}$ cannot be cosy. ∎

PROPOSITION 3.2. — Let $(\Omega, \mathcal{A}, \mathsf{P}, (\mathcal{F}_t)_{t \in T})$ be a filtered probability space. Suppose given a strictly increasing sequence $(t_n)_{n \leqslant 0}$ in T (that is, $t_{n-1} < t_n$), a sequence $(\varepsilon_n)_{n < 0}$ in T such that $\sum_n \varepsilon_n < \infty$, and an $\mathbb{R}^{-\mathbb{N}}$-valued random vector $(U_n)_{n \leqslant 0}$, with diffuse law, such that U_n is $\mathcal{F}_{t_n}$-measurable for each n and U_0 takes only finitely many values.

Assume that for any filtered probability space $(\bar{\Omega}, \bar{\mathcal{A}}, \bar{\mathsf{P}}, \bar{\mathcal{F}})$ and for any two filtrations $\mathcal{F}'$ and $\mathcal{F}''$ isomorphic to $\mathcal{F}$ and immersed in $\bar{\mathcal{F}}$, one has for each $n < 0$

$$\bar{\mathsf{P}}[U'_{n+1} = U''_{n+1} | \bar{\mathcal{F}}_{t_n}] \leqslant \varepsilon_n \quad \text{on the event } \{U'_n \neq U''_n\}$$

(where U'_n and U''_n denote the copies of U_n in the σ-fields $\mathcal{F}'_\infty$ and $\mathcal{F}''_\infty$). Then $\mathcal{F}$ is not cosy.

PROOF. — If $\mathcal{F}'$ and $\mathcal{F}''$ are isomorphic to $\mathcal{F}$ and immersed in $\bar{\mathcal{F}}$, we know that

$$\mathbb{1}_{\{U'_n \neq U''_n\}} \bar{\mathsf{P}}[U'_{n+1} \neq U''_{n+1} | \bar{\mathcal{F}}_{t_n}] \geqslant \mathbb{1}_{\{U'_n \neq U''_n\}} (1 - \varepsilon_n) .$$

by induction on n, this implies

$$\mathbb{1}_{\{U'_n \neq U''_n\}} \bar{\mathsf{P}}[U'_{n+1} \neq U''_{n+1}, \ldots, U'_0 \neq U''_0 | \bar{\mathcal{F}}_{t_n}] \geqslant \mathbb{1}_{\{U'_n \neq U''_n\}} (1 - \varepsilon_{-1}) \ldots (1 - \varepsilon_n)$$

and a fortiori

$$(3.1) \qquad \bar{\mathsf{P}}[U'_0 \neq U''_0 | \bar{\mathcal{F}}_{t_n}] \geqslant \gamma \quad \text{on the event } \{U'_n \neq U''_n\} ,$$

where $\gamma > 0$ denotes the value of the convergent infinite product $\prod_{n < 0} (1 - \varepsilon_n)$.

To establish non-cosiness, we shall apply Proposition 3.1 with $U = U_0$. So suppose $\mathcal{F}'$ and $\mathcal{F}''$ are two filtrations isomorphic to $\mathcal{F}$, separate and immersed in some $\bar{\mathcal{F}}$. As the law of $(U_n)_{n \leqslant 0}$ is diffuse, the separation assumption gives $\bar{\mathsf{P}}[U'_n \neq U''_n$ for some $n \leqslant 0] = 1$, and there exists an $m \leqslant 0$ such that

$$\bar{\mathsf{P}}[U'_n \neq U''_n \text{ for some } n \in \{m, m+1, \ldots, 0\}] \geqslant \tfrac{1}{2} .$$

Call N the smallest n in $\{m, m+1, \ldots, 0\}$ such that $U'_n \neq U''_n$ (if there is one). The random variable T equal to t_N if N exists and to $+\infty$ else. is an $\bar{\mathcal{F}}$-stopping time. that verifies $\bar{\mathsf{P}}[T < \infty] \geqslant \tfrac{1}{2}$ and $U'_n \neq U''_n$ on $\{T = t_n\}$. The minoration (3.1) gives

$$\bar{\mathsf{P}}[U'_0 \neq U''_0 | \bar{\mathcal{F}}_T] \geqslant \gamma \quad \text{on } \{T < \infty\} ,$$

whence $\bar{\mathsf{P}}[U'_0 \neq U''_0] \geqslant \bar{\mathsf{P}}[U'_0 \neq U''_0, T < \infty] \geqslant \tfrac{1}{2} \gamma$; and Proposition 3.1 applies. ∎

As shown by Tsirelson [T 97], a Brownian filtration is always cosy. His proof works just as well with our definition, and shows more generally that the filtration generated by a Gaussian process is always cosy (see [BE 99]). It is easy to verify that a filtration immersed into a cosy filtration is itself cosy.

DEFINITIONS. — A filtration $(\mathcal{F}_t)_{t \in -\mathbb{N}}$ is standard if it is generated by an independent sequence $(V_t)_{t \in -\mathbb{N}}$ of random variables with diffuse laws.

A filtration is substandard if it is isomorphic to a filtration immersed in a standard filtration.

For instance, a filtration generated by an independent sequence $(V'_t)_{t \in -\mathbb{N}}$ of random variables is always substandard; indeed, up to isomorphism, it is possible to consider V'_t as given by $V'_t = f_t(V_t)$, where V_t are independent and diffuse; and in this case, the natural filtration of V' is immersed in that of V.

Clearly, all standard filtrations are isomorphic to each other. A standard filtration is generated by an independent sequence of Gaussian random variables, so it is always cosy; consequently, all substandard filtrations are cosy too.

Proposition 2.3 shows that the filtration generated by Example 2.1 satisfies the criterion for non-cosiness formulated in Proposition 3.2. So it is not cosy, and a fortiori not substandard.

4. — A continuous time-change for Brownian motion

THEOREM 4.1. — On some $(\Omega, \mathcal{A}, \mathbf{P})$, let B be a Brownian motion and $\mathcal{F}$ its natural filtration. Given any $\varepsilon > 0$, there exists a family $(T_t)_{t \geqslant 0}$ of stopping times such that

(i) $T_0 = 0$;

(ii) almost all functions $t \mapsto T_t$ are smooth, increasing, with derivative $\dfrac{dT_t}{dt}$ verifying
$$\left| \frac{dT_t}{dt} - 1 \right| \leqslant \varepsilon \, ;$$

(iii) the time-changed filtration $\mathcal{G}$ defined by $\mathcal{G}_t = \mathcal{F}_{T_t}$ is not cosy.

We start proving the theorem. From now on, ε, Ω, $\mathcal{A}$, $\mathbf{P}$, B and $\mathcal{F}$ are fixed. The construction of the time-change T_t will use a small lemma, notationally complicated but actually quite elementary.

LEMMA 4.2. — The data are an integer $p \geqslant 2$ and an interval $I = [a, b]$ (with $a < b$). There exist p increasing bijections $\sigma_1, \ldots, \sigma_p$ from I onto itself, p numbers $s_1 < \ldots < s_p$ in I, and an interval $J \subset I$ with non-empty interior, such that

(i) each σ_q is smooth, is identity in a neighbourhood of a and b, and its derivative satisfies
$$\left| \frac{d\sigma_q}{dt} - 1 \right| \leqslant \varepsilon \, ;$$

(ii) for $1 \leqslant r \leqslant q \leqslant p$, $\sup J \leqslant \sigma_r(s_q)$; for $1 \leqslant q < r \leqslant p$, $\inf J \geqslant \sigma_r(s_q)$.

PROOF OF LEMMA 4.2. — Let ϕ be a C^∞ function equal to 1 on the interval $[\frac{3a+b}{4}, \frac{a+3b}{4}]$ and with compact support included in the open interval (a, b). Setting $\alpha = \varepsilon / (p \sup |\phi'|)$, the functions $\sigma_q(t) = t - \alpha q \, \phi(t)$ clearly satisfy (i).

Put $s_q = \frac{a+b}{2} + q\alpha$. Using $\sup |\phi'| \geqslant 4/(b-a)$ and $\varepsilon < 1$, one easily sees that $\frac{a+b}{2} < s_1 < \ldots < s_p < \frac{a+3b}{4}$, implying $\phi(s_q) = 1$ and $\sigma_r(s_q) = \frac{a+b}{2} + (q-r)\alpha$. So (ii) holds with $J = [\frac{a+b}{2} - \alpha, \frac{a+b}{2}]$. ∎

The next lemma describes the elementary bricks to be used in the construction. If J is an interval $[s,t]$ with $0 \leqslant s < t$. B^J will denote the normalized Brownian increment $(B_t - B_s)/\sqrt{t-s}$. which is $N(0.1)$-distributed.

LEMMA 4.3. — *Given an interval $I = [a.b]$ with $0 < a < b$. an integer $p \geqslant 2$. and a bounded, Borel function f defined on $\mathbb{R}$. let $\sigma_1, \ldots, \sigma_p$, $s_1, \ldots, s_p$ and J be as in Lemma 4.2: let Q be an $\mathcal{F}_a$-measurable r.v. with values in $\{1, \ldots, p\}$: set $T_t = \sigma_Q(t)$ and $R = f(B^J)$.*

(i) *For each $t \in I$. the random variable T_t is an $\mathcal{F}$-stopping time: the function $t \mapsto T_t$ from I to I is smooth, with derivative ε-close to 1.*

(ii) *For each $q \in \{1, \ldots, p\}$. there exists an $\mathcal{F}_{T_{s_q}}$-measurable r.v. R_q, equal to R on the event $\{Q \leqslant q\}$.*

(iii) *For each $q \in \{1, \ldots, p\}$ and every bounded, Borel ϕ, on the event $\{Q > q\}$ one has $\mathsf{E}[\phi \circ R | \mathcal{F}_{T_{s_q}}] = \mathsf{E}[\phi \circ R]$.*

The meaning of (ii) and (iii) is that, at time T_{s_q}, R is already known if $Q \leqslant q$, but still completely unknown if $Q > q$.

PROOF OF LEMMA 4.3. — (i) Since Q is $\mathcal{F}_a$-measurable. so is T_t too: as $T_t \geqslant a$, it is a stopping time. For fixed ω. the function $t \mapsto T_t(\omega)$ is one of the σ_q's constructed in Lemma 4.2. so its derivative is close to 1.

(ii) On the event $\{Q \leqslant q\}$. Lemma 4.2 (ii) gives $\sup J \leqslant \sigma_Q(s_q) = T_{s_q}$, so R_q can be defined by

$$R_q = \begin{cases} f(B^J) & \text{if } \sup J \leqslant T_{s_q}; \\ 0 & \text{if } \sup J > T_{s_q}. \end{cases}$$

(iii) On the event $\{Q > q\}$. Lemma 4.2 (ii) gives $\inf J \geqslant \sigma_Q(s_q) = T_{s_q}$, so on this event J is equal to the random interval

$$K = \begin{cases} J & \text{if } T_{s_q} \leqslant \inf J \\ [T_{s_q}, T_{s_q}+1] & \text{if } T_{s_q} > \inf J \end{cases}$$

and R to the random variable $S = f(B^K)$. But the Markov property at time T_{s_q} implies that B^K is independent of $\mathcal{F}_{T_{s_q}}$, with law $N(0,1)$; so, on the $\mathcal{F}_{T_{s_q}}$-event $\{Q > q\}$, one can write $\mathsf{E}[\phi \circ R | \mathcal{F}_{T_{s_q}}] = \mathsf{E}[\phi \circ S | \mathcal{F}_{T_{s_q}}] = \mathsf{E}[\phi \circ R]$. ∎

PROOF OF THE THEOREM. — Put $I_n = [2^n, 2^{n+1}]$; when n ranges over $\mathbb{Z}$, the intervals I_n form a subdivision of $(0, \infty)$. Choose a sequence $(p_n)_{n \in \mathbb{Z}}$ of integers such that $p_n \geqslant 2$ and $\sum_{n \leqslant 0} 1/p_n < \infty$. For each $n \in \mathbb{Z}$, Lemma 4.2 applied to I_n and p_n gives p_n bijections σ_q^n from I_n to itself, p_n numbers $s_q^n \in I_n$ and a sub-interval $J_n \subset I_n$. Choose some functions $f_n : \mathbb{R} \to \{1, \ldots, p_{n+1}\}$ such that the image of $N(0,1)$ by f_n is the uniform law on $\{1, \ldots, p_{n+1}\}$, and define $Q_n = f_n(B^{J_n})$; this random variable is $\mathcal{F}_{2^{n+1}}$-measurable and uniformly distributed on $\{1, \ldots, p_{n+1}\}$.

Set $T_0 = 0$ and for $t \in I_n$ let

$$T_t = \sigma_{Q_{n-1}}^n(t) .$$

At this point, it is worth interrupting the proof for a minute, to compare this formula with the discrete formula (2.1). The pointer Q_{n-1} depends only on the behaviour of B in the interval J_{n-1}; it tells us which of the σ_q will be used to time-change the interval I_n. According to Lemma 4.2, the image $\sigma_q^{-1}(J_n)$ of J_n by the chosen time-change will be included in one of the intervals $[s_r, s_{r+1}]$, and this interval is also completely determined by the pointer Q_{n-1}. The rôle of the lotteries $R_{n,q}$ is played by the behaviour of B on those intervals $[s_r, s_{r+1}]$; as they are disjoint intervals, the lotteries are independent. And the definition $Q_n = f_n(B^{J_n})$ says that the choice of the next pointer depends only on the result of the current lottery.

We resume proving the theorem. Since Q_{n-1} is $\mathscr{F}_{2^n}$-measurable. Lemma 4.3 (i) tells us that T_t is a stopping time, depends smoothly upon t, and that its derivative dT_t/dt is ε-close to 1. It remains to prove that the time-changed filtration $\mathscr{G}_t = \mathscr{F}_{T_t}$ is not cosy; this will be done by applying Proposition 3.2 to $\mathscr{G}$, with $t_n = 2^n$ and $U_n = Q_{n-1}$.

As $\mathscr{G}_{2^n} = \mathscr{F}_{2^n}$, Q_{n-1} is $\mathscr{G}_{2^n}$-measurable. As the J_n's are disjoint, the Q_n's are independent, and, for a deterministic sequence $(q_n)_{n \leqslant 0}$, the estimation

$$\mathsf{P}\big[(Q_n)_{n\leqslant 0} = (q_n)_{n\leqslant 0}\big] = \prod_{n\leqslant 0} \frac{1}{p_{n+1}} \leqslant \prod_{n\leqslant 0} \frac{1}{2} = 0$$

shows that the law of $(Q_n)_{n\leqslant 0}$ is diffuse.

To obtain non-cosiness, we shall show that, for any filtered probability space $(\bar{\Omega}, \bar{\mathcal{A}}, \bar{\mathsf{P}}, \mathcal{H})$ and any two filtrations $\mathscr{G}'$ and $\mathscr{G}''$ isomorphic to $\mathscr{G}$ and immersed in $\mathcal{H}$, one has for every $n \in \mathbb{Z}$

$(*) \qquad \bar{\mathsf{P}}\big[Q'_n = Q''_n \,|\, \mathcal{H}_{2^n}\big] = \dfrac{1}{p_{n+1}} \qquad$ on the event $\{Q'_{n-1} \neq Q''_{n-1}\}$;

since $\sum\limits_{n\leqslant 0} 1/p_n$ converges, Proposition 3.2 will then apply, with $\varepsilon_n = 1/p_n$.

So n is now fixed, and, to simplify the notations, we shall write I, J, p, f, s_q instead of I_n, J_n, p_n, f_n, s_q^n. We shall also set $a = 2^n$ and substitute Q for Q_{n-1} and R for Q_n; so $T_t = \sigma_{Q_{n-1}}^n(t)$ becomes $T_t = \sigma_Q(t)$, $Q_n = f_n(B^{J_n})$ becomes $R = f(B^J)$, and we may freely use Lemma 4.3.

Supposing $\mathscr{G}'$ and $\mathscr{G}''$ are two filtrations isomorphic to $\mathscr{G}$ and immersed in some $\mathcal{H}$, Assertion $(*)$ can now be written

$$\bar{\mathsf{P}}\big[R' = R'' \,|\, \mathcal{H}_a\big] = \frac{1}{p_{n+1}} \qquad \text{on the event } \{Q' \neq Q''\}.$$

As $\mathscr{G}'$ and $\mathscr{G}''$ play the same role, it suffices by symmetry to establish this equality on the event $\{Q' < Q''\}$. So we may fix q in $\{1, \dots, p\}$ and work on the event $A = \{Q' = q, \; Q'' > q\}$. The event $\{Q' = q\}$ is in $\mathscr{G}'_a$, hence also in $\mathcal{H}_a$; similarly $\{Q'' > q\}$ is in $\mathcal{H}_a$, and their intersection A is in $\mathcal{H}_a$ too. By isomorphic transfer, the following two facts are obtained from Lemma 4.3 (ii) and (iii):

a) There exists a $\mathscr{G}'_{s_q}$-measurable r.v. R'_q equal to R' on $\{Q' \leqslant q\}$; a fortiori, R'_q is $\mathcal{H}_{s_q}$-measurable and equal to R' on A.

b) For $1 \leqslant r \leqslant p_{n+1}$, $\bar{\mathsf{P}}\big[R'' = r \,|\, \mathscr{G}''_{s_q}\big] = \bar{\mathsf{P}}\big[R'' = r\big]$ on the event $\{Q'' > q\}$; since R'' is $\mathscr{G}''_\infty$-measurable, $\mathscr{G}''$ immersed in $\mathcal{H}$, and R uniformly distributed, this implies $\bar{\mathsf{P}}\big[R'' = r \,|\, \mathcal{H}_{s_q}\big] = 1/p_{n+1}$ on $\{Q'' > q\}$, and a fortiori on A.

For $r \in \{1, \dots, p_{n+1}\}$, we may write

$$\mathbb{1}_A \, \bar{\mathsf{P}}\big[R' = R'' = r \,|\, \mathcal{H}_{s_q}\big] = \mathbb{1}_A \, \bar{\mathsf{P}}\big[A, \; R' = R'' = r \,|\, \mathcal{H}_{s_q}\big]$$

$$= \mathbb{1}_A \, \bar{\mathsf{P}}\big[A, \; R'_q = R'' = r \,|\, \mathcal{H}_{s_q}\big] = \mathbb{1}_A \, \mathbb{1}_{\{R'_q = r\}} \, \bar{\mathsf{P}}\big[R'' = r \,|\, \mathcal{H}_{s_q}\big]$$

$$= \mathbb{1}_A \, \mathbb{1}_{\{R' = r\}} \, \frac{1}{p_{n+1}} \, .$$

Summing over all r's from 1 to p_{n+1} gives $\mathbb{1}_A \, \bar{\mathsf{P}}\big[R' = R'' \,|\, \mathcal{H}_{s_q}\big] = \mathbb{1}_A \, \dfrac{1}{p_{n+1}}$; and, as $A \in \mathcal{H}_a$, applying $\bar{\mathsf{E}}\big[\;\,|\, \mathcal{H}_a\big]$ to both sides establishes the claim. ∎

REMARK. — Theorem 4.1 can be restated in terms of laws of martingales. Recall that every continuous martingale M with $M_0 = 0$ and $[M, M]_\infty = \infty$ can be written as a time-changed Brownian motion: $M = B_{[M,M]}$, where B is some Brownian motion and $[M, M]$ the quadratic variation of M. The sigma-field $\sigma(M)$ generated by M always contains $\sigma(B)$; when $\sigma(M) = \sigma(B)$, each $[M, M]_t$ is a stopping time for the filtration of B, and one says that M is *pure*; whether M is pure or not depends only on its law. (For more on pure martingales, see for instance Section V.4 of [RY 91].)

Call M *very pure* if it is pure and if the time-change $t \mapsto [M, M]_t$ that makes it Brownian is absolutely continuous and strictly increasing. If B, $(T_t)_{t \geqslant 0}$ and $\mathcal{G}$ are as in Theorem 4.1, the martingale $M_t = B_{T_t}$ is a very pure martingale, whith a non-cosy (and a fortiori non-Brownian) filtration $\mathcal{G}$.

5. — References

[BE 99] S. Beghdadi-Sakrani & M. Émery. On certain probabilities equivalent to coin-tossing, d'après Schachermayer. In this volume.

[DFST 96] L. Dubins, J. Feldman, M. Smorodinsky & B. Tsirelson. Decreasing sequences of σ-fields and a measure change for Brownian motion. *Ann. Prob. 24*, 882–904, 1996.

[RY 91] D. Revuz & M. Yor. Continuous Martingales and Brownian Motion. Springer, 1991.

[T 97] B. Tsirelson. Triple points: From non-Brownian filtrations to harmonic measures. *GAFA, Geom. funct. anal. 7*, 1096–1142, 1997.

[V 73] A. M. Vershik. Approximation in measure theory. *Doctor Thesis*. Leningrad 1973. Expanded and updated english version: The theory of decreasing sequences of measurable partitions. *St. Petersburg Math. J. 6*, 705–761, 1995.

[vW 83] H. von Weizsäcker. Exchanging the order of taking suprema and countable intersections of σ-algebras. *Ann. Inst. Henri Poincaré 19*, 91–100, 1983.

[W 91] D. Williams. Probability with Martingales. Cambridge University Press, 1991.

The Existence of a Multiple Spider Martingale in the Natural Filtration of a Certain Diffusion in the Plane

Shinzo Watanabe, Kyoto University

Introduction

The notion of spider martingales (martingales-araignées) with n rays, $n = 2, 3, \ldots, \infty$, has been introduced by Yor ([Y]) by generalizing Walsh's Brownian motions. A spider martingale with 2 rays is essentially a continuous local martingale and, on the other hand, a non-trivial spider martingale with $n \geq 3$ rays is called a *multiple spider martingale*. By the recent works by Tsirelson ([T]) and Barlow, Emery, Knight, Song and Yor ([BEKSY]), it has been recognized that a multiple spider martingale plays an important role in distinguishing a filtration from a Brownian filtration or, more generally, from a filtration which is homomorphic to a Brownian filtration; that is, any filtration which is homomorphic to a Brownian filtration can not contain a multiple spider martingale. In other words, as a noise generating the randomness of probability models, multiple spider martingales could sometimes provide us with a more useful information than a usual martingale could do. So it seems important to study, for a given stochastic process, if there exists a multiple spider martingale or not in its natural filtration. For the convenience of readers, we give in Section 1 a brief survey on the isomorphism problem of filtrations in connection with spider martingales.

The filtration of a smooth diffusion, "smooth" in the sense that it can be obtained as a strong solution of an Itô's stochastic differential equation (SDE) driven by a Wiener process, can not contain a multiple spider martingale because, as a strong solution of SDE, the natural filtration of the diffusion is homomorphic to the Brownian filtration generated by the driving Wiener process. On the other hand, the natural filtration of a Walsh's Brownian motion on $n \geq 3$ rays is a typical (and, indeed, a trivial) example of a filtration containing a multiple spider martingale. In Section 2, as a main purpose of this note, we give a less trivial example of a diffusion process on the plane $\mathbf{R}^2$ whose natural filtration contains a multiple spider martingale. This diffusion process has been studied by Ikeda and Watanabe ([IW 1] or [IW 2]) as an example of diffusions whose infinitesimal generators are not differential operators in the classical sense.

1 The isomorphism problem of filtrations

As we said in Introduction, we give here a summary of recent important results on the isomorphism problem of filtrations by Tsirelson ([T]) and Barlow, Emery, Knight, Song and Yor ([BEKSY]). No proofs are given. The reader is recommended to refer to [T] and [BEKSY] for proofs and more details.

As usual, by a *filtration* $\mathbf{F} = (F(t))_{t \in [0,\infty)}$ on a complete probability space (Ω, F, P), we mean an increasing family of sub σ-fields of F satisfying the usual conditions, that is, it is right-continuous and $F(0)$ contains all P-null sets. We set $F(\infty) = \bigvee_{t \in [0,\infty)} F(t)$. Let $\mathbf{F}$ be a filtration on (Ω, F, P) and $\mathbf{F}'$ be another filtration on (Ω', F', P').

Definition 1.1. *By a* morphism π *from* $\mathbf{F}$ *to* $\mathbf{F}'$, *denoted by* $\pi : \mathbf{F} \to \mathbf{F}'$, *we mean a map*

$$\pi_* : L^0(\Omega'; F'(\infty)) \longrightarrow L^0(\Omega; F(\infty))$$

satisfying the following conditions (i)~(iii). ($L^0(\Omega; F(\infty))$, *or* $L^0(F(\infty))$, *stands for the real vector space formed of all* $F(\infty)$*-measurable real random variables on* Ω.)

(i) *For any* $X_1, \ldots, X_n \in L^0(\Omega'; F'(\infty))$,

$$([X_1, \ldots, X_n], \ P') \stackrel{d}{=} ([\pi_*(X_1), \ldots, \pi_*(X_n)], \ P).$$

(ii) *For any* $X_1, \ldots, X_n \in L^0(\Omega'; F'(\infty))$ *and* $f : \mathbf{R}^n \to \mathbf{R}$ *which is Borel measurable,*

$$\pi_*[f(X_1, \ldots, X_n)] = f(\pi_*(X_1), \ldots, \pi_*(X_n)).$$

(iii) *For any* $X \in L^1(\Omega'; F'(\infty))$, *(then, obviously* $\pi_*(X) \in L^1(\Omega; F(\infty))$,*)*

$$\pi_*[E(X|F'(t)] = E[\pi_*(X)|F(t)], \quad \text{for all} \quad t \geq 0.$$

A morphism is also called a homomorphism; *we say that* $\mathbf{F}'$ *is* homomorphic *to* $\mathbf{F}$ *if there exists a morphism* π *from* $\mathbf{F}$ *to* $\mathbf{F}'$.

Note that the map π_* is obviously one-to-one and *non-anticipative* in the sense that, for every $t \geq 0$, $\pi_*(X)$ is $F(t)$-measurable if X is $F'(t)$-measurable.

Definition 1.2. *A morphism* π *from* $\mathbf{F}$ *to* $\mathbf{F}'$ *is called an* isomorphism *from* $\mathbf{F}$ *to* $\mathbf{F}'$ *if* $\pi_* : L^0(\Omega'; F'(\infty)) \longrightarrow L^0(\Omega; F(\infty))$ *is onto. We say that* $\mathbf{F}'$ *is* isomorphic *to* $\mathbf{F}$ *or,* $\mathbf{F}$ *and* $\mathbf{F}'$ *are isomorphic, if there exists an isomorphism* π *from* $\mathbf{F}$ *to* $\mathbf{F}'$.

Remark 1.1. *Since the map* π_* *is one-to-one, we can say equivalently as follows:* $\mathbf{F}$ *and* $\mathbf{F}'$ *are isomorphic if and only if there exists a morphism* π *from* $\mathbf{F}$ *to* $\mathbf{F}'$ *and a morphism* π' *from* $\mathbf{F}'$ *to* $\mathbf{F}$ *such that* $\pi' \circ \pi = \text{id}$ *and* $\pi \circ \pi' = \text{id}$, *i.e.,* $\pi_* \circ \pi'_* = \text{id}_*$ *on* $L^0(\Omega; F(\infty))$ *and* $\pi'_* \circ \pi_* = \text{id}_*$ *on* $L^0(\Omega'; F'(\infty))$.

Generally, for two filtrations $\mathbf{F} = (F(t))$ and $\mathbf{G} = (G(t))$ on the same probability space (Ω, F, P), we denote $\mathbf{G} \subset \mathbf{F}$ if $G(t) \subset F(t)$ for all $t \geq 0$.

A probability space (Ω, F, P) endowed with a filtration $\mathbf{F}$ is called a *filtered probability space* and is denoted by $\{(\Omega, F, P), \mathbf{F}\}$.

Definition 1.3. *For an* $\mathbf{F}$*-adapted stochastic process* $X = (X(t))$ *on* $\{(\Omega, F, P), \mathbf{F}\}$ *and an* $\mathbf{F}'$*-adapted stochastic process* $Y = (Y(t))$ *on* $\{(\Omega', F', P'), \mathbf{F}'\}$, *we say that* Y *has a* canonical representation *by* X *if the natural filtration* $\mathbf{F}^Y(\subset \mathbf{F}')$ *of* Y *is homomorphic to the natural filtration* $\mathbf{F}^X(\subset \mathbf{F})$ *of* X.

Definition 1.4. *If, in Def. 1.3,* $\mathbf{F}^X$ *and* $\mathbf{F}^Y$ *are isomorphic, then we say that* Y *has a* properly canonical representation *by* X.

We denote by $\mathcal{M}(\mathbf{F})$ the space of all locally square-integrable $\mathbf{F}$-martingales $M = (M(t))$ with $M(0) = 0$.

Proposition 1.1. *(1) Y has a canonical representation by X if and only if*

$$\exists\, Y' \stackrel{d}{=} Y \quad \text{such that} \quad \mathbf{F}^{Y'} \subset \mathbf{F}^X \quad \text{and} \quad \mathcal{M}(\mathbf{F}^{Y'}) \subset \mathcal{M}(\mathbf{F}^X).$$

(2) Y has a properly canonical representation by X if and only if

$$\exists\, Y' \stackrel{d}{=} Y \quad \text{such that} \quad \mathbf{F}^{Y'} = \mathbf{F}^X, \quad (\text{then, obviously, } \mathcal{M}(\mathbf{F}^{Y'}) = \mathcal{M}(\mathbf{F}^X)).$$

Remark 1.2. *By Proposition 1.1, we can see clearly that the notions of the canonical and properly canonical representations exactly correspond to Hida's ([H]) (in the case of linear representations of Gaussian processes by a Wiener process) and Nisio's ([N]) (in the case of nonlinear representations of general stochastic processes by a Wiener process).*

Remark 1.3. *We say that a map $\pi_* : L^0(F'(\infty)) \longrightarrow L^0(F(\infty))$ is a morphism in the weak sense if it satisfies the same conditions as in Def. 1.1 in which (iii) is replaced by: (iii)' If $X \in L^1(F'(\infty))$ is $F'(t)$-measurable, then $\pi_*(X)$ is $F(t)$-measurable for every $t \geq 0$.*

The existence of a weak morphism corresponds, in the case of stochastic processes, to the condition: $\exists\, Y' \stackrel{d}{=} Y$ such that $\mathbf{F}^{Y'} \subset \mathbf{F}^X$. In such a case, we say that Y has a non-anticipative representation by X. However, this notion is very weak compared to that of canonical or properly canonical representation. Indeed, denoting by $BM^0(m)$ an m-dimensional standard Brownian motion starting at 0, if $X = BM^0(m)$ and $Y = BM^0(n)$, then a non-anticipative representation of Y by X exists for any m and n. However, a canonical representation of Y by X exists if and only if $n \leq m$, and a properly canonical representation of Y by X exists if and only if $n = m$. (These facts follow immediately from Theorem 1.1 and its Corollary given below since the multiplicity of the natural filtration of $X = BM^0(m)$ is m.)

In the problem of existence or non-existence of canonical and properly canonical representations for stochastic processes, or more generally, existence and non-existence of homomorphisms and isomorphisms for filtrations, a useful and well-known invariant is the *multiplicity* or the *rank* of filtrations (cf. Davis-Varaiya ([DV]), Skorohod [S], cf. also, Motoo-Watanabe ([MW]), Kunita-Watanabe ([KW])).

Let $\mathbf{F} = (F(t))$ be a filtration on $(\Omega, \mathcal{F}, P)$. We assume that the filtration is separable in the sense that the Hilbert space $\mathbf{L}_2(\Omega, F(\infty), P)$ is separable.

Theorem 1.1.

(1) There exist $M_1, M_2, \ldots \in \mathcal{M}(\mathbf{F})$ such that

$$\langle M_i, M_j \rangle = 0 \quad \text{if} \quad i \neq j, \quad \langle M_1 \rangle \gg \langle M_2 \rangle \gg \cdots$$

and every $M \in \mathcal{M}(\mathbf{F})$ can be represented as a sum of stochastic integrals for some $\mathbf{F}$-predictable processes Φ_i as

$$M = \sum_i \int \Phi_i dM_i.$$

If $N_1, N_2, \ldots$ is another such sequence, then

$$\langle M_1 \rangle \approx \langle N_1 \rangle, \quad \langle M_2 \rangle \approx \langle N_2 \rangle, \quad \cdots.$$

Such a system $M_1, M_2, \cdots$ is called a basis of $\mathcal{M}(\mathbf{F})$. Here as usual, $\langle M, N \rangle$ for $M, N \in \mathcal{M}^2(\mathcal{F})$ is the quadratic covariational process, $\langle M \rangle = \langle M, M \rangle$ and. $\gg$ and $\approx$ denote the absolute continuity and the equivalence of increasing processes, respectively.

In particular, the cardinal of a basis is an invariant of the filtration $\mathbf{F}$ which we call the multiplicity *of $\mathbf{F}$ and denote by $\mathrm{mult}(\mathbf{F})$.*

(2) *Let $\mathbf{F}' \subset \mathbf{F}$ be a sub-filtration of $\mathbf{F}$ and suppose $\mathcal{M}(\mathbf{F}') \subset \mathcal{M}(\mathbf{F})$. Let $\{M_i\}$ and $\{M_i'\}$ be the basis of $\mathcal{M}(\mathbf{F})$ and $\mathcal{M}(\mathbf{F}')$, respectively. Then*

$$\langle M_1' \rangle \ll \langle M_1 \rangle, \quad \langle M_2' \rangle \ll \langle M_2 \rangle, \quad \cdots.$$

In particular, $\mathrm{mult}(\mathbf{F}') \leq \mathrm{mult}(\mathbf{F})$.

Corollary 1.1. *If $\mathbf{F}'$ is homomorphic to $\mathbf{F}$, then $\mathrm{mult}(\mathbf{F}') \leq \mathrm{mult}(\mathbf{F})$.*

Let $\mathcal{M}^c(\mathbf{F})$ be the totality of continuous elements of $\mathcal{M}(\mathbf{F})$. Then the property that $\mathcal{M}^c(\mathbf{F}) = \mathcal{M}(\mathbf{F})$, is an invariant for the existence of homomorphisms: If $\mathcal{M}^c(\mathbf{F}) = \mathcal{M}(\mathbf{F})$ and $\mathbf{F}'$ is homomorphic to $\mathbf{F}$, then we must have $\mathcal{M}^c(\mathbf{F}') = \mathcal{M}(\mathbf{F}')$.

The notion of the multiplicity of filtrations is useful to distinguish various filtrations. However, it is by no means complete; in fact, we have several examples of filtrations $\mathbf{F}$ such that $\mathcal{M}^c(\mathbf{F}) = \mathcal{M}(\mathbf{F})$ with a single base M_1 such that $\langle M_1 \rangle(t) = t$ and $\mathbf{F}$ is not a natural filtration of $BM^0(1)$. An example was given by Dubins, Feldman, Smorodinsky and Tsirelson ([DFST]) and recently, a conjecture of Barlow, Pitman and Yor ([BPY]) that the natural filtration of a Walsh's Brownian motion on $n \geq 3$ rays is not a Brownian filtration has been finally settled affirmatively by Tsirelson ([T]). For this, Tsirelson introduced another invariant notion for filtrations, the notion of *cosiness* of filtrations. We would formulate this notion as follows:

Definition 1.5. *A family $\mathbf{F}_\alpha = (F_\alpha(t)), \alpha \in [0,1]$, of filtrations on (Ω, F, P) is called a T-system (Tsirelson system) if it satisfies the following properties:*

(1) *There exists a filtration $\widehat{\mathbf{F}}$ such that, for every $\alpha \in [0,1]$, $\mathbf{F}_\alpha \subset \widehat{\mathbf{F}}$ and $\mathcal{M}(\mathbf{F}_\alpha) \subset \mathcal{M}(\widehat{\mathbf{F}})$ so that the injection $i_* : L^0(F_\alpha(\infty)) \to L^0(\widehat{F}(\infty))$ satisfies the conditions (i)$\sim$(iii) of Def. 1.1.*

(2) *For every $\alpha \in (0,1]$, there exists $0 < \rho(\alpha) < 1$ such that, for all $M \in \mathcal{M}(\mathbf{F}_0) \subset \mathcal{M}(\widehat{\mathbf{F}})$ and $N \in \mathcal{M}(\mathbf{F}_\alpha) \subset \mathcal{M}(\widehat{\mathbf{F}})$, the following holds:*

$$|\langle M, N \rangle(t)| \leq \rho(\alpha)\sqrt{\langle M \rangle(t) \cdot \langle N \rangle(t)}, \quad \forall\, t \geq 0, \ a.s..$$

(3) *$\forall\, \alpha \in (0,1]$, $\exists$ isomorphism $\pi_\alpha : \mathbf{F}_\alpha \to \mathbf{F}_0$, i.e., $(\pi_\alpha)_* : L^0(F_0(\infty)) \to L^0(F_\alpha(\infty))$ such that*

$$\|X - (\pi_\alpha)_*(X)\|_2 \to 0 \quad as \quad \alpha \to 0$$

for all $X \in L^2(F_0(\infty))$. Note that, by (1), $X \in L^2(F_0(\infty)) \subset L^2(\widehat{F}(\infty))$ and $(\pi_\alpha)_*(X) \in L^2(F_\alpha(\infty)) \subset L^2(\widehat{F}(\infty))$.

A typical example is that induced from a family of Brownian filtrations as follows:

Example 1.1. *On a suitable probability space* (Ω, F, P), *we set up a* $BM^0(2d)$ *as* $B = (B(t)) = (B_1(t), B_2(t))$ *where* B_1, B_2 *are two mutually independent* $BM^0(d)$'s. *Set*

$$W_\alpha(t) = \sqrt{1 - \alpha^2} B_1(t) + \alpha B_2(t), \quad \alpha \in [0, 1].$$

Then W_α *is a* $BM^0(d)$ *for all* α. *Set* $\widehat{\mathbf{F}} = \mathbf{F}^B$ *and* $\mathbf{F}_\alpha = \mathbf{F}^{W_\alpha}$. *Then,* $\{(\Omega, F, P), \{\mathbf{F}_\alpha\}, \widehat{\mathbf{F}}\}$ *is a T-system.*

Indeed, if $(\mathbf{W}^d, F(\mathbf{W}^d), \mu^d)$ is the d-dimensional Wiener space, i.e.,

$$\mathbf{W}^d = \{w \in \mathcal{C}([0, \infty) \to \mathbf{R}^d) \mid w(0) = 0\}$$

and μ^d is the d-dimensional Wiener measure defined on the σ-field $F(\mathbf{W}^d)$ of μ^d-measurable sets, then for every $X \in L^0(F_0(\infty))$, there exists a unique $\widetilde{X} \in L^0(F(\mathbf{W}^d))$ such that $X(\omega) = \widetilde{X}(W_0(\omega))$. Define $(\pi_\alpha)_* : L^0(F_0(\infty)) \to L^0(F_\alpha(\infty))$ by $(\pi_\alpha)_*(X) = \widetilde{X}(W_\alpha(\omega))$. (2) can be deduced, by taking $\rho(\alpha) = \sqrt{1 - \alpha^2}$, from the martingale representation theorem for $\mathcal{M}(\mathbf{F}_\alpha)$ and the relation $\langle W_0, W_\alpha \rangle(t) = \sqrt{1 - \alpha^2} \cdot t \cdot I$, I being $d \times d$-identity matrix. Finally, (3) can be deduced from the relation

$$\|X - (\pi_\alpha)_*(X)\|_2 = \|\widetilde{X} - T_t\widetilde{X}\|_{L^2(\mu^d)}, \quad X \in L^2(F_0(\infty)).$$

Here, $1 - \alpha^2 = e^{-2t}$ and T_t is the Ornstein-Uhlenbeck semigroup on the Wiener space.

Definition 1.6. *A filtration* $\mathbf{F}$ *on* (Ω, F, P) *is said to be cosy if there exist a T-system* $\{(\Omega', F', P'), \{\mathbf{F}'_\alpha\}, \widehat{\mathbf{F}}'\}$ *and a morphism* $\pi : \mathbf{F}'_0 \to \mathbf{F}$; *that is,* $\mathbf{F}$ *is homomorphic to* $\mathbf{F}'_0$.

From this definition and Example 1.1, we can easily deduce the following proposition:

Proposition 1.2. *(1) The Brownian filtration, i.e., the natural filtration of a* $BM^0(d)$, *for any dimension* d, *is cosy.*

(2) If $\mathbf{F}$ *is cosy and* $\mathbf{F}'$ *is homomorphic to* $\mathbf{F}$, *then* $\mathbf{F}'$ *is cosy.*

The notion of spider martingales (martingales-araignées) has been introduced by Yor ([Y], p.110). We follow the definition given in [BEKSY]: Before proceeding, we give some notions and notations. Let $n \geq 2$ and $\mathbf{E}$ be a real vector space of $n - 1$ dimension. Let $\mathbf{U} = \{u_1, \ldots, u_n\}$ be a set of n nonzero vectors in $\mathbf{E}$ such that $\mathbf{U}$ spans the whole space and $\sum_{k=1}^n u_k = 0$. Let

$$\mathbf{T}(= \mathbf{T}(\mathbf{U})) := \bigcup_{k=1}^n \{\lambda u_k \mid \lambda \in [0, \infty)\} \subset \mathbf{E}.$$

$\mathbf{T}$ is called a *web* (une toile d'araignée) of n-rays. When $n = 2$, then $\mathbf{U} = \{u_1, u_2 = -u_1\}$ and

$$\mathbf{T} = \{\lambda u_1 \mid \lambda \in [0, \infty)\} \cup \{\lambda u_1 \mid \lambda \in (-\infty, 0]\} = \{\lambda u_1 \mid \lambda \in \mathbf{R}\} \cong \mathbf{R},$$

so that a web of 2 rays is essentially a real line.

Definition 1.7. *A spider martingale is a* $\mathbf{T}(\subset \mathbf{E})$*-valued continuous local martingale* $M = (M(t))$ *with* $M(0) = 0$ *for some web* $\mathbf{T}$ *in* $\mathbf{E}$. *If* $\mathbf{T}$ *is a web of n rays, then M is said a spider martingale with n rays.*

Thus a spider martingale with 2 rays is essentially a continuous local martingale M with $M(0) = 0$.

Definition 1.8. *A nontrivial spider martingale with $n \geq 3$ rays is called a* multiple *spider martingale.*

When martingales are referred to a filtration $\mathbf{F}$, we say an $\mathbf{F}$-spider martingale.

We denote by $WBM^x(n; p_1, \ldots, p_n)$ a Walsh's Brownian motion (cf. [BPY]) on n rays from the origin with the rate of excursions at the origin given by $p_1, \ldots, p_n$, $(p_i > 0, \sum_{i=1}^n p_i = 1)$ and starting at x. A Walsh's Brownian motion $WBM^0(n; 1/n, \ldots, 1/n)$ is a typical example of a spider martingale with n rays and, indeed, a spider martingale with n rays is essentially obtained from a $WBM^0(n; 1/n, \ldots, 1/n)$ by a time change.

Theorem 1.2. *(Tsirelson 1996, cf. [T], [BEKSY]). If there exists a multiple $\mathbf{F}$-spider martingale in a filtration $\mathbf{F}$, then $\mathbf{F}$ is not cosy. In particular, such a filtration can not be homomorphic to a Brownian filtration in any dimension. In other words, when a filtration $\mathbf{F}$ is the natural filtration of a stochastic process $X = (X(t))$ and a multiple $\mathbf{F}$-spider martingale exists, then X can not have a canonical representation by a Wiener process in any dimension.*

It is not difficult to see that a Walsh's Brownian motion $WBM^x(n; p_1, \ldots, p_n)$ contains a multiple spider martingale with n rays in its natural filtration. Also, the following result was obtained by Barlow et al. (cf. [BEKSY]).

Theorem 1.3. *If $\mathbf{F}$ is a filtration which is homomorphic to the natural filtration of a $WBM^x(n; p_1, \ldots, p_n)$ and if $m > n$, there does not exist any multiple $\mathbf{F}$-spider martingale with m rays.*

We recall the operation of *time change* on filtrations (cf. [IW 3], p.102). Given a filtration $\mathbf{F}$ on (Ω, F, P), we mean, by a *process of time change* with respect to $\mathbf{F}$, an $\mathbf{F}$-adapted increasing process $A = (A(t))$ such that, $A(0) = 0$, $t \to A(t)$ is continuous, strictly increasing and $\lim_{t \uparrow \infty} A(t) = \infty$, almost surely. If A is a process of time change with respect to $\mathbf{F}$, then, for each $t \geq 0$, $A^{-1}(t) = \inf\{u | A(u) = t\}$ is an $\mathbf{F}$-stopping time and the σ-field $F(A^{-1}(t)) := F^{(A)}(t)$ is defined as usual. Then, we have a filtration $\mathbf{F}^{(A)} = (F^{(A)}(t))$ and, with respect to which, the increasing process $A^{-1} = (A^{-1}(t))$ is a process of time change. We can easily see that $\mathbf{F} = \{\mathbf{F}^{(A)}\}^{(A^{-1})}$: More generally, if A is a process of time change with respect to $\mathbf{F}$ and B is a process of time change with respect to $\mathbf{F}^{(A)}$, then $C = B \circ A = \{C(t) := B(A(t))\}$ is a process of time change with respect to $\mathbf{F}$ and $\mathbf{F}^{(C)} = \{\mathbf{F}^{(A)}\}^{(B)}$. Also, the following proposition can be easily deduced:

Proposition 1.3. *If $\pi : \mathbf{F} \to \mathbf{F}'$ is a morphism (i.e., $\mathbf{F}'$ is homomorphic to $\mathbf{F}$) and $A' = (A'(t))$ is a process of time change with respect to $\mathbf{F}'$, then the process $A = (A(t))$ defined by $A(t) = \pi_*(A'(t))$ is a process of time change with respect to $\mathbf{F}$ and the same map $\pi_* : L^0(F'(\infty)) \to L^0(F(\infty))$ induces the homomorphism $\pi : \mathbf{F}^{(A)} \to \mathbf{F}'^{(A')}$.*

Remark 1.4. *The property of cosiness of filtrations is not invariant under the time change; indeed, as we shall see, the filtration of a sticky Brownian motion, which is a time change of a Brownian filtration, is not cosy.*

The following strengthens a little Theorem 1.2.

Theorem 1.4. *If there exists an F-multiple spider martingale in a filtration F, then F can not be homomorphic to any time change of a cosy filtration, in particular, to any time change of a Brownian filtration in any dimension.*

Indeed, if **F** is homomorphic to a time change $\mathbf{G}^{(A)}$ of a cosy filtration **G**, then $\mathbf{G}^{(A)}$ contains a $\mathbf{G}^{(A)}$-multiple spider martingale and hence, **G** contains a **G**-multiple spider martingale. However, this contradicts with Theorem 1.2.

Finally, we introduce the notion of the *direct product* of filtrations. Given filtrations $\mathbf{F}^{(i)}$ on $(\Omega^{(i)}, F^{(i)}, P^{(i)})$, $i = 1, \ldots, m$, define their direct product $\otimes_{i=1}^m \mathbf{F}^{(i)}$, as a filtration on the product probability space $(\Pi_{i=1}^m \Omega^{(i)}, \otimes_{i=1}^m F^{(i)}, \otimes_{i=1}^m P^{(i)})$, by

$$\otimes_{i=1}^m \mathbf{F}^{(i)} = (\otimes_{i=1}^m F^{(i)}(t))_{t \geq 0},$$

where $\otimes_{i=1}^m F^{(i)}(t)$ is the usual product σ-field. Then the following propositions are easily deduced:

Proposition 1.4. *If* $\pi_i : \mathbf{F}^{(i)} \to \mathbf{F}'^{(i)}$, $i = 1, \ldots, m$ *are morphisms, then there exists a unique morphism*

$$\otimes_{i=1}^m \pi_i : \otimes_{i=1}^m \mathbf{F}^{(i)} \to \otimes_{i=1}^m \mathbf{F}'^{(i)}$$

such that $(\otimes_{i=1}^m \pi_i)_*(\otimes_{i=1}^m X_i) = \otimes_{i=1}^m [(\pi_i)_*(X_i)]$ *for* $X_i \in L^0(\Omega'^{(i)}; F'^{(i)}(\infty))$.

Here, as usual, $[\otimes_{i=1}^m X_i](\omega_1', \ldots, \omega_m') = \Pi_{i=1}^m X_i(\omega_i')$, $(\omega_1', \ldots, \omega_m') \in \Pi_{i=1}^m \Omega'^{(i)}$.

Proposition 1.5. *If* $\{(\Omega^{(i)}, F^{(i)}, P^{(i)}), \{\mathbf{F}_\alpha^{(i)}\}, \widehat{\mathbf{F}}^{(i)}\}$, $i = 1, \ldots, m$, *are T-systems, then* $\{(\Pi_{i=1}^m \Omega^{(i)}, \otimes_{i=1}^m F^{(i)}, \otimes_{i=1}^m P^{(i)}), \{\otimes_{i=1}^m \mathbf{F}_\alpha^{(i)}\}, \otimes_{i=1}^m \widehat{\mathbf{F}}^{(i)}\}$ *is a T-system.*

Corollary 1.2. *If* $\mathbf{F}^{(i)}$ *are cosy for all* i, *then their direct product* $\otimes_{i=1}^m \mathbf{F}^{(i)}$ *is cosy.*

2 The existence of a multiple spider martingale in a diffusion on the plane

In the plane $\mathbf{R}^2$, let $L_1, \ldots, L_n$ be n different straight half-lines (rays) starting at the origin 0. Let $\mathbf{e}^{(k)} = (e_1^k, e_2^k)$, $k = 1, \ldots, n$, be the unit direction vector of L_k, respectively. Let $\mathcal{D}_0$ be the space of all C^∞-functions on the plane $\mathbf{R}^2$ with a compact support and vanishing also in a neighborhood of the origin 0. Define the following bilinear form for $f, g \in \mathcal{D}_0$;

$$\mathcal{E}(f, g) = \frac{1}{2} \int_{\mathbf{R}^2} \sum_{i=1}^2 \frac{\partial f(x)}{\partial x^i} \frac{\partial g(x)}{\partial x^i} dx + \sum_{k=1}^n \int_{L_k} \sum_{i,j=1}^2 e_i^k e_j^k \frac{\partial f(x)}{\partial x^i} \frac{\partial g(x)}{\partial x^j} d\mu_k(x) \qquad (1)$$

where $d\mu_k$ is the (one-dimensional) Lebesgue measure on the half-line L_k. Then, setting $D = \mathbf{R}^2 \setminus \{0\}$, $\mathcal{E}(f, g)$ with domain $\mathcal{D}_0$ is a closable Markovian form on $L^2(D; dx)$ and its closure is a regular and local Dirichlet form. Hence, by the general theory of

Dirichlet forms ([FOT]), there corresponds a unique diffusion process $\mathbf{X} = \{X(t), P_x\}$ on D. Locally, the sample paths of this diffusion can be constructed by a skew product of two mutually independent $BM(1)$'s so that the diffusion is precisely defined for every starting point in D. As is proved in [IW 1] or [IW 2], we have that $P_x(\zeta < \infty) = 1$ and $P_x(\lim_{t \uparrow \zeta} X(t) = 0) = 1$ for every $x \in D$, where ζ is the lifetime of $\mathbf{X}$ and the terminal point (cemetery) can be identified with the origin 0. $\mathbf{X}$ is a symmetric diffusion with respect to the Lebesgue measure dx on D and it possesses the continuous transition density $p(t, x, y)$, $(t, x, y) \in (0, \infty) \times D \times D$, so that $p(t, x, y) = p(t, y, x)$.

We say that a continuous function $u(x)$ on D is $\mathbf{X}$-*harmonic* if $u(x) = E_x[u(X(\sigma_{U^c}))]$ for every $x \in D$ and every bounded neighborhood U of x such that $\overline{U} \subset D$, where $\sigma_{U^c} = \inf\{t | X(t) \notin U\}$. $u(x)$ is $\mathbf{X}$-harmonic if and only if, writing $L_k^o = L_k \cap D$,

(i) $u(x)$ is continuous in D,

(ii) $u(x)$ is harmonic in the usual sense in the open set $D \setminus \{\cup_{k=1}^n L_k\}$,

(iii) for each $k = 1, \ldots, n$

$$u|_{L_k^o} \in C^2(L_k^o) \quad \text{and} \quad 2\frac{\partial^2 u}{\partial \xi^2}(\xi, 0) = \frac{\partial u}{\partial \eta}(\xi, 0+) - \frac{\partial u}{\partial \eta}(\xi, 0-)$$

where we introduce a local coordinate (ξ, η) of $y \in U$, in a sufficiently small neighborhood U of $x \in L_k^o$, by $y - x = \xi \cdot \mathbf{e}^{(k)} + \eta \cdot \mathbf{e}^{(k)\perp}$; $(\mathbf{e}^{(k)\perp} = (-e_2^k, e_1^k)$: the unit vector perpendicular to $\mathbf{e}^{(k)}$.)

It was shown in [IW 1] or [IW 2] that, for each $k = 1, \ldots, n$, there exists a unique bounded $\mathbf{X}$-harmonic function $u_k(x)$ such that

$$\lim_{x \to 0, x \in L_j^o} u_k(x) = \delta_{jk}, \quad j = 1, \ldots, n.$$

It satisfies $0 < u_k(x) < 1$ and, furthermore, every bounded $\mathbf{X}$-harmonic function $u(x)$ can be expressed as

$$u(x) = \sum_{k=1}^n c_k u_k(x), \quad c_k \in \mathbf{R},$$

the expression being unique because $c_k = \lim_{x \to 0, x \in L_k^o} u(x)$. In particular,

$$\sum_{k=1}^n u_k(x) \equiv 1.$$

If we set

$$\Xi^{(k)} = \{X(t) \to 0 \text{ as } t \uparrow \zeta \text{ tangentially along } L_k\}, \quad k = 1, \ldots, n,$$

(for the precise meaning of "tangentially along," cf. [IW 1] or [IW 2]), then

$$u_k(x) = P_x(\Xi^{(k)}), \quad x \in D, \quad k = 1, \ldots, n.$$

For each $k = 1, \ldots, n$, $u_k(x)$ is an **X**-excessive function and we can define the u_k-subprocess $\mathbf{X}^{(k)} = (X(t), P_x^{(k)})$, i.e., the diffusion on D obtained from **X** by the transformation by the multiplicative functional (cf. [FOT], Chap. 6.3),

$$M(t) = 1_{\{\zeta > t\}} \cdot \frac{u_k(X(t))}{u_k(X(0))}.$$

This process satisfies

$$P_x^{(k)}(\Xi^{(k)}) = 1 \tag{2}$$

for all $x \in D$. For $j = 1, \ldots, n$, $u_j(x)u_k(x)^{-1}$ is an $\mathbf{X}^{(k)}$-excessive function and, by the symmetry of **X**, the measure $u_j(x)u_k(x)dx$ is $\mathbf{X}^{(k)}$-excessive measure. Then, we can construct the $\mathbf{X}^{(k)}$-*Markovian measure* $\mathbf{N}_{jk}$, called also the *approximate process* or *quasi-process*, associated to $\mathbf{X}^{(k)}$-excessive measure $u_j(x)u_k(x)dx$, cf. e.g. Weil ([We]): $\mathbf{N}_{jk}$ is a σ-finite measure on the path space

$$\mathcal{W} = \{ w \in C([0, \infty) \to \mathbf{R}^2) \mid w(0) = 0, \exists \, \sigma(w) \in (0, \infty) \text{ such that } $$
$$ w(t) \in \mathbf{R}^2 \setminus \{0\} \text{ for } t \in (0, \sigma(w)) \text{ and } w(t) = 0 \text{ for } t \geq \sigma(w) \}$$

endowed with σ-field $\mathcal{B}(\mathcal{W})$ generated by Borel cylinder sets, uniquely determined by the following properties:

(i)

$$\int_0^\infty dt \int_{\mathcal{W}} f(w(t)) 1_{\{\sigma(w) > t\}} \mathbf{N}_{jk}(dw) = \int_D f(x) u_j(x) u_k(x) dx, \quad f \in C_0(D),$$

(ii) for $t > 0$, $E \in \mathcal{B}(D)$ and $U \in \mathcal{B}(\mathcal{W})$,

$$\mathbf{N}_{jk}(\{ w \mid w(t) \in E, \theta_t(w) \in U \}) = \int_{\mathcal{W}} P_{w(t)}^{(k)}(X \in U) 1_{\{w(t) \in E\}} \mathbf{N}_{jk}(dw), \tag{3}$$

where $\theta_t(w)$ is the shifted path: $\theta_t(w)(s) = w(t + s)$.

Since **X** is symmetric, we can deduce the following property under the time reversal:

$$\mathbf{N}_{jk}\{T^{-1}(U)\} = \mathbf{N}_{kj}(U), \quad U \in \mathcal{B}(\mathcal{W}), \quad j, k = 1, \ldots, n \tag{4}$$

where $T : \mathcal{W} \to \mathcal{W}$ is the time reversal operator:

$$(Tw)(t) = \begin{cases} w(\sigma(w) - t), & 0 \leq t \leq \sigma(w), \\ 0, & t \geq \sigma(w). \end{cases}$$

If we set, for $k = 1, \ldots, n$,

$$\tilde{\Xi}^{(k)} = \{w \in \mathcal{W} \mid w(t) \to 0 \text{ as } t \uparrow \sigma(w) \text{ tangentially along } L_k\}$$

and

$$\tilde{\Pi}^{(k)} = T^{-1}(\tilde{\Xi}^{(k)}) = \{w \in \mathcal{W} \mid w(t) \text{ starts at } 0 \text{ tangentially along } L_k\},$$

then, obviously, $\tilde{\Xi}^{(1)}, \cdots, \tilde{\Xi}^{(n)}$ are mutually disjoint and so are also $\tilde{\Pi}^{(1)}, \cdots, \tilde{\Pi}^{(n)}$. From (2), (3) and (4), we can deduce the following:

Proposition 2.1.

$$\mathbf{N}_{jk}(\mathcal{W} \setminus \{\tilde{\Pi}^{(j)} \cup \tilde{\Xi}^{(k)}\}) = 0, \quad j, k = 1, \dots, n.$$

If we set

$$\mathbf{N}_j = \sum_{k=1}^{n} \mathbf{N}_{jk}, \quad j = 1, \dots, n, \tag{5}$$

then $\mathbf{N}_j$ is the $\mathbf{X}$-Markovian measure associated to $\mathbf{X}$-excessive measure $u_j(x)dx$.

Now, the possible extension of $\mathbf{X}$ to a diffusion on the whole plane can be obtained by applying Itô's theory of excursion point processes (cf. [I]).

Theorem 2.1. *([IW 1] or [IW 2].) An extension* $\widehat{\mathbf{X}} = (X(t), \widehat{P}_x)$, *for which the origin* 0 *is not a trap, is determined by the nonnegative parameters* $p_1, \dots, p_n$ *and* m *such that* $\sum_{k=1}^{n} p_k = 1$. $m = 0$ *if and only if* $\int_0^{\infty} 1_{\{0\}}(X(t))dt = 0$ *a.s. with respect to* $\widehat{P}_x$ *for every* $x \in \mathbf{R}^2$. $\widehat{\mathbf{X}}$ *is symmetric with respect to some measure on* $\mathbf{R}^2$ *if and only if* $p_1 = \dots = p_n = 1/n$ *and, then, a symmetrizing measure is given by* $m(dx) = dx + m \cdot \delta_{\{0\}}(dx)$. *In this case, the corresponding Dirichlet form is the closure on* $L^2(m(dx))$ *of the* $\mathcal{E}(f, g)$ *given by (1) with the domain* $C_0^{\infty}(\mathbf{R}^2)$.

The sample paths of $\widehat{\mathbf{X}}$ starting at the origin 0 can be constructed as follows. Let $\mathbf{N} = \sum_{k=1}^{m} p_k \mathbf{N}_k$ which is a σ-finite measure on $(\mathcal{W}, \mathcal{B}(\mathcal{W}))$ with infinite total mass. We set up a Poisson point process p on the state space $\mathcal{W}$ with the characteristic measure $\mathbf{N}$ (cf. [I] or [IW 3], p.43 and p.123-130). Note that each sample of p is a point function $p : \mathbf{D}_p \in t \mapsto p_t \in \mathcal{W}$, where the domain $\mathbf{D}_p$ of p is a countable subset of $(0, \infty)$. Set

$$A(t) = mt + \sum_{s \in \mathbf{D}_p, s \leq t} \sigma(p_s).$$

Then, it is a càdlàg increasing process with stationary independent increments and $A(0) = 0$. Since $\mathbf{N}(\mathcal{W}) = \infty$, $t \mapsto A(t)$ is strictly increasing and $\lim_{t \uparrow \infty} A(t) = \infty$, a.s. Hence, for each $t \geq 0$, there exists unique $s \geq 0$ such that $A(s-) \leq t \leq A(s)$. $s \in \mathbf{D}_p$ if and only if $A(s) > A(s-)$. Set, for each $t \in [0, \infty)$,

$$X(t) = \begin{cases} p_s(t - A(s-)), & s \in \mathbf{D}_p, A(s-) \leq t \leq A(s), \\ 0, & t = A(s-) = A(s). \end{cases}$$

Then, $\widehat{\mathbf{X}} = (X(t))$ is the sample path starting at 0 of the diffusion which is the extension of $\mathbf{X}$ corresponding to the parameters $p_1, \dots, p_n$ and m.

Let $\mathbf{A} = \{k \in \{1, \dots, n\} | p_k > 0\}$. Let $\mathbf{F} = \mathbf{F}^{\widehat{\mathbf{X}}}$ be the natural filtration of the diffusion $\widehat{\mathbf{X}}$ constructed above.

Theorem 2.2. *Assume that the set* $\mathbf{A}$ *contains* $l \geq 2$ *elements. Then there exists a non-trivial* $\mathbf{F}$-*spider martingale with* l *rays. Hence, if* $\mathbf{A}$ *contains* $l \geq 3$ *elements, there exists a multiple* $\mathbf{F}$-*spider martingale so that the filtration* $\mathbf{F}$ *is not cosy.*

Proof. Let $\widehat{\mathbf{X}} = (X(t))$ be the diffusion constructed above and $\mathbf{F} = \{F(t)\}$ be the natural filtration of $\widehat{\mathbf{X}}$. For $t > 0$, set

$$g(t) = \sup\{s \in [0, t] \mid X(s) = 0\}.$$

Then, $g(t)$ is an **F**-honest time and, by Proposition 2.1 and an excursion theory, we deduce

$$F(g(t)+) = F(g(t)) \bigvee \{\Theta_k, k \in \mathbf{A}\},$$

where

$$\Theta_k = [\theta_{g(t)} X \in \tilde{\Pi}^{(k)}], \quad k \in \mathbf{A}.$$

The existence of a non-trivial **F**-spider martingale with l rays follows from a general result in [BEKSY].

Or, we can give a more direct construction of a non-trivial **F**-spider martingale by piecing out some part of each excursion by the method given in [Wat], (the collection of excursions is the point process p from which we have constructed the process $\widehat{\mathbf{X}}$).

Multidimensional extensions of Theorem 2.2 are of course possible. We give a typical example in the case of a three dimensional diffusion process. We define a diffusion $\mathbf{X}$ on $D = \mathbf{R}^3 \setminus \{0\}$ similarly as above by the following Dirichlet form: Let $\Pi_j, j = 1, \ldots, m$, be m different planes in $\mathbf{R}^3$, each passing through the origin 0, and let $L_k, k = 1, \ldots, n$, be n different half lines, each, starting at the origin and lying on some plane Π_j. Let $\mathcal{D}_0$ be, as above, the space of all C^∞-functions with a compact support and vanishing also in a neighborhood of origin. Let $D(u, v)$, $u, v \in \mathcal{D}_0$, be the usual Dirichlet integral on $\mathbf{R}^3$, $D_{\Pi_j}(u, v)$ be the two-dimensional Dirichlet integral for $u|_{\Pi_j}, v|_{\Pi_j}$ on Π_j (by regarding Π_j as two-dimensional Euclidean space by the imbedding) and $D_{L_k}(u, v)$ be the one-dimensional Dirichlet integral for $u|_{L_k}, v|_{L_k}$ on L_k. For positive constants $\mu_j, j = 1, \ldots, m$ and $\nu_k, k = 1, \ldots, n$, define a bilinear form on $\mathcal{D}_0$ by

$$\mathcal{E}(u, v) = \frac{1}{2} D(u, v) + \sum_{j=1}^{m} \mu_j D_{\Pi_j}(u, v) + \sum_{k=1}^{n} \nu_k D_{L_k}(u, v), \quad u, v \in \mathcal{D}_0.$$

Then, it is a closable Markovian form on $L^2(D; dx)$ and its closure is a regular Dirichlet form. Therefore, there corresponds a unique diffusion $\mathbf{X}$ on D with a finite life time. We can obtain similar results as above: the space of bounded $\mathbf{X}$-harmonic functions are n-dimensional and the possible extensions of $\mathbf{X}$ as diffusions on $\mathbf{R}^3$ are determined in exactly the same way as Theorem 2.1. Also, Theorem 2.2 is valid in the same way: Namely, if an extension $\widehat{\mathbf{X}}$ which corresponds to nonnegative parameters $p_1, \ldots, p_n$ and m, is such that $\#\{ k \mid p_k > 0 \} = l$, then the natural filtration of $\widehat{\mathbf{X}}$ contains a multiple spider martingale with l rays.

3 An application to sticky Brownian motions

Here we apply Theorem 2.2 to show the non-cosiness of the filtration of one-dimensional Brownian motion which is sticky at the origin 0.

For given $c \geq 0$, $\rho \geq 0$ with $c + \rho > 0$, consider the following stochastic differential equation for a continuous **F**-semimartingale $X = (X(t))$ on $\mathbf{R}$ on a filtered probability space $\{(\Omega, F, P), \mathbf{F}\}$:

$$d[X(t) \vee 0] = 1_{\{X(t) > 0\}} \cdot dB(t) + c \cdot d\phi(t), \quad d[X(t) \wedge 0] = 1_{\{X(t) < 0\}} \cdot dB(t) - c \cdot d\phi(t) \quad (6)$$

where $B(t)$ is an **F**-Wiener process with $B(0) = 0$, $\phi(t)$ is an **F**-adapted, continuous increasing process with $\phi(0) = 0$ such that

$$\phi(t) = \int_0^t 1_{\{X(s)=0\}} \cdot d\phi(s), \quad \int_0^t 1_{\{X(s)=0\}} \cdot ds = \rho\phi(t). \tag{7}$$

Given $x \in \mathbf{R}$, a solution $X = (X(t))$ with $X(0) = x$ exists on a suitable filtered probability space and it is unique in the law sense ([IW 3]). When $c = 0$, then the solution $X(t)$ is the Brownian motion $x + B(t)$ stopped at $\sigma_0 = \min\{u \mid x + B(u) = 0\}$. When $\rho = 0$, the solution is irrelevant to c and coincides with the Brownian motion $x + B(t)$. In these two extreme cases, the natural filtration $\mathbf{F}^X$ of the solution X is either homomorphic or isomorphic to the Brownian filtration $\mathbf{F}^B$ so that it is a cosy filtration.

So we assume $c > 0, \rho > 0$ and, replacing $c\phi$ by ϕ and ρ by ρ/c, we can always assume $c = 1$ in the equation (6). In the following, we assume that $X(0) = 0$, for simplicity, and denote the solution by X_ρ.

Theorem 3.1. *The natural filtration $\mathbf{F}^{X_\rho}$ of X_ρ is not cosy.*

Proof. Without loss of generality, we may assume $\rho = 1$. Let $X^{(1)}$ and $X^{(2)}$ be independent copies of X_1 and define a diffusion $\mathbf{X} = (X(t))$ in the plane $\mathbf{R}^2$ by $\mathbf{X} = (X^{(1)}, X^{(2)})$. Then, $\mathbf{F}^{\mathbf{X}} = \mathbf{F}^{X_1} \otimes \mathbf{F}^{X_1}$. Let L_1, L_2, L_3, L_4 be the positive part of x-axis, the positive part of y-axis, the negative part of x-axis and the negative part of y-axis, respectively, so that $\mathbf{L} = L_1 \cup L_2 \cup L_3 \cup L_4$ coincides with the union of x- and y-axes. Let

$$A(t) = \int_0^t 1_{\{\mathbf{R}^2\backslash\mathbf{L}\}}(X(s))ds.$$

Then, we easily deduce that $t \mapsto A(t)$ is strictly increasing and $\lim_{t\uparrow\infty} A(t) = \infty$, a.s. so that $A = (A(t))$ is a process of time change. Define $\widehat{X}(t) = X(A^{-1}(t))$ and $\widehat{\mathbf{X}} = ((\widehat{X}(t))$. Then, the natural filtration $\mathbf{F}^{\widehat{\mathbf{X}}}$ is just the time change $(\mathbf{F}^{\mathbf{X}})^{(A)}$. A key observation is that the diffusion process $\widehat{\mathbf{X}}$ is a particular case of diffusions given in Theorem 2.1: It is the case of $n = 4$ with L_1, L_2, L_3, L_4 given above and, $m = 0$, $p_1 = p_2 = p_3 = p_4 = 1/4$, cf. [IW 1], p.118. Hence, by Theorem 2.2, the filtration $\mathbf{F}^{\widehat{\mathbf{X}}}$ contains an $\mathbf{F}^{\widehat{\mathbf{X}}}$-multiple spider martingale with 4 rays. Since the filtration $\mathbf{F}^{\mathbf{X}}$ is obtained from the filtration $\mathbf{F}^{\widehat{\mathbf{X}}}$ by a time change as $\mathbf{F}^{\mathbf{X}} = \{\mathbf{F}^{\widehat{\mathbf{X}}}\}^{(A^{-1})}$, it also contains an $\mathbf{F}^{\mathbf{X}}$-multiple spider martingale with 4 rays. By Theorem 1.2, we can conclude that the filtration $\mathbf{F}^{\mathbf{X}} = \mathbf{F}^{X_1} \otimes \mathbf{F}^{X_1}$ is not cosy. Now, the non-cosiness of the filtration $\mathbf{F}^{X_1}$ follows from Corollary 1.2.

Remark 3.1. *Recently, J. Warren ([War]) proved directly that the natural filtration of a reflecting sticky Brownian motion is not cosy. His result is stronger than ours because there is a homomorphism from the filtration of a bilateral sticky Brownian motion to a reflecting sticky Brownian motion. However, we can give the following argument from which we can also deduce that the natural filtration of a reflecting sticky Brownian motion is not cosy.*

Let $X^{(1)}$, $X^{(2)}$ and $X^{(3)}$ be independent copies of X_1 and define a diffusion process $\mathbf{X} = (X(t))$ in $\mathbf{R}^3$ by $\mathbf{X} = (X^{(1)}, X^{(2)}, X^{(3)})$. Let Π_1, Π_2, Π_3 be coordinate planes in $\mathbf{R}^3$ and let Π be their union. Let

$$A(t) = \int_0^t 1_{\{\mathbf{R}^3\backslash\Pi\}}(X(s))ds \quad \text{and} \quad \widehat{X}(t) = X(A^{-1}(t)).$$

Then, the process $\widehat{\mathbf{X}} = ((\widehat{X}(t))$ is exactly a kind of diffusions discussed in Section 2 as multi-dimensional extensions of Theorem 2.2: It is the case that $m = 3$, Π_1, Π_2, Π_3 are coordinate planes as above and, $n = 6$, $L_1, \ldots, L_6$ are six half coordinate axes each starting at the origin. Furthermore, $\mu_j = \nu_k \equiv 1$, and $\widehat{\mathbf{X}}$ corresponds to parameters $p_1 = \ldots = p_6 = 1/6$ and $m = 0$. Hence, we can conclude that the natural filtration $\mathbf{F}^{\widehat{\mathbf{X}}}$ of $\widehat{\mathbf{X}}$ contains a multiple $\mathbf{F}^{\widehat{\mathbf{X}}}$-spider martingale with 6 rays. Then, the natural filtration $\mathbf{F}^{\mathbf{X}} = \mathbf{F}^{X_1} \otimes \mathbf{F}^{X_1} \otimes \mathbf{F}^{X_1}$ of $\mathbf{X}$ contains also a multiple $\mathbf{F}^{\mathbf{X}}$-spider martingale with 6 rays because $\mathbf{F}^{\mathbf{X}}$ is obtained from $\mathbf{F}^{\widehat{\mathbf{X}}}$ by a time change. From this, we can deduce that the filtration $\mathbf{F} = \mathbf{F}' \otimes \mathbf{F}' \otimes \mathbf{F}'$, where $\mathbf{F}'$ is the natural filtration of $|X_1| = (|X_1(t)|)$, has a multiple $\mathbf{F}$-spider martingale with 3 rays. Therefore, $\mathbf{F}$ is not cosy by Theorem 1.2 and hence, by Corollary 1.2, the filtration $\mathbf{F}'$, which is the natural filtration of the reflecting sticky Brownian motion $|X_1|$, is not cosy.

References

[BEKSY] M. T. Barlow, M. Émery, F. B. Knight, S. Song et M. Yor, Autour d'un théorème de Tsirelson sur des filtrations browniennes et non browniennes, *Séminaire de Probabilités XXXII*, **LNM 1686**, Springer, Berlin (1998), 264-305

[BPY] M. T. Barlow, J. W. Pitman and M. Yor, On Walsh's Brownian motions, *Séminaire de Probabilités XXIII*, **LNM 1372**, Springer, Berlin (1989), 275-293

[DV] M. H. Davis and P. Varaiya, The multiplicity of an increasing family of σ-fields, Annals of Probab., **2**(1974), 958-963

[DFST] L. Dubins, J. Feldman, M. Smorodinsky and B. Tsirelson, Decreasing sequences of σ-fields and a measure change for Brownian motion, Annals of Probab., **24**(1996), 882-904

[EY] M. Émery et M. Yor, Sur un théorème de Tsirelson relatif à des mouvements browniens corrélés et à la nullité de certains temps locaux, *Séminaire de Probabilités XXXII*, **LNM 1686**, Springer, Berlin (1998), 306-312

[FOT] M. Fukushima, Y. Oshima and M. Takeda, *Dirichlet Forms and Symmetric Markov Processes*, Walter de Gruyter, Berlin-New York, 1994

[H] T. Hida, Canonical representation of Gaussian processes and their applications, Mem. Colleg. Sci. Univ. Kyoto, A, **33**(1960), 109-155

[IW 1] N. Ikeda and S. Watanabe, *The Local Structure of Diffusion Processes*, (*Kakusan-Katei no Kyokusho Kôzô*), Seminar on Probab. **Vol. 35**, Kakuritsuron Seminar, 1971 (in Japanese)

[IW 2] N. Ikeda and S. Watanabe, The local structure of a class of diffusions and related problems, *Proc. 2nd. Japan-USSR Symp*, **LNM 330**, Springer, Berlin (1973), 124-169

[IW 3] N. Ikeda and S. Watanabe, *Stochastic Differential Equations and Diffusion Processes*, Second Edition, North-Holland/Kodansha, Amsterdam/Tokyo, 1988

[I] K. Itô, Poisson point processes attached to Markov processes, in *Kiyosi Itô, Selected Papers*, Springer, New York (1987), 543-557, Originally published in Proc. Sixth Berkeley Symp. Math. Statist. Prob. III(1970), 225-239

[KW] H. Kunita and S. Watanabe, On square integrable martingales, *Nagoya Math Jour.* **30**(1967), 209-245

[MW] M. Motoo and S. Watanabe, On a class of additive functionals of Markov processes, J. Math. Kyoto Univ. **4**(1965), 429-469

[N] M. Nisio, Remark on the canonical representation of strictly stationary processes, *J. Math. Kyoto Univ.* **1**(1961), 129-146

[RY] D. Revuz and M. Yor, *Continuous Martingales and Brownian Motion*, Springer, Berlin, 1991

[S] A. B. Skorokhod, Random processes in infinite dimensional spaces, *Proc. ICM. 1986, Berkeley*, Amer. Math. Soc., 163-171 (in Russian).

[T] B. Tsirelson, Triple points: From non-Brownian filtrations to harmonic measures, Geom. Func. Anal. **7**(1997), 1096-1142

[War] J. Warren, On the joining of sticky Brownian motion, *in this volume.*

[Wat] S. Watanabe, Construction of semimartingales from pieces by the method of excursion point processes, Ann. Inst. Henri Poincaré **23**(1987), 293-320

[We] M. Weil, Quasi-processus, *Séminaire de Probabilités IV*, **LNM 124**, Springer, Berlin (1970), 216-239

[Y] M. Yor, *Some Aspects of Brownian Motion*, Part **II**, Lectures in Math. ETH Zürich, Birkhäuser, 1997

A REMARK ON TSIRELSON'S
STOCHASTIC DIFFERENTIAL EQUATION

M. Émery and W. Schachermayer

ABSTRACT. — Tsirelson's stochastic differential equation is called "celebrated and mysterious" by Rogers and Williams [16]. This note aims at making it a little more celebrated and a little less mysterious.

Using a deterministic time-change, we translate the study of Tsirelson's equation into the study of "eternal" Brownian motion on the circle. This allows us to show that the filtration generated by any solution of Tsirelson's equation is also generated by some Brownian motion (which, however, *cannot* be the Brownian motion driving the equation, because the equation has no strong solution).

Introduction

The so-called innovation problem is a remarkable phenomenon in the theory of filtered probability spaces; see for instance § 5.4 of von Weizsäcker [24]. When the answer to the innovation problem is negative, some kind of creation of information occurs. This may happen in discrete, or continuous time (by discrete time, we refer to processes parametrized by $\mathbf{Z}$).

In discrete time, a paradigmatic example, well-known in ergodic theory, is obtained from an independent sequence $(U_n)_{n \in \mathbf{Z}}$ of random variables uniformly distributed on $\{-1, 1\}$. Call $(\mathcal{G}_n)_{n \in \mathbf{Z}}$ the natural filtration of U and $(\mathcal{F}_n)_{n \in \mathbf{Z}}$ the filtration generated by the "innovations" $V_n = U_{n-1}U_n$ $(= U_n/U_{n-1})$. Both $\mathcal{G}_{-\infty}$ and $\mathcal{F}_{-\infty}$ are trivial (by Kolmogorov's zero-one law), and both filtrations have V_n as innovations: V_n is independent of $\mathcal{F}_{n-1}$ (resp. $\mathcal{G}_{n-1}$), and, together with it, generates $\mathcal{F}_n$ (resp. $\mathcal{G}_n$). Yet, the filtration $\mathcal{G}$ strictly contains $\mathcal{F}$, because each random variable U_n is independent of $\mathcal{F}_\infty$. More precisely, in the filtration $\mathcal{F}$, the process U is observed up to multiplication by an unknown factor ± 1 only. This example has been independently discovered by several authors; the earliest reference we know is Vershik [22], see also von Weizsäcker [24] and Exercise 4.12 in Williams [25].

In this example, the filtrations $\mathcal{G}$ and $\mathcal{F}$ are isomorphic, because they are generated by processes $(U_n)_{n \in \mathbf{Z}}$ and $(V_n)_{n \in \mathbf{Z}}$ with the same law.

In continuous time, there is an example very similar to the above discrete-time example. Replace the discrete time $\mathbf{Z}$ by the line $\mathbf{R}$ and the state space $\{-1, 1\}$ by the circle $\mathbf{T} = \mathbf{R}/\mathbf{Z}$. Consider a $\mathbf{T}$-valued Brownian motion $(U_t)_{t \in \mathbf{R}}$: each U_t is uniformly distributed on $\mathbf{T}$ and U moves Brownianly. For each $t \in \mathbf{R}$, call $\mathcal{G}_t$ the σ-algebra generated by all the past positions U_s for $s \leqslant t$ and $\mathcal{F}_t$ the σ-algebra generated by all the past innovations $U_t - U_s$ for $s \leqslant t$. As above, each U_t is independent of $\mathcal{F}_\infty$; this should be rather obvious on an intuitive level (and a formal proof will be provided below). As above, both filtrations $\mathcal{G}$ (generated by U) and $\mathcal{F}$ have trivial tail σ-fields at $-\infty$. By analogy with the previous example, a natural question is whether both filtrations are isomorphic. More precisely, it is easy to see that, after transforming the time-axis $\mathbf{R}$ into $(0,\infty)$ by the time-change $\log t \mapsto t$, $\mathcal{F}$ becomes the natural filtration of some real Brownian motion (first part of Proposition 3). The question becomes, is the time-changed filtration $(\mathcal{G}_{\log t})_{t\geqslant 0}$ also generated by some Brownian motion? The next section will show that the answer is positive (second part of Proposition 3).

Another continuous-time example pertaining to the innovation problem is Tsirelson's stochastic differential equation ([21]). This equation is of the form $dX_t = dB_t + g\bigl(t, X([0,t])\bigr)\,dt$, where B is a Brownian motion and g a bounded function of t and of the past of X up to t, given by some explicit formula involving fractional parts. Tsirelson has shown in [21] that solutions X exist and all have the same law, but cannot be adapted to the natural filtration of B. Our third section will establish that, as far as filtrations are concerned, this situation is the same as the preceding one: given such a pair (X, B) on the time-interval $[0,\infty)$, the natural filtrations of X and B become, after a deterministic time-change from $[0,\infty)$ to $[-\infty,\infty)$, those of a T-valued Brownian motion and its innovations (Proposition 4). As a consequence, the natural (non time-changed) filtration of every solution X to Tsirelson's stochastic differential equation is generated by some Brownian motion (Corollary 2).

This result should be appreciated in the context of the recent literature; in the remarkable paper [7], Dubins, Feldman, Smorodinsky and Tsirelson construct a variant of Tsirelson's equation whose solution X *generates a non-Brownian* filtration. The question remained open, whether Tsirelson's original equation also has this property. As this note shows, the answer is no; in retrospect, this makes the—technically very involved—construction of [7] still more interesting. For further recent pertinent examples of Brownian and non-Brownian filtrations, we refer to [11], [9], [10], [18], [20], [2], [5], [17], [23], [3], [8]. Most of these examples feature the above-mentioned phenomenon of creation of information: there is a decreasing sequence of σ-fields, all of them containing some common information, but with trivial intersection. And Tsirelson's equation does exhibit this phenomenon, so much so, that it inspired Rogers and Williams to comment in [16]: "somehow, magically, this independent random variable has appeared from somewhere! Indeed, it really has appeared from thin air, because [...] it is not present at time 0!"

Throughout this note, the usual hypotheses are in force: the probability spaces are complete, the filtrations contain all negligible events and are right-continuous.

Circular Brownian motions

The circle $\mathbf{R}/\mathbf{Z}$ is called T; the symbol $\overset{1}{=}$ means equality modulo 1 (between two real numbers, or between a real number and an element of T). For $x \in \mathbf{R}$ or $x \in \mathsf{T}$, $\{x\}$ denotes the fractional part of x, that is, the real number $\{x\}$ such that $0 \leqslant \{x\} < 1$ and $\{x\} \overset{1}{=} x$. If U is a continuous, T-valued process, we shall distinguish between the increment $U_t - U_s$, which takes its values in T, and the real random variable $\int_s^t dU_r$, defined by the following three (lifting) conditions: it depends continuously on t, it vanishes for $t = s$, and $\int_s^t dU_r \overset{1}{=} U_t - U_s$.

DEFINITION. — Given a filtration $\mathcal{H} = (\mathcal{H}_t)_{t \in \mathbf{R}}$, a T-valued process $(U_t)_{t \in \mathbf{R}}$ will be called a *circular Brownian motion for* $\mathcal{H}$ (abbreviated $\mathcal{H}$-CBM) if it is continuous and adapted to $\mathcal{H}$ and if for each $s \in \mathbf{R}$, the process $t \mapsto \int_s^t dU_r$, defined on the interval $[s, \infty)$, is a real-valued Brownian motion for the filtration $(\mathcal{H}_t)_{t \in [s,\infty)}$. (Equivalently. U is a Markov process for $\mathcal{H}$, with transition probabilities the Brownian semigroup on the circle.)

If $\mathcal{H}$ equals the natural filtration $\mathcal{G}$ of U, we shall simply say that U is a CBM. Note that this notion only depends on the law of U. It is easy to see that circular Brownian motions exist: as the uniform probability on T is invariant for the Brownian semigroup, a stationary, T-valued Brownian motion with the uniform law at each fixed time is easy to construct. Uniqueness in law of CBM is less straightforward. but follows (among other possible proofs) from Proposition 1 below.

We have phrased the definition of a CBM U with an arbitrary filtration $\mathcal{H}$ (instead of only considering the natural filtration $\mathcal{G}$ of U), as we shall encounter situations where $\mathcal{G}$ is strictly contained in another filtration $\mathcal{H}$. It then follows from the predictable representation property of (usual) Brownian motion that a CBM U is an $\mathcal{H}$-CBM if and only if every $\mathcal{G}$-martingale is an $\mathcal{H}$-martingale.

Associated to a CBM U are two filtrations: its natural filtration, generated by the past values U_t, and the innovation filtration, generated by the past increments $U_t - U_s$ or by their liftings $\int_s^t dU_r$ (which amounts to the same). Clearly, the innovation filtration is included in the natural filtration; Proposition 1 will show that it is always strictly smaller.

All CBM's have the same law, as shown by the next proposition, whose proof is borrowed from Proposition (6.13) of Stroock and Yor [19] (see also § IX.3 of [15] and § V.18 of [16]).

PROPOSITION 1. — *Let U be a CBM, with innovation filtration $\mathcal{F} = (\mathcal{F}_t)_{t \in \mathbf{R}}$. For each $t \in \mathbf{R}$ the random variable U_t is uniformly distributed on T and independent of the σ-field $\mathcal{F}_\infty$ (generated by all increments).*

PROOF. — Call $\mathcal{G} = (\mathcal{G}_t)_{t \in \mathbf{R}}$ the natural filtration of U and for $p \in \mathbf{Z}$ denote by e_p the character on T defined by $e_p(u) = \exp(2\pi i p x)$ if $u \in \mathbf{T}$, $x \in \mathbf{R}$ and $u \stackrel{.}{=} x$. For $s < t$, $\int_s^t dU_r$ is independent of $\mathcal{G}_s$, with law $\mathcal{N}(0, t-s)$, so

$$\mathbf{E}[e_p(U_t)|\mathcal{G}_s] = \mathbf{E}[e_p(U_s)\, e_p(U_t - U_s)|\mathcal{G}_s] = e_p(U_s)\, \mathbf{E}[e_p(U_t - U_s)|\mathcal{G}_s]$$
$$= e_p(U_s)\, \mathbf{E}\big[\exp\big(2\pi i p \int_s^t dU_r\big) \mid \mathcal{G}_s\big] = e_p(U_s)\, \exp\big[-\tfrac{1}{2}4\pi^2 p^2(t-s)\big] .$$

This implies

$$\big|\mathbf{E}[e_p(U_t)]\big| = \big|\mathbf{E}[\mathbf{E}[e_p(U_t)|\mathcal{G}_s]]\big| \leqslant \mathbf{E}[|\mathbf{E}[e_p(U_t)|\mathcal{G}_s]|] = \exp\big[-\tfrac{1}{2}4\pi^2 p^2(t-s)\big] ,$$

and, by letting s tend to $-\infty$, $\mathbf{E}[e_p(U_t)] = 0$ for $p \neq 0$. As a consequence, by the Stone-Weierstraß theorem, U_t has a uniform law on T.

For each s, the σ-field $\mathcal{F}^s = \sigma(U_t - U_s, t \geqslant s)$ of increments after s is independent of $\mathcal{G}_s$ and a fortiori of U_s. Thus, for $s < t$ and $p \neq 0$,

$$\mathbf{E}[e_p(U_t)|\mathcal{F}^s] = \mathbf{E}[e_p(U_s)|\mathcal{F}^s]\, e_p(U_t - U_s) = \mathbf{E}[e_p(U_s)]\, e_p(U_t - U_s) = 0 ,$$

since we have just seen that $\mathbf{E}[e_p(U_s)] = 0$. Now, $\bigvee_{s \in \mathbf{R}} \mathcal{F}^s = \mathcal{F}_\infty$, and, when $s \downarrow -\infty$, $\mathbf{E}[e_p(U_t)|\mathcal{F}^s] \to \mathbf{E}[e_p(U_t)|\mathcal{F}_\infty]$ by martingale convergence; so $\mathbf{E}[e_p(U_t)|\mathcal{F}_\infty] = 0$ too, and U_t is not only uniform, but also independent of $\mathcal{F}_\infty$. ∎

DEFINITIONS. — A *regular time-change* is an increasing (deterministic) bijection a from $\mathbf{R}$ to $(0, \infty)$ such that both a and its inverse a^{-1} are absolutely continuous functions.

A filtration $(\mathcal{F}_t)_{t \geqslant 0}$ will be called *Brownian* if it is the natural filtration of some real Brownian motion $(B_t)_{t \geqslant 0}$ issued from the origin. (Only one-dimensional Brownian filtrations will be considered. so we simply call them Brownian.)

PROPOSITION 2 AND DEFINITION. — *Let $(\mathcal{F}_t)_{t \in \mathbf{R}}$ be a filtration indexed by $\mathbf{R}$. The following are equivalent:*

(i) *for some regular time-change a, the filtration $(\mathcal{F}'_t)_{t \geqslant 0}$ defined by $\mathcal{F}'_t = \mathcal{F}_{a^{-1}(t)}$ for $t > 0$ and by right-continuity for $t = 0$ is Brownian;*

(ii) *for every regular time-change a, the filtration $(\mathcal{F}'_t)_{t \geqslant 0}$ defined by $\mathcal{F}'_t = \mathcal{F}_{a^{-1}(t)}$ for $t > 0$ and by right-continuity for $t = 0$ is Brownian.*

When these conditions are met, the filtration $(\mathcal{F}_t)_{t \in \mathbf{R}}$ is called Brownian.

For instance, if $(B_t)_{t \geqslant 0}$ is a real Brownian motion issued from the origin, the stationary Ornstein-Uhlenbeck process $(X_t)_{t \in \mathbf{R}}$ defined by $X_t = e^{-t/2} B_{e^t}$ generates a Brownian filtration on $\mathbf{R}$; and conversely, every Brownian filtration on $\mathbf{R}$ is the natural filtration of such a stationary Ornstein-Uhlenbeck process.

PROOF OF PROPOSITION 2. — We have to show (i) $\Rightarrow$ (ii); this reduces to checking that, if a and b are regular time-changes, the homeomorphism $\phi = a \circ b^{-1}$ from $(0, \infty)$ to itself has the following property: if B is a Brownian motion on $[0, \infty)$, started at 0, the process $X_t = B_{\phi(t)}$ generates a Brownian filtration. As ϕ and its inverse are absolutely continuous, the Lebesgue derivative ψ of ϕ is almost everywhere defined and strictly positive. The process $X = B \circ \phi$ is a gaussian martingale, with quadratic variation $\phi(t)$. The martingale $\beta_t = \int_0^t \psi^{-1/2}(s)\, dX_s$ is a Brownian motion; as $X_t = \int_0^t \psi^{1/2}(s)\, d\beta_s$, both X and β generate the same filtration. ∎

DEFINITIONS. — A *chopping sequence* is a sequence $(t_k)_{k \in \mathbf{Z}}$ of real numbers such that $t_k < t_{k+1}$, $\lim_{k \to -\infty} t_k = -\infty$ and $\lim_{k \to +\infty} t_k = +\infty$.

Given a chopping sequence $(t_k)_{k \in \mathbf{Z}}$, a *chopped Brownian motion* (respectively a T-*valued chopped Brownian motion*) is a càdlàg process Z whose restriction Z^k to each interval $[t_k, t_{k+1})$ is a real Brownian motion (respectively a T-valued Brownian motion) started from 0 at time t_k, all the Z^k's being independent.

LEMMA 1. — *The filtration generated by a chopped Brownian motion, or by a* T-*valued chopped Brownian motion, is Brownian.*

REMARK. — The converse also holds (but we shall not need it): every Brownian filtration indexed by $\mathbf{R}$ is the natural filtration of a chopped Brownian motion. More generally, if some càdlàg process Z generates a Brownian filtration, every Brownian filtration is generated by a process with the same law as Z.

PROOF OF LEMMA 1. — A T-valued chopped Brownian motion V clearly generates the same filtration as the real chopped Brownian motion Z defined by $Z_t = \int_{t_k}^t dV_s$ for $t \in [t_k, t_{k+1})$; so it suffices to prove the lemma for real processes.

The chopping sequence $(t_k)_{k \in \mathbf{Z}}$ and the chopped Brownian motion Z are given. Introduce a sequence $(c_k)_{k \in \mathbf{Z}}$ of strictly positive constants such that $\sum_{k < 0} c_k (t_{k+1} - t_k) < \infty$ and $\sum_{k \geqslant 0} c_k (t_{k+1} - t_k) = \infty$; define a regular time-change by

$$a(t) = \int_{-\infty}^{t} \sum_{k \in \mathbb{Z}} c_k \, \mathbb{1}_{(t_k, t_{k+1}]}(s) \, ds$$

and a process $(B_t)_{t>0}$ by

$$B_{a(t)} = \sqrt{c_k} \, Z_t + \sum_{\ell \leqslant k} \sqrt{c_{\ell-1}} \, Z_{t_\ell -} \qquad \text{for } t_k \leqslant t < t_{k+1}.$$

By scaling, $B_t^k = B_t - B_{a(t_k)} = \sqrt{c_k} \, Z_{a^{-1}(t)}^k$ is a Brownian motion on the interval $[a(t_k), a(t_{k+1}))$; and these Brownian motions B^k are independent. So the series in the definition of B is convergent and B is a Brownian motion. As

$$Z_t = \frac{1}{\sqrt{c_k}} \left(B_{a(t)} - B_{a(t_k)} \right) \qquad \text{for } t_k \leqslant t < t_{k+1},$$

the natural filtrations of Z and B are time-changed from each other, and Z generates a Brownian filtration. ∎

PROPOSITION 3. — *The natural filtration $\mathcal{G}$ and the innovation filtration $\mathcal{F}$ of a CBM are Brownian.*

PROOF. — Let $U = (U_t)_{t \in \mathbb{R}}$ be a CBM, $\mathcal{G}$ its natural filtration, and $\mathcal{F}$ its innovation filtration. Choose any chopping sequence $(t_k)_{k \in \mathbb{Z}}$; for instance, $t_k = k$ is a possible choice. The chopped Brownian motion

$$Z_t = \sum_{k \in \mathbb{Z}} \mathbb{1}_{[t_k, t_{k+1})}(t) \int_{t_k}^{t} dU_r$$

generates $\mathcal{F}$, so $\mathcal{F}$ is Brownian by Lemma 1.

The proof for $\mathcal{G}$ is less straightforward. In fact, it involves a key idea of the present paper, a certain coupling lemma (for a vivid presentation of the use of coupling in the theory of Markov processes, see Diaconis [6]).

LEMMA 2. — *If $(X_t)_{t \geqslant 0}$ is a T-valued Brownian motion defined on some $(\Omega, \mathcal{A}, \mathbf{P}, (\mathcal{H}_t)_{t \geqslant 0})$, there exists a T-valued Brownian motion $(Y_t)_{t \geqslant 0}$, defined on the same $(\Omega, \mathcal{A}, \mathbf{P}, (\mathcal{H}_t)_{t \geqslant 0})$, satisfying the following four properties:*

(i) $Y_0 = 0 \in \mathsf{T}$;

(ii) *Y is independent of X_0;*

(iii) *both processes $(X_t)_{t \geqslant 0}$ and $(X_0, Y_t)_{t \geqslant 0}$ generate the same filtration;*

(iv) *calling S the $\mathcal{H}$-stopping time $\inf \{t : X_t = Y_t\}$, one has $\mathbf{P}[S \geqslant t] \leqslant 1/(4t)$ and $Y = X$ on $[\![S, \infty[\![$.*

PROOF OF LEMMA 2. — Introduce the $\mathcal{H}$-stopping time $S = \inf \{t : X_t + X_t = X_0\}$. The process

$$Y_t = \begin{cases} X_0 - X_t & \text{if } t \leqslant S \\ X_t & \text{if } t \geqslant S. \end{cases}$$

is continuous by definition of S; it starts from 0, is $\mathcal{H}$-adapted and verifies

$$\int_0^t dY_s = \int_0^t (-\mathbb{1}_{[\![0,S]\!]} + \mathbb{1}_{]\!]S,\infty[\![})(s) \, dX_s,$$

so it is a T-valued $\mathcal{H}$-Brownian motion. As it starts from 0, it is independent of $\mathcal{H}_0$, whence (ii). Replacing $\mathcal{H}$ by the natural filtration of X shows that (X_0, Y) is adapted to that filtration; to show (iii), it suffices to reconstruct X from X_0 and Y. This is easy: S is also the first time when $Y + Y = X_0$, and X is equal to $X_0 - Y$ up to S, and equal to Y from S on. Clearly, S is the same as the one defined in (iv). Last, to establish the estimate $\mathbf{P}[S \geqslant t] \leqslant 1/(4t)$ in (iv), define

$$T = \inf \left\{ t \; : \; \left| \int_0^t dX_s \right| = \tfrac{1}{2} \right\} .$$

When t ranges from 0 to T, $X_t - X_0$ visits all points of one of the two arcs linking 0 and $\tfrac{1}{2}$ in the circle $\mathbb{T}$; so during this time-interval $(X_t - X_0) + (X_t - X_0)$ assumes all possible values on the circle, in particular the value $-X_0$. This implies $S \leqslant T$, whence the majoration $\mathbf{P}[S \geqslant t] \leqslant \mathbf{P}[T \geqslant t]$. Now the Brownian estimate $\mathsf{E}[T] = \mathsf{E}\big[\big(\int_0^T dX_s\big)^2\big] = (\pm 1/2)^2 = 1/4$ yields $\mathbf{P}[T \geqslant t] \leqslant \mathsf{E}[T]/t = 1/(4t)$. ∎

END OF THE PROOF OF PROPOSITION 3. — To show that the natural filtration $\mathcal{G}$ of U is Brownian, it suffices by Lemma 1 to exhibit a $\mathbb{T}$-valued chopped Brownian motion V that generates $\mathcal{G}$. To this end, choose any chopping sequence $(t_k)_{k \in \mathbb{Z}}$ such that

$$\sum_{k < 0} \frac{1}{t_{k+1} - t_k} < \infty \, ;$$

for instance, $t_k = k^3$ is a possible choice. For each k, Lemma 2 applied to the $\mathbb{T}$-valued Brownian motion $(U_t)_{t \geqslant t_k}$ provides us with a $\mathbb{T}$-valued Brownian motion V^k defined on $[t_k, \infty)$, issued from 0 at time t_k, verifying condition (iii) of Lemma 2, and equal to U after some $\mathcal{G}$-stopping time S_k such that $\mathbf{P}[S_k - t_k \geqslant t] \leqslant 1/(4t)$. We shall establish that the process $(V_t)_{t \in \mathbb{R}}$ equal to V^k on $[t_k, t_{k+1})$ is a $\mathbb{T}$-valued chopped Brownian motion and generates the filtration $\mathcal{G}$. Clearly, V is $\mathcal{G}$-adapted.

To see that V is a $\mathbb{T}$-valued chopped Brownian motion, we only have to show that the processes $(V_t^k)_{t \in [t_k, t_{k+1})}$ are independent; it suffices to establish that V^k is independent of $\mathcal{G}_{t_k}$. This can be obtained by writing, for a real-valued, bounded Borel functional f,

$$\mathsf{E}[f(V^k)|\mathcal{G}_{t_k}] \overset{(1)}{=} \mathsf{E}\big[f \circ \phi_k((U_t)_{t \geqslant t_k}) \,\big|\, \mathcal{G}_{t_k}\big] \overset{(2)}{=} \mathsf{E}\big[f \circ \phi_k((U_t)_{t \geqslant t_k}) \,\big|\, U_{t_k}\big]$$

$$= \mathsf{E}[f(V^k)|U_{t_k}] \overset{(3)}{=} \mathsf{E}[f(V^k)] \, ,$$

where (1) stems from the fact that V^k is a functional of $(U_t)_{t \geqslant t_k}$ (by Property (iii) of Lemma 2), (2) from the Markov property of U, and (3) from the independence of V^k and U_{t_k} (by Property (ii) of Lemma 2).

It remains to see that V generates $\mathcal{G}$, or equivalently that U is adapted to V. By Property (iii) of Lemma 2, there are some adapted Borel functionals ψ_k such that

$$(U_t)_{t \in [t_k, t_{k+1})} = \psi_k\big(U_{t_k}, (V_t)_{t \in [t_k, t_{k+1})}\big) \qquad \text{a. s. for each } k.$$

(Adaptedness means that for $t \in [t_k, t_{k+1})$, the restriction of $\psi_k(x, v)$ to $[t_k, t]$ is a function of x and of the restriction of v to $[t_k, t]$.) Using those ψ's, it is possible for each $k \in \mathbb{Z}$ to define inductively a $\mathbb{T}$-valued, càdlàg, V-adapted process $(U_t^k)_{t \in \mathbb{R}}$ by

$$U_t^k = 0 \qquad\qquad\qquad\qquad \text{for } t \in (-\infty, t_k),$$

$$(U_t^k)_{t \in [t_k, t_{k+1})} = \psi_k\big(V_{t_k -}, (V_t)_{t \in [t_k, t_{k+1})}\big) \, ;$$

$$(U_t^k)_{t \in [t_\ell, t_{\ell+1})} = \psi_\ell\big(U_{t_\ell -}^k, (V_t)_{t \in [t_\ell, t_{\ell+1})}\big) \qquad \text{for } \ell > k.$$

(If the left-limit $U_{t_\ell -}^k$ is not defined, put for instance $U_t^k = 0$ for $t \geqslant t_\ell$.) On the event $E_k = \{S_{k-1} < t_k\}$, one has $V_{t_k -} = U_{t_k -} = U_{t_k}$; by definition of ψ_k, this gives

$U^k = U$ on $[t_k, t_{k+1}) \times E_k$. Now if, for some $\ell \geqslant k$, U^k is equal to U on the set $[t_\ell, t_{\ell+1}) \times E_k$, then $U^k_{t_{\ell+1}-} = U_{t_{\ell+1}-} = U_{t_{\ell+1}}$ on E_k and so. by definition of $\psi_{\ell+1}$, U^k agrees with U on the next interval $[t_{\ell+1}, t_{\ell+2}) \times E_k$. Consequently, U^k and U are equal on $[t_k, \infty) \times E_k$.

Now take $t = t_{k+1} - t_k$ in the estimate $\mathbb{P}[S_k - t_k \geqslant t] \leqslant 1/(4t)$ and use the condition on the t_k's. to get $\sum_{k<0} \mathbb{P}[E_k^c] < \infty$. The Borel-Cantelli lemma then says that $\limsup_{k \to -\infty} E_k^c$ is negligible. and E_k converges almost surely to Ω when $k \to -\infty$. So $[t_k, \infty) \times E_k$ tends to $\mathbb{R} \times \Omega$, and consequently U^k converges (stationarily) to U. As each U^k is V-adapted, so is also U. ∎

REMARK (not used in the sequel). — When $\mathbb{T}$ is replaced by a compact, connected Riemannian manifold $\mathbb{M}$, it remains true that Brownian motion in $\mathbb{M}$ indexed by $\mathbb{R}$ has a Brownian filtration (but the innovation filtration can no longer be defined, unless $\mathbb{M}$ is endowed with some extra structure, e.g. Lie group or symmetric space). The definition of a Brownian filtration must of course be modified. with d-dimensional Bronian motions instead of real ones (where d is the dimension of $\mathbb{M}$). We have not found a simple proof of this fact; M. Arnaudon establishes it in an appendix to this note by an argument using a mirror-coupling and some hypoelliptical diffusions. But the case when $\mathbb{M}$ is a d-dimensional sphere is much simpler (the mirror-coupling can be defined so as to avoid all cut-locus difficulties); as an illustration, we show here how that case can be dealt with. and refer to Arnaudon's appendix [1] for the general case.

It suffices to extend Lemma 2 to $\mathbb{M}$, the rest of the proof carries over almost verbatim. There are only two small differences: first, in (i) the point $0 \in \mathbb{T}$ has to be replaced with an arbitrary origin $O \in \mathbb{M}$; second, in (iv), any estimate of the form $\mathbb{P}[S \geqslant t] \leqslant f(t)$ will do, provided $f(t) \to 0$ when $t \to \infty$ (the condition $\sum_{k<0}(t_{k+1}-t_k)^{-1} < \infty$ in the proof will become $\sum_{k<0} f(t_{k+1}-t_k) < \infty$).

To prove the extended Lemma 2, consider $\mathbb{M}$ as the subset of $\mathbb{R}^{d+1}$ made of all points at distance R from the origin $0 \in \mathbb{R}^{d+1}$; for $x \in \mathbb{M}$ call H_x the perpendicular bissector of the segment $[Ox]$ and define the mirror-map Φ_x as the symmetry with respect to the hyperplane H_x; Φ_x is an isometry of $\mathbb{M}$ exchanging O and x (when $x = O$, just take $\Phi_x = \mathrm{Id}_\mathbb{M}$). It then suffices to put $S = \inf \{t : X_t \in H_{X_0}\}$ (take $S = 0$ if $X_0 = O$) and $Y_t = \Phi_{X_0} X_t$ if $t \leqslant S$, $Y_t = X_t$ if $t \geqslant S$. This Y is a $\mathbb{M}$-valued $\mathcal{X}$-Brownian motion satisfying (in $\mathbb{R}^{d+1}$)

$$Y_t = O + \int_0^t (\mathbb{1}_{[\![0,S]\!]} \Phi_{X_0} + \mathbb{1}_{]\!]S,\infty[\!]} \mathrm{Id})(dX_s) ,$$

whence properties (i), (ii) and (iii). To establish the estimate (iv), it suffices to show that, if E denotes a fixed hemisphere and T the first hitting time of the boundary ∂E by a Brownian motion in E, then $\mathbb{P}^x[T \geqslant t]$ tends to 0 uniformly in $x \in E$ when $t \to \infty$. But the function $g(t,x) = \mathbb{P}^x[T \geqslant t]$ is continuous on $(0,\infty) \times \overline{E}$ (it is the solution to the heat equation $\partial g/\partial t = \frac{1}{2}\Delta g$ with boundary conditions $g = 1$ for $t = 0$ and $g = 0$ on ∂E, and for each fixed x, it decreases to zero when $t \to \infty$; so uniformity in x is a consequence of Dini's lemma.

COROLLARY 1. — Let $(\Omega, \mathcal{A}, \mathbb{P}, (\mathcal{H}_t)_{t \in \mathbb{R}})$ be a filtered probability space. There exists an $\mathcal{H}$-CBM if and only if for some (or equivalently for every) regular time-change a, the filtration $(\mathcal{H}'_t)_{t \geqslant 0}$ defined by $\mathcal{H}'_t = \mathcal{H}_{a^{-1}(t)}$ admits an $\mathcal{H}'$-Brownian motion (i.e., some $\mathcal{H}'$-martingale is a Brownian motion).

PROOF. — If $U = (U_t)_{t \in \mathbb{R}}$ is an $\mathcal{H}$-CBM, its natural filtration $\mathcal{G}$ is Brownian by Proposition 3, and the time-changed filtration $\mathcal{G}'$ is generated by a Brownian motion B with $B_0 = 0$. As U is a Markov process for $\mathcal{H}$, every $\mathcal{G}$-martingale is an $\mathcal{H}$-martingale, and, by time-change, every $\mathcal{G}'$-martingale is an $\mathcal{H}'$-martingale. So B is an $\mathcal{H}'$-Brownian motion.

Conversely, if B is an $\mathcal{H}'$-Brownian motion, call $\mathcal{G}'$ the natural filtration of $B - B_0$ and notice that every $\mathcal{G}'$-martingale is an $\mathcal{H}'$-martingale. The time-change transforms $\mathcal{G}'$ into the natural filtration $\mathcal{G}$ of some CBM U; as every $\mathcal{G}$-martingale is an $\mathcal{H}$-martingale, U is also an $\mathcal{H}$-CBM. ∎

Tsirelson's stochastic differential equation

In 1975, Tsirelson [21] has constructed the first example of a stochastic differential equation

$$(T) \qquad dX_t = dB_t + g[t, (X_s, s \leqslant t)] \, dt \qquad X_0 = 0$$

having the following properties: $B = (B_t)_{t \geqslant 0}$ is a Brownian motion, g is a bounded, measurable function of the past $(X_s, 0 \leqslant s \leqslant t)$ of the solution $X = (X_t)_{t \geqslant 0}$, (T) has some (weak) solutions, but no solution X can be adapted to the filtration generated by B.

Tsirelson's equation has been extensively studied; see for instance [4], [14], [19], [13], [26], [27], [12].

To define g, Tsirelson introduces a sequence $(t_k)_{k \leqslant 0}$ of instants verifying $t_k < t_{k+1}$ and $\lim_{k \to -\infty} t_k = 0$. The function g he considers is given by

$$g[t, (X_s, s \leqslant t)] = \sum_{k < 0} \left\{ \frac{X_{t_k} - X_{t_{k-1}}}{t_k - t_{k-1}} \right\} \mathbb{1}_{(t_k, t_{k+1}]}(t)$$

(recall that $\{x\}$ denotes the fractional part of x).

A solution to (T) is a system $(\Omega, \mathcal{A}, \mathbf{P}, \mathcal{H}, B, X)$, where $(\Omega, \mathcal{A}, \mathbf{P}, \mathcal{H})$ is a filtered probability space, B is an $\mathcal{H}$-Brownian motion started at 0, and X is $\mathcal{H}$-adapted and verifies (T). Given such a solution, if $\mathcal{G}$ denotes the sub-filtration of $\mathcal{H}$ generated by X, then B is clearly adapted to $\mathcal{G}$, so it is also a $\mathcal{G}$-Brownian motion, and $(\Omega, \mathcal{A}, \mathbf{P}, \mathcal{G}, B, X)$ is a solution too. The aim of this section is to establish that $\mathcal{G}$ is generated by some real Brownian motion started at the origin. This will be done by reducing the problem to the one addressed in Proposition 3; to do so, we shall need some notation.

Fix once and for all the sequence $(t_k)_{k \leqslant 0}$ and by this choice the function g. Define two functions f and α on $(0, \infty)$ by

$$f(t) = \sum_{k < 0} \frac{1}{t_{k+1} - t_k} \, \mathbb{1}_{(t_k, t_{k+1}]}(t) + \mathbb{1}_{(t_0, \infty)}(t) \qquad \alpha(t) = \int_{t_0}^{t} f(s)^2 \, ds \, .$$

As

$$\int_0^{t_0} f(s)^2 \, ds = \sum_{k < 0} \frac{1}{t_{k+1} - t_k} = +\infty \qquad \text{and} \qquad \int_{t_0}^{\infty} f(s)^2 \, ds = \int_{t_0}^{\infty} ds = +\infty \, ,$$

α is a homeomorphism from $(0, \infty)$ to $\mathbb{R}$; its inverse a is a regular time-change.

PROPOSITION 4. — *Fix the probability space* $(\Omega, \mathcal{A}, \mathbf{P})$.

a) *If, on* $(\Omega, \mathcal{A}, \mathbf{P})$, $(\mathcal{H}, B, X)$ *is a solution to* (T), *define a* T-*valued process* $(U_t)_{t \in (0, \infty)}$ *by*

$$
U_t \doteq \begin{cases} \dfrac{B_t - B_{t_k}}{t_{k+1} - t_k} + \dfrac{X_{t_k} - X_{t_{k-1}}}{t_k - t_{k-1}} & \text{for } t \in (t_k, t_{k+1}] \\[2ex] B_t - B_{t_0} + \dfrac{X_{t_0} - X_{t_{-1}}}{t_0 - t_{-1}} & \text{for } t > t_0 \end{cases}
$$

and time-change both U *and* $\mathcal{H}$ *by* a: *for* $t \in \mathbf{R}$, *set* $\tilde{U}_t = U_{a(t)}$ *and* $\tilde{\mathcal{H}}_t = \mathcal{H}_{a(t)}$. *The process* $\tilde{U}$ *is a* $\tilde{\mathcal{H}}$-*CBM.*

b) *Conversely, if* $(\tilde{\mathcal{H}}_t)_{t \in \mathbf{R}}$ *is a filtration on* $(\Omega, \mathcal{A}, \mathbf{P})$ *and if* $(\tilde{U}_t)_{t \in \mathbf{R}}$ *is an* $\tilde{\mathcal{H}}$-*CBM, time-change them to a filtration* $(\mathcal{H}_t)_{t \geqslant 0}$ *and a process* $(U_t)_{t > 0}$ *by* $\mathcal{H}_t = \tilde{\mathcal{H}}_{\alpha(t)}$ *and* $U_t = \tilde{U}_{\alpha(t)}$; *for* $t > 0$ *set*

$$
B_t = \int_0^t \frac{1}{f(s)} \, dU_s \; ; \qquad X_t = B_t + \int_0^t \sum_{k < 0} \{U_{t_k}\} \, \mathbb{1}_{(t_k, t_{k+1}]}(s) \, ds \; .
$$

Then $(\mathcal{H}, B, X)$ *is a solution to* (T).

c) *The two maps defined in* a) *and* b) *are inverse to each other: they establish a bijection between the solutions* $(\mathcal{H}, B, X)$ *to* (T) *and the CBM's* $(\tilde{\mathcal{H}}, \tilde{U})$.

d) *Given a corresponding pair* $(\mathcal{H}, B, X) \longleftrightarrow (\tilde{\mathcal{H}}, \tilde{U})$, *call* $\mathcal{G}$ *the natural filtration of* X, $\mathcal{F}$ *that of* B, $\tilde{\mathcal{G}}$ *that of* $\tilde{U}$ *and* $\tilde{\mathcal{F}}$ *the innovation filtration of* $\tilde{U}$. *These filtrations are time-changed from one another:*

$$
\tilde{\mathcal{G}}_t = \mathcal{G}_{a(t)} \qquad \mathcal{G}_t = \tilde{\mathcal{G}}_{\alpha(t)} \; ; \qquad \tilde{\mathcal{F}}_t = \mathcal{F}_{a(t)} \qquad \mathcal{F}_t = \tilde{\mathcal{F}}_{\alpha(t)} \; .
$$

In particular, $\mathcal{H}$ *is the natural filtration of* X *if and only if* $\tilde{\mathcal{H}}$ *is the natural filtration of* $\tilde{U}$.

REMARK. — This long statement can be summarized as follows. For fixed $(\Omega, \mathcal{A}, \mathbf{P})$, there exists a bijection between the solutions $(\mathcal{H}, B, X)$ to (T) on the one hand and the pairs $(\tilde{\mathcal{H}}, \tilde{U})$ where $\tilde{U}$ is a $\tilde{\mathcal{H}}$-CBM on the other hand, with the following properties: The time-changes a and α exchange the filtrations $\mathcal{H}$ and $\tilde{\mathcal{H}}$; they also exchange the natural filtrations $\mathcal{G}$ of X and $\tilde{\mathcal{G}}$ of $\tilde{U}$, as well as the natural filtration $\mathcal{F}$ of B and the innovation filtration $\tilde{\mathcal{F}}$ of $\tilde{U}$.

So the process X corresponds to $\tilde{U}$ while B corresponds to the innovations of $\tilde{U}$. Remark that the triple $(\mathcal{H}, B, X)$ is redundant, for B is a functional of X; consequently the proposition could be rephrased so as to state a correspondence between the pairs $(\mathcal{H}, X)$ and $(\tilde{\mathcal{H}}, \tilde{U})$; this would look more symmetric.

PROOF OF PROPOSITION 4. — a) If $(\mathcal{H}, B, X)$ is a solution to (T), for each $k < 0$ one has

$$
X_{t_{k+1}} - X_{t_k} = B_{t_{k+1}} - B_{t_k} + \int_{t_k}^{t_{k+1}} g[t, (X_s, s \leqslant t)] \, dt
$$

$$
= B_{t_{k+1}} - B_{t_k} + \left\{ \frac{X_{t_k} - X_{t_{k-1}}}{t_k - t_{k-1}} \right\} (t_{k+1} - t_k) \; ;
$$

dividing both sides by $t_{k+1} - t_k$ and working modulo 1 to strip off the braces yields

$$
\frac{X_{t_{k+1}} - X_{t_k}}{t_{k+1} - t_k} \stackrel{1}{=} \frac{B_{t_{k+1}} - B_{t_k}}{t_{k+1} - t_k} + \frac{X_{t_k} - X_{t_{k-1}}}{t_k - t_{k-1}} \; .
$$

Considered as elements of T. the right-hand side is $U_{t_{k+1}}$ and the left-hand one is the limit of U_t when t tends to t_{k+1} from above. This shows that the process U is continuous at point t_{k+1} and hence everywhere, and gives a meaning to $\int_s^t dU_r$.

For $s \leqslant t$ with s and t in the same interval $(t_k, t_{k+1}]$ or (t_0, ∞), the definition of U implies

$$\int_s^t dU_r = \int_s^t f(r)\, dB_r .$$

By additivity, this formula remains valid for all pairs $s \leqslant t$ in $(0, \infty)$; notice that the right-hand side is an $\mathcal{H}$-martingale in t on the interval $[s, \infty)$. By time-change, one gets, for $s \leqslant t$ in $\mathbf{R}$,

$$\int_s^t d\tilde{U}_r = \int_{a(s)}^{a(t)} f(r)\, dB_r$$

and the right-hand side is an $\tilde{\mathcal{H}}$-Brownian motion since $\int_{a(s)}^{a(t)} f^2(r)\, dr = t - s$. As $\tilde{U}$ is adapted to $\tilde{\mathcal{H}}$, it is an $\tilde{\mathcal{H}}$-CBM.

b) For each $s > 0$, the process $(B_t)_{t \geqslant s}$ is a continuous $\mathcal{H}$-martingale, with quadratic variation

$$\int_s^t \frac{1}{f^2(r)}\, d[U, U]_r = \int_{a(s)}^{a(t)} \frac{1}{f^2(\alpha(r))}\, dr = t - s ,$$

hence a Brownian motion. Consequently, $(B_t)_{t \geqslant 0}$ is an $\mathcal{H}$-Brownian motion, starting at $B_0 = 0$.

Since $0 \leqslant \{U_{t_k}\} < 1$, the integral in the definition of X is convergent. For $k \leqslant 0$,

$$X_{t_k} - X_{t_{k-1}} = B_{t_k} - B_{t_{k-1}} + \{U_{t_{k-1}}\}(t_k - t_{k-1})$$

$$= \int_{t_{k-1}}^{t_k} \frac{1}{f(r)}\, dU_r + \{U_{t_{k-1}}\}(t_k - t_{k-1})$$

$$= (t_k - t_{k-1})\left[\int_{t_{k-1}}^{t_k} dU_r + \{U_{t_{k-1}}\}\right] ,$$

so $\dfrac{X_{t_k} - X_{t_{k-1}}}{t_k - t_{k-1}} \overset{1}{=} U_{t_k}$, wherefrom $\{U_{t_k}\} = \left\{\dfrac{X_{t_k} - X_{t_{k-1}}}{t_k - t_{k-1}}\right\}$ and X verifies (T).

c) We first show that the composed map $(\mathcal{H}, B, X) \mapsto (\tilde{\mathcal{H}}, \tilde{U}) \mapsto (\mathcal{H}', B', X')$ is identity: $(\mathcal{H}', B', X') = (\mathcal{H}, B, X)$. That $\mathcal{H}' = \mathcal{H}$ is trivial: two inverse time-changes cancel. Then, B' is defined by $dB'_t = dU_t/f(t)$ and U verifies $dU_t = f(t)\, dB_t$, giving $B' = B$. Last, the right-continuity of U at t_k yields $U_{t_k} \overset{1}{=} (X_{t_k} - X_{t_{k-1}})/(t_k - t_{k-1})$. whence $dX'_t = dB_t + g[t, (X_s, s{\leqslant}t)]\, dt = dX_t$, giving $X' = X$.

Now the other way round: $(\tilde{\mathcal{H}}, \tilde{U}) \mapsto (\mathcal{H}, B, X) \mapsto (\tilde{\mathcal{H}}', \tilde{U}')$ is identity too. Again, $\tilde{\mathcal{H}}' = \tilde{\mathcal{H}}$ is trivial. Time-change $\tilde{U}$ and $\tilde{U}'$ to get a U and a U'. For $t \in [t_k, t_{k+1})$ (with the convention $t_1 = \infty$),

$$U'_t \overset{1}{=} \int_{t_k}^t f(s)\, dB_s + \frac{X_{t_k} - X_{t_{k-1}}}{t_k - t_{k-1}} = \int_{t_k}^t dU_s + \frac{X_{t_k} - X_{t_{k-1}}}{t_k - t_{k-1}} .$$

But we have seen in the proof of b) that $\dfrac{X_{t_k} - X_{t_{k-1}}}{t_k - t_{k-1}} \overset{1}{=} U_{t_k}$, so $U'_t = U_t$.

d) As X and B are $\mathcal{G}$-adapted, the definition given in a), how to obtain U from B and X, shows that U is $\mathcal{G}$-adapted. Conversely, the definitions given in b), how to obtain B and X from U. show that X is adapted to the natural filtration of U. So U has $\mathcal{G}$ as natural filtration: by time-change, this gives the relation between $\mathcal{G}$ and $\widetilde{\mathcal{G}}$.

The formulae $dB = (1/f)\,dU$ and $dU = f\,dB$ show that the increments of U generate the filtration $\mathcal{F}$: by time-change, this gives the relation between $\mathcal{F}$ and $\widetilde{\mathcal{F}}$. ∎

Proposition 4 reduces the study of Tsirelson's equation to that of CBM's. Transferring to (T) what we know about CBM's gives the following statement (where only the last sentence is new):

COROLLARY 2. — *Solutions* $(\Omega, \mathcal{A}, \mathbf{P}, \mathcal{H}, B, X)$ *to Tsirelson's equation exist. More precisely, on a given* $\left(\Omega, \mathcal{A}, \mathbf{P}, (\mathcal{H}_t)_{t \geqslant 0}\right)$, *a solution* (B, X) *exists if and only if there exists an* $\mathcal{H}$-*Brownian motion. The law of a solution* (B, X) *depends only on the sequence* $(t_k)_{k \leqslant 0}$. *Given any solution, the natural filtration of* X *is Brownian.*

PROOF. — If $(\mathcal{H}, B, X)$ is a solution, B is an $\mathcal{H}$-Brownian motion. Conversely, if an $\mathcal{H}$-Brownian motion exists, the filtration $(\widetilde{\mathcal{H}}_t)_{t \in \mathbb{R}}$ defined by $\widetilde{\mathcal{H}}_t = \mathcal{H}_{a(t)}$ contains some CBM $\widetilde{U}$ by Corollary 1 and a solution $(\mathcal{H}, \widetilde{B}, X)$ exists by Proposition 4.b.

The rest of the corollary is straightforward from Propositions 3 and 4. ∎

REMARKS. — a) If the reader does not care about CBM's and is only interested in knowing that the filtration of X is Brownian, the proofs given above can be shortened.

First, in Proposition 4, only a) and d) are needed, so half of the computations can be dispensed with.

More important, a long detour we have taken can be bypassed. Our proof consisted in time-changing (T) to get a CBM; and the proof that the filtration of a CBM is Brownian was done by time-changing back the CBM to work on $\mathbb{R}_+$ (with a time-change provided by Lemma 1). By choosing in the proof of Proposition 3 the same (modulo the time-change) sequence $(t_k)_{k \leqslant 0}$ that is used to define g, it is possible to show directly that the filtration of X is Brownian, with a proof quite similar to that of Proposition 3, that uses the time-changed CBM U featuring in Proposition 4 a).

The reason for this detour was that we found it instructive, on an intuitive level, that the "mysterious" Tsirelson example may be one-to-one translated into Brownian motion on the circle, an object which to us does not seem mysterious at all.

b) After this work was completed, we learned that A. Vershik proved a long time ago the following (unpublished) result: When taken at the subdivision times t_n, a solution X to (T) generates a *standard* discrete filtration, a necessary condition for the continuous-time filtration to be Brownian. We find it striking that his (combinatorial) method of proof also involves coupling two (discrete) processes.

This standardness result was independently rediscovered by M. Malric (also unpublished).

References

[1] M. Arnaudon. Appendice à l'exposé précédent : La filtration naturelle du mouvement brownien indexé par **R** dans une variété compacte. In this volume.

[2] M.T. Barlow, M. Émery, F.B. Knight, S. Song & M. Yor. Autour d'un théorème de Tsirelson sur des filtrations browniennes et non browniennes. *Séminaire de Probabilités XXXII*. Lecture Notes in Mathematics 1686, Springer 1998.

[3] S. Beghdadi-Sakrani & M. Émery. On certain probabilities equivalent to coin-tossing, d'après Schachermayer. In this volume.

[4] V. E. Beneš. Non existence of strong non-anticipating solutions to SDE's; implication for functional DE's, filtering and control. *Stoch. Proc. Appl. 17*, 243–263, 1977.

[5] B. De Meyer. Une simplification de l'argument de Tsirelson sur le caractère non-brownien des processus de Walsh. In this volume.

[6] P. Diaconis. From shuffling cards to walking around the building: An introduction to modern Markov chain theory. *Documenta Mathematica. Extra volume ICM 1998. I* 47–64, 1998.

[7] L. Dubins, J. Feldman, M. Smorodinsky & B. Tsirelson. Decreasing sequences of σ-fields and a measure change for Brownian motion. *Ann. Prob. 24*, 882–904, 1996.

[8] M. Émery & W. Schachermayer. Brownian filtrations are not stable under equivalent time-changes. In this volume.

[9] J. Feldman. ε-close measures producing non-isomorphic filtrations. *Ann. Prob. 24*, 912–915, 1996.

[10] J. Feldman & M. Smorodinsky. Simple examples of non-generating Girsanov processes. *Séminaire de Probabilités XXXI*, Lecture Notes in Mathematics 1655, Springer 1997.

[11] J. Feldman & B. Tsirelson. Decreasing sequences of σ-fields and a measure change for Brownian motion. II *Ann. Prob. 24*, 905–911, 1996.

[12] J. Kallsen. A stochastic differential equation with a unique (up to indistinguishability) but not strong solution. In this volume.

[13] J.-F. Le Gall and M. Yor. Sur l'équation stochastique de Tsirelson. *Séminaire de Probabilités XVII*, Lecture Notes in Mathematics 986, Springer 1983.

[14] R. S. Liptser & A. N. Shiryaev. Statistics of Random Processes I. Springer, 1977.

[15] D. Revuz & M. Yor. Continuous Martingales and Brownian Motion. Springer, 1991.

[16] L.C.G. Rogers and D. Williams. Diffusions, Markov Processes, and Martingales. Volume 2: Itô Calculus. Wiley, 1987.

[17] W. Schachermayer. On certain probabilities equivalent to Wiener measure, d'après Dubins, Feldman, Smorodinsky and Tsirelson. In this volume.

[18] M. Smorodinsky. Processes with no standard extension. *Israel J. Math.*, to appear.

[19] D.W. Stroock & M. Yor. On extremal solutions of martingale problems. *Ann. Sci. École Norm. Sup. 13*, 95–164, 1980.

[20] B. Tsirelson. Triple points: From non-Brownian filtrations to harmonic measures. *GAFA, Geom. funct. anal. 7*, 1096–1142, 1997.

[21] B. S. Tsirel'son. An example of a stochastic differential equation having no strong solution. *Theor. Prob. Appl. 20*, 427–430, 1975.

[22] A. M. Vershik. Approximation in measure theory. *Doctor Thesis*, Leningrad 1973. Expanded and updated english version: The theory of decreasing sequences of measurable partitions. *St. Petersburg Math. J. 6*, 705–761, 1995.

[23] J. Warren. On the joining of sticky Brownian motion. In this volume.

[24] H. von Weizsäcker. Exchanging the order of taking suprema and countable intersections of σ-algebras. *Ann. Inst. Henri Poincaré 19*, 91–100. 1983.

[25] D. Williams. Probability with Martingales. Cambridge University Press. 1991.

[26] M. Yor. De nouveaux résultats sur l'équation de Tsirel'son. *C. R. Acad. Sci., Paris. Sér. I 309*. 511–514, 1989.

[27] M. Yor. Tsirel'son's equation in discrete time. *Probab. Theory Relat. Fields 91*. 135–152. 1992.

Appendice à l'exposé précédent :
La filtration naturelle du mouvement brownien
indexé par ℝ
dans une variété compacte

M. Arnaudon

Résumé. — Étant donnés deux points x et y d'une variété riemannienne compacte M de dimension $d \geq 2$ et un mouvement brownien X issu de x, on prouve qu'il existe un mouvement brownien issu de y, qui rencontre presque sûrement X et qui engendre la même filtration. Avec une démonstration analogue à celle de l'exposé précédent, on en déduit que la filtration naturelle du mouvement brownien indexé par ℝ et à valeurs dans M est, à un changement de temps régulier près, égale à la filtration naturelle d'un mouvement brownien indexé par $ℝ_+$, à valeurs dans $ℝ^d$ et issu de 0.

1. Introduction

La filtration naturelle du mouvement brownien à valeurs dans le cercle $ℝ/ℤ$ et *indexé par la droite* ℝ est une filtration brownienne au sens où tout changement de temps régulier (c'est-à dire déterministe, absolument continu ainsi que son inverse, et transformant ℝ en $]0, \infty[$), en fait une filtration brownienne au sens usuel. Ce résultat est établi dans l'exposé précédent (proposition 3) par un argument de couplage de browniens sur le cercle. Le but de cet appendice est de l'étendre au cas où le cercle est remplacé par une variété riemannienne compacte de dimension supérieure ou égale à deux.

Nos énoncés seront repérés par des lettres, les énoncés numérotés renvoyant à l'exposé précédent.

Nous nous donnons donc une variété M de classe C^∞, riemannienne compacte, connexe, sans bords, de dimension finie d supérieure ou égale à 2. La définition des « circular Brownian motions » de l'exposé précédent se généralise sans difficulté :

DÉFINITION. — *Un processus $U = (U_t)_{t \in ℝ}$ indexé par ℝ et à valeurs dans M est un mouvement brownien pour une filtration $\mathcal{H} = (\mathcal{H}_t)_{t \in ℝ}$ si pour tout s fixé, la restriction de U à $[s, \infty[$ est un mouvement brownien dans M pour la filtration $(\mathcal{H}_t)_{t \geq s}$.*

Un tel processus est aussi un mouvement brownien pour sa filtration naturelle ; c'est un processus stationnaire et toutes les variables aléatoires U_t ont la loi uniforme sur M, c'est-à-dire la probabilité $(1/V)\rho$, où ρ est la mesure riemannienne sur M et la constante de normalisation V est le volume $\rho(M)$.

Bien que cette propriété ne soit pas utilisée dans la suite, rappelons brièvement comment on peut l'établir :

Soit μ_t la loi de U_t. Pour $s < t$, l'équation de Chapman-Kolmogorov donne $\mu_t(dy) = \int_{x \in M} \mu_s(dx)\, p_{t-s}(x, y)\, \rho(dy)$, où p_t est le noyau de la chaleur sur $M \times M$ (avec la normalisation des probabilistes : l'opérateur de la chaleur est $\frac{1}{2}\Delta - \partial_t$) ; la théorie des équations aux dérivées partielles elliptiques dit que $p_t(x, y)$ est continu (et même C^∞) en (t, x, y) sur $]0, \infty[\times M \times M$ (voir par exemple Davies [D]). En particulier, $p_1(x, y)$ est continu sur $M \times M$, et, par compacité, borné par une constante C. L'équation de Chapman-Kolmogorov entraîne que la fonction continue $u_s(y) = \int \mu_{s-1}(dx)\, p_1(x, y)$ est une densité de μ_s par rapport à ρ. Cette fonction vérifie

$$\|u_s\|_{L^2} = \left\| \int \mu_{s-1}(dx)\, p_1(x, \cdot) \right\|_{L^2} \le \int \mu_{s-1}(dx)\, \|p_1(x, \cdot)\|_{L^2} \le C.$$

Mais, par compacité et connexité, les fonctions harmoniques sont constantes, et la projection dans L^2 de u_t sur l'espace des fonctions harmoniques est la constante $1/V$. La propriété de trou spectral du laplacien (voir par exemple Berger, Gauduchon et Mazet [B,G,M]) donne

$$\| u_t - (1/V) \|_{L^2} \le e^{-\lambda_1(t-s)} \| u_s - (1/V) \|_{L^2} \le \left(C + \|1/V\|_{L^2} \right) e^{-\lambda_1(t-s)} ,$$

où $\lambda_1 > 0$ est la première valeur propre non nulle de $-\frac{1}{2}\Delta$. En faisant tendre s vers $-\infty$, on obtient $u_t = 1/V$.

Appelons filtration brownienne à d dimensions (indexée par $\mathbb{R}_+$) la filtration naturelle d'un mouvement brownien dans $\mathbb{R}^d$ issu de 0, et filtration brownienne à d dimensions indexée par $\mathbb{R}$ toute filtration qui s'en déduit par un changement de temps régulier. Étendons de même à $\mathbb{R}^d$ et à M en fixant une origine $O \in M$, la définition des mouvements browniens tronçonnés (« chopped Brownian motions »). Il est clair que la proposition 2 reste vraie lorsque l'on y remplace « filtration brownienne » par « filtration brownienne à d dimensions ». De même, on peut énoncer l'analogue du lemme 1 : la filtration engendrée par un mouvement brownien tronçonné dans $\mathbb{R}^d$, ou par un mouvement brownien tronçonné dans M, est une filtration brownienne à d dimensions. La preuve est identique, et le passage de $\mathbb{R}^d$ à M se fait à l'aide du développement stochastique, après avoir fixé une isométrie $\mathbb{R}^d \to T_O M$.

Pour démontrer que la filtration naturelle de tout mouvement brownien indexé par $\mathbb{R}$ et à valeurs dans M est une filtration brownienne de dimension d (proposition E), nous allons procéder de façon analogue à l'exposé précédent. Comme dans le lemme 2, nous commencerons par établir un résultat de couplage dans une même filtration de mouvements browniens issus de points différents (proposition A). Il faudra distinguer le comportement au voisinage du cutlocus (lemme B) et le comportement en dehors de ce voisinage (lemme D).

Contrairement à la première moitié de la proposition 3, nous n'étendrons pas aux variétés la seconde moitié. En effet, lorsque M n'a aucune structure particulière (telle que groupe de Lie ou espace symétrique), nous ne voyons pas comment définir la filtration des innovations de façon intrinsèque, sans faire intervenir une structure supplémentaire, par exemple le choix (non canonique !) d'un champ mesurable de repères.

Dans la suite, la différentielle d'Itô d'une semimartingale X à valeurs dans M sera notée $d_\Gamma X$. Elle admet une décomposition canonique en partie martingale et partie à variation finie, qui seront notées respectivement $d_m X$ et $\check{d}_\Gamma X$. En coordonnées locales, si $dX^i = dN^i + dA^i$ où N^i est une martingale locale réelle et A^i un processus à variation finie, et si on désigne par Γ^k_{ij} les symboles de Christoffel de la connexion, on a

$$d_\Gamma X = \left(dX^i + \frac{1}{2} \Gamma^i_{jk} d{<}X^j, X^k{>} \right) \frac{\partial}{\partial x^i},$$

$$d_m X = dN^i \frac{\partial}{\partial x^i} \quad \text{et} \quad \check{d}_\Gamma X = \left(dA^i + \frac{1}{2} \Gamma^i_{jk} d{<}X^j, X^k{>} \right) \frac{\partial}{\partial x^i}.$$

Formellement, $d_\Gamma X$, $d_m X$ et $\check{d}_\Gamma X$ sont des vecteurs tangents au point X (voir [M]). Si M' est une autre variété avec connexion, si e est une section C^1 du fibré $TM' \otimes (TM)^*$ au-dessus de $M \times M'$ (pour $(x,y) \in M \times M'$, $e(x,y)$ est une application linéaire de $T_x M$ dans $T_y M'$) et X est une semimartingale dans M, alors l'équation d'Itô

$$d_\Gamma Y = e(X,Y) d_\Gamma X$$

a une unique solution de condition initiale Y_0, à durée de vie éventuellement finie (voir par exemple [E1], [E2], [A,T]).

2. Couplage de mouvements browniens dans une variété riemannienne compacte

Il est bien connu qu'étant donnés deux points de M, il existe une filtration et deux mouvements browniens dans cette filtration, à valeurs dans M, issus chacun de l'un des points, et qui se rencontrent presque sûrement (voir [K] et [C]). La proposition qui suit est plus exigeante : elle demande que les filtrations naturelles des deux mouvements browniens soient égales.

PROPOSITION A. — *Il existe une constante $\alpha > 0$ (ne dépendant que de M) telle que pour tout $y \in M$ et pour tout mouvement brownien $(X_t)_{t \geq 0}$ dans M, il existe un mouvement brownien $(Y_t)_{t \geq 0}$ dans M vérifiant les quatre propriétés suivantes:*

(i) $Y_0 = y$;

(ii) Y est indépendant de X_0 ;

(iii) *les processus $(X_t)_{t \geq 0}$ et $(X_0, Y_t)_{t \geq 0}$ engendrent la même filtration* ;

(iv) *notant S le temps d'arrêt $\inf\{t, \ X_t = Y_t\}$, on a pour tout $n \in \mathbb{N}^*$*

$$\mathbb{P}[S \leq n] > 1 - (1 - \alpha)^n$$

et $Y = X$ sur $[S, \infty[$. En particulier, $\mathbb{P}[S < \infty] = 1$.

Pour établir cette proposition, on utilise deux lemmes. Le premier permet de construire Y au voisinage du cutlocus. La notation $\beta((x,y),r)$ désigne la boule centrée en $(x,y) \in M \times M$ et de rayon $r > 0$.

LEMME B. — *Il existe trois constantes $\varepsilon_h > 0$, $\varepsilon > 0$ et $\alpha_h > 0$ ne dépendant que de M, telles que pour toute filtration $\mathcal{F} = (\mathcal{F}_t)_{t \geq 0}$, pour toute v.a. $\mathcal{F}_0$-mesurable $\mathscr{Y}$ à valeurs dans M et pour tout $\mathcal{F}$-mouvement brownien $(X_t)_{t \geq 0}$, il existe un mouvement brownien $(Y_t)_{t \geq 0}$ vérifiant les quatre propriétés suivantes:*

(i) $Y_0 = \mathscr{Y}$;

(ii) conditionnellement à Y_0, Y est indépendant de X_0;

(iii) les processus $(X_t, Y_0)_{t \geq 0}$ et $(X_0, Y_t)_{t \geq 0}$ engendrent la même filtration;

(iv) notant $T = \inf\{t > 0, (X_t, Y_t) \notin \beta((X_0, Y_0), \varepsilon_h)\}$ et pour un ouvert A de $M \times M$, $T^A = \inf\{t \geq 0, (X_t, Y_t) \notin A\}$, si A est de mesure inférieure à ε, alors

$$\mathbb{P}\left[T^A < T \wedge 1/2\right] > \alpha_h.$$

La démonstration va reposer sur la construction d'une diffusion hypoelliptique (X, Y) dans $M \times M$, dont les deux composantes X et Y sont des mouvements browniens, et telle que les trois processus X, Y et (X, Y) aient une même filtration naturelle. Pour cela, on a besoin du lemme suivant.

LEMME C. — *Soit $(a, b) \in M \times M$. Il existe un voisinage ouvert $\mathscr{B}$ de (a, b) et d champs de vecteurs $W_1, \ldots, W_d$ définis sur $\mathscr{B}$ vérifiant les propriétés suivantes:*

(i) notant $W_i = (A_i, B_i)$, pour tout $(x, y) \in \mathscr{B}$, $(A_1(x, y), \ldots, A_d(x, y))$ est une base orthonormale de $T_x M$ et $(B_1(x, y), \ldots, B_d(x, y))$ est une base orthonormale de $T_y M$;

(ii) pour tout $(x, y) \in \mathscr{B}$, $Lie(W_1, \ldots, W_d)(x, y) = T_{(x,y)}M \times M$, où $Lie(W_1, \ldots, W_d)$ désigne l'algèbre de Lie engendrée par $W_1, \ldots, W_d$.

Démonstration du lemme C. — On considère tout d'abord le cas $d = 3$. Soit $(a, b) \in M \times M$, et soit $(A_1(a), A_2(a), A_3(a))$ une base orthonormale de $T_a M$. Pour x^1, x^2, x^3 au voisinage de 0, on la transporte parallèlement le long de $t \mapsto \exp(t A_1(a))$ de $t = 0$ à $t = x^1$, puis le long de $t \mapsto \exp(t A_2(\exp(x^1 A_1(a))))$ de $t = 0$ à $t = x^2$, puis le long de $t \mapsto \exp(t A_3(\exp(x^2 A_2(\exp(x^1 A_1(a))))))$ de $t = 0$ à $t = x^3$. On obtient ainsi des coordonnées (x^1, x^2, x^3) centrées en a. On effectue la même construction au voisinage de b, et on note C_1, C_2 et C_3 les champs de vecteurs obtenus, (y^1, y^2, y^3) les coordonnées. En a, les dérivées covariantes $\nabla_{A_i} A_j$ s'annulent, ainsi que $\nabla_{A_1} \nabla_{A_1} A_2$ et $\nabla_{A_1} \nabla_{A_2} A_1$ Par conséquent, les crochets $[A_i, A_j]$ et $[A_1, [A_1, A_2]]$ s'annulent aussi. Il en va de même en b avec les vecteurs C_i. On définit ensuite au voisinage de (a, b)

$$W_1 = (A_1, \cos x^1 \cos x^2 C_1 + \sin x^1 \cos x^2 C_2 + \sin x^2 C_3),$$

$$W_2 = (A_2, -\sin x^1 C_1 + \cos x^1 C_2),$$

$$W_3 = (A_3, -\cos x^1 \sin x^2 C_1 - \sin x^1 \sin x^2 C_2 + \cos x^2 C_3).$$

La propriété (i) est facile à vérifier. D'après les remarques ci-dessus sur l'annulation des crochets, au point (a,b), $\Big(W_1,\ W_2,\ W_3,\ [W_1,W_2],\ [W_1,[W_1,W_2]],\ [W_2,W_3])\Big)$ est égal à

$$\Big((A_1,C_1),\ (A_2,C_2),\ (A_3,C_3),\ (0,-C_1-C_3),\ (0,-C_2),\ (0,-C_1)\Big).$$

C'est une famille génératrice de $T_{(a,b)}M \times M$. Par continuité, on déduit que $\Big(W_1,\ W_2,\ W_3,\ [W_1,W_2],\ [W_1,[W_1,W_2]],\ [W_2,W_3])\Big)(x,y)$ engendre $T_{(x,y)}M \times M$ pour (x,y) dans un voisinage de (a,b), d'où la propriété (ii).

Pour les autres valeurs de d, on conserve la construction des A_i et des C_i en remplaçant les 3 transports parallèles par d transports parallèles. Si d est pair, on construit les W_j par couples en posant

$$W_{2i+1} = (A_{2i+1}, \cos x^{2i+1} C_{2i+1} + \sin x^{2i+1} C_{2i+2})$$

et

$$W_{2i+2} = (A_{2i+2}, -\sin x^{2i+1} C_{2i+1} + \cos x^{2i+1} C_{2i+2}).$$

Avec les mêmes calculs que plus haut, on vérifie alors qu'au point (a,b),

$$\Big(W_{2i+1},\ W_{2i+2},\ [W_{2i+1},W_{2i+2}],\ [W_{2i+1},[W_{2i+1},W_{2i+2}]]\Big)$$

est égal à

$$\Big((A_{2i+1},C_{2i+1}),\ (A_{2i+2},C_{2i+2}),\ (0,-C_{2i+1}),\ (0,-C_{2i+2}),\Big).$$

Si d est de la forme $2n+3$, on construit n couples comme ci-dessus et les trois derniers W_j comme en dimension 3, en remplaçant les indices j par $2n+j$. Dans tous les cas, (i) est facile à vérifier et $\mathrm{Lie}(W_1,\ldots,W_d)(a,b) = T_{(a,b)}M \times M$, donc pour tout (x,y) dans un voisinage $\mathscr{B}$ de (a,b), $\mathrm{Lie}(W_1,\ldots,W_d)(x,y) = T_{(x,y)}M \times M$, d'où la propriété (ii). $\square$

Démonstration du lemme B. — On recouvre $M \times M$ par un nombre fini, disons n, d'ouverts $\mathscr{B}_1$, ..., $\mathscr{B}_n$ définis comme l'ouvert $\mathscr{B}$ du lemme C, puis on définit ε_h tel que toute boule $\beta((x,y),\varepsilon_h)$ ait son adhérence incluse dans l'un des ouverts du recouvrement.

On se place sur l'événement $\big\{\beta((X_0,Y_0),\varepsilon_h) \subset \mathscr{B}_1\big\}$. On se contente de construire (X,Y) jusqu'au premier temps de sortie T de $\beta((X_0,Y_0),\varepsilon_h)$, un prolongement après ce temps ne pose pas de difficulté.

Pour $(x,y) \in \mathscr{B}_1$, on note $W(x,y)$ l'application linéaire $\mathbb{R}^d \to T_{(x,y)}M \times M$ qui à $(\zeta^1,\ldots,\zeta^d)$ associe $\sum_{i=1}^d \zeta^i W_i(x,y)$. On a $W(x,y) = (A(x,y),B(x,y))$, où $A(x,y)$ est une isométrie de $\mathbb{R}^d$ dans T_xM, et $B(x,y)$ est une isométrie de $\mathbb{R}^d$ dans T_yM. On définit le mouvement brownien Y par $Y_0 = \mathscr{Y}$ et

$$d_\Gamma(X,Y) = W(X,Y)A(X,Y)^{-1}d_\Gamma X.$$

Si on note $W = (A, B)$, les processus X et Y satisfont les équations différentielles

$$d_\Gamma Y = B(X,Y)A(X,Y)^{-1}d_\Gamma X$$

et

$$d_\Gamma X = A(X,Y)B(X,Y)^{-1}d_\Gamma Y,$$

dont les coefficients sont de classe C^∞, donc (X^T, Y_0) et (X_0, Y^T) engendrent la même filtration. Ainsi on a (iii). Et comme Y est une martingale dans la filtration engendrée par (X_0, Y) et est aussi un mouvement brownien, on obtient (ii) grâce à la propriété de Markov. Pour établir l'inégalité de (iv), on peut conditionner par $(X_0, Y_0) = (x, y)$. Quitte à remplacer A par $A \cap \beta((x,y), \varepsilon_h)$, on peut supposer que A est inclus dans $\beta((x,y), \varepsilon_h)$. Comme $\mathrm{Lie}(W_1, \ldots, W_d)$ engendre $TM \times TM$ en tout point, le processus (X, Y) est une diffusion hypoelliptique sur $[0, T[$. On veut ensuite utiliser un résultat de majoration des densités des diffusion hypoelliptiques dans $\mathbb{R}^{2d}$. Pour cela, à l'aide d'une carte et quitte à réduire la taille de $\mathscr{B}_1$, on identifie $\mathscr{B}_1$ à un ouvert de $\mathbb{R}^{2d}$ et on prolonge les champs de vecteurs à tout $\mathbb{R}^{2d}$ en des champs dont toutes les dérivées sont bornées, et on en rajoute éventuellement qui s'annulent sur $\mathscr{B}_1$ afin que la condition d'hypoellipticité de Hörmander soit réalisée. On dispose grâce à ces champs de vecteurs d'une diffusion hypoelliptique (X', Y') sur $\mathbb{R}^{2d}$, issue de (x, y) qui coïncide avec (X, Y) jusqu'au premier temps de sortie de $\mathscr{B}_1$. Or pour tout $\eta > 0$, il existe $t_\eta > 0$ tel que si $t \in]0, t_\eta[$, la probabilité de transition de $Z = (X', Y')$ vérifie $p_t(z, z') \leq e^{\frac{\eta}{2t}} e^{-\frac{\mathrm{dist}^2(z, z')}{2t}}$ ([L], théorème 1 ; puisqu'on cherche une majoration, on peut prendre pour dist la distance euclidienne qui est plus petite que la distance hypoelliptique). Pour t fixé suffisamment petit (en particulier $t < 1/2$), la densité est donc majorée dans la boule de rayon ε_h par un nombre $M(t)$, et par conséquent la probabilité de sortir d'un ouvert quelconque de mesure ε avant l'instant t est minorée par $\alpha = 1 - \varepsilon M(t)$. On peut choisir t suffisamment petit pour que Z reste dans $\beta((x,y), \varepsilon_h)$ avant l'instant t avec une grande probabilité α', et ensuite ε suffisamment petit pour que α soit proche de 1. En posant $\alpha_h = \alpha + \alpha' - 1$, on a la propriété voulue pour le processus Z. Mais comme Z^T et $(X, Y)^T$ coïncident, on a le résultat recherché. $\square$

On note Δ la diagonale de $M \times M$, et $\mathscr{C} = \cup_{x \in M}(\{x\} \times \mathscr{C}(x))$, où $\mathscr{C}(x) \subset M$ est la réunion du cutlocus et du "conjugate locus" de x. L'ensemble $\mathscr{C}$ est un compact de $M \times M$ de mesure nulle.

Avec le lemme suivant, on construit Y lorsque (X, Y) est en dehors d'un petit voisinage de $\mathscr{C}$.

LEMME D. — *Soit $W_\mathscr{C}$ un voisinage ouvert de $\mathscr{C}$. Il existe $\alpha_c > 0$ ne dépendant que de M, tel que pour toute filtration $\mathcal{F} = (\mathcal{F}_t)_{t \geq 0}$, pour toute v.a. $\mathcal{F}_0$-mesurable $\mathscr{Y}$ à valeurs dans M et pour tout $\mathcal{F}$-mouvement brownien $(X_t)_{t \geq 0}$ tels que p.s. $(X_0, \mathscr{Y}) \in W_\mathscr{C}^c$, il existe un mouvement brownien $(Y_t)_{t \geq 0}$ vérifiant les quatre propriétés suivantes:*

(i) $Y_0 = \mathscr{Y}$;

(*ii*) *conditionnellement à* Y_0, Y *est indépendant de* X_0 ;

(*iii*) *les processus* $(X_t, Y_0)_{t \geq 0}$ *et* $(X_0, Y_t)_{t \geq 0}$ *engendrent la même filtration* ;

(*iv*) *notant* $T = \inf\{t > 0, (X_t, Y_t) \in \mathscr{C}\}$ *et* $S = \inf\{t \geq 0, (X_t, Y_t) \in \Delta\}$, *on a*

$$\mathbb{P}[S < T \wedge 1/2] > \alpha_c.$$

Démonstration du lemme D. — Si on remplace $W_{\mathscr{C}}^c$ par un voisinage V_Δ arbitrairement petit de la diagonale de $M \times M$ et le temps $1/2$ par $1/4$, alors ce résultat peut se déduire de [C] théorème 1. Il nous suffit donc de construire un Y tel qu'avec une probabilité strictement positive α'_c, le processus (X, Y) issu de $(X_0, \mathscr{Y}) \in W_{\mathscr{C}}^c$ atteigne V_Δ avant d'atteindre le cutlocus, et avant le temps $1/4$. On construira Y à partir de X avec un couplage miroir. La clef de la démonstration consistera à trouver des coordonnées dans $M \times M$ telles que la première coordonnée soit la distance de x à y et telles que lorsqu'on exprime (X, Y) à l'aide des autres coordonnées, le mouvement brownien directeur soit indépendant du mouvement brownien directeur utilisé pour la distance de X à Y. Cette propriété n'est pas utile dans la démonstration de Cranston.

Soit SM le fibré unitaire, i.e. la sous-variété de TM formée des vecteurs de norme 1. Si $(x, y) \notin \mathscr{C}$, soit $m(x, y)$ le milieu de la géodésique minimisante reliant x à y, $2\rho(x, y)$ la distance de x à y et $u(x, y) := \dfrac{1}{\rho(x, y)} \exp_{m(x,y)}^{-1} y$. L'application

$$M \times M \backslash (\mathscr{C} \cup \Delta) \to \mathbb{R}_+^* \times SM$$
$$(x, y) \mapsto (\rho(x, y), u(x, y))$$

est un difféomorphisme local d'application réciproque

$$(\rho, u) \mapsto (\exp(-\rho u), \exp(\rho u)).$$

Il existe $\varepsilon' > 0$ vérifiant la propriété suivante. Si $(x, y) \in W_{\mathscr{C}}^c$, alors $\mathscr{C}^c$ contient l'ouvert $T(x, y) =$

$$\{(x', y') \in M \times M, \ \mathrm{dist}(u(x', y'), u(x, y)) < \varepsilon' \ \mathrm{et} \ \varepsilon' < \rho(x', y') < \rho(x, y) + \varepsilon'\}.$$

Il suffit de construire (X, Y) jusqu'au temps de sortie de $T(X_0, Y_0)$, il sera ensuite facile de prolonger Y après ce temps. On va définir le couplage miroir et montrer que le mouvement brownien Y construit à partir de X avec un couplage miroir répond à la question. Pour cela, quitte à diminuer ε', il suffit de montrer qu'avec une probabilité strictement positive, le processus (X, Y) sort de $T(X_0, Y_0)$ pour la première fois au moment où $\rho(X, Y)$ atteint ε'. On notera F cet événement.

Si $(x, y) \in \mathscr{C}^c$, on désigne par $t \mapsto \gamma(t, x, y)$ $(t \in [0, 1])$ la géodésique minimisante qui vérifie $\gamma(0, x, y) = x$ et $\gamma(1, x, y) = y$, et pour $\jmath_0 \in T_x M$, $\jmath_1 \in T_y M$, on note $J(t, \jmath_0, \jmath_1) = T\gamma(x, y)(\jmath_0, \jmath_1)$ où $T\gamma$ est la dérivée par rapport à (x, y), $\dot{J}(t, \jmath_0, \jmath_1) = \frac{\nabla}{dt} J(t, \jmath_0, \jmath_1)$. On désigne par $\tau(x, y, \cdot)$ le transport parallèle de $T_x M$ vers $T_y M$ le long de $t \mapsto \gamma(t, x, y)$, $p(x, y, \cdot)$ la projection orthogonale d'un vecteur de T_xM parallèlement à $\dot{\gamma}(0,x,y)$ où $\dot{\gamma}$ est la dérivée par rapport à t, $\mathscr{M}(x,y,\cdot) = \tau(x,y,2p(x,y,\cdot) - I(x))$ l'application miroir ($I(x): T_xM \to T_xM$ est l'application identité), et on pose $u'(x,y) = 2\rho(x,y)u(x,y)$.

On a alors $m(x,y) = \gamma(1/2,x,y)$, $u'(x,y) = \dot{\gamma}(1/2,x,y)$, et pour $v \in T_xM$,

$$J\left(\frac{1}{2}, v, \mathscr{M}(x,y,v)\right) = J\left(\frac{1}{2}, p(x,y,v), \tau(x,y,p(x,y,v))\right)$$

et

$$\dot{J}\left(\frac{1}{2}, v, \mathscr{M}(x,y,v)\right) = \dot{J}\left(\frac{1}{2}, p(x,y,v), \tau(x,y,p(x,y,v))\right)$$
$$- 2\tau(x, m(x,y), v - p(x,y,v)).$$

On ne s'intéresse à (X,Y) qu'avant le temps de sortie T' de $T(X_0,Y_0)$. Le processus Y issu de $\mathscr{Y}$ est défini en résolvant l'équation d'Itô

$$d_\Gamma Y = \mathscr{M}(X,Y,d_\Gamma X).$$

Comme X satisfait l'équation

$$d_\Gamma X = \mathscr{M}(Y,X,d_\Gamma Y),$$

les processus $(X^{T'}, Y_0)$ et $(X_0, Y^{T'})$ engendrent la même filtration. On obtient ainsi (iii). De plus, Y est une martingale dans la filtration engendrée par (X_0,Y) et un mouvement brownien, donc on obtient (ii) grâce à la propriété de Markov.

Pour établir (iv), on peut conditionner par $(X_0,Y_0) = (x,y)$. Posant $U_t = u(X_t,Y_t)$, $U'_t = u'(X_t,Y_t)$, $M_t = m(X_t,Y_t)$ et $\rho_t = \rho(X_t,Y_t)$, on obtient pour les parties martingales

$$d_m M = J\left(\frac{1}{2}, d_m X, d_m Y\right)$$
$$= J\left(\frac{1}{2}, p(X,Y,d_m X), \tau(X,Y,p(X,Y,d_m X))\right),$$

$$\nabla_m U' = \dot{J}\left(\frac{1}{2}, d_m X, d_m Y\right)$$
$$= \dot{J}\left(\frac{1}{2}, p(X,Y,d_m X), \tau(X,Y,p(X,Y,d_m X))\right)$$
$$- 2\tau(X, M, d_m X - p(X,Y,d_m X)),$$

et

$$d_m \rho = \langle U, -\tau(X, M, d_m X - p(X,Y,d_m X)) \rangle. \tag{1}$$

Comme

$$\nabla_m U = \frac{1}{2\rho} \nabla_m U' - \frac{d_m \rho}{\rho} U,$$

on obtient

$$\nabla_m U = \frac{1}{2\rho_t} j\left(\frac{1}{2}, p(X,Y,d_m X), \tau(X,Y,p(X,Y,d_m X)) \right).$$

ce qui donne en définitive

$$
\begin{aligned}
d_m U = {} & h_U J\left(\frac{1}{2}, p(X,Y,d_m X), \tau(X,Y,p(X,Y,d_m X)) \right) \\
& + v_U \frac{1}{2\rho_t} j\left(\frac{1}{2}, p(X,Y,d_m X), \tau(X,Y,p(X,Y,d_m X)) \right)
\end{aligned}
\tag{2}
$$

où h_u (resp. v_u) : $T_{\pi(u)}M \to T_u TM$ désigne le relèvement horizontal (resp. vertical). De (1) on déduit que ρ_t peut s'écrire

$$d\rho_t = -dB_t + b(\rho_t, U_t)dt$$

où B est un mouvement brownien réel et $|b|$ est borné par une constante $C > 0$ qui ne dépend que de M et ε'. De (2) on déduit que dans des coordonnées centrées en $u(x, y)$,

$$dU_t^i = \sigma^i(\rho_t, U_t)dB_t' + c^i(\rho_t, U_t)dt,$$

où B' est un mouvement brownien de dimension $d - 1$ indépendant de B, σ et c sont bornés ainsi que leurs dérivées d'ordre 1, par une constante qui ne dépend que de M et ε'. Pour un temps t_0 suffisamment petit (en particulier $t_0 < 1/4$), la probabilité pour U de ne pas sortir d'un voisinage de U_0 de rayon ε' est supérieure à α_c'' proche de 1, ne dépendant que de M et ε'. D'autre part, avec une probabilité α_c''' strictement positive (éventuellement très petite), on a pour tout $t \in [0, t_0]$,

$$\rho_0 - B_t + Ct \le \rho_0 - \frac{\rho_0}{t_0}t + \varepsilon'.$$

Les deux événements étant indépendants, leur intersection a une probabilité supérieure à $\alpha_c' = \alpha_c'' \alpha_c'''$. Et comme

$$\rho_t \le \rho_0 - B_t + Ct$$

pour tout t inférieur au temps de sortie de $T(x, y)$, cette intersection est incluse dans F. Ceci achève la démonstration. □

Démonstration de la proposition A. — Soit $(X_t)_{t \ge 0}$ un mouvement brownien. On choisit un voisinage ouvert $W_{\mathscr{C}}$ de $\mathscr{C}$ de mesure inférieure au nombre ε donné par le lemme B.

On va définir par récurrence une suite de temps d'arrêt $(R_k)_{k \ge 0}$ et construire le processus Y sur les intervalles $[R_k, R_{k+1}]$.

On pose $R_0 = 0$. On a $(X, Y)_{R_0} = (X_0, y)$.

Soit $k \ge 0$. On suppose que R_j est construit pour $j \le k$ et que Y est construit jusqu'au temps d'arrêt R_k.

Si $Y_{R_k} = X_{R_k}$, alors on pose $Y_t = X_t$ pour $t \geq R_k$, $S = R_k$ et $R_n = R_k$ pour tout $n \geq k$.

Si $(X, Y)_{R_k} \in W_{\mathscr{C}}^c$, on construit Y par le lemme D (avec un couplage miroir) jusqu'au temps

$$R_{k+1} = \left(R_k + \frac{1}{2} \right) \wedge \inf \left\{ t > R_k, \ (X, Y)_t \in \mathscr{C} \cup \Delta \right\}.$$

Si $(X, Y)_{R_k} \in W_{\mathscr{C}}$, on construit Y par le lemme B ((X, Y) est alors une diffusion hypoelliptique) jusqu'au temps

$$R_{k+1} = \left(R_k + \frac{1}{2} \right) \wedge \inf \left\{ t > R_k, \ (X, Y)_t \notin W_{\mathscr{C}} \cap \beta((X, Y)_{R_k}, \varepsilon_h) \right\}$$

où ε_h est défini dans le lemme B (les champs de vecteurs servant à construire la diffusion hypoelliptique sont définis dans l'ouvert $\mathscr{B}_j$ de la preuve du lemme B, j étant le plus petit indice tel que $\beta((X, Y)_{R_k}, \varepsilon_h) \subset \mathscr{B}_j$).

On a ainsi construit (X, Y) jusqu'à un temps d'arrêt $R_{k+1} \leq 1/2 + R_k$. On peut donc recommencer la procédure en remplaçant k par $k+1$.

Partant de $W_{\mathscr{C}}^c$, le couplage a lieu en temps inférieur à $1/2$ avant de sortir de $\mathscr{C}^c$ avec une probabilité supérieure à α_c (lemme D). Partant de $W_{\mathscr{C}}$, (X, Y) entre dans $W_{\mathscr{C}}^c$ en temps inférieur à $1/2$ avec une probabilité supérieure à α_h (lemme B en prenant $A = W_{\mathscr{C}}$). On choisit $\alpha = \alpha_h \alpha_c$. Soit $(\mathcal{F}_t)_{t \geq 0}$ la filtration engendrée par X. Ainsi, pour $n \geq 0$, à l'aide d'un conditionnement par rapport à $\mathcal{F}_{R_n}$, on obtient $\mathbb{P}(S > R_{n+2}) < (1 - \alpha) \mathbb{P}(S > R_n)$, donc pour tout $n \geq 1$, $\mathbb{P}(S \leq R_{2n}) > 1 - (1 - \alpha)^n$. Comme $R_{2n} \leq n$, cela donne $\mathbb{P}(S \leq n) > 1 - (1 - \alpha)^n$, et le processus Y ainsi construit est bien défini pour tout $t \geq 0$ et est un mouvement brownien dans la filtration engendrée par X.

La détermination de X à partir de Y et X_0 se fait de la même façon, donc X est un mouvement brownien dans la filtration engendrée par (X_0, Y) ; cela donne (iii). De plus, Y est une martingale dans la filtration engendrée par (X_0, Y) et un mouvement brownien, donc on obtient (ii) grâce à la propriété de Markov. □

On peut maintenant établir le résultat principal.

PROPOSITION E. — *Soit M une variété riemannienne de classe C^∞, compacte et connexe, de dimension $d \geq 2$. La filtration naturelle de tout mouvement brownien indexé par $\mathbb{R}$ et à valeurs dans M est une filtration brownienne de dimension d.*

Démonstration. — On fixe une origine $O \in M$. On procède comme dans la preuve de la proposition 3 en apportant les modifications décrites dans la remarque qui suit cette preuve, et en remplaçant le lemme 2 par la proposition A. Comme fonction $t \mapsto f(t)$ qui majore $t \mapsto \mathbb{P}(S \geq t)$, on peut choisir $f(t) = (1 - \alpha)^{E(t)}$ où $E(t)$ est la partie entière de t. □

RÉFÉRENCES

[A,T]   Arnaudon (M.), Thalmaier (A.) — *Stability of stochastic differential equations in manifolds*, Séminaire de probabilités XXXII, Lecture Notes in mathematics 1686, 1998. p. 188–214.

[B,G,M]   Berger (M.), Gauduchon (P.), Mazet (E) — *Le spectre d'une variété riemannienne.* Lecture Notes in Mathematics, t. **194**, 1971.

[C]   Cranston (M.) — *Gradient estimates on manifolds using coupling*, Journal of Functional Analysis. t. **99**, 1991, p. 110–124.

[D]   Davies (E.B.) — *Heat Kernels and Spectral Theory.* — Cambridge Tracts in Mathematics 92, Cambridge University Press, 1990.

[E1]   Emery (M.) — *Stochastic calculus in manifolds.* — Springer, 1989.

[E2]   Emery (M.) — *On two transfer principles in stochastic differential geometry*, Séminaire de Probabilités XXIV, Lecture Notes in Mathematics, Vol 1426, Springer, 1989, p. 407–441.

[K]   Kendall (W.S.) — *Nonnegative Ricci curvature and the Brownian coupling property*, Stochastics, t. **19**, 1986, p. 111–129.

[L]   Léandre (R.) — *Majoration en temps petit de la densité d'une diffusion dégénérée*, Probability theory and Related Fields, t. **74**, 1987, p. 289–294.

[M]   Meyer (P.A.) — *Géométrie stochastique sans larmes,*, Séminaire de Probabilités XV, Lecture Notes in Mathematics, Vol 850, Springer, 1981.

A Stochastic Differential Equation with a Unique (up to Indistinguishability) but not Strong Solution

Jan Kallsen

Abstract

Fix a filtered probability space $(\Omega, \mathcal{F}, (\mathcal{F}_t)_{t \geq 0}, P)$ and a Brownian motion B on that space and consider any solution process X (on Ω) to a stochastic differential equation (SDE) $dX_t = f(t, X) \, dB_t + g(t, X) \, dt$ (1). A well-known theorem states that pathwise uniqueness implies that the solution X to SDE (1) is strong, i.e., it is adapted to the P-completed filtration generated by B. Pathwise uniqueness means that, on *any* filtered probability space carrying a Brownian motion and for any initial value, SDE (1) has at most one (weak) solution. We present an example that if we only assume that, for any initial value, there is at most one solution process on the *given* space $(\Omega, \mathcal{F}, (\mathcal{F}_t)_{t \geq 0}, P))$, we can no longer conclude that the solution X is strong.

1 Introduction

Consider the following stochastic differential equation (SDE)

$$X_t = X_0 + \int_0^t f(s, X) \, dB_s + \int_0^t g(s, X) \, ds, \tag{1.1}$$

where $f, g : \mathbb{R}^+ \times \mathbb{C}(\mathbb{R}^+) \to \mathbb{R}$ are predictable mappings and B denotes Brownian motion ($\mathbb{C}(\mathbb{R}^+) := \{f : \mathbb{R}^+ \to \mathbb{R} : f \text{ continuous}\}$ denotes Wiener space and predictability is defined as in Revuz & Yor (1994), IX,§1). There are at least two fundamentally different concepts of approaching SDE (1.1).

Firstly, one can start with a filtered probability space $(\Omega, \mathcal{F}, (\mathcal{F}_t)_{t \geq 0}, P)$ and a Brownian motion B on that space. SDE (1.1) is then interpreted as an equation only for processes defined on Ω and by B one always refers to the same Brownian motion on Ω. *Existence and uniqueness of a solution* means in this context that, for any initial value X_0, there is (up to indistinguishability) exactly one solution process on Ω satisfying Equation (1.1). This concept is applied e.g. by Protter (1992) and it easily extends to arbitrary semimartingales as driving processes.

Alternatively, one may regard SDE (1.1) independently of a fixed underlying probability space and a fixed Brownian motion. In this context, SDE (1.1) has a *(weak) solution* whenever there is a probability space and two processes X and B on that space such that B is a Brownian motion and Equation (1.1) holds for this particular choice. Here, the space $(\Omega, \mathcal{F}, (\mathcal{F}_t)_{t \geq 0}, P)$ and the Brownian motion are part of the solution. *Pathwise uniqueness* holds if, for any two solutions $(\Omega, \mathcal{F}, (\mathcal{F}_t)_{t \geq 0}, P, (X, B))$ and $(\tilde{\Omega}, \tilde{\mathcal{F}}, (\tilde{\mathcal{F}}_t)_{t \geq 0}, \tilde{P}, (\tilde{X}, \tilde{B}))$ with $(\Omega, \mathcal{F}, (\mathcal{F}_t)_{t \geq 0}, P) = (\tilde{\Omega}, \tilde{\mathcal{F}}, (\tilde{\mathcal{F}}_t)_{t \geq 0}, \tilde{P})$, $B = \tilde{B}$, and $X_0 = \tilde{X}_0$, the solutions X and $\tilde{X}$ are indistinguishable. The concept of weak solutions is discussed in many books (see e.g. Revuz & Yor (1994), Karatzas & Shreve (1991)). Clearly, a solution on a fixed space is always a weak solution. Also, *pathwise uniqueness* implies *uniqueness on a fixed space*. For a thorough account of both viewpoints see Jacod (1979).

Following Revuz & Yor (1994) we call a (weak) solution $(\Omega, \mathcal{F}, (\mathcal{F}_t)_{t \geq 0}, P, (X, B))$ *strong* if X is adapted to the P-completed filtration generated by the driving Brownian motion B. A well-known theorem due to Yamada & Watanabe (cf. Revuz & Yor (1994), Theorem IX.1.7; for a generalization to SDE's involving random measures see Jacod (1979), Théorème 14.94) states that pathwise uniqueness implies that any (weak) solution to SDE (1.1) is strong.

Now, consider the following situation. Starting from a fixed probability space $(\Omega, \mathcal{F}, (\mathcal{F}_t)_{t \geq 0}, P)$ and a fixed Brownian motion B, we are given a solution X to SDE (1.1) and we know that X is (up to indistinguishability) the only solution on that space starting in X_0. Is it, in general, true that X is a strong solution? (Note that we do not assume pathwise uniqueness, as pathwise uniqueness involves weak solutions on other spaces as well.) We give an example that the answer is *no*. More precisely, we present a SDE having no strong solution, having exactly one solution (for a fixed initial value) on some probabilty space and more than one solution on others. The example will be closely related to Tsirel'son's SDE (cf. Revuz & Yor (1994), p. 373).

We use the following notation: $[\cdot]$ denotes the integer part of a real number, λ is Lebesgue measure. For random variables U, V we write P^U, $P^{U|V}$, $P^{U|V=v}$ for the distribution (under P) of U, the conditional distribution of U given V, the factorisation of the conditional distribution of U given V, respectively. π_1 and $\pi_2 : \mathbb{R}^2 \to \mathbb{R}$ denote the projection on the first and the second coordinate.

2   The example

Consider the SDE

$$X_t = X_0 + B_t + \int_0^t \tau(s, X) \, ds, \tag{2.1}$$

where B stands for standard Brownian motion and $\tau : \mathbb{R}^+ \times \mathbb{C}(\mathbb{R}^+) \to \mathbb{R}$ is defined by

$$\tau(t, \omega) := \begin{cases} \alpha\left(\left\{\frac{\omega(t_k) - \omega(t_{k-1})}{t_k - t_{k-1}}\right\}\right) & \text{for } t_k < t \leq t_{k+1}, \\ 0 & \text{for } t = 0 \text{ or } t > 1, \end{cases}$$

where $\{x\}$ denotes x modulo 1, the function α is defined by $\alpha(x) := x 1_{[0,1/2)}(x) + (x + 1/4) 1_{[1/2,3/4)}(x) + (x - 1/4) 1_{[3/4,1)}(x)$, and $(t_k)_{k \in -\mathbb{N}}$ is a strictly increasing sequence of numbers such that $t_0 = 1$ and $\lim_{k \to -\infty} t_k = 0$.

As for Tsirel'son's example (where we have the identity instead of α) τ is predictable and bounded and a weak solution $(\Omega, \mathcal{F}, (\mathcal{F}_t)_{t \geq 0}, P, (X, B))$ to SDE (2.1) with $X_0 = 0$ exists (see e.g. Revuz & Yor (1994), Theorem IX.1.11).

By $(\mathcal{F}_t^B)_{t \geq 0}$ and $(\mathcal{F}_t^X)_{t \geq 0}$ we denote the P-completed natural filtrations of B resp. X. Let $(\tilde{X}_t)_{t \geq 0}$ be another weak solution defined on the same filtered probability space, with respect to the same Brownian motion B, and with $\tilde{X}_0 = X_0 = 0$. If we set for $t_k < t \leq t_{k+1}$

$$\eta_t := \frac{X_t - X_{t_k}}{t - t_k}, \quad \tilde{\eta}_t := \frac{\tilde{X}_t - \tilde{X}_{t_k}}{t - t_k}, \quad \varepsilon_t := \frac{B_t - B_{t_k}}{t - t_k},$$

we have that for $t_k < t \le t_{k+1}$

$$X_t = B_t + \sum_{l \le k} \alpha(\{\eta_{t_{l-1}}\})(t_l - t_{l-1}) + \alpha(\{\eta_{t_k}\})(t - t_k) \tag{2.2}$$

and hence

$$\eta_t = \varepsilon_t + \alpha(\{\eta_{t_k}\}), \tag{2.3}$$

and accordingly for $\tilde{X}$ and $\tilde{\eta}$.

Now the following statements hold:

1. For any $0 < s < t \le 1$ we have $\mathcal{F}_t^X = \sigma(\{\eta_s\}) \vee \mathcal{F}_t^B$.

2. For any $k \in -\mathbb{N}$ there is a probability measure ρ on $[0, 1/4)$ and constants $c_1, c_2 \ge 0$ with $c_1 + c_2 = 1$ (independent of k) such that the distribution of $(\{\eta_{t_k}\}, \{\tilde{\eta}_{t_k} - \eta_{t_k}\})$ is

$$\lambda|_{[0,1)} \otimes \left(\rho|_{(0,1/4)} * \frac{1}{4}(\epsilon_0 + \epsilon_{1/4} + \epsilon_{1/2} + \epsilon_{3/4}) \right.$$
$$\left. + c_1 \rho|_{\{0\}} + c_2 \rho|_{\{0\}} * \frac{1}{3}(\epsilon_{1/4} + \epsilon_{1/2} + \epsilon_{3/4}) \right),$$

   where the asterisk denotes convolution and ϵ_a is the Dirac measure in a.

3. For any $k \in -\mathbb{N}$ the random vector $(\{\eta_{t_k}\}, \{\tilde{\eta}_{t_k}\})$ is independent of $\mathcal{F}_1^B$.

4. X is not strong.

Since $(\Omega, \mathcal{F}, (\mathcal{F}_t^X)_{t \ge 0}, P, (X, B))$ is a weak solution of SDE (2.1), let us assume $(\mathcal{F}_t)_{t \ge 0} = (\mathcal{F}_t^X)_{t \ge 0}$ for the following. Then we have in addition:

5. For any $k \in -\mathbb{N}$ there is a measureable mapping $\beta : \mathbb{R} \to \mathbb{R}$ such that $\{\tilde{\eta}_{t_k}\} = \beta(\{\eta_{t_k}\})$ P-a.s.

6. $(X_t)_{t \ge 0}$ and $(\tilde{X}_t)_{t \ge 0}$ are indistinguishable.

7. On $(\Omega, \mathcal{F}, (\mathcal{F}_t^X)_{t \ge 0}, P)$ and for any $a \in \mathbb{R}$, the process $X^a := X + a$ is (up to indistinguishability) the unique solution to SDE (2.1) starting at a in $t = 0$, but it is not strong.

Remark. Statement 7 can be strengthened in that, for any $T > 0$ (and for any fixed initial value), there is no other process on that space solving SDE (2.1) on $[0, T]$.

3 Proofs

Proof of Statement 1.
The $\supset$-inclusion follows from the definitions and from Equation (2.2). Since (2.3) implies $\{\eta_{t_k}\} = \alpha(\{\eta_t - \varepsilon_t\})$ for $t_k < t \le t_{k+1}$, the inclusion "$\subset$" follows easily from Equation (2.2).

Proof of Statement 2.

We will proceed in four steps.

Step 1: Definition of several Markov kernels
We start by defining mappings

$$S^y : [0,4) \to [0,4), \quad x \mapsto \alpha(\{x\}) + [4\{\alpha(\{y + [x]/4 + x\}) - \alpha(\{x\})\}]$$

for any $y \in [0, 1/4)$. For $k = 0, \ldots, 15$ we set

$$A_k := \{(x,y) : y \in [0, 1/4), x \in [k/4, (k+1)/4 - y)\},$$

$$B_k := \{(x,y) : y \in [0, 1/4), x \in [(k+1)/4 - y, (k+1)/4)\}.$$

With this notation, we have the following simple graphical representation of the mapping $[0,4) \times [0, 1/4) \to [0, 4, 0, 1/4), (x, y) \mapsto (S^y(x), y)$. The image of

$\frac{1}{4}$	B_0	B_1	B_2	B_3	B_4	B_5	B_6	B_7	B_8	B_9	B_{10}	B_{11}	B_{12}	B_{13}	B_{14}	B_{15}
0	A_0	A_1	A_2	A_3	A_4	A_5	A_6	A_7	A_8	A_9	A_{10}	A_{11}	A_{12}	A_{13}	A_{14}	A_{15}

(with $0, 1, 2, 3, 4$ along the horizontal axis)

under this mapping is

$\frac{1}{4}$	B_0	B_5	B_{11}	B_6	B_8	B_1	B_3	B_{10}	B_4	B_9	B_7	B_2	B_{12}	B_{13}	B_{15}	B_{14}
0	A_0	A_1	A_3	A_2	A_4	A_9	A_{15}	A_{10}	A_{12}	A_5	A_7	A_{14}	A_8	A_{13}	A_{11}	A_6

(with $0, 1, 2, 3, 4$ along the horizontal axis).

Consider, for example, $(x, y) \in B_7$, i.e., $x \in [2 - y, 2)$. It follows

$$
\begin{aligned}
S^y(x) &= \alpha(\{x\}) + [4\{\alpha(\{y + [x]/4 + x\}) - \alpha(\{x\})\}] \\
&= x - 1 - 1/4 + [4\{\alpha(\{y + 1/4 + x\}) - (x - 1 - 1/4)\}] \\
&= x - 5/4 + [4\{y + 1/4 + x - 2 - (x - 5/4)\}] \\
&= x - 5/4 + 2 = x + 3/4.
\end{aligned}
$$

Hence, B_7 is shifted in x-direction by $3/4$ on B_{10}.

Now, we define Markov kernels K^y, $L^{y,b}$ (for any fixed $y \in [0, 1/4), b \in \mathbb{R}$) from $[0, 4)$ to $[0, 4)$ as follows:

$$L^{y,b}(x, A) := \epsilon_{\{b - \alpha(\{x\})\} + [S^y(x)]}(A) \text{ for } x \in [0, 4), A \in \mathcal{B}([0, 4))$$

and

$$K^y(x, A) := \int_A \kappa^y(x, x') \, dx' \text{ for } x \in [0, 4), A \in \mathcal{B}([0, 4))$$

with

$$\kappa^y(x, x') := \sum_{n \in \mathbb{Z}} \phi(\{x'\} + n - \alpha(\{x\})) 1_{\{0\}}([S^y(x)] - [x']),$$

where ϕ denotes the density of the standard normal distribution.

Lemma 3.1 *For $x \in [0, 1)$, we have $K^0(x, [1, 4)) = 0$. For $x \in [1, 4)$, we have $K^0(x, [0, 1))$* $= 0$.

Proof. Since $[S^0(x)] = 0$ for $x \in [0,1)$, we have $\kappa^0(x, x') = 0$ for $x \in [0,1)$, $x' \in [1,4)$, hence $K^0(x, [1,4)) = \int_{[1,4)} \kappa^0(x, x') \, dx' = 0$ for $x \in [0,1)$. From the graphical representation of S^y one observes $[S^0(x)] \neq 0$ for $x \in [1,4)$, hence $\kappa^0(x, x') = 0$ for $x \in [1,4)$, $x' \in [0,1)$. It follows $K^0(x, [0,1)) = \int_{[0,1)} \kappa^0(x, x') \, dx' = 0$ for $x \in [1,4)$. $\qquad \square$

Therefore, we can define Markov kernels K_1 from $[0,1)$ to $[0,1)$ and K_2 from $[1,4)$ to $[1,4)$ by

$$K_1(x, A) := K^0(x, A) \quad \text{for } x \in [0,1), A \in \mathcal{B}([0,1));$$
$$K_2(x, A) := K^0(x, A) \quad \text{for } x \in [1,4), A \in \mathcal{B}([1,4)).$$

Step 2: Fixed points of the Markov kernels defined in Step 1

Notation. Let I be an interval.

1. For any Markov kernel K from I to I, we denote the corresponding Markov operator $\mathcal{M}^1(I) \to \mathcal{M}^1(I)$ again by K (i.e., $KQ : A \mapsto \int_I K(x, A)Q(dx)$ for $Q \in \mathcal{M}^1(I) := \{Q : Q \text{ probability measure on } I\}$, $A \in \mathcal{B}(I)$).

2. We set $\mathcal{D}_I := \{g \in L^1(I) : g \geq 0, \int_I g \, d\lambda = 1\}$. If a Markov kernel K from I to I has a transition density $\kappa : I \times I \to \mathbb{R}^+$ (i.e., $K(x, A) = \int_A \kappa(x, x') \, dx'$), then we denote the mapping $L^1(I) \to L^1(I)$, $g \mapsto \kappa g$ with

$$(\kappa g)(x') := \int_I \kappa(x, x') g(x) \, dx,$$

   also by κ. Observe that $\kappa|_{\mathcal{D}_I} \subset \mathcal{D}_I$.

3. Powers of a transition density $\kappa : I \times I \to \mathbb{R}^+$ shall be defined recursively by $\kappa^1(x, x') := \kappa(x, x')$ and $\kappa^{n+1}(x, x') := \int_I \kappa(x'', x') \kappa^n(x, x'') \, dx''$.

Lemma 3.2 *1. For any $y \in [0, 1/4)$, $b \in \mathbb{R}$, the distribution $\frac{1}{4}\lambda|_{[0,4)} \in \mathcal{M}^1([0,4))$ is a fixed point of the Markov operators K^y and $L^{y,b}$.*

2. $\lambda|_{[0,1)} \in \mathcal{M}^1([0,1))$ is a fixed point of the Markov operator K_1.

3. $\frac{1}{3}\lambda|_{[1,4)} \in \mathcal{M}^1([1,4))$ is a fixed point of the Markov operator K_2.

4. For any $b \in \mathbb{R}$ and any $c_1, c_2 > 0$ with $c_1 + c_2 = 1$, the distribution $c_1 \lambda|_{[0,1)} + c_2 \frac{1}{3}\lambda|_{[1,4)} \in \mathcal{M}^1([0,4))$ is a fixed point of the Markov operator $L^{0,b}$.

Proof.

1. Fix $y \in [0, 1/4)$, $b \in \mathbb{R}$. For any $A \in \mathcal{B}([0,4))$, we have

$$
\begin{aligned}
K^y\left(\frac{1}{4}\lambda|_{[0,4)}\right)(A) &= \frac{1}{4} \int_{[0,4)} \int_A \kappa^y(x, x') \, dx' \, dx \\
&= \frac{1}{4} \int_A \int_{(S^y)^{-1}([[x'],[x']+1))} \sum_{n \in \mathbb{Z}} \phi(\{x'\} + n - \alpha(\{x\})) \, dx \, dx'
\end{aligned}
$$

$$= \frac{1}{4} \int_A \int_{[[x'],[x']+1)} \sum_{n \in \mathbb{Z}} \phi(\{x'\} + n - \alpha(\{S^y(x'')\})) \, dx'' \, dx'$$

$$= \frac{1}{4} \int_A \int_{[[x'],[x']+1)} \sum_{n \in \mathbb{Z}} \phi(\{x'\} + n - \{x''\}) \, dx'' \, dx'$$

$$= \frac{1}{4} \int_A dx' = \frac{1}{4} \lambda(A),$$

where the third equation follows from the fact that λ is invariant under $S^y = (S^y)^{-1}$ (because S^y is a permutation of the intervals A_k^y, B_k^y, $k = 0, \dots, 15$), and the fourth equation follows from $\{S^y(x'')\} = \alpha(\{x''\})$ for any $x'' \in [0,4)$. Similarly, we obtain for any $A \in \mathcal{B}([0,4))$

$$L^{y,b}\left(\frac{1}{4}\lambda|_{[0,4)}\right)(A) = \frac{1}{4}\lambda(A).$$

2., 3., and 4. follow along the same lines (Observe that $S^0(x) \in [0,1)$ for $x \in [0,1)$ and $S^0(x) \in [1,4)$ for $x \in [1,4)$). $\qquad \square$

Step 3: Convergence of iterates of the Markov kernels defined in Step 1

Lemma 3.3 *Let I be an interval, K a Markov kernel from I to I defined by a transition density $\kappa : I \times I \to \mathbb{R}$, and suppose that there are $j \in \mathbb{N}$, $s > 0$ such that $\kappa^j(x,x') > s$ for any $x, x' \in I$. Further assume that $\tilde{g}$ is a fixed point of $\kappa|_{\mathcal{D}_I}$. Then we have*

$$\sup_{g \in \mathcal{D}_I} \|\kappa^n g - \tilde{g}\|_{L^1(I)} \to 0 \text{ for } n \to \infty.$$

Proof. Since $\kappa^n g - \tilde{g} = \kappa^n(g - \tilde{g})$ and $\|\kappa^{n+1}(g - \tilde{g})\|_{L^1(I)} = \|\kappa(\kappa^n(g - \tilde{g}))\|_{L^1(I)} \le \|\kappa^n(g - \tilde{g})\|_{L^1(I)}$ for $g \in \mathcal{D}_I$, $n \in \mathbb{N}$ (cf. Lasota & Mackey (1985), Prop. 3.1.1), it suffices to show that $\|\kappa^j h\|_{L^1(I)} \le (1 - \lambda(I)s)\|h\|_{L^1(I)}$ for any $h \in L^1(I)$ with $\int_I h \, d\lambda = 0$.

Let $h \in L^1(I) \setminus \{0\}$ with $\int_I h \, d\lambda = 0$, and denote $c := \|h^+\|_{L^1(I)} = \|h^-\|_{L^1(I)}$. For any $g \in \mathcal{D}_I$, $x' \in I$, we have $\kappa^j g(x') = \int_I \kappa^j(x,x')g(x) \, dx \ge s$, hence $(\kappa^j g - s)^- = 0$. By $h^+/c \in \mathcal{D}_I$, $h^-/c \in \mathcal{D}_I$, it follows $\kappa^j(h^+/c) \ge s$, $\kappa^j(h^-/c) \ge s$. Therefore,

$$\|\kappa^j(h^+/c) - s\|_{L^1(I)} = \int_I (\kappa^j(h^+/c)(x) - s) \, dx = 1 - \lambda(I)s,$$

and accordingly $\|\kappa^j(h^-/c) - s\|_{L^1(I)} = 1 - \lambda(I)s$. Together, we obtain

$$\begin{aligned}
\|\kappa^j h\|_{L^1(I)} &= \|c(\kappa^j(h^+/c) - s) - c(\kappa^j(h^-/c) - s)\|_{L^1(I)} \\
&\le c(\|\kappa^j(h^+/c) - s\|_{L^1(I)} + \|\kappa^j(h^-/c) - s\|_{L^1(I)}) \\
&= 2c(1 - \lambda(I)s) = (1 - \lambda(I)s)\|h\|_{L^1(I)}.
\end{aligned}$$

$\qquad \square$

In order to apply the preceding lemma to the kernels K^y, K_1, K_2, we state

Lemma 3.4 *1. Let $y \in (0, 1/4)$. There is a $s > 0$ such that, for any $x, x' \in [0,4)$, we have $(\kappa^y)^3(x,x') > s$.*

2. *There is a $s > 0$ such that, for any $x, x' \in [0, 1)$, we have $\kappa^0(x, x') > s$ and, for any $x, x' \in [1, 4)$, we have $(\kappa^0)^3(x, x') > s$.*

Proof. Since the mapping $[0, 1] \times [0, 1] \to \mathbb{R}$, $(u, v) \mapsto \sum_{n \in \mathbb{Z}} \phi(u + n - v)$, is positive and continuous, it has a lower bound $m > 0$. Hence, we have $\kappa^y(x, x') \geq m$ for any $y \in [0, 1/4)$ and any $x, x' \in [0, 4)$ with $[S^y(x)] = [x']$.

For $y \in [0, 1/4)$, $k = 0, \ldots, 15$ define the sets

$$A_k^y := \{x : (x, y) \in A_k\}, \ B_k^y := \{x : (x, y) \in B_k\}.$$

In the following cases (among others), we have $\kappa^y(x, x') \geq m$:

$x \in (S^y)^{-1}([0, 1))$	, $x' \in B_1^y$	$x \in B_1^y$	, $x' \in [1, 2)$	$x \in B_5^y$	, $x' \in [0, 1)$
$x \in (S^y)^{-1}([1, 2))$	, $x' \in A_4^y$	$x \in A_4^y$	, $x' \in [1, 2)$	$x \in A_4^y$	, $x' \in [1, 2)$
$x \in (S^y)^{-1}([2, 3))$	, $x' \in A_9^y$	$x \in A_9^y$	, $x' \in [1, 2)$	$x \in A_5^y$	, $x' \in [2, 3)$
$x \in (S^y)^{-1}([3, 4))$	, $x' \in A_{15}^y$	$x \in A_{15}^y$	, $x' \in [1, 2)$	$x \in A_6^y$	, $x' \in [3, 4)$.

1. Fix $y \in [0, 1/4)$. There is a $\delta > 0$ such that $\lambda(A_k^y) > \delta$, $\lambda(B_k^y) > \delta$ for $k = 0, \ldots, 15$. Define $s := m^3 \delta^2$ and observe that, for $x' \in [0, 1)$,

$$\int_{[1,2)} \kappa^y(v, x') \, dv \geq \int_{B_5^y} m \, dv \geq m\delta,$$

and accordingly for $x' \in [1, 2)$, $[2, 3)$, $[3, 4)$ (with A_4^y, A_5^y, A_6^y instead of B_5^y). It follows for $x \in (S^y)^{-1}([0, 1))$, $x' \in [0, 4)$:

$$
\begin{aligned}
(\kappa^y)^3(x, x') &= \int_{[0,4)} \int_{[0,4)} \kappa^y(x, u)\kappa^y(u, v)\kappa^y(v, x') \, du \, dv \\
&\geq \int_{[0,4)} \int_{B_1^y} m\kappa^y(u, v)\kappa^y(v, x') \, du \, dv \\
&\geq \int_{B_1^y} du \int_{[1,2)} m^2 \kappa^y(v, x') \, dv \\
&\geq m^3 \delta^2,
\end{aligned}
$$

and accordingly for $x \in (S^y)^{-1}([1, 2))$, $x \in (S^y)^{-1}([2, 3))$, $x \in (S^y)^{-1}([3, 4))$ (with A_4^y, A_9^y, A_{15}^y instead of B_1^y).

2. Obviously, $\lambda(A_k^0) = 1/4$ for $k = 0, \ldots, 15$. Define $s := \min\{m, m^3/16\}$. For $x \in [0, 1)$, we have $S^0(x) = \alpha(\{x\})$, hence $[S^y(x)] = 0$. Therefore, $\kappa^0(x, x') = \sum_{n \in \mathbb{Z}} \phi(\{x'\} + n - \alpha(\{x\})) \geq m$ for any $x, x' \in [0, 1)$. The second statement follows as in 1. (but this time with $B_1^y = \emptyset = B_5^y$). $\square$

Corollary 3.5 *If we denote the transition densities of K_1, K_2 by κ_1, κ_2 (i.e., $\kappa_1 : [0, 1) \times [0, 1) \to \mathbb{R}$, $\kappa_1(x, x') = \kappa(x, x')$; $\kappa_2 : [1, 4) \times [1, 4) \to \mathbb{R}$, $\kappa_2(x, x') = \kappa(x, x')$), we obtain*

1. $\sup_{g \in \mathcal{D}_{[0,4)}} \|(\kappa^y)^n g - 1/4\|_{L^1([0,4))} \to 0$ *for $n \to \infty$, for any $y \in (0, 1/4)$,*

2. $\sup_{g \in \mathcal{D}_{[0,1)}} \|(\kappa_1)^n g - 1\|_{L^1([0,1))} \to 0$ *for $n \to \infty$,*

3. $\sup_{g \in \mathcal{D}_{[1,4)}} \|(\kappa_2)^n g - 1/3\|_{L^1([1,4))} \to 0$ *for* $n \to \infty$.

Proof. Lemma 3.2, Lemma 3.3, Lemma 3.4. □

Step 4: The joint distribution of $(\{\eta_{t_k}\}, \{\tilde{\eta}_{t_k} - \eta_{t_k}\})$

Define the mapping $\psi : [0,1) \times [0,1) \to [0,4) \times [0,1/4)$ by $(x', y') \mapsto (x' + [4y'], \{4y'\}/4)$. ψ is a bijection with converse $\psi^{-1} : [0,4) \times [0,1/4) \to [0,1) \times [0,1)$, $(x,y) \mapsto (\{x\}, y + [x]/4)$. Further we define, for any probability measure Q on $[0,4) \times [0,1/4)$, the Markov kernels $\bar{K}(Q)$ and, for any $b \in \mathbb{R}$, $\bar{L}^b(Q)$ from $[0,1/4)$ to $[0,4)$ by

$$\bar{K}(Q)(y,A) := (K^y Q^{\pi_1|\pi_2=y})(A) = \int K^y(x,A)\, Q^{\pi_1|\pi_2=y}(dx),$$

$$\bar{L}(Q)(y,A) := (L^{y,b} Q^{\pi_1|\pi_2=y})(A) = \int L^{y,b}(x,A)\, Q^{\pi_1|\pi_2=y}(dx)$$

for any $y \in [0,1/4)$, $A \in \mathcal{B}([0,4))$. One easily checks that $\bar{K}(Q)$, $\bar{L}^b(Q)$ are indeed Markov kernels. For any $k \in -\mathbb{N}$, we denote by μ_k the distribution of $\psi(\{\eta_{t_k}\}, \{\tilde{\eta}_{t_k} - \eta_{t_k}\})$.

Lemma 3.6 *For any* $k \in -\mathbb{N}$, *we have* $\mu_k = \bar{K}(\mu_{k-1}) \otimes \mu_{k-1}^{\pi_2}$.

Proof. For $k \in -\mathbb{N}$ let $(U_k, V_k) := \psi(\{\eta_{t_k}\}, \{\tilde{\eta}_{t_k} - \eta_{t_k}\})$. Then we have

$$
\begin{aligned}
(U_k, V_k) &= \psi(\{\eta_{t_k}\}, \{\tilde{\eta}_{t_k} - \eta_{t_k}\}) \\
&= \psi(\{\varepsilon_{t_k} + \alpha(\{\eta_{t_{k-1}}\})\}, \{\alpha(\{\tilde{\eta}_{t_{k-1}}\}) - \alpha(\{\eta_{t_{k-1}}\})\}) \\
&= (\{\varepsilon_{t_k} + \alpha(\{U_{k-1}\})\} + [4\{\alpha(\{U_{k-1} + V_{k-1} + [U_{k-1}]/4\}) - \alpha(\{U_{k-1}\})\}], \\
&\qquad \{4\{\alpha(\{\tilde{\eta}_{t_{k-1}}\}) - \alpha(\{\eta_{t_{k-1}}\})\}\}/4) \\
&= (\{\varepsilon_{t_k} + \alpha(\{U_{k-1}\})\} + [S^{V_{k-1}}(U_{k-1})], \{4\{\tilde{\eta}_{t_{k-1}} - \eta_{t_{k-1}}\}\}/4) \\
&= (\{\varepsilon_{t_k} + \alpha(\{U_{k-1}\})\} + [S^{V_{k-1}}(U_{k-1})], V_{k-1}).
\end{aligned}
\tag{3.1}
$$

Since ε_{t_k} is independent of $\mathcal{F}_{t_{k-1}}$ and $N(0,1)$-distributed, we have for any $A \in \mathcal{B}([0,4) \times [0,1/4))$

$$
\begin{aligned}
\mu_k(A) &= \iint \mathbf{1}_A(\{w + \alpha(\{u\})\} + [S^v(u)], v)\, \mu_{k-1}(d(u,v))\phi(w)\, dw \\
&= \iint \mathbf{1}_A(\{w'\} + [S^v(u)], v)\phi(w' - \alpha(\{u\}))\, \mu_{k-1}(d(u,v))\, dw' \\
&= \iint_{[0,1)} \sum_{n \in \mathbb{Z}} \mathbf{1}_A(w' + [S^v(u)], v)\phi(w' + n - \alpha(\{u\}))\, dw'\, \mu_{k-1}(d(u,v)) \\
&= \iint_{[[S^v(u)],[S^v(u)]+1)} \sum_{n \in \mathbb{Z}} \mathbf{1}_A(w'', v)\phi(\{w''\} + n - \alpha(\{u\}))\, dw''\, \mu_{k-1}(d(u,v)) \\
&= \iiint_{\mathbb{R}} \mathbf{1}_A(w'', v) \sum_{n \in \mathbb{Z}} \phi(\{w''\} + n - \alpha(\{u\})) \mathbf{1}_{\{0\}}([w''] - [S^v(u)])\, dw''\, \mu_{k-1}(d(u,v)) \\
&= \iiint_{[0,4)} \mathbf{1}_A(w'', v)\kappa^v(u, w'')\, dw''\, \mu_{k-1}^{\pi_1|\pi_2=v}(du)\, \mu_{k-1}^{\pi_2}(dv) \\
&= (\bar{K}(\mu_{k-1}) \otimes \mu_{k-1}^{\pi_2})(A).
\end{aligned}
$$

□

Lemma 3.7 *Fix $k \in -\mathbb{N}$. Then we have:*

1. $\mu_k^{\pi_2}$ *does not depend on k. We denote this distribution by ρ.*

2. $\mu_k^{\pi_1|\pi_2=y} = \frac{1}{4}\lambda|_{(0,4)}$ *ρ-a.s. for $y \in (0,1/4)$. There are constants $c_1, c_2 \geq 0$ with $c_1 + c_2 = 1$ and such that $\mu_k^{\pi_1|\pi_2=0} = c_1\lambda|_{[0,1)} + c_2\frac{1}{3}\lambda|_{[1,4)}$ ρ-a.s. In addition, c_1, c_2 are independent of k.*

3. μ_k *does not depend on k. We write $\mu := \mu_k$.*

Proof.

1. This follows by induction from Lemma 3.6.

2. Since, by Lemma 3.6, $\mu_{k'}^{\pi_1|\pi_2=y}(A) = \int_A \int \kappa^y(x, x') \mu_{k'-1}^{\pi_1|\pi_2=y}(dx) dx'$ for ρ-almost all $y \in [0, 1/4)$, $A \in \mathcal{B}([0,4))$, we conclude that $\mu_{k'}^{\pi_1|\pi_2=y}$ has a Lebesgue density $g_{k'}^y \in \mathcal{D}_{[0,4)}$ for any $k' \in -\mathbb{N}$.

It suffices to show:

$$\|g_k^y - 1/4\|_{L^1((0,4))} = 0 \text{ for } \rho\text{-a.a. } y \in (0, 1/4)$$

and, if $\rho(\{0\}) > 0$, then there are $c_1, c_2 \geq 0$ with $c_1 + c_2 = 1$ such that

$$\left\|g_k^0 - \left(c_1 1_{[0,1)} + c_2\frac{1}{3}1_{[1,4)}\right)\right\|_{L^1((0,4))} = 0.$$

By Lemma 3.6 and induction, one has that for any $l \in \mathbb{N}$:

$$g_k^y = (\kappa^y)^l g_{k-l}^y \ \lambda\text{-a.s. for } \rho\text{-a.a. } y \in (0, 1/4)$$

and

$$g_k^0(\cdot) = (\kappa^y)^l g_{k-l}^0 = (\kappa_1)^l g_{k-l}^0|_{[0,1)} 1_{[0,1)}(\cdot) + (\kappa_2)^l g_{k-l}^0|_{[1,4)} 1_{[1,4)}(\cdot) \ \lambda\text{-a.s.},$$

hence

$$g_k^0(\cdot)|_{[0,1)} = (\kappa_1)^l g_{k-l}^0|_{[0,1)} \ \lambda\text{-a.s.}, \ \ g_k^0(\cdot)|_{[1,4)} = (\kappa_2)^l g_{k-l}^0|_{[1,4)} \ \lambda\text{-a.s.} \quad (3.2)$$

Let $\varepsilon > 0$ and choose $l \in \mathbb{N}$ big enough to ensure

$$\sup_{g \in \mathcal{D}_{[0,4)}} \|(\kappa^y)^l g - 1/4\|_{L^1((0,4))} < \varepsilon \text{ for } y \in (0, 1/4)$$

and

$$\sup_{g \in \mathcal{D}_{[0,1)}} \|(\kappa_1)^l g - 1\|_{L^1((0,1))} < \varepsilon, \quad \sup_{g \in \mathcal{D}_{[1,4)}} \|(\kappa_2)^l g - 1/3\|_{L^1((1,4))} < \varepsilon.$$

Then we obtain

$$\|g_k^y - 1/4\|_{L^1((0,4))} = \|(\kappa^y)^l g_{k-l}^y - 1/4\|_{L^1((0,4))} < \varepsilon \text{ for } \rho\text{-a.a. } y \in (0, 1/4).$$

We define real numbers $c_1 := \mu_{k-l}^0([0,1)) = \int_{[0,1)} g_{k-l}^0(x)\,dx$ and $c_2 := \mu_{k-l}^0([1,4)) = \int_{[1,4)} g_{k-l}^0(x)\,dx$. By Equation (3.2), c_1, c_2 are independent of l. Since $\frac{1}{c_1} g_{k-l}^0|_{[0,1)} \in \mathcal{D}_{[0,1)}$ and $\frac{1}{c_2} g_{k-l}^0|_{[1,4)} \in \mathcal{D}_{[1,4)}$, we obtain

$$
\left\| g_k^0 - \left(c_1 1_{[0,1)} + c_2 \frac{1}{3} 1_{[1,4)} \right) \right\|_{L^1([0,4))}
$$

$$
= \left\| g_k^0 - c_1 \right\|_{L^1([0,1))} + \left\| g_k^0 - c_2 \frac{1}{3} \right\|_{L^1([1,4))}
$$

$$
= c_1 \left\| (\kappa_1)^l \left(\frac{1}{c_1} g_{k-l}^0 \right) - 1 \right\|_{L^1([0,1))} + c_2 \left\| (\kappa_2)^l \left(\frac{1}{c_2} g_{k-l}^0 \right) - \frac{1}{3} \right\|_{L^1([1,4))}
$$

$$
< c_1 \varepsilon + c_2 \varepsilon = \varepsilon.
$$

3. This follows immediately from 1. and 2. □

Corollary 3.8 ρ, c_1, c_2 *satisfy Statement 2 in Section 2.*

Proof. Fix $k \in -\mathbb{N}$. By ν we denote the distribution of $(\{\eta_{t_k}\}, \{\tilde{\eta}_{t_k} - \eta_{t_k}\})$, i.e., $\nu = \psi^{-1}(\mu)$. Then we have for any $A \in \mathcal{B}([0,1) \times [0,1))$:

$$
\nu(A) = \mu(\psi(A))
$$

$$
= \mu(A \cap ([0,1) \times [0,1/4))) + \mu((A \cap ([0,1) \times [1/4,1/2))) + (1,-1/4))
$$

$$
+ \mu((A \cap ([0,1) \times [1/2,3/4))) + (2,-1/2))
$$

$$
+ \mu((A \cap ([0,1) \times [3/4,1))) + (3,-3/4))
$$

$$
= \left(\frac{1}{4} \lambda|_{[0,1)} \otimes \mu^{\pi_2}|_{(0,1/4)} \right)(A) + \left(c_1 \lambda|_{[0,1)} \otimes \mu^{\pi_2}|_{\{0\}} \right)(A)
$$

$$
+ \left(\frac{1}{4} \lambda|_{[0,1)} \otimes (\mu^{\pi_2}|_{(0,1/4)} * \epsilon_{1/4}) \right)(A) + \left(c_2 \frac{1}{3} \lambda|_{[0,1)} \otimes (\mu^{\pi_2}|_{\{0\}} * \epsilon_{1/4}) \right)(A)
$$

$$
+ \left(\frac{1}{4} \lambda|_{[0,1)} \otimes (\mu^{\pi_2}|_{(0,1/4)} * \epsilon_{1/2}) \right)(A) + \left(c_2 \frac{1}{3} \lambda|_{[0,1)} \otimes (\mu^{\pi_2}|_{\{0\}} * \epsilon_{1/2}) \right)(A)
$$

$$
+ \left(\frac{1}{4} \lambda|_{[0,1)} \otimes (\mu^{\pi_2}|_{(0,1/4)} * \epsilon_{3/4}) \right)(A) + \left(c_2 \frac{1}{3} \lambda|_{[0,1)} \otimes (\mu^{\pi_2}|_{\{0\}} * \epsilon_{3/4}) \right)(A)
$$

$$
= \lambda|_{[0,1]} \otimes \left(\rho|_{(0,1/4)} * \frac{1}{4}(\epsilon_0 + \epsilon_{1/4} + \epsilon_{1/2} + \epsilon_{3/4}) \right.
$$

$$
\left. + c_1 \rho|_{\{0\}} + c_2 \rho|_{\{0\}} * \frac{1}{3}(\epsilon_{1/4} + \epsilon_{1/2} + \epsilon_{3/4}) \right)(A).
$$

□

Proof of Statement 3.
Fix $k \in -\mathbb{N}$. It suffices to show that $\psi(\{\eta_{t_k}\}, \{\tilde{\eta}_{t_k} - \eta_{t_k}\})$ is independent of $(B_t - B_{t_{k-l}})_{t_{k-l} < t \leq 1}$ for any $l \in \mathbb{N}$. This follows from

Lemma 3.9 *For any* $l \in \mathbb{N}$, μ *is a version of* $P^{\psi(\{\eta_{t_k}\},\{\tilde{\eta}_{t_k}-\eta_{t_k}\})|(B_t-B_{t_{k-l}})_{t_{k-l}<t\leq 1}}$.

Proof. We prove the lemma by induction on l. For $l = 0$ this holds, because $(B_t - B_{t_k})_{t_k < t \leq 1}$ is independent of $\mathcal{F}_{t_k}$. Assume that the lemma holds for $l - 1 \in \mathbb{N}$ (regardless of k). By Equation (3.1), by assumption, and by Lemmas 3.7.2 and 3.2, we have for any $A \in \mathcal{B}([0, 4) \times [0, 1/4))$:

$$P^{\psi(\{\eta_{t_k}\}, \{\tilde{\eta}_{t_k} - \eta_{t_k}\}) | (B_t - B_{t_{k-l}})_{t_{k-l} < t \le 1}}(A)$$

$$= \int 1_A(\{\varepsilon_{t_k} + \alpha(\{u\})\} + [S^v(u)], v) \, P^{\psi(\{\eta_{t_{k-1}}\}, \{\tilde{\eta}_{t_{k-1}} - \eta_{t_{k-1}}\}) | (B_t - B_{t_{k-l}})_{t_{k-l} < t \le 1}}(d(u, v))$$

$$= \int 1_A(\{\varepsilon_{t_k} + \alpha(\{u\})\} + [S^v(u)], v) \, \mu(d(u, v))$$

$$= \iiint 1_A(u', v) \, \epsilon_{\{\varepsilon_{t_k} + \alpha(\{u\})\} + [S^v(u)]}(du') \, \mu^{\pi_1 | \pi_2 = v}(du) \, \mu^{\pi_2}(dv)$$

$$= (\bar{L}^{\varepsilon_{t_k}}(\mu) \otimes \mu^{\pi_2})(A)$$

$$= \mu(A)$$

$\square$

Proof of Statement 4.
This follows easily from 3. and the fact that the distribution of $\{\eta_{t_k}\}$ is $\lambda|_{[0,1)}$ and hence not degenerate.

Proof of Statement 5.
We have

$$\{\tilde{\eta}_{t_k}\} = E(\{\tilde{\eta}_{t_k}\} | \mathcal{F}_1^X) = E(\{\tilde{\eta}_{t_k}\} | \sigma(\{\eta_{t_k}\}) \vee \mathcal{F}_1^B) = E(\{\tilde{\eta}_{t_k}\} | \sigma(\{\eta_{t_k}\})) \; P\text{-a.s.},$$

where the third equality follows from Statement 3 (cf. Bauer (1991), Satz 15.5).

Proof of Statement 6.
Observe that

$$\epsilon_{\{\beta(\{\eta_{t_k}\}) - \{\eta_{t_k}\}\}} = P^{\{\beta(\{\eta_{t_k}\}) - \{\eta_{t_k}\}\} | \{\eta_{t_k}\}}$$

$$= P^{\{\tilde{\eta}_{t_k} - \eta_{t_k}\} | \{\eta_{t_k}\}}$$

$$= \left(\rho|_{(0,1/4)} * \frac{1}{4}(\epsilon_0 + \epsilon_{1/4} + \epsilon_{1/2} + \epsilon_{3/4}) \right.$$

$$\left. + c_1 \rho|_{\{0\}} + c_2 \rho|_{\{0\}} * \frac{1}{3}(\epsilon_{1/4} + \epsilon_{1/2} + \epsilon_{3/4}) \right).$$

This is only possible if $\rho|_{(0,1/4)} = 0$, $c_2 = 0$ and hence $\{\tilde{\eta}_{t_k}\} = \beta(\{\eta_{t_k}\}) = \{\eta_{t_k}\}$ P-a.s. By Equations (2.3), (2.2), and according equations for $\tilde{X}$, we conclude that $\{\tilde{\eta}_{t_l}\} = \{\eta_{t_l}\}$ P-a.s. for any $l \in -\mathbb{N}$ and therefore $\tilde{X}_t = X_t$ P-a.s. for any $t \ge 0$.

Proof of Statement 7.
X^a is obviously a solution to SDE (2.1). By 4., it is not strong. For any solution $\tilde{X}^a$ starting at a, the process $\tilde{X}^a - a$ is a solution starting at 0 and hence indistinguishable from X. Thus, $\tilde{X}^a$ is indistinguishable from X^a.

Proof of the remark.
The whole proof works analogously if all processes are restricted to $[0, T]$ for any $T > 0$.

References

Bauer, H., (1991). *Wahrscheinlichkeitstheorie*, 4th edn. De Gruyter, Berlin.

Jacod, J., (1979). *Calcul Stochastique et Problèmes de Martingales*, Lecture Notes in Mathematics, vol. 714. Springer, Berlin.

Karatzas, I. and Shreve, S., (1991). *Brownian Motion and Stochatic Calculus*, 2nd edn. Springer, New York.

Lasota, A. and Mackey, M., (1985). *Probabilistic Properties of Deterministic Systems*. Cambridge University Press, Cambridge.

Protter, P., (1992). *Stochastic Integration and Differential Equations*, 2nd edn. Springer, Berlin.

Revuz, D. and Yor, M., (1994). *Continuous Martingales and Brownian Motion*, 2nd edn. Springer, Berlin.

Tsirel'son, B., (1975). An Example of a Stochastic Differential Equation Having No Strong Solution. *Theory of Probability and Applications* 20, 416-418.

Some Remarks on the Uniform Integrability of Continuous Martingales

Koichiro TAKAOKA

In this article we show a property on the tails of the supremum and the quadratic variation of real-valued continuous (local) martingales, and furthermore use the property to give a characterization of uniform integrable martingales. Our result refines or generalizes the main theorems of the following three papers: Azéma-Gundy-Yor [1], Elworthy-Li-Yor [2], and the continuous martingale version of Galtchouk-Novikov [4]. The present article is also closely related to a recent paper of Elworthy-Li-Yor [3].

We should mention two more works on related topics. H. Sato [7] gave a result on the uniform integrability of stochastically continuous additive martingales. Concerning exponential local martingales, see Kazamaki's book [5] and its references to earlier papers.

The author obtained the idea of using the technique in Step 4 of the proof of our main theorem when he attended a course on mathematical finance given by Professor Freddy Delbaen in Tokyo, February–March 1998. The author also would like to thank Professor Marc Yor and Professor Kohei Uchiyama for their helpful comments.

Our result is the following

Theorem. *Let $M = (M_t)_{t \in \mathbf{R}_+}$ be a real-valued continuous local martingale, with $M_0 = 0$, on a certain filtered probability space satisfying the usual conditions. Assume $M_\infty \overset{\text{def}}{=} \lim_{t \to \infty} M_t$ exists a.s. and $E[\,|M_\infty|\,] < \infty$. Then both*

$$\ell \overset{\text{def}}{=} \lim_{\lambda \to \infty} \lambda\, P\big[\, \sup_t |M_t| > \lambda \,\big] \quad and \quad \sigma \overset{\text{def}}{=} \lim_{\lambda \to \infty} \lambda\, P\big[\, \langle M \rangle_\infty^{1/2} > \lambda \,\big]$$

exist in $\mathbf{R}_+ \cup \{\infty\}$, and satisfy

$$\ell = \sqrt{\frac{\pi}{2}}\, \sigma = \sup_{U \in T(M)} E\big[\,|M_U|\,\big] - E\big[\,|M_\infty|\,\big],$$

where $T(M)$ is the set of all reducing stopping times for M. Furthermore, M is a uniformly integrable martingale if and only if $\ell = \sigma = 0$.

Remarks.

1. The expression $\sup_{U \in T(M)} E\big[\,|M_U|\,\big]$ is less complicated than it looks since, as we will see later in this paper,

$$\sup_{U \in \mathcal{T}(M)} E\big[|M_U|\big] = \sup_{U \in \mathcal{T}} E\big[|M_U|\big] = E[L_\infty],$$

where $\mathcal{T}$ is the set of all stopping times and L_t is the local time of M at the origin up to time t.

2. Azéma-Gundy-Yor [1] proved the equivalence

"M is a u.i. martingale " $\Leftrightarrow$ "ℓ exists and $\ell = 0$ " $\Leftrightarrow$ "σ exists and $\sigma = 0$ "

under the assumptions that M is a martingale and that $\sup_t E\big[|M_t|\big] < \infty$.

3. The continuous martingale version of the main theorem of Galtchouk-Novikov [4] shows that, under the same assumptions as ours, $E[M_\infty] = 0$ is implied by a condition weaker than "σ exists and $\sigma = 0$ " and stronger than "$\liminf_{\lambda \to \infty} \lambda P\big[\langle M \rangle_\infty^{1/2} > \lambda\big] = 0$. "

4. Theorem 1 of Elworthy-Li-Yor [2] and Lemma 1 of Galtchouk-Novikov [4] use a variant of the Tauberian theorem to prove that if M is bounded below (actually in a somewhat more general setting), then the two limits ℓ and σ exist and satisfy $\ell = \sqrt{\frac{\pi}{2}}\sigma = -E[M_\infty]$. Note that in this case our result agrees with theirs. Indeed, if $(M_t^-)_t$ is of class (D), then $M_{T_n}^- \to M_\infty^-$ in L^1 for every sequence T_n in $\mathcal{T}(M)$ increasing to ∞ a.s., so we find that

$$\begin{aligned}
\sup_{U \in \mathcal{T}(M)} E\big[|M_U|\big] - E\big[|M_\infty|\big] &= \lim_{n \to \infty} E\big[|M_{T_n}|\big] - E\big[|M_\infty|\big] \\
&= 2 \lim_{n \to \infty} E\big[M_{T_n}^-\big] - E\big[|M_\infty|\big] \\
&= 2 E\big[M_\infty^-\big] - E\big[|M_\infty|\big] \\
&= -E\big[M_\infty\big]
\end{aligned}$$

(for the validity of the first equality see Lemma below).

5. The assumption $E\big[|M_\infty|\big] < \infty$ is essential for our theorem. See Azéma-Gundy-Yor [1] for an example of how things would go wrong without this hypothesis.

The rest of this paper is devoted to the proof of our theorem. We need the following easy lemma.

Lemma. *For every sequence $(T_n)_{n=1}^\infty$ in $\mathcal{T}(M)$ increasing to ∞ a.s., we have*

$$\lim_{n \to \infty} E\big[|M_{T_n}|\big] = \sup_n E\big[|M_{T_n}|\big] = \sup_{U \in \mathcal{T}(M)} E\big[|M_U|\big] = \sup_{U \in \mathcal{T}} E\big[|M_U|\big],$$

where $\mathcal{T}$ is the set of all stopping times.

Proof. For every stopping time U, observe that $U \wedge T_n \in \mathcal{T}(M)$ and that $M_U = \lim_{n \to \infty} M_{U \wedge T_n}$ a.s. Hence, by Fatou's lemma,

$$E\big[|M_U|\big] \le \lim_{n \to \infty} E\big[|M_{U \wedge T_n}|\big] \le \lim_{n \to \infty} E\big[|M_{T_n}|\big]. \qquad \square$$

Proof of the main theorem. We divide the proof into six steps.

Step 1. We first show the existence of the limit ℓ and the equality $\ell = \sup_{U \in T(M)} E[|M_U|] - E[|M_\infty|]$. For $\lambda > 0$, define the stopping time

$$T_\lambda \overset{\text{def}}{=} \inf\{t : |M_t| > \lambda\}; \qquad (\inf \emptyset \overset{\text{def}}{=} \infty)$$

then $T_\lambda \in T(M)$ and

$$E[|M_{T_\lambda}|] = \lambda P[\sup_t |M_t| > \lambda] + E[|M_\infty|; \sup_t |M_t| \le \lambda].$$

Here the left-hand side increases with λ, and the second term on the right-hand side converges to $E[|M_\infty|]$ $(< \infty)$ as $\lambda \to \infty$. Therefore the limit of the first term on the right-hand side also exists. The desired equality also follows from this together with the above Lemma.

Step 2. For the proof of the equivalence "M is a u.i. martingale " $\Leftrightarrow$ "$\ell = 0$", we make the following observation:

$$\ell = 0 \quad \Leftrightarrow \quad \sup_{U \in T(M)} E[|M_U|] = E[|M_\infty|] \quad \text{(by Step 1)}$$

$$\Leftrightarrow \quad \lim_{n \to \infty} E[|M_{T_n}|] = E[|M_\infty|] \text{ for every sequence } T_n \text{ in } T(M)$$

increasing to ∞ a.s. (by Lemma)

$$\Leftrightarrow \quad \lim_{n \to \infty} M_{T_n} = M_\infty \text{ in } L^1 \text{ for every sequence } T_n \text{ in } T(M)$$

increasing to ∞ a.s. (since $E[|M_\infty|] < \infty$)

$$\Leftrightarrow \quad M \text{ is a u.i. martingale.}$$

Step 3. Next we show the inequality

$$\ell \le \sqrt{2\pi} \liminf_{\lambda \to \infty} \lambda P[\langle M \rangle_\infty^{1/2} > \lambda],$$

which gives the implication "$\ell = \infty$" $\Rightarrow$ "σ exists and $\sigma = \infty$." We apply an argument similar to the proof of the main theorem of Galtchouk-Novikov [4]. For $x > 0$, define the stopping time

$$S_x \overset{\text{def}}{=} \inf\{t : M_t > x\}. \qquad (\inf \emptyset \overset{\text{def}}{=} \infty)$$

Since $(M_{t \wedge S_x})_t$ is a continuous local martingale bounded above, it is proved in Elworthy-Li-Yor [2] and Galtchouk-Novikov [4] that

$$\sqrt{\tfrac{\pi}{2}} \lim_{\lambda \to \infty} \lambda P[\langle M \rangle_{S_x}^{1/2} > \lambda] = E[M_{S_x}]$$

(see Remark 4 after the statement of our Theorem). Also, observe that

$$E[M_{S_x}] = x P[\sup_t M_t > x] + E[M_\infty; \sup_t M_t \le x],$$

and thus

$$\sqrt{\tfrac{\pi}{2}} \liminf_{\lambda \to \infty} \lambda P\big[\langle M\rangle_\infty^{1/2} > \lambda\big]$$

$$\geq \sqrt{\tfrac{\pi}{2}} \lim_{\lambda \to \infty} \lambda P\big[\langle M\rangle_{S_x}^{1/2} > \lambda\big]$$

$$= x P\big[\sup_t M_t > x\big] + E\big[M_\infty\,;\, \sup_t M_t \leq x\big].$$

Likewise, replacing $(M_t)_t$ with $(-M_t)_t$ we have

$$\sqrt{\tfrac{\pi}{2}} \liminf_{\lambda \to \infty} \lambda P\big[\langle M\rangle_\infty^{1/2} > \lambda\big]$$

$$\geq x P\big[\inf_t M_t < -x\big] - E\big[M_\infty\,;\, \inf_t M_t \geq -x\big].$$

Therefore

$$\sqrt{2\pi} \liminf_{\lambda \to \infty} \lambda P\big[\langle M\rangle_\infty^{1/2} > \lambda\big]$$

$$\geq x\Big\{P\big[\sup_t M_t > x\big] + P\big[\inf_t M_t < -x\big]\Big\}$$

$$+ E\big[M_\infty\,;\, \sup_t M_t \leq x\big] - E\big[M_\infty\,;\, \inf_t M_t \geq -x\big]$$

$$\geq x P\big[\sup_t |M_t| > x\big]$$

$$+ E\big[M_\infty\,;\, \sup_t M_t \leq x\big] - E\big[M_\infty\,;\, \inf_t M_t \geq -x\big],$$

and by letting $x \to \infty$ we get the desired inequality.

Step 4. In this step we make some preparations for the proof of the existence of σ (in Step 5) and the equality $\sqrt{\tfrac{\pi}{2}}\sigma = \sup_{U \in T(M)} E\big[|M_U|\big] - E\big[|M_\infty|\big]$ (in Step 6). By virtue of Step 3, we need to consider only the case $\ell < \infty$ (or equivalently $\sup_{U \in T(M)} E\big[|M_U|\big] < \infty$), which we will assume for the rest of the proof. Define the local martingale $N = (N_t)_{t \in \mathbf{R}_+}$ by

$$N_t \stackrel{\text{def}}{=} -\int_0^t \text{sgn}\,(M_s)\, dM_s.$$

Note that $\langle N\rangle. = \langle M\rangle.$ It also follows from Tanaka's formula that

$$|M_t| = -N_t + L_t, \qquad t \geq 0, \quad \text{a.s.},$$

where L_t is the local time of M at the origin up to time t. Since this is the Doob-Meyer decomposition of the local submartingale $(|M_t|)_t$, we see that

$$\forall U \in T(M), \qquad E\big[|M_U|\big] = E\big[L_U\big],$$

and hence

$$E\big[L_\infty\big] = \sup_{U \in T(M)} E\big[|M_U|\big] < \infty.$$

Furthermore, it follows from the Skorohod equation argument that

$$L_t = \sup_{u \in [0,t]} N_u, \qquad t \geq 0, \quad \text{a.s.}$$

Therefore

$$E\big[\,\sup_t N_t\,\big] \;<\; \infty,$$

which is crucial for the rest of the proof.

Step 5. With the assumption in Step 4 and the notations there assumed, we will here prove

$$\limsup_{\lambda\to\infty} \lambda P\big[\langle N\rangle_\infty^{1/2} > \lambda\big] \;\le\; \lim_{x\to\infty}\lim_{\lambda\to\infty} \lambda P\big[\langle N\rangle_{S_{x'}}^{1/2} > \lambda\big],$$

where

$$S_{x}' \;\overset{\text{def}}{=}\; \inf\{\,t \,:\, N_t > x\,\}. \qquad (\inf\emptyset \overset{\text{def}}{=} \infty)$$

Note that this implies the existence of the limit σ since it is trivial that

$$\liminf_{\lambda\to\infty} \lambda P\big[\langle N\rangle_\infty^{1/2} > \lambda\big] \;\ge\; \lim_{x\to\infty}\lim_{\lambda\to\infty} \lambda P\big[\langle N\rangle_{S_{x'}}^{1/2} > \lambda\big].$$

It suffices to show that, for each fixed $0 < a < 1$,

$$\limsup_{\lambda\to\infty} \lambda P\big[\langle N\rangle_\infty^{1/2} > \lambda\big] \;\le\; \frac{1}{a}\lim_{x\to\infty}\lim_{\lambda\to\infty} \lambda P\big[\langle N\rangle_{S_{x'}}^{1/2} > \lambda\big].$$

For $x > 0$, we have

$$P\big[\langle N\rangle_\infty^{1/2} > \lambda\big] \;\le\; P\big[\langle N\rangle_{S_{x'}}^{1/2} \le a\lambda,\ \langle N\rangle_\infty^{1/2} > \lambda\big] + P\big[\langle N\rangle_{S_{x'}}^{1/2} > a\lambda\big]$$

and hence

$$\limsup_{\lambda\to\infty} \lambda P\big[\langle N\rangle_\infty^{1/2} > \lambda\big] \;\le\; \frac{1}{a}\lim_{\lambda\to\infty} \lambda P\big[\langle N\rangle_{S_{x'}}^{1/2} > \lambda\big]$$

$$+ \sup_\lambda \lambda P\big[\,\langle N\rangle_{S_{x'}}^{1/2} \le a\lambda,\ \langle N\rangle_\infty^{1/2} > \lambda\,\big].$$

Thus it suffices to show

$$\lim_{x\to\infty} \sup_\lambda \lambda P\big[\,\langle N\rangle_{S_{x'}}^{1/2} \le a\lambda,\ \langle N\rangle_\infty^{1/2} > \lambda\,\big] \;=\; 0.$$

Fix $x > 0$ for the moment. For $t \ge 0$, define

$$\tilde{N}_t^{(x)} \overset{\text{def}}{=} N_{S_{x'}+t} - N_{S_{x'}} \quad\text{and}\quad \tilde{\mathcal{F}}_t^{(x)} \overset{\text{def}}{=} \mathcal{F}_{S_{x'}+t}.$$

Note that $\big(\tilde{N}_t^{(x)}\big)_t$ is a continuous local martingale w.r.t. the filtration $\big(\tilde{\mathcal{F}}_t^{(x)}\big)_t$. Also, observe that

$$\sup_\lambda \lambda P\big[\,\langle N\rangle_{S_{x'}}^{1/2} \le a\lambda,\ \langle N\rangle_\infty^{1/2} > \lambda\,\big]$$

$$\le\; \sup_\lambda \lambda P\big[\,\langle \tilde{N}^{(x)}\rangle_\infty^{1/2} > \sqrt{1-a^2}\,\lambda\,\big]$$

$$=\; \frac{1}{\sqrt{1-a^2}}\sup_\lambda \lambda P\big[\,\langle \tilde{N}^{(x)}\rangle_\infty^{1/2} > \lambda\,\big]$$

$$\le\; \frac{C}{\sqrt{1-a^2}}\sup_\lambda \lambda P\big[\,\sup_t |\tilde{N}_t^{(x)}| > \lambda\,\big], \qquad (*)$$

where the last inequality follows from the well-known good λ inequality (see e.g. §IV.4 of Revuz-Yor [6]), with the constant C universal; in particular, C does not depend on x. Since

$$\forall \lambda > 0, \qquad \lambda P\big[\, \sup_t |\tilde{N}_t^{(x)}| > \lambda \,\big] \;\leq\; E\big[|\tilde{N}_{\tilde{T}_\lambda}^{(x)}|\big],$$

$$(\,\tilde{T}_\lambda \overset{\text{def}}{=} \inf\{t : |\tilde{N}_t^{(x)}| > \lambda\}\,)$$

it follows that

$$(*) \;\leq\; \frac{C}{\sqrt{1-a^2}} \sup_{U \in T(\tilde{N}^{(x)})} E\big[|\tilde{N}_U^{(x)}|\big]$$

$$(\text{ where } T(\tilde{N}^{(x)}) \text{ is defined the same way as } T(M)\,)$$

$$= \; \frac{C}{\sqrt{1-a^2}} \, 2 \sup_{U \in T(\tilde{N}^{(x)})} E\big[\tilde{N}_U^{(x)\,+}\big]$$

$$\leq \; \frac{C}{\sqrt{1-a^2}} \, 2 \, E\big[(\sup_t N_t - x)^+\big].$$

The last expression converges to 0 as $x \to \infty$, since $E\big[\sup_t N_t\big] < \infty$.

Step 6. It remains to prove the equality $\sqrt{\tfrac{\pi}{2}}\,\sigma \;=\; \sup_{U \in T(M)} E\big[|M_U|\big] - E\big[|M_\infty|\big]$.
We assume the notations in the previous two steps. For $x > 0$, the same argument as in Step 3 gives

$$\sqrt{\tfrac{\pi}{2}} \lim_{\lambda \to \infty} \lambda P\big[\langle N \rangle_{S_{x'}}^{1/2} > \lambda\big]$$

$$= x P\big[\sup_t N_t > x\big] + E\big[N_\infty ; \sup_t N_t \leq x\big].$$

Here the first term on the right-hand side converges to 0 as $x \to \infty$, since $E\big[\sup_t N_t\big] < \infty$. The second term converges to

$$E[N_\infty] \;=\; E[L_\infty] - E\big[|M_\infty|\big]$$
$$=\; \sup_{U \in T(M)} E\big[|M_U|\big] - E\big[|M_\infty|\big].$$

Therefore

$$\lim_{x \to \infty} \sqrt{\tfrac{\pi}{2}} \lim_{\lambda \to \infty} \lambda P\big[\langle N \rangle_{S_{x'}}^{1/2} > \lambda\big] \;=\; \sup_{U \in T} E\big[|M_U|\big] - E\big[|M_\infty|\big],$$

which together with Step 5 completes the proof. $\square$

References

[1] Azéma, J., Gundy, R.F., Yor, M.: Sur l'intégrabilité uniforme des martingales continues. *Séminaire de Probabilités XIV*, LNM 784, Springer (1980), pp. 53–61.

[2] Elworthy, K.D., Li, X.M., Yor, M.: On the tails of the supremum and the quadratic variation of strictly local martingales. *Séminaire de Probabilités XXXI*, LNM 1655, Springer (1997), pp. 113–125.

[3] ———: The importance of strictly local martingales; applications to radial Ornstein-Uhlenbeck processes. Submitted to Probab. Theory Relat. Fields.

[4] Galtchouk, L.I., Novikov, A.A.: On Wald's equation. Discrete time case. *Séminaire de Probabilités XXXI*, LNM 1655, Springer (1997), pp. 126–135.

[5] Kazamaki, N.: *Continuous Exponential Martingales and BMO*. LNM 1579, Springer (1994).

[6] Revuz, D., Yor, M.: *Continuous Martingales and Brownian Motion*, Second edition. Springer (1994).

[7] Sato, H.: Uniform integrabilities of an additive martingale and its exponential. *Stochastics and Stochastics Reports* **30** (1990), 163–169.

An alternative proof of a theorem of Aldous concerning convergence in distribution for martingales.

Maurizio Pratelli

We consider regular right continuous stochastic processes $X = (X_t)_{0 \leq t \leq 1}$ defined on the finite time interval $[0, 1]$: let P^X be the distribution of X on the canonical Skorokhod space $\mathsf{D} = \mathsf{D}([0, 1]; \mathsf{R})$ of "càdlàg" paths.

We consider on D, besides the usual Skorokhod topology referred as S–topology (Jacod–Shiryaev is perhaps the best reference for our purposes, see [4]), the "pseudo-path" or MZ–topology: we refer to the paper of Meyer–Zheng ([6]) for a complete account of this rather neglected topology (see also Kurtz [5]).

We will use the notation $X^n \Longrightarrow^S X$ (respectively $X^n \Longrightarrow^{MZ} X$) to indicate that the probabilities P^{X^n} converge strictly to P^X when the space D is endowed respectively with the S– or the MZ–topology. We will write also $X^n \Longrightarrow^{f.d.d.} X$ to indicate that all finite dimensional distributions of $(X_t^n)_{0 \leq t \leq 1}$ converge to those of $(X_t)_{0 \leq t \leq 1}$.

The following theorem holds true:

Theorem. *Let (M^n) be a sequence of martingales, and M a continuous martingale, and suppose that the following integrability condition is satisfied:*

(1) all random variables $\left(\sup_{0 \leq t \leq 1} |M_t^n| \right)$, $n = 1, 2, \ldots$ are uniformly integrable.

Then the following statements are equivalent:

(a) $M^n \Longrightarrow^S M$,

(b) $M^n \Longrightarrow^{f.d.d.} M$,

(c) $M^n \Longrightarrow^{MZ} M$.

The implication $(a) \Longrightarrow (b)$ is quite obvious, since Skorokhod convergence implies convergence of finite dimensional distributions for all continuity points of M (see [4]).

The implications $(b) \Longrightarrow (c)$ is an easy consequence of the results of Meyer-Zheng: in fact the sequence (M^n) is "tight" for the MZ-topology ([6] p. 368) and, if X is a limit process, there exists ([6] p. 365) a subsequence (M^{n_k}) and a set $I \subset [0,1]$ of full Lebesgue measure such that all finite dimensional distributions $(M_t^{n_k})_{t \in I}$ converge to those of $(X_t)_{t \in I}$: necessarily $\mathbf{P}^X = \mathbf{P}^M$.

Aldous (in [2]) gives a proof of the implication $(b) \Longrightarrow (a)$, but (although he does not mention the MZ-topology) the implication $(c) \Longrightarrow (a)$ is more or less implicit in his paper (see [2] p. 591).

The purpose of this paper is to give a proof of the implication $(c) \Longrightarrow (a)$, completely different form the Aldous' original one and strictly in the spirit of the paper of Meyer-Zheng; I hope that this contributes also to a better knowledge of the result of "Stopping times and tightness II" ([2]), which is in my opinion very important and seems to be almost unknown.

The proof will be postponed after some remarks.

Remark 1. I want to point out that Aldous' proof of the implication $(b) \Longrightarrow (a)$ requires the following weaker integrability condition:

(2) *all random variables* M_1^n, $n = 1, 2, \dots$ *are uniformly integrable.*

Condition (2) implies that all r.v. of the form M_T^n, $n = 1, 2, \dots$, with T a natural stopping time for M^n, are uniformly integrable; instead our proof needs a more stringent condition, i.e. that all r.v. of the form M_S^n, $n = 1, 2, \dots$, with S a random variable in $[0, 1]$, are uniformly integrable.

Remark 2. The extension of the Theorem to processes whose time interval is $[0, +\infty)$ is straightforward: in that case the correct hypothesis is that, for every fixed t, the r.v. $\sup_{0 \leq s \leq t} |M_s^n|$, $n = 1, 2, \dots$ are uniformly integrable.

In fact, if the limit function f is continuous, $f_n \to f$ for the S-topology (respectively the MZ-topology) on $\mathbf{D}(\mathbf{R}^+; \mathbf{R})$ if and only if the restrictions of f_n to every finite time interval converge to those of f (for the S- or the MZ-topology).

Remark 3. The Theorem fails to be true if the limit martingale M is not continuous ([2] p. 588), and fails for more general processes, e.g. for supermartingales.

Let indeed T be a Poisson r.v. and put, for every n:

$$X_t^n = \left(I_{\{t \geq T\}} - t \wedge T \right) - n \left((t - T) I_{\{t \geq T\}} \wedge 1 \right) .$$

The processes X^n are supermartingales whose paths converge in measure (but not uniformly) to the paths of the continuous supermartingale $X_t = -(t \wedge T)$.

Remark 4. Suppose that the processes X^n are supermartingales, and consider their Doob-Meyer decompositions $X^n = M^n - A^n$. If separately $M^n \Longrightarrow^{MZ} M$ and the martingale M is continuous, and if $A^n \Longrightarrow^{MZ} A$ and the increasing process A is continuous, then $X^n \Longrightarrow^S X = M - A$ (remark that, for monotone processes, convergence in the MZ-sense to a continuous limit implies convergence for the S-topology).

An application of the latter result can be found in [7], theorem 5.5 .

The proof of the implication $(c) \Longrightarrow (a)$ of the Theorem is rather technical, and will be divided in several steps.

Step 1. Given $\epsilon > 0$, there exists $\delta > 0$ such that, if S is a r.v. with values in $[0, 1]$ and $0 \leq d \leq \delta$:

$$(3) \qquad \mathbf{E}\left[|M_{S+d} - M_S|\right] \leq \epsilon .$$

This is an easy consequence of the path-continuity of the limit process M, and of the integrability of $M^* = \sup_{0 \leq t \leq 1} |M_t|$. Remark that the function $f \to \sup_{0 \leq t \leq 1} |f(t)|$ is lower semi-continuous on $\mathbf{D}$ endowed with the topology of convergence in measure (i.e. the MZ–topology); therefore the integrability of M^* is a consequence of condition (1) of the theorem.

Step 2. Suppose that (a) is false; then the sequence does not verify Aldous' tightness condition ([1] p. 335, see also [4]); therefore there exists $\epsilon > 0$ such that for every $\delta > 0$ it is possible to determine a subsequence n_k and, for every k, a natural stopping time T_k (i.e. a stopping time for the filtration generated by M^{n_k}) and $0 < d_k \leq \delta$ such that

$$(4) \qquad \mathbf{E}^{n_k}\left[|M^{n_k}_{T_k+d_k} - M^{n_k}_{T_k}|\right] \geq \epsilon .$$

(In the sequel, for the sake of simplicity of notations, we will assume that indices have been renamed so that the whole sequence verifies (4)). We choose δ such that, for any r.v. S whatsoever, we also have (step 1) $\mathbf{E}\left[|M_{S+2\delta} - M_S|\right] \leq \frac{\epsilon}{4}$.

Step 3. There exists a random variable T with values in $[0,1]$ such that (M^n, T_n) converge in distribution to (M, T) on the space $\mathbf{D}([0,1], \mathbf{R}^+) \times [0,1]$ equipped with the product topology ($\mathbf{D}$ being equipped with the MZ–topology).

In fact the laws of (M^n, T_n) are evidently tight since the laws of M^n are tight on $\mathbf{D}$ ([6] p. 368); we point out that the limit r.v. T is not a natural stopping time for the stochastic process M (but it can be proved that M is a martingale for the canonical filtration on $\mathbf{D} \times [0,1]$, i.e. the smallest filtration that makes M adapted and T a stopping time).

Step 4. For c and d in $[0, 1]$, we have the inequality

$$(5) \qquad \mathbf{E}^n\left[|M^n_{T_n+\delta+c} - M^n_{T_n-d}|\right] \geq \frac{\varepsilon}{2} .$$

(It is technically convenient to regard each process M as extended to $[-1, 2]$ by putting $M_t = M_0$ for $t < 0$ and $M_t = M_1$ for $t > 1$: this enables us to write $M_{T+\delta}$ instead of $M_{(T+\delta)\wedge 1}$.)

Concerning the inequality (5), firstly we note that

$$\left(M^n_{T_n+d_n} - M^n_{T_n}\right) = \mathbf{E}^n\left[M^n_{T_n+\delta+c} - M^n_{T_n} \,\big|\, \mathcal{F}_{T_n+d_n}\right]$$

and therefore

$$\mathbf{E}^n\left[|M^n_{T_n+\delta+c} - M^n_{T_n}|\right] \geq \mathbf{E}^n\left[|M^n_{T_n+d_n} - M^n_{T_n}|\right] \geq \epsilon .$$

Then we remark that $(T_n - c)$ is not a stopping time, but the r.v. $M^n_{T_n-c}$ is $\mathcal{F}_{T_n}$-measurable: in fact $M^n_{T_n-c}.I_{\{T_n \leq t\}} = M^n_{(T_n \wedge t)-c}.I_{\{T_n \leq t\}}$ and $(T_n \wedge t - c)$ is $\mathcal{F}_t$-measurable.

Let $X = \left(M^n_{T_n+\delta+c} - M^n_{T_n}\right)$, $Y = \left(M^n_{T_n} - M^n_{T_n-c}\right)$ and $\mathcal{G} = \mathcal{F}_{T_n}$: Y is $\mathcal{G}$-adapted and $\mathsf{E}[X|\mathcal{G}] = 0$.

We remark that $\mathsf{E}\left[X^+|\mathcal{G}\right] = \mathsf{E}\left[X^-|\mathcal{G}\right] = \frac{1}{2}\mathsf{E}\left[|X||\mathcal{G}\right]$, and that $|X + Y| \geq X^+.I_{\{Y \geq 0\}} + X^-.I_{\{Y < 0\}}$.

One gets $\mathsf{E}\left[|X + Y||\mathcal{G}\right] \geq \frac{1}{2}\mathsf{E}\left[|X||\mathcal{G}\right]$; and, taking expectations, inequality (5).

Step 5. There exists a subsequence and a set $I \subset [-1, 1]$ of full Lebesgue measure such that the finite dimensional distributions of $\left(M^n_{T_n+t}\right)_{t \in I}$ converge to those of $(M_{T+t})_{t \in I}$.

The proof of this step is a slight modification of the argument given in [6] (p. 364): Dudley's extension of the Skorokhod representation theorem implies that one can find on some probability space $(\Omega, \mathcal{F}, \mathbf{P})$ some random variables (X^n, S_n) and (X, S) with values in $\mathbf{D} \times [0, 1]$ such that the laws of (X^n, S_n) (resp. (X, S)) are equal to those of (M^n, T_n) (resp. (M, T)) and that, for almost all ω, $(X^n(\omega), S_n(\omega))$ converge to $(X(\omega), S(\omega))$: to be accurate, the "paths" $t \rightarrow (X^n_t(\omega))$ converge in measure to the path $t \rightarrow (X_t(\omega))$ and $S_n(\omega)$ converge to $S(\omega)$.

We remark that the Skorokhod theorem cannot be applied directly since $\mathbf{D}$ is not a Polish space ([6] p. 372), but Dudley's extension works well (see [3]).

By substituting X^n with $\text{arctg}(X^n)$, we can suppose that X^n and X are uniformly bounded: therefore we have

$$(6) \qquad \lim_{n \rightarrow \infty} \left(\int_{-1}^{+1} dt \int_\Omega \left|X^n_{T_n(\omega)+t}(\omega) - X_{T(\omega)+t}(\omega)\right| d\mathbf{P}(\omega)\right) = 0 .$$

By taking a subsequence, we find that for every t in a set $I \subset [-1, 1]$ of full Lebesgue measure,

$$(7) \qquad \lim_{n \rightarrow \infty} \int_\Omega \left|X^n_{T_n(\omega)+t}(\omega) - X_{T(\omega)+t}(\omega)\right| d\mathbf{P}(\omega) = 0 .$$

Hence one gets easily the convergence of finite dimensional distributions of $\left(M^n_{T+t}\right)_{t \in I}$.

Step 6. We choose $0 \leq d, c \leq 1$ such that $d + c < \delta$ and that $\left(M^n_{T_n+\delta+c}, M^n_{T_n-d}\right)$ converge in distribution to $(M_{T+\delta+c}, M_{T-d})$; since the r.v. involved are uniformly integrable, the inequality (5) gives in the limit

$$\mathsf{E}\left[|M_{T+\delta+c} - M_{T-d}|\right] \geq \frac{\varepsilon}{2}$$

and finally we have a contradiction.

References

[1] Aldous D.: *Stopping Times and Tightness*. Ann. Prob. **6**, 335–340 (1979)

[2] Aldous D.: *Stopping Times and Tightness II*. Ann. Prob. **17**, 586–595 (1989)

[3] Dudley R.M.: *Distances of Probability measures and Random Variables*. Ann. of Math. Stat. **39**, 1563–1572 (1968)

[4] Jacod J., Shiryaev A.N.: *Limit theorems for stochastic processes*. Berlin, Heidelberg, New York: Springer 1987.

[5] Kurtz T.G.: *Random time changes and convergence in distribution under the Meyer-Zheng conditions* Ann. Prob. **19**, 1010–1034 (1991)

[6] Meyer P.A., Zheng W.A.: *Tigthness criteria for laws of semimartingales*. Ann. Inst. Henri Poincaré Vol. **20** No. 4, 353–372 (1984)

[7] Mulinacci S., Pratelli M.: *Functional convergence of Snell envelopes: Applications to American options approximations*. Finance Stochast. **2**, 311–327 (1998)

A short proof of decomposition of strongly reduced martingales

Michał Morayne and Krzysztof Tabisz

Abstract

A short proof of the following theorem is given: If M is a martingale, $T > 0$ is a stopping time, $M = M^T$ and $E(|M_T|| \mathcal{F}_t) 1_{[t<T]}$ is bounded, then M is a sum of a BMO (and, thus, square-integrable) martingale and a martingale of integrable variation.

The purpose of this note is to give a short proof of P.A. Meyer's theorem ([Me]) stated in the abstract. Although the proof given here is very much in the spirit of that of [Me] it seems to be simpler and shorter (in particular, we do not use potentials and Riesz decomposition in the proof). The proof presented here reduces to a sequence of easy inequalities. A shortcut has been possible because of the fact that the dual predictable projection of a reduced process is reduced by the same stopping time.

Let $\mathbf{R}_+ = [0, \infty)$. Let $(\mathcal{F}_t, t \in \mathbf{R}_+)$ be a fixed right-hand side continuous complete filtration. We shall consider martingales with respect to this filtration, assuming always that they are CADLAG (i.e. that almost all their trajectories are right-hand side continuous and have left-hand side finite limits). For a process X by X_∞ we denote $\lim_{t \to \infty} X_t$, when such a limit exists a.s.

BMO denotes the class of those uniformly integrable martingales M for which $E M_\infty^2 < \infty$ and there exists a constant c such that for each stopping time S the following inequality is satisfied: $E((M_\infty - M_{S-})^2 | \mathcal{F}_S) < c$. $\mathcal{A}$ denotes the class of the processes of integrable variation i.e. $A \in \mathcal{A}$ if $E \operatorname{Var}|_0^\infty A_t < \infty$.

For a stopping time T we shall put $[[0, T]] = \{(\omega, t) : t \leq T(\omega)\}$, $[[T, \infty)) = \{(\omega, t) : t \geq T(\omega)\}$, $((T, \infty)) = \{(\omega, t) : t > T(\omega)\}$ and $[[T]] = \{(\omega, t) : T(\omega) = t\}$.

We say that a process X is *reduced* by a stopping time T if $X^T = X$, where $X_t^T = X_{\min(t,T)}$. If, in addition, $T > 0$ and for some constant C we have $\sup_{t<T} E(|X_T|| \mathcal{F}_t) < C$ a.s. we say that X is *strongly reduced* by T.

We are going to prove the following theorem ([Me], Chap. IV, 8, p. 294 and Chap. V, 5c, p. 335).

The work of the first author was supported in part by KBN Grant 2 P03A 01813.

This article was written in part when the first author was visiting the Department of Mathematics, University of Louisville, Kentucky, USA.

AMS Subject Classification: 60 G 44.

Key words and phrases: martingale, decomposition.

Theorem. If $T > 0$ is a stopping time and M is a martingale strongly reduced by T, then M can be expressed as a sum $M = N + A$, where $N \in BMO$ and $A \in \mathcal{A}$.

The following lemma is crucial for our proof.

Lemma. Let $T > 0$ be a stopping time, $\Phi \in L_1$, Φ be $\mathcal{F}_T$ - measurable, $E(|\Phi||\mathcal{F}_t)$ be strongly reduced by T, U be the dual predictable projection of the process $\Phi 1_{[[T,\infty))}$. Then there exists a constant c such that for each stopping time S

$$E((U_\infty - U_{S-})^2|\mathcal{F}_S) < c.$$

Proof. First we shall show that there exists such a constant c that for each stopping time S

$$E((U_\infty - U_S)^2|\mathcal{F}_S) < c. \tag{1}$$

Let W be the dual predictable projection of the process $|\Phi|1_{[[T,\infty))}$. Let $P_t = E(|\Phi||\mathcal{F}_t)$. We have for a set $B \in \mathcal{F}_S$:

$$E((U_\infty - U_S)^2 1_B) \le E((W_\infty - W_S)^2 1_B) \le 2E \int_0^\infty 1_{((S,\infty))} 1_B (W_t - W_S) dW_t$$

$$= 2E \int_0^\infty 1_{((S,\infty))} 1_B (W_t - W_S) d(|\Phi|1_{[[T,\infty))})_t = 2E(1_B |\Phi|(W_T - W_S)) =$$

$$2E \int_0^\infty 1_{((S,\infty))} 1_B |\Phi| dW_t = 2E \int_0^\infty 1_{((S,\infty))} 1_B P_{t-} dW_t$$

$$(= 2E \int_0^\infty 1_{((S,\infty))} 1_B P_{t-} d(|\Phi|1_{[[T,\infty))})_t \le 2E(P_{T-}|\Phi|) \le 2CE|\Phi| < \infty)$$

$$\le 2CE(1_B 1_B (W_\infty - W_S)) \le 2C(P(B))^{1/2}(E(1_B (W_\infty - W_S)^2))^{1/2}.$$

To get the second and the third equality and the next inequality we used the fact that the process W is reduced by T. The last step was possible by virtue of Schwarz inequality.

Comparing the second and the last term of the sequence of (in)equalities above and repeating the first inequality we get

$$E((U_\infty - U_S)^2 1_B) \le 4C^2 P(B)$$

and this gives (1).

We still need to show that all the jumps of the process U are uniformly bounded. First let us notice that the process U, being predictable, possibly jumps only at countable number of graphs of predictable stopping times. So let us assume that S is a predictable stopping time which means that there exists a nondecreasing sequence of stopping times, S_n such that $S_n < S$ on the set $[S > 0]$ and $\lim_n S_n = S$. Let $B \in \mathcal{F}_S$. Let $R_t = E(1_B|\mathcal{F}_t)$. We have:

$$E(1_B |U_S - U_{S-}|) \le E(1_B (W_S - W_{S-})) = E \int_0^\infty 1_{[[S]]} 1_B dW_t =$$

$$E \int_0^\infty \mathbf{1}_{[[S]]} R_t - d(|\Phi| \mathbf{1}_{[[T,\infty]]})_t = E(\mathbf{1}_{[S=T]} R_{S-} |\Phi|) = EE(\mathbf{1}_{[S=T]} R_{S-} |\Phi| | \mathcal{F}_{S-}) =$$

$$E(R_{S-} E(\mathbf{1}_{[S=T]} |\Phi| | \mathcal{F}_{S-})) = E(R_{S-} \lim_n E(\mathbf{1}_{[S=T]} |\Phi| | \mathcal{F}_{S_n})) \leq$$

$$E(R_{S-} \sup_n (\mathbf{1}_{[S_n<T]} P_{S_n})) \leq C E R_{S-} = C P(B).$$

This finishes the proof.

Now we can give a proof of Theorem.

Proof of Theorem. Let U now denote the dual predictable projection of $M_T \mathbf{1}_{[[T,\infty))}$. The decomposition of M is defined as in [Me] as $M = N + A$, where $N = M - M_T \mathbf{1}_{[[T,\infty))} + U$ and $A = M_T \mathbf{1}_{[[T,\infty))} - U$ and we apply Lemma to get $N \in BMO$. It is obvious that $A \in \mathcal{A}$. This finishes the proof.

References

[Me] P.A. Meyer, *Un cours sur les integrales stochastiques*, Séminaire de Probabilités X, Lecture Notes in Mathematics 511, Berlin, Heidelberg, New York 1976.

Some remarks on L^∞, H^∞ and BMO

Peter Grandits

1 Introduction

In [1] C. Dellacherie, P.A. Meyer and M. Yor proved that L^∞ is neither closed nor dense in BMO, except in trivial cases (i.e. if the underlying filtration is constant). The same is true for H^∞ (c.f. [3] section 2.6 and [5]). So one may ask, whether it is possible to find a martingale $X \in BMO$, which has a best approximation in L^∞ resp. in H^∞, i.e.

$$\inf_{Z \in L^\infty} ||X - Z||_{BMO} = ||X - \bar{Z}||_{BMO} \text{ for some } \bar{Z} \in L^\infty$$

resp.

$$\inf_{Z \in H^\infty} ||X - Z||_{BMO} = ||X - \bar{Z}||_{BMO} \text{ for some } \bar{Z} \in H^\infty.$$

It is easy to see that this is equivalent to the question: does there exist a martingale $X \in BMO$ s.t.

$$\inf_{Z \in L^\infty} ||X - Z||_{BMO} = ||X||_{BMO}$$

resp.

$$\inf_{Z \in H^\infty} ||X - Z||_{BMO} = ||X||_{BMO}.$$

holds? R. Durrett poses this problem for L^∞ in [2], p. 214, and he conjectures a solution for X. We show in this paper that a discrete time analogue of Durrett's example works, but in continuous time it does not. In the case of H^∞ we provide a class of processes (including Durrett's example), for which $\bar{Z} = 0$ is indeed the best approximation in H^∞. Note that for the negative result in L^∞ we work with the norm $|| \cdot ||_{BMO_1}$, as the problem was posed by Durrett in this way. For the positive result in H^∞ we use $|| \cdot ||_{BMO_2}$, which seems to be more natural in this case.

2 Notations and Preliminaries

We denote by BMO the space of continuous martingales X on a given stochastic basis $(\Omega, \mathcal{F}, (\mathcal{F}_t)_{t=0}^\infty, P)$, satisfying the "usual conditions" of completeness and right continuity, for which the the following equivalent norms are finite

$$||X||_{BMO_1} = \sup_T \{ ||E[|X_\infty - X_T| \,|\, \mathcal{F}_T]||_\infty \} = \sup_T \left\{ \left(\frac{E[|X_\infty - X_T|]}{P[T < \infty]} \right) \right\}$$

$$\|X\|_{BMO_2} = \sup_T \{\|E[(X_\infty - X_T)^2|\mathcal{F}_T]^{\frac{1}{2}}\|_\infty\} = \sup_T \left\{ \left(\frac{E[(X_\infty - X_T)^2]}{P[T < \infty]} \right)^{\frac{1}{2}} \right\} =$$

$$\sup_T \{\|E[\langle X \rangle_\infty - \langle X \rangle_T|\mathcal{F}_T]^{\frac{1}{2}}\|_\infty\}.$$

Here T runs through all stopping times. In the present context H^∞ denotes the space of continuous martingales M on $(\Omega, \mathcal{F}, (\mathcal{F}_t)_{t=0}^\infty, P)$ s.t.

$$\|M\|_{H^\infty} = \text{ess sup } \langle M \rangle_\infty^{\frac{1}{2}} < \infty$$

holds. We also use the following standard notation. If M is a martingale and T a stopping time, we denote by M^T the martingale stopped at time T, i.e.

$$M_t^T = M_{t \wedge T}$$

and by $^T M$ the martingale started at time T, i.e.

$$^T M = M - M^T.$$

The next easy lemma is maybe folklore, but for the convenience of the reader we provide a proof.

Lemma 2.1 *Let X be in BMO and R an arbitrary stopping time. Then we have*

$$\|^R X\|_{BMO_2} \leq \|X\|_{BMO_2}.$$

Proof: We prove that $\|\int H dX\|_{BMO_2} \leq \|X\|_{BMO_2}$, if H is previsible with $|H| \leq 1$, which immediately implies the assertion of the lemma.

$$\|\int H dX\|_{BMO_2}^2 = \sup_T \{\|E[(\int H dX \rangle_\infty - \langle \int H dX \rangle_T|\mathcal{F}_T]\|_\infty\} =$$

$$\sup_T \{\|E[\int_T^\infty H^2 d\langle X \rangle|\mathcal{F}_T]\|_\infty\} \leq \sup_T \{\|E[\int_T^\infty d\langle X \rangle|\mathcal{F}_T]\|_\infty\} =$$

$$\sup_T \{\|E[\langle X \rangle_\infty - \langle X \rangle_T|\mathcal{F}_T]\|_\infty\} = \|X\|_{BMO_2}^2 \qquad \square$$

3 The case L^∞ - a discrete time example

We give in this section an example of a discrete-time process, for which $\bar{Z} = 0$ is indeed the best approximation in L^∞, if we use the space bmo_1 (c.f. [4]) as an analogue to BMO_1 in the continuous-time setting. Let $(W_n)_{n=0}^\infty$ be a standard random walk on $(\Omega, \mathcal{F}, (\mathcal{F}_n)_{n=0}^\infty, P)$ with natural filtration, i.e. $P[\{\Delta W_n = 1\}] = P[\{\Delta W_n = -1\}] = \frac{1}{2}$ and $W_0 = 0$. Let T be the stopping time $T = inf\{n|\Delta W_n = -1\}$, and $B_n = W_n^T = W_{T \wedge n}$. This is a discrete-time analogue of the continuous martingale, which we consider in section 4, and which was suggested by Durrett in [2].

Denoting the bmo_1-norm by $\|\cdot\|_*$, an easy calculation gives

$$\|B\|_* = \sup_S \|E[|B_\infty - B_S||\mathcal{F}_S]\|_\infty = \sup_{k \in N_0} \|E[|B_\infty - B_S||B_S = k]\|_\infty =$$

$$\|E[|B_\infty - B_S||B_S = 0]\|_\infty = E[|B_\infty|] = \sum_{r=-1}^\infty |r|2^{-(r+2)} = 1,$$

where the supremum is taken over all stopping times S and N_0 denotes the set $\{0, 1, 2, ...\}$. We denote by $L^\infty(\Omega, \mathcal{F}, (\mathcal{F}_n)_{n=0}^\infty, P)$ the space of all bounded martingales with respect to the given filtration. Our claim is

Proposition 3.1

$$\inf_{Z \in L^\infty} \|B - Z\|_* = 1,$$

i.e. $\hat{Z} = 0$ is the best approximation in L^∞ of B.

Proof: We shall show that assuming the existence of a $Z \in L^\infty$, which fulfills $\|B - Z\|_* = \alpha < 1$, leads to a contradiction.

As the definition of the bmo_1-norm is invariant with respect to an additive constant, we assume that $Z \geq 0$ holds. Furthermore the function $f(t) = \|B - tZ\|_*$ is a continuous convex function with $f(0) = 1$ and $f(1) = \alpha$. Therefore we may assume w.l.o.g. that $\|Z\|_\infty < \frac{1}{4}$ holds, and we remain with

$$Z_\infty = a_k \qquad \text{on } C_k \text{ for } k = -1, 0, 1, ...,$$

where

$$0 \leq a_k \leq \frac{1}{4} \tag{1}$$

holds, and the atoms C_k are defined by $C_k = \{B_\infty = k\}$. Since the filtration is given by

$$\mathcal{F}_n = \{C_{-1}, C_0, ..., C_{n-2}, (C_{n-1} \cup C_n \cup ...)\} \qquad n = 0, 1, 2, ...,$$

and $P[C_k] = 2^{-(k+2)}$, one can easily calculate $Z_n = E[Z_\infty | \mathcal{F}_n]$ for $n = 0, 1, 2, ...$

$$Z_n = a_k \qquad \text{on } C_k \text{ for } k = -1, 0, 1, ..., n-2$$

$$Z_n = \sum_{r=n-1}^\infty a_r 2^{-(r+2-n)} =: \gamma_{n-1} \qquad \text{on } (C_{n-1} \cup C_n \cup ...)$$

Hence we get for $n = 0, 1, 2, ...$

$$B_\infty - B_n = Z_\infty - Z_n = 0 \qquad \text{on } C_{-1} \cup C_0 \cup ... \cup C_{n-2},$$

resp.

$$B_\infty - B_n = s - n \qquad \text{on } C_s \text{ for } s = n-1, n, n+1, ...$$

and

$$Z_\infty - Z_n = a_s - \gamma_{n-1} \qquad \text{on } C_s \text{ for } s = n-1, n, n+1, ...$$

As the supremum over all stopping times in the definition of the bmo_1-norm can be replaced by a supremum over all fixed times n, we calculate $E[|B_\infty - B_n - Z_\infty + Z_n| \,|\, \mathcal{F}_n]$, which is 0 on $C_{-1}, C_0, ..., C_{n-2}$ and

$$\sum_{s=-1}^\infty |s - a_{n+s} + \gamma_{n-1}| 2^{-(s+2)}$$

on $C_{n-1} \cup C_n \cup, ...$

Using eq. (1) and our assumption $\|B - Z\|_* = \alpha < 1$, we conclude that

$$(a_{n-1} - \gamma_{n-1})\frac{1}{2} + |a_n - \gamma_{n-1}|(\frac{1}{2})^2 + \sum_{s=1}^{\infty}(-a_{n+s} + \gamma_{n-1})2^{-(s+2)} \leq -\rho := \alpha - 1$$

has to hold for $n = 0, 1, 2, \dots$. We now distinguish two cases.

Case 1: $-a_n + \gamma_{n-1} \geq 0$
A simple calculation gives

$$a_{n-1} \leq \gamma_{n-1} - \rho.$$

Case 2: $-a_n + \gamma_{n-1} < 0$
In this case we get $-\gamma_{n-1} + \frac{2}{3}a_{n-1} + \frac{1}{3}a_n \leq -\frac{2}{3}\rho$. This inequality and our assumption in case 2 allow us to conclude that $a_{n-1} < a_n$ has to hold, and we finally get

$$a_{n-1} < \gamma_{n-1} - \frac{2}{3}\rho.$$

Denoting now $\sigma = \frac{2}{3}\rho > 0$, we can combine case 1 and case 2, which yields

$$-a_{n-1} + \gamma_{n-1} > \sigma \qquad n = 0, 1, 2, \dots$$

or

$$-\frac{1}{2}a_{n-1} + \sum_{s=n}^{\infty} a_s 2^{-(s-n+2)} > \sigma \qquad n = 0, 1, 2, \dots. \tag{2}$$

Defining $A = \sup_{s=-1,0,\dots} a_s$, implies the existence of an M, s.t. $a_M > A - \sigma$, and we infer that

$$-\frac{1}{2}a_M + \sum_{s=M+1}^{\infty} a_s 2^{-(s-M+1)} < \frac{\sigma}{2}$$

holds, which is a contradiction to eq. (2). $\qquad \square$

4 The case L^∞ - a continuous time example

In contrast to the discrete case it seems to be not so easy to find a martingale in BMO in continuous time, which has a best approximation $\bar{Z} = 0$ in L^∞. It is shown in this section that the - in some sense - natural guess of Durrett [2] of a martingale, which is quasi-stationary, in a sense to be defined later, does not work. However, it will be shown in section 5 that this quasistationarity is sufficient to guarantee a best approximation $\bar{Z} = 0$ in H^∞.

Let $(W_t)_{t=0}^{\infty}$ be a standard Brownian motion on $(\Omega, \mathcal{F}, (\mathcal{F}_t)_{t=0}^{\infty}, P)$. As in [2] we define $R_0 = 0$, $R_n = \inf\{t > R_{n-1} : |W_t - W_{R_{n-1}}| > 1\}$, $N = \inf\{n : W_{R_n} - W_{R_{n-1}} = -1\}$ and finally $X_t = W_{t \wedge R_N}$. The following formula is valid for $a \in (-1, 1)$ (c.f. [2],p. 208)

$$\|X\|_{BMO_1} = \sup_T \|E[|X_\infty - X_T| \,|\, \mathcal{F}_T]\|_\infty = \sup_{a \in (-1,1)} E[|X_\infty - X_T| \,|\, X_T = a] =$$

$$\sup_{a \in (-1,1)} 1_{(-1,0]}(a)(1 - a^2) + 1_{(0,1]}(a)\frac{(a+1)(2-a)}{2} = \frac{9}{8}.$$

Our claim is now

Proposition 4.1

$$\inf_{Z \in L^\infty} \|X - Z\|_{BMO_1} < \frac{9}{8},$$

where $L^\infty = L^\infty(\Omega, \mathcal{F}, \mathcal{F}_t, P)$ is the space of continuous bounded martingales.

This answers negatively the question posed by Durrett in Ex. 1 of sect. 7.7 in [2].
Proof: In order to prove the proposition some further notation is needed. We define

$$
\begin{aligned}
A_r &= \{\omega : X_\infty = r\} & r &= -1, 0, 1, \ldots \\
S_n &= \inf\{t > R_{n-1} : X_t - X_{R_{n-1}} = \tfrac{1}{2}\} & n &= 1, 2, 3, \ldots,
\end{aligned}
$$

where we use the convention $\inf \emptyset = \infty$. Furthermore we need

$$
\begin{aligned}
A_r^b &= A_r \cap \{S_{r+2} = \infty\} \\
A_r^g &= A_r \cap \{S_{r+2} < \infty\}, \quad r = -1, 0, 1, \ldots
\end{aligned}
$$

and finally

$$
M_t = \begin{cases} 0 & R_{n-1} \le t < S_n \\ 1 & S_n \le t < R_n, \end{cases} \quad n = 1, 2, 3, \ldots.
$$

The process M indicates, whether X has reached the value $X_{R_{n-1}} + \frac{1}{2}$ in the stochastic interval $[[R_{n-1}, R_n[[$ or not, and is essential for the calculation of the conditional expectations occurring in the sequel.

A straightforward but lengthy application of the optional stopping theorem yields the following table of conditional probabilities, which we will need later on.

$r = 0, 1, 2, \ldots$	$-1 < a \le \frac{1}{2}$	$\frac{1}{2} < a < 1$
$P[A_{-1}^b \mid X_T = a, M_T]$	$\frac{1-2a}{3}(1 - M_T)$	0
$P[A_{-1}^g \mid X_T = a, M_T]$	$\frac{a+1}{6}(1 - M_T) + \frac{1-a}{2} M_T$	$\frac{1-a}{2}$
$P[A_r^b \mid X_T = a, M_T]$	$\frac{1+a}{6}\frac{1}{2^r}$	$\frac{1+a}{6}\frac{1}{2^r}$
$P[A_r^g \mid X_T = a, M_T]$	$\frac{1+a}{12}\frac{1}{2^r}$	$\frac{1+a}{12}\frac{1}{2^r}$

We define now a bounded continuous martingale Z, which gives a better approximation of X than the trivial approximation $Z = 0$:

$$
\begin{aligned}
Z_\infty &= \delta 1_{\bigcup_{r=-1}^\infty A_r^b} \\
Z_t &= E[Z_\infty \mid \mathcal{F}_t]
\end{aligned}
$$

with $\delta > 0$. This yields

$$Z_T = \delta P[S_N = \infty \mid \mathcal{F}_T].$$

Again it suffices to consider $a \in (-1, 1)$. Since $\|Z\|_{BMO_1} \le \delta$ holds, we only have to show that

$$E[|X_\infty - X_T - Z_\infty + Z_T| \mid X_T = a, M_T] < E[|X_\infty - X_T| \mid X_T = a] = \frac{(a+1)(2-a)}{2}$$

holds for $a \in [\frac{1}{4}, \frac{3}{4}]$ uniformly in M_T, and then to choose δ small enough.

Using again the optional stopping theorem, an easy calculation gives the following table.

$r = -1, 0, 1, \ldots$	X_∞	X_T	Z_∞	Z_T	
				$-1 < a < \frac{1}{2}$	$\frac{1}{2} \le a < 1$
A_r^b	r	a	δ	$\delta(\frac{2-a}{3}(1 - M_T) + \frac{a+1}{3} M_T)$	$\delta \frac{a+1}{3}$
A_r^g	r	a	0	"	"

Putting things together we arrive - after a lot of algebra - at

$$E[|X_\infty - X_T - Z_\infty + Z_T||X_T = a, M_T] =$$

$$\begin{cases} \frac{(a+1)(2-a)}{2} + \frac{\delta}{6}(M_T(a^2 - 1) + (1 - M_T)(-a^2 - a)) & \frac{1}{4} \le a \le \frac{1}{2} \\ \frac{(a+1)(2-a)}{2} + \frac{\delta}{6}(a^2 - 1) & \frac{1}{2} \le a \le \frac{3}{4}, \end{cases}$$

which clearly proves our assertion. $\square$

5 The case H^∞

In this section we introduce a class of processes for which $\bar{Z} = 0$ is indeed the best approximation in H^∞. This class includes also the example of section 4. We start with definitions.

Definition 5.1 *Let X be in BMO. Then we call a stopping time T proper for X, if $P[\langle X \rangle_T < \langle X \rangle_\infty] > 0$.*

Definition 5.2 *A process X in BMO has the property QS (quasi-stationary), if for each proper stopping time T for X, we can find another proper stopping time $S \ge T$ $P - a.s.$ for X, s.t. $^S X 1_{\{^S X \ne 0\}}/P[\{^S X \not\equiv 0\}] \sim X$ hold. Here $\sim$ stands for equality in law.*

Our next lemma shows that - not very surprisingly - for QS processes the BMO-norm "does not decline", no matter when the process is started.

Lemma 5.1 *Let X be in BMO with the property QS. Then for all proper stopping times R we have $\|^R X\|_{BMO_2} = \|X\|_{BMO_2}$*

Proof: Let U be a proper stopping time s.t. $U \ge R$ $P-a.s.$ and $^U X 1_{\{^U X \ne 0\}}/P[\{^U X \not\equiv 0\}] \sim X$ hold. We get

$$\|^R X\|_{BMO_2}^2 = \sup_T \frac{E[(^R X_\infty - ^R X_T)^2]}{P[T < \infty]} = \sup_T \frac{E[(X_\infty - X_{T \vee R})^2]}{P[T < \infty]} \ge$$

$$\sup_{T \ge R} \frac{E[(X_\infty - X_T)^2]}{P[T < \infty]} \ge \sup_{T \ge U} \frac{E[(X_\infty - X_T)^2]}{P[T < \infty]} = \|X\|_{BMO_2}^2.$$

The reverse inequality follows from Lemma 2.1. $\square$

Using a result proved by W. Schachermayer in [5], which characterizes the distance of a given martingale to H^∞ in $\|\cdot\|_{BMO_2}$, we get our final result.

Theorem 5.1 *Let X be in BMO with the property QS. Then we have*

$$\inf_{Z \in H^\infty} \|X - Z\|_{BMO_2} = \|X\|_{BMO_2}$$

Proof: Assuming the contrary, namely

$$\inf_{Z \in H^\infty} \|X - Z\|_{BMO_2} < \|X\|_{BMO_2},$$

yields, by applying Theorem 1.1 of [5], a finite increasing sequence of stopping times

$$0 = T_0 \leq T_1 \leq \ldots \leq T_N \leq T_{N+1} = \infty$$

s.t.

$$\|{}^{T_n}X^{T_{n+1}}\|_{BMO_2} < \|X\|_{BMO_2} \qquad n = 0, \ldots, N$$

(Without loss of generality we may assume that T_N is a proper stopping time for X.) In particular we find

$$\|{}^{T_N}X\|_{BMO_2} < \|X\|_{BMO_2},$$

which is a contradiction to Lemma 5.1. □

Acknowledgement: Support by "Fonds zur Förderung der wissenschaftlichen Forschung in Österreich" (Project Nr. P11544) is gratefully acknowledged.

References

[1] C. Dellacherie, P.A. Meyer and M. Yor, Sur certaines propriétés des espaces de Banach H^1 et BMO. Séminaire de probabilités XII, Lecture notes in Math. 649,98-113.

[2] R. Durrett, Brownian motion and martingales in analysis, Wadsworth, Belmont, Calif. 1984.

[3] N. Kazamaki, Continuous exponential martingales and BMO, Lecture notes in mathematics 1579, Springer 1994.

[4] R. Long, Martingale spaces and inequalities, Peking University Press 1993.

[5] W. Schachermayer, A characterisation of the closure of H^∞ in BMO, Séminaire de probabilités XXX, Lecture notes in Math. 1626, 344-356.

A Bipolar Theorem for $L^0_+(\Omega, \mathcal{F}, \mathbb{P})$

W. Brannath and W. Schachermayer

Abstract. A consequence of the Hahn-Banach theorem is the classical bipolar theorem which states that the bipolar of a subset of a locally convex vector space equals its closed convex hull.

The space $L^0(\Omega, \mathcal{F}, \mathbb{P})$ of real-valued random variables on a probability space $(\Omega, \mathcal{F}, \mathbb{P})$ equipped with the topology of convergence in measure fails to be locally convex so that — a priori — the classical bipolar theorem does not apply. In this note we show an analogue of the bipolar theorem for subsets of the positive orthant $L^0_+(\Omega, \mathcal{F}, \mathbb{P})$, if we place $L^0_+(\Omega, \mathcal{F}, \mathbb{P})$ in duality with itself, the scalar product now taking values in $[0, \infty]$. In this setting the order structure of $L^0(\Omega, \mathcal{F}, \mathbb{P})$ plays an important role and we obtain that the bipolar of a subset of $L^0_+(\Omega, \mathcal{F}, \mathbb{P})$ equals its closed, convex and solid hull.

In the course of the proof we show a decomposition lemma for convex subsets of $L^0_+(\Omega, \mathcal{F}, \mathbb{P})$ into a "bounded" and a "hereditarily unbounded" part, which seems interesting in its own right.

1. The Bipolar Theorem

Let $(\Omega, \mathcal{F}, \mathbb{P})$ be a probability space and denote by $L^0(\Omega, \mathcal{F}, \mathbb{P})$ the vector space of (equivalence classes of) real-valued measurable functions defined on $(\Omega, \mathcal{F}, \mathbb{P})$ which we equip with the topology of convergence in measure (see [KPR 84], chapter II, section 2). Recall the wellknown fact (see, e.g.,[KPR 84], theorem 2.2) that, for a diffuse measure $\mathbb{P}$, the topological dual of $L^0(\Omega, \mathcal{F}, \mathbb{P})$ is reduced to $\{0\}$ so that there is no counterpart to the duality theory, which works so nicely in the context of locally convex spaces (compare [Sch 67], chapter IV).

By $L^0_+(\Omega, \mathcal{F}, \mathbb{P})$ we denote the positive orthant of $L^0(\Omega, \mathcal{F}, \mathbb{P})$, i.e.,

$$L^0_+(\Omega, \mathcal{F}, \mathbb{P}) = \{f \in L^0(\Omega, \mathcal{F}, \mathbb{P}), f \geq 0\}.$$

The research of this paper was financially supported by the Austrian Science Foundation (FWF) under grant SFB#10 ('Adaptive Information Systems and Modelling in Economics and Management Science')

1980 *Mathematics Subject Classification* (1991 *Revision*). Primary: 62B20; 28A99; 26A20; 52A05; 46A55 Secondary: 46A40; 46N10; 90A09.

Key words and phrases. Convex sets of measurable functions; Bipolar theorem; bounded in probability; hereditarily unbounded.

We may consider the dual pair of convex cones $\langle L^0_+(\Omega, \mathcal{F}, \mathbb{P}), L^0_+(\Omega, \mathcal{F}, \mathbb{P}) \rangle$ where we define the scalar product $\langle f, g \rangle$ by

$$\langle f, g \rangle = \mathbb{E}[fg], \qquad f, g \in L^0_+(\Omega, \mathcal{F}, \mathbb{P}).$$

Of course, this is not a scalar product in the usual sense of the word as it may assume the value $+\infty$. But the expression $\langle f, g \rangle$ is a welldefined element of $[0, \infty]$ and the application $(f, g) \longrightarrow \langle f, g \rangle$ has — mutatis mutandis — the obvious properties of a bilinear function.

The situation is similar to the one encountered at the very foundation of measure theory: to overcome the difficulty that $\mathbb{E}[f]$ does not make sense for a general element $f \in L^0(\Omega, \mathcal{F}, \mathbb{P})$ one may either restrict to elements $f \in L^1(\Omega, \mathcal{F}, \mathbb{P})$ or to elements $f \in L^0_+(\Omega, \mathcal{F}, \mathbb{P})$, admitting in the latter case the possibility $\mathbb{E}[f] = +\infty$. In the present note we adopt this second point of view.

1.1 DEFINITION. We call a subset $C \subseteq L^0_+$ *solid*, if $f \in C$ and $0 \le g \le f$ implies that $g \in C$. The set C is said to be *closed in probability* or simply *closed*, if it is closed with respect to the topology of convergence in probability.

1.2 DEFINITION. For $C \subseteq L^0_+$ we define the *polar* C^0 of C by

$$C^0 = \{g \in L^0_+ : \mathbb{E}[fg] \le 1, \quad \text{for each } f \in C\}$$

1.3 Bipolar Theorem. For a set $C \subseteq L^0_+(\Omega, \mathcal{F}, \mathbb{P})$ the polar C^0 is a closed, convex, solid subset of $L^0_+(\Omega, \mathcal{F}, \mathbb{P})$.

The bipolar

$$C^{00} = \{f \in L^0_+ : \mathbb{E}[fg] \le 1, \quad \text{for each } g \in C^0\}$$

is the smallest closed, convex, solid set in $L^0_+(\Omega, \mathcal{F}, \mathbb{P})$ containing C.

To prove theorem 1.3 we need a decomposition result for convex subsets of L^0_+ we present in the next section. The proof of theorem 1.3 will be given in section 3.

We finish this introductory section by giving an easy extension of the bipolar theorem 1.3 to subsets of L^0 (as opposed to subsets of L^0_+). Recall that, with the usual definition of *solid* sets in vector lattices (see [Sch 67], chapter V, section 1), a set $D \subset L^0$ is defined to be *solid* in the following way.

1.4 DEFINITION. A set $D \subset L^0$ is solid, if $f \in D$ and $h \in L^0$ with $|h| \le |f|$ implies $h \in D$.

Note that a set $D \subset L^0$ is solid if and only if the set of its absolut values $|D| = \{|h| : h \in D\} \subset L^0_+$ form a solid subset of L^0_+ as defined in 1.1 and $D = \{h \in L^0 : |h| \in |D|\}$. Hence the second part of theorem 1.3 implies:

1.5 Corollary. Let $C \subset L^0$ and $|C| = \{|f| : f \in C\}$. Then the smallest closed, convex, solid set in L^0 containing C equals $\{f \in L^0 : |f| \in |C|^{00}\}$.

PROOF. Let D' be the smallest closed, convex, solid set in L^0_+ containing $|C|$ and $D = \{f : |f| \in D'\}$. One easily verifies that D is the smallest closed, convex and solid subset of L^0 containing C. Applying theorem 1.3 to $|C|$, we obtain that $D' = |C|^{00}$, which implies that $D = \{f \in L^0 : |f| \in |C|^{00}\}$. □

For more detailed results in the line of corollary 1.5 concerning more general subsets of L^0 we refer to [B 97].

2. A Decomposition Lemma for Convex Subsets of $L_+^0(\Omega, \mathcal{F}, \mathbb{P})$

Recall that a subset of a topological vector space X is *bounded* if it is absorbed by every zero-neighborhood of X ([Sch 67], Chapter I, Section 5). In the case of $L^0(\Omega, \mathcal{F}, \mathbb{P})$ this amounts to the following well-known concept.

2.1 DEFINITION. A subset $C \subseteq L^0(\Omega, \mathcal{F}, \mathbb{P})$ is *bounded in probability* if, for $\varepsilon > 0$, there is $M > 0$ such that

$$\mathbb{P}[|f| > M] < \varepsilon, \qquad \text{for } f \in C.$$

We now introduce a concept which describes a strong form of unboundedness in $L^0(\Omega, \mathcal{F}, \mathbb{P})$.

2.2 DEFINITION. A subset $C \subseteq L^0(\Omega, \mathcal{F}, \mathbb{P})$ is called *hereditarily unbounded in probability on a set* $A \in \mathcal{F}$, if, for every $B \in \mathcal{F}, B \subseteq A, \mathbb{P}[B] > 0$ we have that $C|_B = \{f\chi_B : f \in C\}$ fails to be a bounded subset of $L^0(\Omega, \mathcal{F}, \mathbb{P})$.

We now are ready to formulate the decomposition result:

2.3 Lemma. *Let C be a convex subset of $L_+^0(\Omega, \mathcal{F}, \mathbb{P})$. There exists a partition of Ω into disjoint sets $\Omega_u, \Omega_b \in \mathcal{F}$ such that*

(1) The restriction $C|_{\Omega_b}$ of C to Ω_b is bounded in probability.

(2) C is hereditarily unbounded in probability on Ω_u.

The partition $\{\Omega_u, \Omega_b\}$ is the unique partition of Ω satisfying (1) and (2) (up to null sets). Moreover

(3) If $\mathbb{P}[\Omega_b] > 0$ we may find a probability measure Q_b equivalent to the restriction $\mathbb{P}|_{\Omega_b}$ of $\mathbb{P}$ to Ω_b such that C is bounded in $L^1(\Omega, \mathcal{F}, Q_b)$. In fact, we may choose Q_b such that $\frac{dQ_b}{d\mathbb{P}}$ is uniformly bounded.

(4) For $\varepsilon > 0$ there is $f \in C$ s.t.

$$\mathbb{P}[\Omega_u \cap \{f < \varepsilon^{-1}\}] < \varepsilon.$$

(5) Denote by D the smallest closed, convex, solid set containing C. Then D has the form

$$D = D|_{\Omega_b} \oplus L_+^0|_{\Omega_u},$$

where $D|_{\Omega_b} = \{u\chi_{\Omega_b} : u \in D\}$ and $L_+^0|_{\Omega_u} = \{v\chi_{\Omega_u} : v \in L_+^0(\Omega, \mathcal{F}, \mathbb{P})\}$.

PROOF. Noting that the lemma holds true for C iff it holds true for the solid hull of C we may assume w.l.g. that C is solid and convex.

We now use a standard exhausting argument to obtain Ω_u. Denote by $\mathcal{B}$ the family of sets $B \in \mathcal{F}, \mathbb{P}[B] > 0$, verifying

$$\text{for } \varepsilon > 0 \text{ there is } f \in C, \text{ s.t. } \mathbb{P}[B \cap \{f < \varepsilon^{-1}\}] < \varepsilon.$$

Note that $\mathcal{B}$ is closed under countable unions: indeed, for $(B_n)_{n=1}^\infty$ is $\mathcal{B}$ and $\varepsilon > 0$, find elements $(f_n)_{n=1}^\infty$ in C such that

$$\mathbb{P}[B_n \cap \{f_n < 2^n\varepsilon^{-1}\}] < 2^{-n}\varepsilon.$$

Then, by the convexity and solidity of C

$$F_N = \sum_{n=1}^{N} 2^{-n} f_n$$

is in C and, for N large enough,

$$\mathbb{P}[B \cap \{F_N < \varepsilon^{-1}\}] < \varepsilon.$$

Hence there is a set of maximal measure in $\mathcal{B}$, which we denote by Ω_u and which is unique up to null-sets. Let $\Omega_b = \Omega \backslash \Omega_u$.

(1) and (3): If $\mathbb{P}[\Omega_b] = 0$ assertions (1) and (3) are trivially satisfied; hence we may assume that $\mathbb{P}[\Omega_b] > 0$. We want to verify (3). Note, since C is a solid subset of L_+^0, the convex set $C' = C \cap L^1(\Omega, \mathcal{F}, \mathbb{P}|_{\Omega_b})$ is dense in C with respect to the convergence in probability $\mathbb{P}|_{\Omega_b}$; hence, by Fatou's Lemma, it is enough to find a probability measure $Q_b \sim \mathbb{P}|_{\Omega_b}$ such that C' is bounded in $L^1(Q_b)$. To this end we apply Yan's theorem ([Y 80], theorem 2) to C'. For convex, solid subsets C' of $L_+^1(\mathbb{P}|_{\Omega_b})$, this theorem states, that the following two assertions are equivalent:

(i) for each $A \in \mathcal{F}$ with $\mathbb{P}|_{\Omega_b}[A] = \mathbb{P}[\Omega_b \cap A] > 0$, there is $M > 0$ such that $M\chi_A$ is not in the $L^1(\Omega, \mathcal{F}, \mathbb{P}|_{\Omega_b})$-closure of C';

(ii) there exists a probability measure Q_b equivalent to $\mathbb{P}|_{\Omega_b}$ such that C' is a bounded subset of $L_+^1(\Omega, \mathcal{F}, Q_b)$. In addition, we may choose Q_b such that $\frac{dQ_b}{d\mathbb{P}}$ is uniformly bounded.

Assertion (i) is satisfied because otherwise we could find a subset $A \in \mathcal{F}, A \subset \Omega_b, \mathbb{P}[A] > 0$ belonging to the family $\mathcal{B}$, in contradiction to the construction of Ω_u above.

Hence assertion (ii) holds true which implies assertion (3) of the lemma. Obviously (3) implies assertion (1).

(2) and (4): As Ω_u is an element of $\mathcal{B}$ we infer that (4) holds true which in turn implies (2).

(5): Obviously $D \subset D|_{\Omega_b} \oplus L_+^0|_{\Omega_u}$. To show the reverse inclusion let $f = v + w$ with $v \in D|_{\Omega_b}$ and $w \in L_+^0|_{\Omega_u}$. We have to show that $f \in D$. Property (2) implies that, for every $n \in \mathbb{N}$, we find an $f_n \in C$ such that $\mathbb{P}[\{f_n \le n^2\} \cap \Omega_u] \le (1/n)$. Since $h_n = (1 - (1/n))v + (1/n)(f_n \wedge (n\,w)) \in D$ and $h_n \to v + w$ in probability, it follows that $f \in D$.

According to (2), C is unbounded in probability in $L^0(\Omega, \mathcal{F}, \mathbb{P}|_B)$ for each $B \subseteq \Omega_u$ with $\mathbb{P}[B] > 0$; the uniqueness of the decomposition $\Omega = \Omega_u \cup \Omega_b$ (up to null sets) with respect to the assertions (1) and (2) immediately follows from this. $\square$

3. The Proof of the Bipolar Theorem 1.3

To prove the first assertion of theorem 1.3 fix a set $C \subset L_+^0(\Omega, \mathcal{F}, \mathbb{P})$ and note that the convexity and solidity of C^0 are obvious and the closedness of C^0 follows from Fatou's lemma.

To prove the second assertion of the theorem denote by D the intersection of all closed, convex and solid sets in L_+^0 containing C. Clearly D is closed, convex and solid, which implies the inclusion $D \subset C^{00}$. We have to show that $C^{00} \subseteq D$.

Using assertion (5) of lemma 2.3 we may decompose Ω into $\Omega = \Omega_b \cup \Omega_u$ such that $D = D|_{\Omega_b} \oplus L^0_+|_{\Omega_u}$ and (if $\mathbb{P}[\Omega_b] > 0$) we find a probability measure Q_b supported by Ω_b and equivalent to the restriction $\mathbb{P}|_{\Omega_b}$ of $\mathbb{P}$ to Ω_b such that D is bounded in $L^1(\Omega, \mathcal{F}, Q_b)$ (assertion (2)).

Now suppose that there is $f_0 \in C^{00} \setminus D$ and let us work towards a contradiction. Let $f_b = f_0 \chi_{\Omega_b}$ denote the restriction of f_0 to Ω_b. It is enough to show that f_b is in D. Let us denote by $D_b = \{f \chi_{\Omega_b} : f \in D\}$ the restriction of D to Ω_b and by

$$\tilde{D}_b = D_b - L^1_+(\Omega, \mathcal{F}, Q_b) = \{h \in L^1(\Omega, \mathcal{F}, Q_b)\} : \exists f \in D_b \text{ s.t. } h \le f, Q_b - \text{a.s.}\}$$

the set of elements of $L^1(Q_b)$ dominated by an element of D_b. It is straightforward to verify that D_b and $\tilde{D}_b$ are $L^1(Q_b)$-closed, convex subsets of $L^1_+(Q_b)$ and $L^1(Q_b)$ respectively, and that D_b is bounded in $L^1_+(Q_b)$.

To show that f_b is contained in D (equivalently in D_b or in $\tilde{D}_b$) it suffices to show that $f_b \wedge M$ is in D_b, for each $M \in \mathbb{R}_+$. Indeed, by the $L^1(Q)$-boundedness and $L^1(Q)$-closedness of D_b this will imply that $f_b = L^1(Q) - \lim_{M \to \infty} f_b \wedge M$ is in D.

So we are reduced to assuming that f_b is an element of $L^1(Q_b)$ which is not an element of $\tilde{D}_b$. Now we may apply a version of the Hahn-Banach theorem (the separation theorem [Sch 67], theorem 9.2) to the Banach space $L^1(Q_b)$ to find an element $g \in L^\infty(Q_b)$ such that

$$\mathbb{E}[f_b g] > 1 \text{ while } \mathbb{E}[fg] \le 1, \text{ for } f \in \tilde{D}_b.$$

As $\tilde{D}_b$ contains the negative orthant of $L^1(Q_b)$ we conclude that $g \ge 0$. Considering g as an element of $L^0_+(\Omega, \mathcal{F}, \mathbb{P})$ by letting g equal zero on Ω_u we therefore have that $g \in C^0$ and the first inequality above implies that $f_b \notin C^{00}$ and so that $f \notin C^{00}$, a contradiction finishing the proof. $\square$

4. Notes and Comments

4.1 Note: Our motivation for the formulation of the bipolar theorem 1.3 above comes from Mathematical Finance: in the language of this theory there often comes up a duality relation between a set of contingent claims and a set of state price densities, i.e., Radon-Nikodym derivatives of absolutely continuous martingale measures. In this setting it turns out that $L^0(\Omega, \mathcal{F}, \mathbb{P})$ often is the natural space to work in (as opposed to $L^p(\Omega, \mathcal{F}, \mathbb{P})$ for some $p > 0$), as it remains unchanged under the passage from $\mathbb{P}$ to an equivalent measure Q (while $L^p(\Omega, \mathcal{F}, \mathbb{P})$ does change, for $0 < p < \infty$). We refer, e.g., [DS 94] for a general exposition of the above described duality relations and to [KS 97] for an applications of the bipolar theorem 1.3.

4.2 Note: Lemma 2.3 may be viewed as a variation of theorem 1 in [Y 80], which is a result based on previous work of Mokobodzki (as an essential step in Dellacherie's proof of the semimartingale characterization theorem due to Bichteler and Dellacherie; see [Me 79] and [Y 80]). The proof of Yan's theorem is a blend of a Hahn-Banach and an exhaustion argument (see, e.g., [S 94] for a presentation of this proof and [Str 90], [S 94] for applications of Yan's theorem to Mathematical

Finance) In fact, these arguments have their roots in the proof of the Halmos-Savage theorem [HS 49] and the theorems of Nikishin and Maurey [N 70], [M 74].

4.3 Note: In the course of the proof of lemma 2.3 we have shown that a convex subset C of $L^0_+(\Omega, \mathcal{F}, \mathbb{P})$ is hereditarily unbounded in probability on a set $A \in \mathcal{F}$ iff, for $\varepsilon > 0$, there is $f \in C$ with

$$\mathbb{P}[A \cap \{f < \varepsilon^{-1}\}] < \varepsilon.$$

which seems a fact worth noting in its own right.

4.4 Note: Notice that by theorem 1.3 the bipolar C^{00} of a given set $C \subset L^0_+$, although originally defined with respect to $\mathbb{P}$, does not change if we replace $\mathbb{P}$ by an equivalent measure $\mathbb{Q}$. This may also be seen directly (without applying theorem 1.3) in the following way: If $\mathbb{Q} \sim \mathbb{P}$ are equivalent probability measures and $h = d\mathbb{Q}/d\mathbb{P}$ is the Radon-Nikodym derivative of $\mathbb{Q}$ with respect to $\mathbb{P}$, then the polar $C^0(\mathbb{Q})$ of a given convex set $C \subset L^0_+$ with respect to $\mathbb{Q}$ equals $C^0(\mathbb{Q}) = h^{-1} \cdot C^0(\mathbb{P})$, where $C^0(\mathbb{P})$ is the dual of C with respect to $\mathbb{P}$. On the other hand $\mathbb{E}_{\mathbb{P}}[f\,g] = \mathbb{E}_{\mathbb{P}}[f\,h\,h^{-1}\,g] = E_{\mathbb{Q}}[f\,h^{-1}\,g]$ for all $g \in L^0_+$ and therefore the polar $C^{00}(\mathbb{Q})$ of $C^0(\mathbb{Q})$ (defined with respect to $\mathbb{Q}$) coincides with the polar $C^{00}(\mathbb{P})$ of $C^0(\mathbb{P})$ (defined with respect to $\mathbb{P}$).

References

[B 97]. W. Brannath, *No Arbitrage and Martingale Measures in Option Pricing*, Dissertation. University of Vienna (1997).

[DS 94]. F. Delbaen, W. Schachermayer, *A General Version of the Fundamental Theorem of Asset Pricing*, Math. Annalen **300** (1994), 463 — 520.

[HS 49]. Halmos. P.R., Savage, L.J. (1949), *Application of the Radon-Nikodym Theorem to the Theory of Sufficient Statistics*, Annals of Math. Statistics 20. 225–241..

[KS 97]. D. Kramkov, W. Schachermayer, *A Condition on the Asymptotic Elasticity of Utility Functions and Optimal Investment in Incomplete Markets*, Preprint (1997).

[KPR 84]. N.J. Kalton, N.T. Peck, J.W. Roberts, *An F-space Sampler*. London Math. Soc. Lecture Notes **89** (1984).

[M 74]. B. Maurey, *Théorèmes de factorisation pour les opérateurs linéaires à valeurs dans un espace L^p*, Astérisque **11** (1974).

[Me 79]. P.A., Meyer, *Caractérisation des semimartingales, d'après Dellacherie*, Séminaire de Probabilités XIII, Lect. Notes Mathematics **721** (1979), 620 — 623.

[N 70]. E.M. Nikishin, *Resonance theorems and superlinear operators*. Uspekhi Mat. Nauk **25**, Nr. 6 (1970), 129 — 191.

[S 94]. W. Schachermayer, *Martingale measures for discrete time processes with infinite horizon*, Math. Finance **4** (1994), 25 — 55.

[Sch 67]. Schaefer, H.H. (1966), *Topological Vector Spaces*, Springer Graduate Texts in Mathematics.

[Str 90]. Stricker, C., *Arbitrage et lois de martingale*, Ann. Inst. Henri. Pincaré **Vol. 26, no. 3** (1990), 451–460.

[Y 80]. J. A. Yan, *Caractérisation d' une classe d'ensembles convexes de L^1 ou H^1*. Séminaire de Probabilités XIV, Lect. Notes Mathematics **784** (1980), 220–222.

Barycentre canonique pour un espace métrique à courbure négative

ES-SAHIB Aziz & HEINICH Henri

Résumé

Pour une variable aléatoire X intégrable à valeurs dans un espace (M, d) métrique complet séparable et à courbure négative, nous définissons un barycentre de X. Ce point, $b(X)$, appartient à l'ensemble des espérances au sens de Doss de X et ne dépend que de la loi de la variable. De plus si X et Y sont deux variables intégrables, alors $d(b(X), b(Y)) \leq E[d(X, Y)]$.
Nous étudions le problème de cohérence (loi des grands nombres) pour ce barycentre et nous montrons un théorème ergodique.
Puis nous remplaçons l'espérance de Doss par celle de Herer puis par celle d'Émery et Mokobodzki.

Abstract

For X an integrable random variable with values in a complete separable metric space (M, d) with negative curvature, we define a point $b(X)$ called barycenter of X which depends only on the law of X and belongs to the set of Doss expectation of X. Moreover for two integrable variables we have $d(b(X), b(Y)) \leq E(d(X, Y))$. We study the coherence problem: strong law of larges numbers for this barycenter and an ergodic theorem is given.
In the end we change Doss expectation for the Herer one and for the Émery-Mokobodzki one.

Introduction

La notion d'espérance pour une variable aléatoire à valeurs dans un espace métrique M trouve ses premières formulations dans [8] et [5]. Les propriétés de cette espérance se développent dans [2], [6], [9], [11]. Par la suite Herer introduit en [10] et [12] une autre définition et s'intéresse aux espaces à courbure négative. Plus récemment le cas M variété a été abordé par [1] et [7].

Ces différentes approches ont en commun que l'espérance d'une variable aléatoire est un sous ensemble fermé en général non réduit à un point. Ceci soulève des difficultés, en particulier pour obtenir une loi forte des grands nombres, car il faut considérer des convergences de fermés par exemple au sens de Hausdorff ou de Wijsman *c.f.* [15]. Cette dernière référence contient aussi d'autres versions de la loi forte des grands nombres.

Lorsque $d^2(\cdot, x)$ est strictement convexe, d'autres auteurs définissent classiquement le barycentre $b(X)$ d'une variable aléatoire X intégrable, par $b(X) = \underset{a}{\text{Argmin}}\ E[d^2(a, X)]$. L'espérance est alors un singleton. Il en est de même pour la définition adoptée par [13] pour les variétés.

Notre but est d'obtenir de manière "canonique", pour une v.a X intégrable à valeurs dans M aussi général que possible, un point appelé barycentre de X (ou espérance de X).

Nous allons centrer notre travail sur l'espérance au sens de Doss du fait de sa simplicité et de l'adaptation des méthodes introduites à d'autres notions d'espérances. Rappelons, que pour un espace de Banach séparable, la notion d'espérance de Doss et celle de Bochner coïncident [3].

Plan du travail

• La première partie établit l'existence d'un barycentre canonique avec des hypothèses minimales.
- En I-1 nous rappelons les outils nécessaires.
- Dans la partie I-2 nous définissons avec la loi des grands nombres et de manière canonique, pour une v.a. X (ou une probabilité μ) intégrable, un point $b(X)$ (ou $b(\mu)$), appartenant à l'ensemble des espérances de Doss de X. Ce barycentre vérifie la propriété fondamentale :

$$d(b(X), b(Y)) \leq E[d(X, Y)] \ .$$

- Nous donnons dans I-3 un moyen d'obtenir le barycentre précédent sans utiliser la convergence presque sûre. Un exemple est abordé par simulation.
- La partie I-4 est consacrée au problème de cohérence sous deux aspects. Nous montrons d'abord que le barycentre obtenu à partir de la loi empirique converge p.s. vers le barycentre de la loi initiale (supposée intégrable). Nous donnons ensuite une condition nécessaire et suffisante pour que les barycentres des lois empiriques forment une martingale. Pour clore cette partie nous montrons que le théorème ergodique demeure dans ce cadre.

• Dans la partie II nous remplaçons l'espérance de Doss par celle de Herer puis par celle d'Émery-Mokobodzki.

I - 1 Définitions et notations

Dans toute la suite M est un espace métrique séparable complet muni de sa tribu borélienne. La distance de deux points x et y de M est notée xy.

Une variable aléatoire (v.a.) X définie sur un espace de probabilité $(\Omega, \mathcal{A}, P)$ et à valeurs dans M, est dite *intégrable*, si pour un point $a \in M$ (et donc pour tout point) la v.a.r. Xa est intégrable.

Pour une v.a. X intégrable, l'ensemble $\{E[X]\} = \{m | m \in M, \; am \leq E[aX],$ pour tout $a \in M\}$ est l'*espérance* de X au sens de Doss, nous écrirons aussi $\{E[\mu]\}$ pour μ probabilité (intégrable) sur M. L'ensemble $\{E[X]\}$ ne dépend que de la loi de X et est fermé.

L'espace M est dit *convexe*, respectivement *strictement convexe*, si pour toute probabilité $\mu = \dfrac{1}{2}(\delta_x + \delta_y)$, $\{E[\mu]\}$ est non vide, respectivement est un singleton, noté $b(x, y)$.

Nous utiliserons aussi une notion plus forte : un espace M convexe est dit à *coubure négative* si, $\forall (x_1, x_2, y_1, y_2) \in M^4$, $\forall u \in \{E[\frac{1}{2}(\delta_{x_1} + \delta_{x_2})]\}$ et tout $v \in \{E[\frac{1}{2}(\delta_{y_1} + \delta_{y_2})]\}$, on a $uv \leq \dfrac{1}{2}(x_1 y_1 + x_2 y_2)$. Un tel espace est strictement convexe (prendre $y_i = x_i$) et la condition de courbure négative s'écrit :

pour tout point $(x_1, x_2, y_1, y_2) \in M^4$, $b(x_1, x_2) b(y_1, y_2) \leq \dfrac{1}{2}(x_1 y_1 + x_2 y_2)$.

Soit $\mathcal{F}$ une sous tribu de $\mathcal{A}$, on dit qu'une v.a. Y à valeurs dans M est une *moyenne* (ou *espérance*) *conditionnelle* de X relativement à $\mathcal{F}$ si

(i) Y est $\mathcal{F}$-mesurable

(ii) $\forall a \in M$, $aY \leq E^{\mathcal{F}}[aX]$ p.s. (espérance conditionnelle d'une v.a.r.).

On note $\{E^{\mathcal{F}}[X]\}$ l'ensemble des v.a. Y vérifiant (i) et (ii).

Une suite $(Y_n)_n$ adaptée à une filtration monotone $(\mathcal{F}_n)_n$ est une *martingale* si $Y_n \in \{E^{\mathcal{F}_n}[Y_m]\}$ pour $m = n + 1$ dans le cas d'une filtration croissante et pour $m = n - 1$ lors d'une filtration décroissante.

I - 2   Barycentre d'une v.a. intégrable.

Dans un premier temps nous définissons un barycentre intermédiaire pour une loi uniforme sur un ensemble $\{x_1, x_2, \cdots, x_k\}$ où les x_i sont des points *non nécessairement distincts* de M.

Proposition 1   *Soit M un espace métrique complet strictement convexe. Pour tout $k > 1$, il existe une application b_k de M^k dans M telle que le point $b_k(x_1, x_2, \cdots, x_k)$, noté aussi $\displaystyle\sum_{i=1}^{k} \frac{1}{k} x_i$, appartient à $\{E[\sum_{i=1}^{k} \frac{1}{k} \delta_{x_i}]\}$ et vérifie*

$$\sum_{i=1}^{k} \frac{1}{k} x_i = \sum_{i=1}^{k} \frac{1}{k} \Big(\sum_{j \leq k, j \neq i} \frac{1}{k-1} x_j \Big).$$

De plus si l'espace est à courbure négative, alors

$$b_k(x_1, x_2, \cdots, x_k) b_k(y_1, y_2, \cdots, y_k) \leq \frac{1}{k} \sum_{1}^{k} x_i y_i,$$

en particulier l'application b_k est continue.

Preuve. L'existence de $b_k(\cdot)$ se fait par récurrence sur k.

Pour $k = 2$ et $\mu = \frac{1}{2}(\delta_x + \delta_y)$, comme M est strictement convexe posons $b_2(x, y) = b(x, y)$.

Par récurrence supposons que, pour tout $l < k$ $(k > 2)$ et tout point $(x_1, x_2, \cdots, x_l) \in M^l$, on ait défini de manière canonique un barycentre $b_l(x_1, x_2, \cdots, x_l) = \sum_{i=1}^{l} \frac{1}{l} x_i$, vérifiant les assertions de la proposition.

Considérons alors un point $(x_1, x_2, \cdots, x_k) \in M^k$.

Définissons, par récurrence, la suite $(x_1^n, x_2^n, \cdots, x_k^n)_{n \geq 0}$, associée à $(x_1, \cdots, x_k)$, par : $x_i^0 = x_i$ et, pour $n \geq 1$, $x_i^n = \sum_{j \leq k, j \neq i} \frac{1}{k-1} x_j^{n-1}$.

Nous allons montrer que les x_i^n convergent, lorsque $n \to \infty$, vers un point indépendant de i et qui sera par définition $b_k(x_1, x_2, \cdots, x_k)$.

À cette fin, considérons l'ensemble $\mathcal{P}_f(A)$ des probabilités dont le support est contenu dans $A \subset M$, cardinal de A fini. L'enveloppe convexe de A est $E_1(A) = \bigcup_{\mu \in \mathcal{P}_f(A)} \{E[\mu]\}$. Notons $E_2(A) = E_1(E_1(A))$ et montrons que ces deux enveloppes convexes coïncident : "$A = \{x_1, \cdots x_k\} \implies E_1(A) = E_2(A)$."

En effet soit $a \in E_2(A)$, on peut écrire $a \in \{E[\sum_{j \in J} \beta_j \delta_{m_j}]\}$ où J fini et $m_j \in \{E[\sum_{i=1}^{k} \alpha_i^j \delta_{x_i}]\}$. Posons $\gamma_i = \sum_{j \in J} \beta_j \alpha_i^j$ et $\mu = \sum_{i=1}^{k} \gamma_i \delta_{x_i}$. Pour $z \in M$ les inégalités : $az \leq \sum_j \beta_j m_j z \leq \sum_j \beta_j (\sum_i \alpha_i^j x_i z) = \sum_{i,j} \beta_j \alpha_i^j x_i z = \sum_i \gamma_i x_i z$, impliquent $a \in \{E[\mu]\}$ et donc l'inclusion $E_2(A) \subset E_1(A)$. La réciproque est triviale.

Remarquons que $E_1(A)$ est fermé.

Retour à la preuve de la proposition 1.

Avec les notations introduites à partir de la suite $(x_1, \cdots, x_k)$, considérons $A_n := \{x_1^n, x_2^n, \cdots, x_k^n\}$ et montrons que la suite $(E_1(A_n))_n$ est décroissante et converge vers un point.

   a) Décroissance

Comme $x_i^{n+1} \in E_1(A_n)$ pour tout i, on a $A_{n+1} \subset E_1(A_n)$ par conséquent $E_1(A_{n+1}) \subset E_2(A_n)$, avec l'égalité précédente, $E_1(A_{n+1}) \subset E_2(A_n) = E_1(A_n)$.

   b) Montrons que les ensembles $E_1(A_n)$ et A_n ont même diamètre.

Le diamètre d'un ensemble A est, par définition, $\mathrm{diam}(A) := \sup_{x,y \in A} xy$. Soient z et u deux points de $E_1(A_n)$, $z \in \{E[\sum_i \gamma_i \delta_{x_i}]\}$, $u \in \{E[\sum_j \beta_j \delta_{y_j}]\}$ où les x_i et les y_j appartiennent à A_n. On a :

$$zu \leq \sum_i \gamma_i x_i u \leq \sup_i x_i u \leq \sup_i \sum_j \beta_j x_i y_j \leq \sup_{i,j} x_i y_j = \mathrm{diam}(A_n).$$

c) Montrons que diam$(E_1(A_n))$ tend vers 0.

De manière équivalente montrons que $L^{(n)} := \sum_{i \neq j} x_i^n x_j^n \to 0$.

Comme $x_i^{n+1} \in \{E[\sum_{j \leq k, j \neq i} \frac{1}{k-1} \delta_{x_j^n}]\}$, on a $x_i^{n+1} x_j^{n+1} \leq (\frac{1}{k-1})^2 \sum_{p \neq i} \sum_{q \neq j} x_p^n x_q^n$.

En écrivant $\sum_{p \neq i} \sum_{q \neq j} x_p^n x_q^n = \sum_{p \neq q} x_p^n x_q^n - \sum_p x_p^n x_j^n - \sum_q x_i^n x_q^n + x_i^n x_j^n$, puis en sommant sur les couples $(i, j), i \neq j$, on arrive à :

$$L^{(n+1)} = \sum_{i \neq j \leq k} x_i^{n+1} x_j^{n+1} \leq \frac{k^2 - 3k + 3}{(k-1)^2} L^{(n)}.$$ L'espace M étant complet et $k > 2$, nous avons la convergence de la suite $E_1(A_n)$ vers un unique point qui est, par définition, $b_k(x_1, x_2, \cdots, x_k)$.

Montrons que $b_k(x_1, x_2, \cdots, x_k) \in \{E[\frac{1}{k} \sum_{i=1}^k \delta_{x_i}]\}$.

On a les inclusions $\{E[\frac{1}{k} \sum_i \delta_{x_i^{n+1}}]\} \subset \{E[\frac{1}{k} \sum_i \delta_{x_i^n}]\} \subset \{E[\frac{1}{k} \sum_i \delta_{x_i}]\}$ et

$\{E[\frac{1}{k} \sum_i \delta_{x_i^n}]\} \subset E_1(A_n)$, à la limite, on obtient la relation souhaitée.

Le premier terme de la suite associée à $(x_1, \cdots, x_k)$ est $(x_1^1, \cdots, x_k^1)$ où $x_i^1 = \sum_{j \leq k, j \neq i} \frac{1}{k-1} x_j$ et, comme $b_k(x_1, \cdots, x_k) = b_k(x_1^1, \cdots, x_k^1)$, la relation liant $b_k(\cdot)$ et $b_{k-1}(\cdot)$ est démontrée.

Pour achever la preuve, considérons le cas d'un espace à courbure négative et montrons, par récurrence, l'inégalité recherchée.

Pour $k = 2$, c'est la définition même de la courbure négative.

Admettons la relation jusqu'au rang $k-1$ et considérons deux points $(x_1, \cdots, x_k)$ et $(y_1, \cdots, y_k)$. En reprenant les notations précédentes,

$x_i^1 y_i^1 = (\sum_{j \neq i} \frac{1}{k-1} x_j)(\sum_{j \neq i} \frac{1}{k-1} y_j)$, et avec l'hypothèse de récurrence,

$x_i^1 y_i^1 \leq \frac{1}{k-1} \sum_{j \neq i} x_j y_j$. Donc $\sum_{i=1}^k x_i^1 y_i^1 \leq \frac{1}{k-1} \sum_{i=1}^k \sum_{j \neq i} x_j y_j = \sum_{i=1}^k x_i y_i$.

En itérant, la suite $(\sum_{i=1}^k x_i^n y_i^n)$ est décroissante. La preuve s'achève en remarquant que $x_i^n \xrightarrow[n \to \infty]{} b_k(x_1, \cdots, x_k)$ et que $y_i^n \xrightarrow[n \to \infty]{} b_k(y_1, y_2, \cdots, y_k)$. □

Remarques.

- Pour toute permutation σ de $\{1 \cdots k\}$, $\sigma \in \Sigma(k)$, $\sum_{i=1}^k \frac{1}{k} x_i = \sum_{i=1}^k \frac{1}{k} x_{\sigma(i)}$.

- Cas d'une variable aléatoire étagée.

Pour X v.a. définie sur $[0, 1]$, à valeurs dans M et de loi uniforme sur $(x_1, x_2, \cdots, x_k) \in M^k$, on écrit $b_k(X) := b_k(x_1, x_2, \cdots, x_k)$.

Lorsque $X = (X_1, \cdots, X_k)$ est une v.a. étagée à valeurs M^k, l'application :

$$\omega \longrightarrow b_k(X(\omega)) = b_k(X_1(\omega), \cdots, X_k(\omega)) = \sum_{i=1}^{k} \frac{1}{k} X_i(\omega) \text{ est une v.a. étagée.}$$

À partir de la suite $\big(b_n(\cdot)\big)_n$ nous allons définir un barycentre canonique, pour une variable aléatoire intégrable, en utilisant une propriété de martingale. Rappelons à cet effet le théorème de convergence p.s. des martingales selon [2] ou [6] :

"*Si les boules fermées bornées de M sont compactes, alors toute martingale (Y_n) pour une filtration décroissante (S_n), est convergente p.s. i.e.* $\lim\limits_{n \to \infty} Y_n$ *existe p.s.*"

Un espace dont les boules fermées et bornées sont compactes est dit "*finiment compact*". Cette condition est en général trop exigeante pour obtenir uniquement la loi forte des grands nombres. Par exemple, elle n'est pas vérifiée dans les espaces de Banach de dimension infinie. C'est pour cela que nous introduisons la définition suivante :

L'espace M est fini-compact *si, pour tout* $(x_1, \cdots, x_n) \in M^n$, *l'ensemble* $E_1(\{x_1, \cdots, x_n\})$ *est compact.*

Énonçons maintenant le résultat central.

Théorème 2 (barycentre d'une v.a. intégrable)

Soient (X_n) une suite de v.a. i.i.d., X_1 intégrable dans M espace métrique complet séparable, fini compact à courbure négative et S_n la tribu des événements dépendant symétriquement de $(X_1, \cdots, X_n)$ et d'une manière quelconque de $(X_{n+1}, \cdots)$. Alors la suite $\Big(Y_n = b_n(X_1, \cdots, X_n) = \sum_{i=1}^{n} \frac{1}{n} X_i\Big)$ est une martingale adaptée à la filtration (S_n), convergeant p.s. vers une constante, $b(X_1)$, qui ne dépend que de la loi de X_1.

De plus, pour X et Y intégrables, on a $b(X) \in \{E[X]\}$ et $b(X)b(Y) \le E[XY]$.

Preuve   Montrons tout d'abord que Y_n est une v.a.

Pour cela construisons une suite de fonctions (f_k) permettant d'approcher les v.a. par des v.a. étagées. Notons (a_k) une suite dense dans M. Définissons $f_k : M \to M$ par $f_k(x) = a_i$ où i est le plus petit entier inférieur à k et tel que $a_i x \le a_n x$, pour tout n, $1 \le n \le k$. On voit, sans difficulté, que $f_k(M) = \{a_1, \cdots a_k\}$ et que $\forall x \in M$, $x f_k(x)$ tend vers 0 et enfin, $x f_k(x) \le a_1 x$.

Ainsi la suite de v.a. étagées $\big(X_n^k = f_k(X_n)\big)_k$ converge pour tout n fixé, ponctuellement et dans $\mathbb{L}^1(M)$, vers X_n, $\mathbb{L}^1(M)$ étant l'espace des v.a. à valeurs dans M et intégrables. Posons $Y_n^k(\omega) = \sum_{i=1}^{n} \frac{1}{n} X_i^k(\omega)$, Y_n^k est une v.a. étagée et la proposition 1 assure que $Y_n^k(\omega) \xrightarrow[k \to \infty]{} Y_n(\omega)$, Y_n est donc une v.a. De plus l'invariance par permutation : $\sum_{1}^{n} \frac{1}{n} x_i = \sum_{1}^{n} \frac{1}{n} x_{\sigma(i)}$, $\sigma \in \Sigma(n)$, montre que Y_n est S_n mesurable.

Afin d'établir que (Y_n) est une martingale, *i.e.* $Y_{n+1} \in \{E^{\mathcal{S}_{n+1}}[Y_n]\}$, montrons la propriété suivante

(P) *Pour toute bijection,* $\sigma \in \Sigma(n+1)$, $\quad \{E^{\mathcal{S}_{n+1}}[\sum_1^n \frac{1}{n} X_i]\} = \{E^{\mathcal{S}_{n+1}}[\sum_1^n \frac{1}{n} X_{\sigma(i)}]\}$.

Cette propriété une conséquence de

(P') *Pour tout* $a \in M$, $\quad E^{\mathcal{S}_{n+1}}[(a \sum_{i=1}^n \frac{1}{n} X_i)] = E^{\mathcal{S}_{n+1}}[(a \sum_{i=1}^n \frac{1}{n} X_{\sigma(i)})]$ *p.s.*

Pour cela prenons une fonction $f : M^{n+1} \longrightarrow \mathbb{R}$ telle que $f(X_1, \cdots, X_{n+1})$ soit $\mathcal{S}_{n+1}$ mesurable. Notons $X = (X_1, \cdots, X_{n+1})$, $X_\sigma = (X_{\sigma(1)}, \cdots, X_{\sigma(n+1)})$ et $g(X) = a \sum_1^n \frac{1}{n} X_i$. Alors, en désignant par P_X la loi du vecteur X, $E[f(X)g(X)] = E_{P_X}[f(x)g(x)] = E[f(X_\sigma)g(X_\sigma)]$. Comme $f(X) = f(X_\sigma)$ *p.s.*, on a $E[f(X)a \sum_1^n \frac{1}{n} X_i] = E[f(X)a \sum_1^n \frac{1}{n} X_{\sigma(i)}]$. C'est la propriété désirée.

Revenons à la propriété de martingale.

La proposition 1 appliquée à la v.a. $\mathcal{S}_{n+1}$ mesurable $Y_{n+1} = \sum_1^{n+1} \frac{1}{n+1} X_i$, permet d'écrire :

$$Y_{n+1} = \sum_{i=1}^{n+1} \frac{1}{n+1} \Big(\sum_{j \leq n+1, j \neq i} \frac{1}{n} X_j \Big) \text{ et } a\, Y_{n+1} \leq \sum_{i \leq n+1} \frac{1}{n+1} \Big(a \sum_{j \leq n+1, j \neq i} \frac{1}{n} X_j \Big).$$

En prenant l'espérance conditionnelle par rapport à $\mathcal{S}_{n+1}$, il vient :

$$a\, Y_{n+1} \leq \sum_{i \leq n+1} \frac{1}{n+1} E^{\mathcal{S}_{n+1}} \Big(a \sum_{j \leq n+1, j \neq i} \frac{1}{n} X_j \Big).$$

Avec la propriété (P'), $\quad E^{\mathcal{S}_{n+1}} \Big(a \sum_{j \leq n+1, j \neq i} \frac{1}{n} X_j \Big) = E^{\mathcal{S}_{n+1}} \Big(a \sum_{j \leq n} \frac{1}{n} X_j \Big)$,

par conséquent, $\quad a Y_{n+1} \leq \sum_{i \leq n+1} \frac{1}{n+1} E^{\mathcal{S}_{n+1}} (a Y_n) = E^{\mathcal{S}_{n+1}} (a Y_n)$.

Par suite, $Y_{n+1} \in \{E^{\mathcal{S}_{n+1}}[Y_n]\}$ et (Y_n) est bien une martingale.

Montrons la convergence presque sûre.

Pour X_1 étagée de loi $\mu = \sum_{i=1}^k \alpha_i \delta_{a_i}$ et $a \in M$, la suite $(a Y_n(\omega))$ est une sous-martingale positive convergeant p.s. Ainsi, pour $a \in D, D$ dénombrable dense dans M, on a la convergence p.s. de $a Y_n$, Comme $Y_n \in E_1(\{a_1, \cdots, a_k\})$, l'hypothèse de compacité de cet ensemble implique alors que la suite $(Y_n(\omega))$ converge p.s. Par le théorème de 0,1 de Kolmogorov, la limite est presque sûrement une constante. Elle est notée provisoirement $b((X_n))$.

Montrons que cette limite ne dépend que de la loi.

On remarque que si X et Y sont intégrables, il en est de même pour XY.

Soit $(Y_n)_n$ une autre suite de v.a. i.i.d. En utilisant la propriété d'invariance par permutation de b_n pour τ et $\sigma \in \Sigma(n)$:

$$b_n(X_1(\omega), \cdots, X_n(\omega))b_n(Y_1(\omega), \cdots, Y_n(\omega)) =$$
$$b_n(X_{\tau(1)}(\omega), \cdots, X_{\tau(n)}(\omega))b_n(Y_{\sigma(1)}(\omega), \cdots, Y_{\sigma(n)}(\omega)).$$

On voit, avec la proposition 1 et la convergence précédente, que

$$b((X_n))b((Y_n)) \leq \varliminf_n \inf_{\sigma, \tau \in \Sigma(n)} \frac{1}{n} \sum_1^n X_{\tau(i)}(\omega) Y_{\sigma(i)}(\omega) \quad p.s. \tag{1}$$

En prenant l'espérance dans l'inégalité (1), on obtient

$$b((X_n))b((Y_n)) \leq E\Big[\varliminf_n \frac{1}{n} \sum_{i=1}^n X_i Y_i\Big] \leq \varliminf_n \frac{1}{n} \sum_{i=1}^n E[X_i Y_i]. \tag{2}$$

Soit $\big((X_n, Y_n)\big)_n$ une suite i.i.d. de même loi que (X, Y). L'inégalité (2) donne

$$b((X_n))b((Y_n)) \leq \varliminf_n \frac{1}{n} \sum_{i=1}^n E[X_i Y_i] = E[XY]. \tag{3}$$

Cas des v.a. étagées.   Supposons que X_1 et Y_1 ont même loi : $\mu = \sum_1^p \alpha_i \delta_{a_i}$.

À chaque n et ω, associons deux permutations τ et σ telles que $(X_{\tau(1)}(\omega), \cdots, X_{\tau(n)}(\omega)) = (a_1, \cdots, a_1, a_2, \cdots, a_{p-1}, a_p, \cdots, a_p)$. Le nombre de répétitions de chaque a_i est $p_i(\omega, n) = \operatorname{card}\{j | j \leq n, X_j(\omega) = a_i\}$. Procédons de même pour $(Y_{\sigma(1)}(\omega), \cdots, Y_{\sigma(n)}(\omega))$ avec $q_i(\omega, n) = \operatorname{card}\{j | j \leq n, Y_j(\omega) = a_i\}$. Notons enfin $C = \sup_{i,j} a_i a_j$. Avec les relations précédentes :

$$b_n(X_1(\omega), \cdots, X_n(\omega))b_n(Y_1(\omega), \cdots, Y_n(\omega)) \leq \frac{C}{n} \sum_1^p |p_i(\omega, n) - q_i(\omega, n)|.$$

Pour presque tout ω, le terme de droite de l'inégalité tend vers 0 quand $n \to \infty$. Ainsi pour une v.a. étagée X, $b((X_n))$ ne dépend que de la loi, la notation $b(X)$ est justifiée.

De plus, avec la partie précédente, pour deux v.a. étagées, on a l'inégalité

$$b(X)b(Y) \leq E[XY]. \tag{4}$$

Cas des v.a. intégrables.   Pour une suite $(X_n)_n$ i.i.d. avec X_1 intégrable, posons $X_n^k = f_k(X_n)$ où les f_k sont les fonctions définies auparavant. Pour chaque k, la suite $(X_n^k)_n$ est i.i.d. et X_1^k étagée. Nous avons vu, dans ces conditions, que la suite $\big(b_n^k(\omega) = b_n(X_1^k(\omega), \cdots, X_n^k(\omega))\big)_n$ converge p.s. vers un point noté b^k.

La suite $(b^k)_k$ est convergente car l'inégalité (4) donne $b^k b^p \leq E[X_1^k X_1^p]$, ce dernier terme tend vers 0. Notons b la limite de la suite $(b^k)_k$.

L'inégalité triangulaire et (4) montrent que, pour $Y_n(\omega) = \sum_{i=1}^n \frac{1}{n} X_i(\omega)$,

$$bY_n(\omega) \le bb_n^k(\omega) + b_n^k(\omega)Y_n(\omega) \le bb_n^k(\omega) + \frac{1}{n}\sum_{i=1}^{n} X_i^k(\omega)X_i(\omega).$$

La suite $(Z_i^k = X_i^k X_i)_i$ est i.i.d. réelle, Z_1^k intégrable, donc $\frac{1}{n}\sum_{i=1}^{n} Z_i^k$ converge p.s. vers $E[X_1^k X_1]$.

Ainsi presque sûrement, $\overline{\lim_n} \, bY_n(\omega) \le \overline{\lim_n} \, bb_n^k(\omega) + E[X_1^k X_1] = bb^k + E[X_1^k X_1]$.

Ceci prouve la convergence p.s. de la suite $\big(Y_n(\omega)\big)_n$ vers $b = \lim_k b(X_1^k)$.

De plus, en reprenant la partie précédente, on voit que les inégalités $(1), (2), (3)$ et (4) demeurent pour les variables intégrables.

Nous avons en définitive montré que, pour toute suite i.i.d. $(X_n)_n$, X_1 intégrable, la suite $(Y_n = b_n(X_1, \cdots, X_n))_n$ converge p.s. vers une constante ne dépendant que de la loi de X_1, $b(X_1) = \lim_k b(X_1^k)$. Enfin, en choisissant $Y = a$ dans la relation (4), on obtient $b(X) \in \{E[X]\}$. $\square$

Remarque : Il se peut que $b_4(x, y, z, z) \ne b_8(x, x, y, y, z, z, z, z)$, et que $b_n \ne b$ pour tout n, ce qui rend difficile, a priori, l'évaluation de $b(x, y, z, z)$.

I - 3 Identification de la limite

Nous allons donner un moyen de "calculer" $b(\mu)$ pour certaines probabilités en contournant la convergence presque sûre.

Soient $\mu = \frac{1}{k}\sum_{i=1}^{k} \delta_{a_i}$, les a_i distincts, et (X_i) une suite de v.a. i.i.d. de loi μ. Définissons la suite $\mathbf{a} = (a_1, \cdots, a_k, a_1, \cdots)$ formée par répétitions successives de la suite finie $(a_1, \cdots, a_k)$ et notons $\mathbf{a}^n$ la suite formée par les n premiers termes de $\mathbf{a}$. Nous allons montrer que

$$\lim_{n \to \infty} b_n(\mathbf{a}^n) = b(\mu).$$

Fixons provisoirement un entier n et soit $n = km + k^*, 0 \le k^* < k$, son écriture modulo k. Pour tout $i \in \{1 \cdots k\}$, notons p_i^n le cardinal de l'ensemble $I_i^n = \{j | 1 \le j \le n, X_j(\omega) = a_i\}$, $p_i^n = \#(I_i^n)$. En ordonnant de manière croissante, on a $I_i^n = \{j_1^i, \cdots, j_{p_i^n}^i\}$. Pour $J = \{1 \cdots m\}$, posons $J_l = (l-1)m+J$, $\tilde{J}_l = \{1 \cdots m \wedge p_l^n\}$, $\tilde{J}_l^n = (l-1)m + \tilde{J}_l$ où $l \in \{1 \cdots k\}$ et $J_{k+1} = \{km+1, \cdots, n\}$. Les J_l, pour $l \in \{1 \cdots k+1\}$, forment une partition de $\{1 \cdots n\}$ et les $\tilde{J}_l^n$ sont disjoints pour $1 \le l \le k$.

Soit $q \in \tilde{J}_l^n$, posons $\sigma(q) = j_q^l$; on a $X_{\sigma(q)}(\omega) = a_l$ et σ est une injection de $\overset{k}{\underset{l=1}{\cup}} \tilde{J}_l^n$ dans $\{1 \cdots n\}$ on peut donc la prolonger en une permutation, notée encore σ. Par ailleurs, $\#\left(\overset{k}{\underset{l=1}{\cup}} \tilde{J}_l^n\right) = \sum_{l=1}^{k} m \wedge p_l^n$, et si $C = \sup_{i,j} a_i a_j$, alors

$$b_n(X_1(\omega), \cdots, X_n(\omega))b_n(\mathbf{a}^n) \le \frac{C}{n}\left(\sum_{l=1}^{n}(|m - p_l^n| + m \wedge p_l^n) + k^*\right).$$

Ainsi, pour presque tout ω, lorsque $n \to \infty$ on obtient le résultat anoncé. $\square$

Ce procédé s'étend naturellement à toute probabilité $\mu = \sum_{i=1}^{k} \alpha_i \delta_{a_i}$ à coefficients α_i rationnels.

Exemple   Dans l'espace formé par l'arbre homogène ternaire à distance unitaire, soient x, y, z les trois sommets à distance 1 d'un sommet donné o. Une simulation montre que le barycentre $b(\mu)$ de la mesure $\mu = \frac{1}{2}\delta_x + \frac{1}{4}\delta_y + \frac{1}{4}\delta_z$ est situé sur la branche joignant o à x et $ob(\mu)$ est voisin de 0,15.

I - 4   Cohérence et théorème ergodique

Le problème soulevé ici est le suivant : le théorème 2 demeure-t-il lorsque l'on remplace $Y_n = b_n(X_1, \cdots, X_n)$ par $Y_n = b(X_1, \cdots, X_n)$? En d'autres termes le barycentre ainsi défini vérifie-t-il la loi forte des grands nombres ?
Le résultat suivant montre que la partie "convergence p.s." reste valable.

Proposition 3 (de cohérence)   *Sous les conditions du théorème 2, la suite* $\left(b\left(\sum_{i=1}^{n} \frac{1}{n} \delta_{X_i(\omega)} \right) \right)_n$ *converge p.s. vers $b(X)$.*

En fait l'existence de $b(\cdot)$ permet d'affaiblir les hypothèses, en particulier de supprimer l'hypothèse "fini compact". Le théorème suivant regroupe la propriété de convergence presque sûre et celle de martingale :

Théorème 4   *Soient M un espace métrique complet séparable et $\mathcal{P}_1$ l'ensemble des probabilités intégrables sur la tribu borélienne de M ($\mu \in \mathcal{P}_1 \Longrightarrow \int ax \, d\mu(x) < \infty$, $a \in M$). Supposons qu'il existe une application $b : \mathcal{P}_1 \longrightarrow M$, telle que $b(\mu)b(\nu) \leq E[XY]$, pour toutes v.a. X et Y de lois respectives μ et ν. Alors $\forall \mu \in \mathcal{P}_1$ et $\forall (X_n)$ i.i.d. de loi μ, la suite $\left(Y_n(\omega) = b\left(\frac{1}{n} \sum_{i=1}^{n} \delta_{X_i(\omega)} \right) \right)$ converge presque sûrement vers $b(\mu)$.*
De plus les deux assertions suivantes sont équivalentes :

(i) $\forall \mu \in \mathcal{P}_1$ *et* $\forall (X_n)_n$ *i.i.d. de loi* μ, $\left(Y_n(\omega) = b\left(\frac{1}{n} \sum_{i=1}^{n} \delta_{X_i(\omega)} \right) \right)$ *est une martingale adaptée à* $\mathcal{S}_n$.
(ii)   *le barycentre $b(\cdot)$ vérifie,* $\forall a \in M$, $\forall n$ *et* $\forall (x_1, \cdots, x_{n+1}) \in M^{n+1}$,

$$ab\left(\frac{1}{n+1} \sum_{i=1}^{n+1} \delta_{x_i} \right) \leq \sum_{i=1}^{n+1} \frac{1}{n+1} ab\left(\sum_{j \leq n, j \neq i} \frac{1}{n} \delta_{x_j} \right).$$

Preuve   Montrons au préalable que Y_n est $\mathcal{S}_n$ mesurable.
Soient $X_1, \cdots X_n$, n variables mesurables par rapport à une tribu $\mathcal{A}$. Avec les fonctions f_k précédentes posons $X_i^k = f_k(X_i)$, on obtient des v.a. étagées $\mathcal{A}$

mesurables, pour tout k et tout $i = 1 \cdots n$. Pour chaque k, il existe donc une partition Π_k finie : $A_1, \cdots, A_p$, dépendant de k et formée d'éléments de $\mathcal{A}$ telle que sur chaque A_j, $X_i^k(\omega) = x_i^k(j)$. Notons les lois empiriques $\mu_n^\omega = \sum_{i=1}^n \frac{1}{n} \delta_{X_i(\omega)}$ et $({}_k\mu)_n^\omega = \sum_{i=1}^n \frac{1}{n} \delta_{X_i^k(\omega)}$. Par hypothèse on a l'inégalité :

$$b(\mu_n^\omega)b(({}_k\mu)_n^\omega) \leq \frac{1}{n}\sum_1^n X_i(\omega)X_i^k(\omega). \text{ Le terme de droite tend vers } 0 \text{ quand}$$

$k \to \infty$, donc $b(({}_k\mu)_n^\omega) \xrightarrow[k\to\infty]{} b(\mu_n^\omega)$. Or, pour $\omega \in A_j$, $b(({}_k\mu)_n^\omega) = b(\sum_{i=1}^n \frac{1}{n}\delta_{x_i^k(j)})$ et par suite, l'application $\omega \longrightarrow b(({}_k\mu)_n^\omega)$ est $\mathcal{A}$ mesurable. À la limite, $b(\mu_n^\omega)$ est $\sigma(X_1, \ldots, X_n)$ mesurable.

Remarquons alors que la probabilité μ_n^ω est invariante pour toute permutation $\tau \in \Sigma(n)$, en effet $\mu_n^\omega = \frac{1}{n}\sum_1^n \delta_{X_{\tau(i)}(\omega)}$. En reprenant le raisonnement précédent, on en déduit que $Y_n(\omega) = b(\mu_n^\omega)$ est $\mathcal{S}_n$ mesurable.

Montrons la convergence presque-sûre.

De manière classique, par exemple [14] theorem 7.1, on a, presque sûrement en ω, la convergence en loi des lois empiriques $\mu_n^\omega = \frac{1}{n}\sum_{i=1}^n \delta_{X_i(\omega)}$ vers μ. Le théorème de représentation de Skohorod, [16] [4], assure de l'existence, pour tout $\omega \notin \mathcal{N}$, $\mathcal{N}$ négligeable, d'une suite $(U_n^\omega(\cdot))_n$ de variables aléatoires définies sur l'espace canonique $([0,1], \mathcal{B}, \lambda)$, de loi μ_n^ω et convergeant p.p. vers une v.a. U^ω de loi μ. Lorsque μ est concentrée sur une partie bornée, ce qui est le cas lorsque μ est la loi de X_1 d'une v.a. bornée, le terme de droite de l'inégalité : $b(\mu_n^\omega)b(\mu) \leq E_\lambda[U_n^\omega(\cdot)U^\omega(\cdot)]$ tend vers 0. Ainsi la première assertion du théorème est prouvée pour X_1 bornée.

Le cas général se déduit de l'inégalité :

$b(\mu)b(\mu_n^\omega) \leq b(\mu)b({}_k\mu) + b({}_k\mu)b(({}_k\mu)_n^\omega) + b(({}_k\mu)_n^\omega)b(\mu_n^\omega)$ où ${}_k\mu$ est la loi de X_1^k.

En effet, $b(({}_k\mu)_n^\omega)b(\mu_n^\omega) \leq \frac{1}{n}\sum_{i=1}^n X_i(\omega)X_i^k(\omega)$ qui tend vers $E[X_1X_1^k]$ lorsque $n \to \infty$, le terme $b(\mu)b({}_k\mu)$ est inférieur à $E[X_1X_1^k]$ qui tend vers 0 quand k infini et enfin, nous venons de le voir, $b({}_k\mu)b(({}_k\mu)_n^\omega)$ tend vers 0 pour tout k.

Établissons maintenant l'équivalence.

Comme dans la propriété (P), pour toute permutation $\tau \in \Sigma(n+1)$, on établit l'égalité

$$\{E^{\mathcal{S}_{n+1}}[b(\mu_n(\omega))]\} = \{E^{\mathcal{S}_{n+1}}[b(\frac{1}{n}\sum_{i=1}^n \delta_{X_{\tau(i)}(\omega)})]\}.$$

Avec la notation $Y^i(\omega) = b(\sum_{j\leq n+1, j\neq i} \frac{1}{n}\delta_{X_j(\omega)})$ alors, pour tout $a \in M$,

(*) $$E^{\mathcal{S}_{n+1}}[aY_n] = E^{\mathcal{S}_{n+1}}[aY^i].$$

Montrons que la condition est nécessaire.

Si $(Y_n)_n$ est une martingale alors pour tout $a \in M$,

$$aY_{n+1} \leq E^{S_{n+1}}[aY_n] = E^{S_{n+1}}[aY^i]. \text{ D'où } aY_{n+1} \leq \sum_1^{n+1} \frac{1}{n+1} E^{S_{n+1}}[aY^i].$$

La v.a.r. $\sum_1^{n+1} \frac{1}{n+1} aY^i = \sum_{i=1}^{n+1} \frac{1}{n+1} ab\left(\sum_{j \leq n+1, j \neq i} \frac{1}{n} \delta_{X_j} \right)$ étant clairement S_{n+1}

mesurable, on obtient $aY_{n+1} \leq \sum_1^{n+1} \frac{1}{n+1} aY^i$. C'est la relation cherchée en

prenant $\mu = \frac{1}{n} \sum_1^n \delta_{x_i}$.

Montrons que la condition est suffisante.

La v.a. $Y_{n+1} = b\left(\sum_1^{n+1} \frac{1}{n+1} \delta_{X_i} \right)$ vérifie alors $aY_{n+1} \leq \sum_1^n \frac{1}{n+1} aY^i$ p. s. Donc

$aY_{n+1} \leq \sum_1^{n+1} \frac{1}{n+1} E^{S_{n+1}}[aY^i]$. Avec la relation $(*)$, il vient $aY_{n+1} \leq E^{S_{n+1}}[aY_n]$

qui est la propriété de martingale. La preuve du théorème est donc achevée. []

Comme application nous allons prouver un théorème ergodique. À cette fin nous supposerons que l'espace de probabilité est l'espace de Lebesgue et, par conséquent, pour toute sous tribu C de la tribu borélienne B, il existe une version régulière de la probabilité conditionnelle sachant C, *i.e.* un noyau N défini sur $\Omega \times B$ tel que $P(\cdot) = \int N(\omega, \cdot) dP(\omega)$ et que $N(\cdot, B)$ soit une v.a.r. C-mesurable pour tout $B \in B$. Ceci permet d'introduire la définition suivante :

Définition

Avec les conditions du théorème 4, soient X une variable aléatoire intégrable à valeurs dans M et C une sous tribu. *L'espérance conditionnelle de X sachant C* est la variable aléatoire $b^C(X)(\omega) = b_{N(\omega, \cdot)}(X)$, barycentre de X par rapport au noyau de conditionnement N relatif à C.

Il est aisé de vérifier que l'application $\omega \longrightarrow b^C(X)(\omega)$ est C mesurable et $b^C(X) \in \{E^C[X]\}$. De plus pour deux variables X et Y intégrables, on a : $b^C(X)(\omega)b^C(Y)(\omega) \leq E^C[XY](\omega)$ *p.s.*

Théorème 5 (Théorème ergodique)

Soit (X_n) une suite stationnaire de v.a. intégrables à valeurs dans M. Alors, sous les conditions du théorème 4 et, en désignant par I la tribu des événements invariants, $b\left(\mu_n^\omega = \frac{1}{n} \sum_1^n \delta_{X_i(\omega)} \right)$ converge p.s. vers $b^I(X_1)(\omega)$.

Preuve   Considérons d'abord le cas où X_1 est étagée. Le théorème de Birkhoff montre que pour presque tout ω et pour toute fonction numérique f continue bornée $\frac{1}{n} \sum_1^n f(X_i(\omega))$ converge vers $E^I[f(X_1)](\omega) = \int_M f(x) \, d\mu_I^\omega(x)$. Le

théorème de représentation de Skohorod assure de l'existence d'une suite de
v.a. $U_n^\omega(\cdot)$ de loi μ_n^ω, convergeant p.s. vers une v.a. $U^\omega(\cdot)$ de loi μ_T^ω.
Le théorème de convergence dominée de Lebesgue montre que

$$b^{\mathcal{I}}(X_1)(\omega) \sum_1^n \frac{1}{n} X_i(\omega) = b(U^\omega)b(U_n^\omega) \le E[U_n^\omega(\cdot)U^\omega(\cdot)] \underset{n\to\infty}{\longrightarrow} 0.$$

Le cas général se fait de manière similaire avec les fonctions f_k. []

II   Autres notions d'espérances

II - 1   Sur l'espérance au sens de Herer

Le but de cette partie est d'obtenir des résultats similaires en remplaçant
l'espérance au sens de Doss, $\{E[X]\}$, par celle de Herer [12], notée $\mathcal{E}_{\mathcal{H}}(X)$.
L'espace de probabilité de référence est, ici, l'espace canonique $([0,1], \mathcal{B}, \lambda)$.
Soient $a_1, \cdots, a_k$ des points distincts de M, affectés de masses α_i, $\sum_1^n \alpha_i = 1$. Une
suite de points $(x_1, \cdots, x_n)$, $n \ge k$ affectés de masses respectives $\beta_i, i = 1 \cdots n$
est dite issue de $\mu = \sum_1^k \alpha_i \delta_{a_i}$ si

$$\{x_1, \cdots, x_n\} = \{a_1, \cdots, a_k\}, \text{ et, pour tout } i, \ \alpha_i = \sum_{j|x_j=a_i} \beta_j \ ; \ i.e. \ \mu = \sum_1^n \beta_j \delta_{x_j}.$$

À partir d'une suite de points $(x_i)_{i=1\cdots n}$ affectés de masses β_i, on construit
un arbre binaire (chaque sommet intermédiaire joint deux points) issu d'un
sommet et dont les extrémités finales sont les x_j. En remontant à partir des
extrémités finales, on affecte à chaque sommet intermédiaire la masse somme
des masses de ses deux extrémités (la masse du sommet initial est 1). De la
même façon, en remontant, le sommet joignant deux points m_l et m_p de masses
respectives γ_l et γ_p est appelé $\dfrac{\gamma_l}{\gamma_l + \gamma_p} m_l + \dfrac{\gamma_p}{\gamma_l + \gamma_p} m_p$.
Le nom du sommet initial dépend de la forme de l'arbre et est identifié à un
point de M.

On note $\mathcal{H}(\mu)$ l'ensemble des points de M, obtenus comme sommets initiaux
d'un arbre binaire du type précédent, lorsque l'on fait varier les paramètres
suivants : n, la suite $(x_1, \cdots x_n)$ avec les masses $(\alpha_i)_{i=1\cdots n}$ issue de μ, et la
forme de l'arbre.

L'espérance de μ au sens de Herer est, par définition, $\mathcal{E}_{\mathcal{H}}(\mu) := \overline{\mathcal{H}(\mu)}$. On
écrit aussi $\mathcal{H}(X)$ et $\mathcal{E}_{\mathcal{H}}(X)$ si X est de loi μ.

Soient $m_1, \cdots, m_p$ p points de $\mathcal{H}(\mu)$. Chaque m_i est le sommet initial
d'un arbre binaire dont les n_i extrémités sont des points que nous notons,
pour simplifier, $x_{n_{i-1}+1}, \cdots, x_{n_{i-1}+n_i}$, avec la convention $n_0 = 0$. La suite
$(x_j)_{n_{i-1}+1 \le j \le n_{i-1}+n_i}$ affectée de masses $(\beta_j^i)_j$ est issue de μ. Construisons un

arbre binaire partant d'un point m et dont les extrémités sont les m_i, $1 \leq i \leq p$. On obtient naturellement, par prolongement, un arbre binaire joignant m aux points $x_1, \cdots, x_k$ où $k = \sum_{1}^{p} n_i$. Affectons chaque m_i de la masse γ_i, $\sum_i \gamma_i = 1$, et chaque x_j, pour $n_{i-1} + 1 \leq j \leq n_{i-1} + n_i$, est affecté de la masse $\gamma_i \beta_j^i$. On obtient alors une suite issue de μ. Le point m s'interprète comme un point de $\mathcal{H}(\mu)$ et aussi comme un point de $\mathcal{H}(\sum_{1}^{p} \gamma_i \delta_{m_i})$. Ceci montre le résultat suivant :

"Soient μ une probabilité discrète sur M et ν une autre probabilité discrète dont le support est contenu dans $\mathcal{H}(\mu)$, alors $\mathcal{H}(\nu) \subset \mathcal{H}(\mu)$."

Ou encore

"Soient X et Y deux v.a. étagées, si Y prend ses valeurs dans $\mathcal{H}(X)$, alors $\mathcal{H}(Y) \subset \mathcal{H}(X)$."

On obtient alors la propriété suivante qui, grâce à la proposition 1, sert de point de départ de l'étude du barycentre canonique relatif à l'espérance de Doss. En adaptant les notations de la partie I-1, posons $\mathcal{H}_1(A) = \bigcup_{\mu \in \mathcal{P}(A)} \mathcal{E}_\mathcal{H}(\mu)$ et $\mathcal{H}_2(A) = \mathcal{H}_1(\mathcal{H}_1(A))$. On a : *si $A = \{x_1, \cdots, x_n\}$, alors $\mathcal{H}_1(A) = \mathcal{H}_2(A)$.* En effet l'inclusion $\mathcal{H}_1(A) \subset \mathcal{H}_2(A)$ est triviale en prenant pour ν une masse de Dirac, et l'inclusion inverse est une conséquence des propriétés ci-dessus.

L'adaptation de la proposition 1 permet d'obtenir l'existence de $b_n(x_1 \cdots x_n)$ comme un point de $\mathcal{E}_\mathcal{H}(\mu = \sum_{1}^{n} \frac{1}{n} \delta_{x_i})$ et, de là, une version du théorème 2 relative à l'espérance au sens de Herer.

II - 2   Sur l'espérance au sens d'Émery et Mokobodzki

Suivant [7] on considère, à la place de l'espace M précédent, un espace V qui est à courbure négative et est un compact géodésiquement convexe d'une variété riemannienne.

Un point $x \in V$ est un *barycentre convexe* de μ probabilité sur V, si pour toute fonction f convexe définie sur un voisinage de V on a $f(x) \leq \mu(f)$, rappelons que f est convexe si sa restriction à toute géodésique est convexe. On note $\{E_{E-M}[\mu]\}$ l'ensemble des barycentres convexes de μ. L'égalité entre les ensembles $\mathcal{E}_1(A = \{x_1, \cdots, x_n\}) = \bigcup_{\mu \in \mathcal{P}(A)} \{E_{E-M}[\mu]\}$ et $\mathcal{E}_2(A) = \mathcal{E}_1(\mathcal{E}_1(A))$ résulte de la proposition 1 de [7].

Montrons que le barycentre $b(\mu)$ appartient à $\{E_{E-M}[\mu]\}$. Établissons d'abord la propriété pour une probabilité discrète $\mu = \frac{1}{k} \sum_{1}^{k} \delta_{x_i}$ et pour $b_k(\mu)$. Pour $k = 2$, l'exemple d) de [7] montre que $b_2(x,y) \in \{E_{E-M}[\frac{1}{2}(\delta_x + \delta_y)]\}$.

Admettons par récurrence, que $f(b_l(x_1, \cdots, x_l)) \leq \frac{1}{l} \sum_1^l f(x_i)$ pour $l \leq k - 1$

et $k > 2$. Il faut établir $f(\sum_1^k \frac{1}{k} x_i) \leq \frac{1}{k} \sum_1^k f(x_i)$. Avec les notations de la

proposition 1 : $x_i^{n+1} = \sum_{j \neq i} \frac{1}{k-1} x_j^n$, on obtient $f(x_i^{n+1}) \leq \frac{1}{k-1} \sum_{j \neq i} f(x_j^n)$.

D'où, en notant $S^n = \sum_{j=1}^k f(x_j^n)$, $S^{n+1} \leq \frac{1}{k-1}(kS^n - \sum_{i=1}^k f(x_i^n)) = S^n$, la suite

(S^n) est décroissante. Par continuité, pour tout $\epsilon > 0$, on a pour n suffisament

grand, $f(b_k(x_1, \cdots, x_k)) \leq \frac{1}{k} S^n \leq \epsilon + \frac{1}{k} \sum_{j=1}^k f(x_j)$, c'est la relation cherchée :

$b_k(\mu) \in \{E_{E-M}[\mu]\}$.

Le passage à $b(\cdot)$ peut se faire par (voir I-3) : $b(\mu = \frac{1}{k} \sum_1^k \delta_{x_i}) = \lim_n b_n(\mathbf{x}^n)$.

Nous avons, pour f convexe continue et bornée et $n = kp$,

$$f(b(\mu)) = \lim_n f(b_n(\mathbf{x}^n)) \leq \lim_n \frac{1}{n} \sum_1^n f(x_i^n) = \frac{1}{k} \sum_1^k f(x_i).$$

Ce qui exprime que $b(\mu) \in \{E_{E-M}[\mu]\}$. On peut aussi, plus rapidement, utiliser
la fermeture de $\{E_{E-M}[\mu]\}$.

Pour le cas général, considérons une suite (X_n) i.i.d. de loi μ intégrable. Le

théorème 2 montre la convergence p.s. de $b(\frac{1}{n} \sum_1^n \delta_{X_i(\omega)})$ vers $b(\mu)$, ce qui donne

la relation cherchée. $\square$

Remarquons aussi que les conditions a) et b) de la définition 1.2 de [13] vérifient
la seconde partie du théorème 4.

Pour conclure, examinons le cas d'un exemple simple.

On considère la loi uniforme $\mu = \frac{1}{3}(\delta_x + \delta_y + \delta_z)$ où x, y et z sont trois points du
plan muni de sa structure euclidienne usuelle. Le barycentre ordinaire coïncide
avec le barycentre exponentiel. Si on modifie la structure en introduisant une
perturbation dans un (petit) voisinage d'un point situé par exemple au milieu
du segment $[x, y^1]$, y^1 étant, conformément à nos notations, le milieu de $[x, z]$,
le barycentre exponentiel ne change pas. Par contre le milieu du segment $[x, z]$
change et le barycentre canonique de μ (pour la nouvelle structure) diffère alors
du barycentre exponentiel.

Références

[1] M. ARNAUDON *Barycentres convexes et approximations des martingales continues dans les variétés.* Séminaire de Probabilités XXIX, Lecture Notes in Mathematics **1613** 70-85, 1997.

[2] V. BENÈS *Martingales on metric spaces.* Theor. Veroyatnost. i Primenen **7**, 81-82 1962.

[3] B. BRU, H. HEINICH, J.C. LOOGIETER *Distance de Lévy et extensions des théorèmes de la limite centrale et de Glivenko-Cantelli.* Pub. Inst. Stat. Univ. de Paris **37**, 29-42, 1993.

[4] J. A. CUESTA, C. A. MATRÁN *Strong convergence of weighted sums of random element through the equivalence of sequences of distributions.* J. Multivariate Anal. **25**, 311-322, 1988.

[5] S. DOSS *Sur la moyenne d'un élément aléatoire dans un espace distancié* Bull. Sci. Math. **73**, 48-72, 1949

[6] S. DOSS *Moyennes conditionnelles et martingales dans un espace métrique* C. R. Acad. Sci. Paris Série I, t.**254**, 3630-3632, 1962

[7] M. ÉMERY, G. MOKOBODZKI *Sur le barycentre d'une probabilité dans une variété* Sémi. Prob. XXV, Lect. Notes in Math. **1485**, 220-233, 1991

[8] M. FRECHET *Les éléments aléatoires de nature quelconque* Ann. Inst. H. Poincaré **14**, 215-310, 1948.

[9] W. HERER *Espérance mathématique au sens de Doss d'une variable aléatoire dans un espace métrique* C. R. Acad. Sci. Paris Série I, t.**302**, 131-134, 1983.

[10] W. HERER *Espérance mathématique d'une variable aléatoire à valeurs dans un espace métrique à courbure négative* C. R. Acad. Sci. Paris Série I, t.**306**, 681-684, 1988 .

[11] W. HERER *Mathematical expectation and martingales of random subsets of a metric spaces* Prob. and Math. Stat. **11**, 291-304, 1991.

[12] W. HERER *Mathematical expectation and strong law of large numbers for random variables with values in a metric space of negative curvative* Prob. and Math. Stat. **13**, 59-70, 1992.

[13] J. PICARD *Barycentres et martingales sur une variété* Ann. Inst. H. Poincaré **30**, 647-702, 1994.

[14] K. R. PARTHASARATHY *Probability Measures on Metric Spaces.* Academic Press 1967.

[15] P. RAYNAUD de FITTE *Théorème ergodique ponctuel et lois fortes des grands nombres pour des points aléatoires d'un espace métrique à courbure négative.* À paraitre dans Annals of Probability.

[16] A. V. SKOHOROD *Limit theorems for stochastic processes* Theory Probab. Appl. **1** 261-290, 1956.

Dualité du problème des marges et ses applications

NACEREDDINE BELILI

Résumé

Cet article présente une synthèse des théorèmes de dualité relatif au problème des marges, ses diverses applications comme le théorème classique de Strassen, la caractérisation de l'ordre stochastique et la représentation des métriques minimales. On y donne une nouvelle preuve du théorème de couplage Goldstein basée sur la représentation de la distance de variation totale.

Abstract

This paper gives a review of duality theorem for marginal problems and its applications like Strassen theorem, representation of stochastic orders and representation of minimal metrics. We give a new proof of Goldstein theorem based on the representation of total variation metric.

1 Introduction et Notations

Soient $(E_i, \mathcal{E}_i, \mu_i)$, pour $i = 1$, 2 deux espaces de probabilités. On dit que μ_1 et μ_2 sont les marges d'une mesure de probabilité γ sur $(E_1 \times E_2, \mathcal{E}_1 \otimes \mathcal{E}_2)$ si

$$\gamma(A_1 \times E_2) = \mu_1(A_1) \quad \text{pour tout} \quad A_1 \in \mathcal{E}_1$$

et

$$\gamma(E_1 \times A_2) = \mu_2(A_2) \quad \text{pour tout} \quad A_2 \in \mathcal{E}_2.$$

On désigne par $\mathcal{M}(E_i)$ l'ensemble de mesures de probabilités sur E_i, et par $\Gamma(\mu_1, \mu_2)$ l'ensemble de toutes les mesures de probabilités sur $E_1 \times E_2$ ayant μ_1 et μ_2 pour lois marginales. Pour $i = 1$, 2, les applications $\pi_i : E_1 \times E_2 \longrightarrow E_i$ désignent les projections canoniques. L'abréviation $\oplus_i g_i$ est utilisée pour $\sum_i g_i \circ \pi_i$.

Si $(\Omega, \mathcal{A})$ est un espace mesurable, alors $f \in \mathcal{A}$ signifie que f est une application réelle $\mathcal{A}$-mesurable. Une mesure de probabilité μ sur $(\Omega, \mathcal{A})$ est dite *parfaite* ou l'espace probabilisé $(\Omega, \mathcal{A}, \mu)$ est *parfait* si pour tout $f \in \mathcal{A}$ on peut trouver un borélien $B_f \subset f(\Omega)$

tel que $\mu(f^{-1}(B_f)) = 1$. L'espace des fonctions μ_i -intégrable $f \in \mathcal{E}_i$ est noté $\mathcal{L}(\mathcal{E}_i, \mu_i)$; en particulier si $\mathcal{E}_i = \mathcal{B}_i$ est la tribu borélienne de E_i, l'ensemble $\mathcal{L}(\mathcal{E}_i, \mu_i)$ sera noté tout simplement $\mathcal{L}(\mu_i)$. On pose $(\mathcal{E}_1 \otimes \mathcal{E}_2)_{\geq} = \{h \in \mathcal{E}_1 \otimes \mathcal{E}_2 : \exists f_i \in \mathcal{L}(\mathcal{E}_i, \mu_i), \ h \geq \oplus_i f_i\}$, et $(\mathcal{E}_1 \otimes \mathcal{E}_2)_{\leq} = \{h \in \mathcal{E}_1 \otimes \mathcal{E}_2 : \exists f_i \in \mathcal{L}(\mathcal{E}_i, \mu_i), \ h \leq \oplus_i f_i\}$.

Définition 1 *Soit h une fonction $\mathcal{E}_1 \otimes \mathcal{E}_2$ -mesurable sur $E_1 \times E_2$. On définit*

$$U(h) \ := \ \sup\left\{ \int_{E_1 \times E_2} h \, d\gamma : \ \gamma \in \Gamma(\mu_1, \mu_2) \right\},$$

$$L(h) \ := \ \inf\left\{ \sum_{i=1}^{2} \int_{E_i} h_i \, d\mu_i : \ h_i \in \mathcal{L}(\mathcal{E}_i, \mu_i) \ et \ h \leq \oplus_i h_i \right\}.$$

Si $E_1 = E_2 = E$ est un espace métrique séparable muni d'une distance d, le problème de transport de Monge-Kantorovich (cf. [39], [31] et [32]) consiste à évaluer la fonctionnelle

$$\mathcal{K}_c(\mu_1, \mu_2) := \inf \left\{ \int_{E \times E} c(x, y) \, \gamma(dx, dy) : \ \gamma \in \Gamma(\mu_1, \mu_2) \right\}, \tag{1}$$

où le "coût" $c(x, y)$ est une fonction mesurable ≥ 0 sur $E \times E$. La fonctionnelle $\mathcal{K}_c(\mu_1, \mu_2)$ est appelée *fonctionnelle de Kantorovich* (resp. *métrique de Kantorovich* si $c = d$). En fait, le théorème de dualité du problème des marges (cf. les théorèmes 1 et 2) est une extension en un certain sens du problème de transport de Monge-Kantorovich. Dans la section 3, on prouvera le théorème de dualité, dans le cas des espaces polonais, pour les fonctions continues, au moyen d'une forme géométrique du théorème de Hahn-Banach. On donnera diverses applications de ce théorème : la caractérisation de l'ordre stochastique, un théorème classique de Strassen, ainsi que la représentation des métriques minimales.

La technique de couplage a été introduite par Doeblin (cf. [11]) pour prouver qu'une chaîne de Markov homogène à espace d'états fini est faiblement ergodique. Sa technique a été raffinée pour obtenir de meilleurs résultats et son idée a été explorée pour d'autres applications (cf. [24], [25], [26], [36] et [55]). Dans la section 4.3, on présentera une preuve du théorème de Goldstein (cf. [24], [55] et [36]), basée sur la représentation de la distance de variation totale.

2 Résultats

Énonçons maintenant le théorème de dualité étudié principalement par Kellerer (cf. [33]), Ramachandran, Rüschendorf (cf. [46]).

Théorème 1 *(cf. [33] et [46]). Soient $(E_i, \mathcal{E}_i, \mu_i)$, pour $i = 1$, 2 deux espaces de probabilités et supposons μ_2 parfaite. Alors on a*

$$U(h) = L(h) \quad pour \ tout \ h \in (\mathcal{E}_1 \otimes \mathcal{E}_2)_{\geq}. \tag{2}$$

Le résultat suivant montre que l' inf dans $L(h)$, égal au sup dans $U(h)$, est atteint.

Proposition 1 *(cf. [33]) Pour tout $h \in (\mathcal{E}_1 \otimes \mathcal{E}_2)_{\leq}$ avec $L(h) < \infty$ alors il existe $h_i \in \mathcal{L}(\mathcal{E}_i, \mu_i)$ telle que*

$$h \leq \oplus_i h_i \quad et \quad L(h) = \sum_{i=1}^{2} \int_{E_i} h_i \, d\mu_i$$

Si on pose

$$I(h) := \inf\left\{ \int_{E_1 \times E_2} h \, d\gamma : \ \gamma \in \Gamma(\mu_1, \mu_2) \right\},$$

$$S(h) := \sup\left\{ \sum_{i=1}^{2} \int_{E_i} h_i \, d\mu_i : \ h_i \in \mathcal{L}(\mathcal{E}_i, \mu_i) \text{ et } h \geq \oplus_i h_i \right\},$$

comme $U(h) = -I(-h)$ et $L(h) = -S(-h)$ le théorème 1 est équivalent au théorème suivant.

Théorème 2 *Soient $(E_i, \mathcal{E}_i, \mu_i)$, pour $i = 1$, 2 deux espaces de probabilités et supposons μ_2 parfaite. Alors on a*

$$I(h) = S(h) \quad \text{pour tout } h \in (\mathcal{E}_1 \otimes \mathcal{E}_2)_{\leq}. \tag{3}$$

On dit qu'un espace probabilisé $(E_2, \mathcal{E}_2, \mu_2)$ est *un espace de dualité* si pour tout espace probabilisé $(E_1, \mathcal{E}_1, \mu_1)$ les propriétés (2) ou (3) sont vérifiées. D'autre part, l'espace $(E_1, \mathcal{E}_1, \mu_1)$ est dit *contenu* dans $(E_2, \mathcal{E}_2, \mu_2)$ si $E_1 \subset E_2$, $\mathcal{E}_1 = \mathcal{E}_2|_{E_1}$ et $\mu_1 = \mu_2^*|_{\mathcal{E}_1}$. Récemment, dans [47] Ramachandran et Rüschendorf donnent une caractérisation des espaces parfaits et ils montrent par le résultat suivant que la réciproque du théorème de dualité reste vraie.

Théorème 3 *(cf.[47]) Soit $(E, \mathcal{E}, \mu)$ un espace probabilisé. Alors les propriétés suivantes sont équivalentes*

(i) *L'espace $(E, \mathcal{E}, \mu)$ est parfait;*

(ii) *Si l'espace $(E, \mathcal{E}, \mu)$ est contenu dans un espace de probabilité $(E_1, \mathcal{E}_1, \mu_1)$, alors $(E_1, \mathcal{E}_1, \mu_1)$ est parfait;*

(iii) *Si l'espace $(E, \mathcal{E}, \mu)$ est contenu dans un espace de probabilité $(E_1, \mathcal{E}_1, \mu_1)$, alors $(E_1, \mathcal{E}_1, \mu_1)$ est un espace de dualité.*

3 Théorème de dualité dans les espaces polonais

On désigne par E un espace polonais muni de sa tribu borélienne $\mathcal{B}$ et par $\mathcal{C}(E)$ l'ensemble des fonctions réelles continues sur E, et on note $\mathcal{C}(E \times E)$ l'ensemble des fonctions réelles continues sur $E \times E$. Par commodité, on désignera désormais par μ et ν deux mesures de $\mathcal{M}(E)$, au lieu de μ_1 et μ_2 utilisés plus haut. Dans cette section, on va donner en détail une preuve du théorème 2 pour les fonctions continues (dans les espaces polonais). Selon Kellerer [33] l'extension aux fonctions boréliennes se fait ensuite en deux temps : d'abord on étend aux fonctions s.c.s. (semicontinue supérieurement) puis, ayant remarqué que la fonctionnelle $I(\cdot)$, convenablement tronquée, est une capacité fonctionnelle de Choquet, on passe des fonctions s.c.s. aux fonctions boréliennes grâce au théorème de capacitabilité (cf. [5] et [9]).

Théorème 4 *Soient μ et ν deux mesures de $\mathcal{M}(E)$. Alors on a*

$$I(h) = S(h) \quad \text{pour tout} \quad h \in \mathcal{C}(E \times E). \tag{4}$$

On montre d'abord le lemme suivant.

Lemme 1 *Soient μ et ν deux mesures de $\mathcal{M}(E)$. Alors on a*

$$I(h) \geq S(h) \quad \text{pour tout } h \in \mathcal{C}(E \times E).$$

DÉMONSTRATION DU LEMME 1.

Soient $\gamma \in \Gamma(\mu, \nu)$ et $f, g \in \mathcal{C}(E)$ telles que $f(x) + g(y) \leq h(x, y)$ pour tout $x, y \in E$. Alors on a

$$\int_E f(x) \, d\mu(x) + \int_E g(x) \, d\nu(x) = \int (f(x) + g(y)) \, d\gamma(x, y) \leq \int_{E \times E} h(x, y) \, d\gamma(x, y)$$

d'où $I(h) \geq S(h)$.

$\square$

DÉMONSTRATION DU THÉORÈME 4.

(a) Supposons E compact. Pour tout $h \in \mathcal{C}(E \times E)$ soit

$$G = \{ g \in \mathcal{C}(E \times E) \; : \; g(x, y) < h(x, y), \; \forall x, \, y \in E \};$$

et soit H l'ensemble des fonctions $c(x, y) = f(x) + g(y)$ pour tout f et g dans $\mathcal{C}(E)$. D'autre part, on pose

$$s(c) = \int_E f \, d\mu + \int_E g \, d\nu \quad \text{pour} \quad c \in H.$$

On vérifie aisément que s est une forme linéaire bien définie. L'ensemble G est convexe, ouvert pour la norme sup ; d'autre part, la forme linéaire s sur H est non identiquement nulle, et bornée supérieurement sur l'ensemble convexe non vide $G \cap H$. En effet, comme $f(x) + g(y) < h(x, y)$ pour tout x et y dans E, alors $s(c) \leq \sup(f) + \sup(g) < \sup(h)$. Alors, d'après le théorème de Hahn-Banach (voir appendice théorème 17), il existe une forme linéaire η qui prolonge s sur $\mathcal{C}(E \times E)$ et telle que

$$\sup_G \eta = \sup_{G \cap H} s.$$

D'après le théorème de représentation de Riesz, il existe une unique mesure positive finie γ sur $E \times E$ telle que

$$\eta(c) = \int_{E \times E} c \, d\gamma \quad \text{pour tout} \quad c \in \mathcal{C}(E \times E).$$

Comme $\eta = s$ sur H, on obtient donc

$$\int_{E \times E} f(x) \, d\gamma(x, y) = \int_E f(x) \, d\mu(x) \quad \text{pour tout} \quad f \in \mathcal{C}(E)$$

et

$$\int_{E \times E} g(x) \, d\gamma(x, y) = \int_E g(x) \, d\nu(x) \quad \text{pour tout} \quad g \in \mathcal{C}(E).$$

Par conséquent $\gamma \in \Gamma(\mu, \nu)$. Par ailleurs,

$$S(h) = \sup_{r \in G \cap H} s(r) = \sup_{t \in G} \eta(t) = \int_{E \times E} h \, d\gamma. \tag{5}$$

Ceci entraîne que $S(h) \geq I(h)$ pour tout $h \in \mathcal{C}(E \times E)$. En utilisant le Lemme 1, on en déduit le résultat du théorème 4 dans le cas (a).

(b) Cas général.

Les mesures μ et ν sont tendues, alors pour tout entier $n > 0$, il existe une partie compacte K_n de E telle que

$$\max\left(\mu(E \setminus K_n),\ \nu(E \setminus K_n)\right) \le \frac{1}{n}.$$

On note a un élément de E, définissons alors deux mesures de probabilités μ_n et ν_n à support compact en posant pour tout $A \in \mathcal{B}$,

$$\mu_n(A) := \mu(A \cap K_n) + \mu(E \setminus K_n)\delta_a(A)$$

et

$$\nu_n(A) := \nu(A \cap K_n) + \nu(E \setminus K_n)\delta_a(A).$$

où δ_a est la mesure de Dirac au point a. La preuve précédente montre qu'il existe une mesure de probabilité γ_n sur $E \times E$ ayant pour lois marginales μ_n et ν_n vérifiant, d'après les définitions de μ_n et ν_n

$$\int_{E \times E} h(x, y)\, d\gamma_n(x, y)$$
$$\le\ \sup\left\{\int_E f\, d\mu + \int_E g\, d\nu\ :\ f(x) + g(y) \le h(x, y),\ \forall x,\ y \in E\right\} + \frac{h(a, a)}{n}.$$

La suite (γ_n) ayant des marges étroitement convergentes est relativement compacte, on peut en extraire une sous-suite étroitement convergente vers une mesure de probabilité γ qui a pour marges μ et ν. D'après le théorème 9 ci-dessous de Rachev-Rüschendorf, il existe un espace de probabilité $(\Omega, \mathcal{A}, P)$ et une suite de variables aléatoires $(X_n)_{n \in \mathbb{N}}$ sur Ω à valeurs dans E telle que la loi de X_n et γ_n pour tout $n = 1, \ldots,$ et la loi de X_0 est γ et de plus X_n converge p.s vers X_0. Par suite en utilisant le lemme de Fatou on obtient,

$$\liminf \int h(x, y)\, d\gamma_n(x, y)\ =\ \liminf \int h(X_n)\, dP \ge \int \liminf h(X_n)\, dP$$
$$=\ \int h(X_0)\, dP = \int h(x, y)\, d\gamma(x, y);$$

et alors

$$\int h(x, y)\, d\gamma(x, y) \le \sup\left\{\int f\, d\mu + \int g\, d\nu\ :\ f \in \mathcal{L}(\mu),\ g \in \mathcal{L}(\nu),\ f \oplus g \le h\right\}.$$

$\square$

On en déduit le corollaire suivant.

Corollaire 1 *Fixons $h \in \mathcal{C}(E \times E)$ et soient $\mu,\ \nu \in \mathcal{M}(E)$. Alors*

1. *Il existe une mesure "minimale" $\gamma \in \Gamma(\mu, \nu)$ telle que*

$$I(h) = \int h\, d\gamma.$$

2. *Il existe une mesure "maximale" $\gamma \in \Gamma(\mu, \nu)$ telle que*

$$U(h) = \int h\, d\gamma.$$

Dans la partie gauche de l'égalité (4), si $h(x,y)$ est une fonction de coût, on obtient la fonctionnelle de Kantorovich. L'existence d'une mesure minimale appelée aussi *mesure optimale* et la détermination de sa structure ont été récemment explicitées par Gangbo et McCann (cf. [21]), sous des conditions de convexité sur la fonction de coût et dans le cas $E = \mathbb{R}^d$.

4 Applications

Une conséquence très importante du théorème de dualité est le théorème de Strassen (cf. [53], [28], [17] et [51]) qui est en fait la source d'inspiration de toute cette étude. Sa forme dual, en un certain sens, a été établie par Gamboa et Cattiaux dans [4] en utilisant des méthodes de grandes déviations.

Théorème 5 *(cf. [33]) Soient E_1 et E_2 deux espaces polonais, $\mathcal{E}_1$ et $\mathcal{E}_2$ deux σ-algèbres de E_1 et E_2. Et désignons par μ et ν deux mesures de probabilités sur E_1 et E_2. Alors, pour tout $B \in \mathcal{E}_1 \otimes \mathcal{E}_2$ on a*

$$U(B) := U(1_B) = \inf \left\{ \mu(B_1) + \nu(B_2) : \ B \subset \bigcup_{i=1}^{2} \pi_i^{-1}(B_i) \right\} \tag{6}$$

et

$$I(B) := I(1_B) = \sup \left\{ \mu(B_1) + \nu(B_2) - 1 : \ B_1 \times B_2 \subset B \right\} \tag{7}$$

Maintenant on va donner des applications concrètes du théorème de dualité.

4.1 Ordre stochastique

Soit $(E, \preceq)$ un espace polonais muni d'une relation de préordre, et supposons que

$$G(E) := \{(x,y) \in E \times E : \ x \preceq y\},$$

soit fermé dans $E \times E$. On désigne par $\mathcal{I}(E)$ le cône des fonctions réelles f bornées et croissantes sur E (i.e., $x \preceq y \Longrightarrow f(x) \leq f(y)$), et par $\mathcal{I}^*(E)$ la famille des sous-ensembles A de E croissants (i.e., la fonction indicatrice de A est croissante). On vérifie immédiatement que

$$A \in \mathcal{I}^*(E) \Longleftrightarrow ((x \in A \ \text{et} \ x \preceq y) \Longrightarrow y \in A).$$

On dit qu'une mesure $\nu \in \mathcal{M}(E)$ *domine stochastiquement* $\mu \in \mathcal{M}(E)$ et on note $\mu \leq_{st} \nu$ si et seulement si

$$\int f \, d\mu \leq \int f \, d\nu \quad \text{pour tout } f \in \mathcal{I}(E).$$

Ce qui est équivalent à

$$\mu(A) \leq \nu(A), \quad \forall A \in \mathcal{I}^*(E).$$

(cf. [36], [30] et [38]). D'après le théorème 5, on obtient une caractérisation de l'ordre stochastique.

Théorème 6 *Soient μ et ν deux mesures de $\mathcal{M}(E)$. Alors on a*

1. *(cf. [33]).*

$$U(G(E)) = 1 - \sup\{\mu(F) - \nu(F) : \ F \ \text{fermé, croissant}\} \tag{8}$$

2. *(cf. [36] et [33]).*

$$\mu \leq_{st} \nu \Longleftrightarrow \exists \gamma \in \Gamma(\mu, \nu) \ avec \ \gamma(G(E)) = 1. \tag{9}$$

4.2 Représentation des métriques minimales

Certaines métriques (pouvant prendre éventuellement la valeur infini) sur l'ensemble des lois de probabilités, très utiles en théorie des probabilités, ont une représentation de "métrique minimale" (cf. [44]).

Soit E un espace polonais muni d'une distance adéquate d et $\mathcal{B}$ sa tribu borélienne. Pour tout $A \subset E$ et $\varepsilon > 0$, le voisinage d'ordre ε de A est défini par

$$A^\varepsilon := \{x \in E : \exists y \in A, \ d(x, y) < \varepsilon\}, \ \text{et on pose } A^0 := \overline{A} \ \text{(l'adhérence de } A\text{)}.$$

4.2.1 Métrique de Lévy-Prokhorov

Par le théorème 5 de Strassen, on obtient

Théorème 7 *(cf. Dudley [14], Théorème 18.2)*
Considérons μ, $\nu \in \mathcal{M}(E)$ et soient $\varepsilon \geq 0$ et $\delta > 0$. Alors les assertions suivantes sont équivalentes

1. *Il existe $\gamma \in \Gamma(\mu, \nu)$ vérifiant $\gamma\{d(x, y) > \varepsilon\} < \delta$;*

2. *on a*

$$\mu(A) \leq \nu(A^\varepsilon) + \delta, \ \textit{pour tout } A \textit{ fermé de } E.$$

Le théorème 7 entraîne une représentation de Strassen de la métrique de Lévy-Prokhorov

$$\Pi(\mu, \nu) := \inf\{\varepsilon : \mu(A) \leq \nu(A^\varepsilon) + \varepsilon, \ \forall A \text{ fermé de } E\}. \tag{10}$$

Définissons pour γ mesure de probabilité sur $E \times E$ la métrique de probabilité de Ky-Fan

$$K(\gamma) := \inf\{\epsilon > 0 : \ \gamma(d(x, y) > \varepsilon) < \varepsilon\} \tag{11}$$

et considérons sa métrique correspondante

$$\widehat{K}(\mu, \nu) := \inf\{K(\gamma) : \ \gamma \in \Gamma(\mu, \nu)\}. \tag{12}$$

Théorème 8 *(cf. Strassen [53] et Dudley [13]) Soient μ et ν deux mesures de probabilités sur E. On a*

$$\widehat{K}(\mu, \nu) = \Pi(\mu, \nu). \tag{13}$$

La distance Π induit la topologie de la convergence faible sur l'ensemble de mesures boréliennes tendues (cf. [15] théorème 11.7.1), i.e.

$$\mu_n \longrightarrow \mu \text{ en loi si et seulement si } \Pi(\mu_n, \ \mu) \longrightarrow 0.$$

Un résultat fondamental est le théorème de représentation presque sûre. La preuve de la deuxième partie utilise principalement le théorème 8.

Théorème 9

*(i) (théorème de représentation presque sûre) (Skorokhod, Strassen, Dudley, Wichura, cf.[14]).
Soit $(\mu_n)_{n\in\mathbb{N}}$ une suite de mesures de probabilités sur E. Alors $\Pi(\mu_n, \mu_0) \to 0$ si et seulement s'il existe un espace de probabilité $(\Omega, \mathcal{A}, P)$ et une suite de variables aléatoires $(X_n)_{n\in\mathbb{N}}$ sur $(\Omega, \mathcal{A})$ à valeurs dans E, telle que $P^{X_n} = \mu_n$, $P^{X_0} = \mu_0$ et $d(X_n, X_0) \to 0$
p.s.*

(ii) (Rachev, Rüschendorf et Schief [45], Dudley [15]). Soient $(\mu_n)_{n\in\mathbb{N}}$ et $(\nu_n)_{n\in\mathbb{N}}$ deux suites tendues de mesures de probabilités sur E. Alors $\Pi(\mu_n, \nu_n) \to 0$ si et seulement s'il existe deux suites de variables aléatoires $(X_n)_{n\in\mathbb{N}}$ et $(Y_n)_{n\in\mathbb{N}}$ sur $(\Omega, \mathcal{A}, P)$ à valeurs dans E telles que $P^{X_n} = \mu_n$, $P^{Y_n} = \nu_n$ et $d(X_n, Y_n) \to 0$ p.s.

4.2.2 ℓ_p-métriques

Pour γ mesure de probabilité sur $E \times E$, définissons la métrique de probabilité suivante

$$\ell_\infty(\gamma) := \operatorname*{ess\,sup}_\gamma d(x, y) = \inf\{\varepsilon > 0 : \gamma(d(x, y) > \varepsilon) = 0\},$$

et considérons sa métrique correspondante définie par

$$\widehat{\ell_\infty}(\mu, \nu) := \inf\{\ell_\infty(\gamma) : \gamma \in \Gamma(\mu, \nu)\},$$

le théorème 7 implique la représentation de la métrique $\widehat{\ell_\infty}$.

Théorème 10 *(cf. [14]. Théorème 18.2)*
Soient μ et ν deux mesures de $\mathcal{M}(E)$. Alors on a

$$\widehat{\ell_\infty}(\mu, \nu) = \inf\{\varepsilon : \mu(A) \leq \nu(A^\varepsilon), \ \forall A \text{ fermé de } E\}. \tag{14}$$

Pour $0 < \lambda < \infty$, les métriques de type Prokhorov et de Ky-Fan sont définies respectivement par

$$\Pi_\lambda(\mu, \nu) := \inf\{\varepsilon : \mu(A) \leq \nu(A^{\lambda\varepsilon}) + \varepsilon, \ \forall A \text{ fermé de } E\}, \tag{15}$$

et

$$K_\lambda(\gamma) := \inf\{\epsilon > 0 : \gamma(d(x, y) > \lambda\varepsilon) < \varepsilon\}, \tag{16}$$

remarquons que

$$\lim_{\lambda \to \infty} \lambda\Pi_\lambda(\mu, \nu) = \inf\{\varepsilon : \mu(A) \leq \nu(A^\varepsilon), \ \forall A \text{ fermé de } E\}.$$

Pour $1 \leq p < \infty$, la ℓ_p-distance entre deux mesures μ et ν de $\mathcal{M}(E)$ est définie par

$$\widehat{\ell_p}(\mu, \nu) := \inf\left\{\left(\int d^p(x, y) \, d\gamma(x, y)\right)^{1/p} : \gamma \in \Gamma(\mu, \nu)\right\}.$$

Le théorème suivant se déduit immédiatement du théorème 4 de dualité.

Théorème 11 *(cf. [43] et [44])*
*Soient μ et ν deux mesures de $\mathcal{M}(E)$ telles que $\int d^p(x, x_0) \, d(\mu + \nu) < \infty$, pour tout $p \geq 1$.
Alors on a*

$$\widehat{\ell_p}(\mu, \nu) = \sup\left\{\left(\int f \, d\mu + \int g \, d\nu\right)^{1/p} : f \in \mathcal{L}(\mu), \ g \in \mathcal{L}(\nu),\right.$$

$$f(x) + g(y) \leq d^p(x, y), \ \forall x, y \in E \Big\}.$$

En particulier, pour $p = 1$ on a

Théorème 12 *(Kantorovich-Rubinstein, cf. [54], [18], [33] et [44]).*
Si $\int d(x, x_0) d(\mu + \nu) < \infty$, alors

$$\widehat{\ell}_1(\mu, \nu) = \sup \left\{ \int f \ d(\mu - \nu) : \ f \in \mathrm{Lip}_1(E) \right\},$$

où

$$\mathrm{Lip}_1(E) := \{ f : E \to I\!\!R : \ |f(x) - f(y)| \leq d(x, y), \ \forall x, \ y \in E \}.$$

Pour $E = I\!\!R$ les ℓ_p-métriques minimales sont explicitement connues (cf. Dall'Aglio [8], Fréchet [19]-[20] et Vallender [56]) : on a

$$\widehat{\ell}_1(\mu, \nu) = \int \mid M(x) - N(x) \mid \ dx$$

$$\widehat{\ell}_p^{\ p}(\mu, \nu) = \int_0^1 |M^{-1}(t) - N^{-1}(t)|^p \ dt, \ p \geq 1,$$

où M et N sont les fonctions de répartitions de μ et ν. Plus généralement, Major [37] montre que si $\int |x| \ d\mu(x) < \infty$ et $\int |x| \ d\nu(x) < \infty$ alors on a

$$\inf \left\{ \int \psi(x - y) \ d\gamma(x, y) : \ \gamma \in \Gamma(\mu, \nu) \right\} = \int_0^1 \psi(M^{-1}(t) - N^{-1}(t)) \ dt,$$

avec ψ une fonction convexe.

Maintenant, on va étudier un cas particulier important de la ℓ_p-métrique quand $E = I\!\!R^d$ et $p = 2$. Soit $|| \cdot ||$ la norme euclidienne dans $I\!\!R^d$ et $< \cdot, \cdot >$ désigne le produit scalaire dans $I\!\!R^d$. Pour une fonction réelle f sur $I\!\!R^d$, semicontinue inférieurement et convexe on désigne par f^* la fonction conjuguée de f définie par

$$f^*(y) := \sup_{x \in I\!\!R^d} \ \{< x, y > -f(x)\};$$

et le *sous-différentiel* de f en x est défini par

$$\partial f(x) := \{ y \in I\!\!R^d : \ f(z) - f(x) \geq < z - x, y >, \ z \in I\!\!R^d \};$$

(cf. Rockafellar [48]). Les éléments de $\partial f(x)$ sont appelés *sous-gradients* de f en x. La ℓ_2-distance de *Lévy-Wasserstein* $W(\mu, \nu)$ entre deux mesures μ et ν de $\mathcal{M}(E)$ est définie par

$$W(\mu, \nu) := \widehat{\ell}_2(\mu, \nu) = \inf \left\{ \left(\int ||x - y||^2 \ \gamma(dx, dy) \right)^{1/2}, \ \gamma \in \Gamma(\mu, \nu) \right\}. \quad (17)$$

Un couple (X, Y) de variables aléatoires sur un espace de probabilité $(\Omega, \mathcal{A}, P)$ est dit *W-couple optimal* (en abrégé W-c.o.) pour (μ, ν) si

$$E||X - Y||^2 := \int ||X - Y||^2 \ dP = W^2(\mu, \nu).$$

Théorème 13

(i) (Knott et Smith [35], Rüschendorf and Rachev [50]). Soient μ et ν deux mesures de $\mathcal{M}(I\!\!R^d)$ telles que $\int \|x\|^2 \, d\mu(x) < \infty$ et $\int \|x\|^2 \, d\nu(x) < \infty$ et X et Y deux variables aléatoires ayant pour lois μ et ν. Alors $E\|X - Y\|^2 = W^2(\mu, \nu)$ si et seulement s'il existe une fonction $f : I\!\!R^d \to I\!\!R$ semicontinue inférieurement et convexe telle que $Y \in \partial f(X)$.

(ii) (Dowson et Landau [12], Olkin et Pukelsheim [40], Givens et Shortt [23]). Considérons μ et ν deux gaussiennes multidimensionnelles ayant respectivement m_μ, m_ν, Σ_μ et Σ_ν pour moyennes et matrices de covariance. Alors on a

$$W^2(\mu, \nu) = \|m_\mu - m_\nu\|^2 + \text{trace}\left(\Sigma_\mu + \Sigma_\nu - 2\left(\Sigma_\mu^{1/2}\Sigma_\nu\Sigma_\mu^{1/2}\right)^{1/2}\right). \tag{18}$$

Remarque 1

(i) Une fonction réelle f continue différentiable sur $I\!\!R^d$ est convexe si et seulement si

$$\varphi(x) := \nabla f \text{ est monotone, i.e. } < x - y, \varphi(x) - \varphi(y) > \geq 0, \ \forall x, y \in I\!\!R^d$$

où ∇f est le gradient de f, (cf. [48], p. 99).
Soient μ et ν deux mesures de $\mathcal{M}(I\!\!R^d)$, et X une variable aléatoire de loi μ. Si φ est le gradient d'une fonction différentiable f, et $\varphi(X)$ a pour loi ν, alors $(X, \varphi(X))$ est un W-couple optimale si et seulement si φ est monotone.

Si φ est une fonction continue et différentiable sur $I\!\!R^d$, alors $(X, \varphi(X))$ est un W-couple optimal si et seulement si

1. $\dfrac{\partial \varphi_j}{\partial x_i} = \dfrac{\partial \varphi_i}{\partial x_j}, \quad \forall i \neq j$, et

2. φ est monotone. (cf. [49]).

Dans le cas d'une fonction linéaire $\varphi(x) = Ax$ (où A est une matrice), le couple $(X, \varphi(X))$ est W-optimal si et seulement si A est symétrique et semidéfinie positive.

En particulier, dans le cas normal multidimensionel avec $\mu = N(0, \Sigma_\mu)$ et $\nu = N(0, \Sigma_\nu)$, si X a pour loi μ et $A = \Sigma_\mu^{-1/2}(\Sigma_\mu^{1/2}\Sigma_\nu\Sigma_\mu^{1/2})^{1/2}\Sigma_\mu^{-1/2}$, alors (X, AX) est un W-c.o pour (μ, ν), (cf. [49]).

(ii) Le problème de simulation de Monte-Carlo est le suivant :
Pour $\mu_i \in \mathcal{M}(I\!\!R)$, comment construire des variables aléatoires $X_i \sim \mu_i$ (i.e., X_i a pour loi μ_i) telles que

$$\text{Var}\left(\sum_{i=1}^n X_i\right) \leq \text{Var}\left(\sum_{i=1}^n Y_i\right) \quad \text{pour tout} \quad Y_i \sim \mu_i.$$

Pour $n = 2$, on a une solution bien connue sous le nom de "antithetic variates" (cf. Hammersley et Handscomb [27]). Et pour $n \geq 2$ on connait quelques solutions du problème dans des cas particuliers (cf. [52]).
Le problème correspondant est de déterminer le minimum de

$$\sum_{i<j} E < X_i, X_j > .$$

D'après le théorème 13, pour $n = 2$ on obtient une caractérisation de la solution du problème : $E < X_1, X_2 > = \min \{E < Y_1, Y_2 > : Y_1 \sim \mu_1, Y_2 \sim \mu_2\}$ si et seulement s'il existe $f : I\!\!R \longrightarrow I\!\!R$ semicontinue inférieurement et convexe telle que $X_2 \in \partial f(-X_1)$. Ce qui est équivalent à $< -X_1, X_2 > = f(-X_1) + f(X_2)$.

(iii) La preuve de la première partie du théorème 13, est établie grâce au théorème de dualité et d'autres ingrédients d'analyse convexe. On trouve dans [42] d'autres résultats pour la ℓ_2-métrique de Lévy-Wasserstein et ses applications pour les problèmes d'approximations.

(iv) Un cas très intéressant du problème de transport de Monge-Kantorovich consiste à évaluer la fonctionnelle $W^2(\mu, \nu)$. Sous une condition de continuité sur la loi μ, Cuesta et Matrán (cf. [6]) ont montré que les W-c.o. sont de la forme $(X, f(X))$ où X est une variable aléatoire de loi μ et f une application monotone adéquate; récemment dans [7], ils ont étudié les propriétés de la fonction f. D'autre part, une détermination explicite de cette fonction f a été faite par Abdellaoui et Heinich (cf. [1] et [2]).

(v) La deuxième propriété du théorème 13 reste vrai sur un espace de Hilbert séparable si μ et ν sont deux gausssiennes (cf.[22] et [7]).

4.3 Couplage maximal et théorème de Goldstein

Dans cette section, on va présenter une nouvelle démonstration du théorème de Goldstein (cf. [24], [55] et [36]).

Une métrique très intéressante associée à la ℓ_p-métrique est la métrique de variation totale définie par

$$\sigma(\mu, \nu) := \sup_{A \in \mathcal{B}} |\mu(A) - \nu(A)|.$$

En effet, σ peut être obtenue comme cas limite de la métrique $\widehat{\ell}_p$ et celle de Prokhorov Π_λ, c'est-à-dire

$$\sigma(\mu, \nu) = \lim_{p \to \infty} \widehat{\ell}_p(\mu, \nu),$$

et

$$\sigma(\mu, \nu) = \lim_{\lambda \to 0} \Pi_\lambda(\mu, \nu),$$

et admettant donc la représentation suivante

$$\sigma(\mu, \nu) = \inf\left\{ \gamma\{(x, y) \in E \times E : x \neq y\} : \ \gamma \in \Gamma(\mu, \nu) \right\}; \tag{19}$$

Maintenant, considérons $A \in \mathcal{B}$ et soit

$$\Delta(A) := \{(x, x) \in E \times E : \ x \in A\} \in \mathcal{B} \otimes \mathcal{B},$$

la diagonale de l'ensemble A. On définit

$$\mu \wedge \nu(A) := \inf\{\mu(A_1) + \nu(A_2) : \ A_1, \ A_2 \in \mathcal{B}, \ A_1 \cup A_2 = A\}.$$

Le résultat suivant nous permet de donner une autre façon d'avoir la représentation (19).

Théorème 14 *Il existe* $\gamma \in \Gamma(\mu, \nu)$ *telle que*

$$\gamma(\Delta(A)) = U(\Delta(A)) = \mu \wedge \nu(A), \quad pour\ tout \quad A \in \mathcal{B}, \tag{20}$$

où

$$U(\Delta(A)) = \sup\left\{ \gamma(\Delta(A)) \ : \ \gamma \in \Gamma(\mu, \nu) \right\}.$$

DÉMONSTRATION DU THÉORÈME 14. L'égalité $U(\Delta(A)) = \mu \wedge \nu(A)$ découle du théorème 5 de Strassen. Par ailleurs, comme $\mu \wedge \nu$ est une mesure sur E, alors U est additive. Alors, on déduit aisément l'existence de $\gamma \in \Gamma(\mu, \nu)$ telle que

$$\gamma(\Delta(A)) = \mu \wedge \nu(A), \quad \text{pour tout} \quad A \in \mathcal{B}.$$

$\square$

Maintenant, si on prend $A = E$, et si on pose $\Delta = \Delta(E)$ pour simplifier les notations, on montre aisément que l'équation (20) implique

$$\sigma(\mu, \nu) = \sup_{A \in \mathcal{B}} |\mu(A) - \nu(A)| = 1 - U(\Delta) = I(E \setminus \Delta). \tag{21}$$

Soient $X = (X_n)_0^\infty$ et $Y = (Y_n)_0^\infty$ deux suites de variables aléatoires à valeurs dans $(E^\infty, \mathcal{B}^\infty)$ et de lois μ et ν respectivement, avec

$$E^\infty := E \times E \times E \times \ldots \quad \text{et} \quad \mathcal{B}^\infty := \mathcal{B} \otimes \mathcal{B} \otimes \mathcal{B} \otimes \ldots.$$

Pour $n \in \overline{\mathbb{N}} = \mathbb{N} \cup \{\infty\}$, définissons le *shift*

$$\theta_n : E^\infty \longrightarrow E^\infty$$

par

$$\theta_n x = \begin{cases} (x_n, x_{n+1}, x_{n+2}, \ldots); & \text{pour } n < \infty, \\ (z, z, z, \ldots); & \text{pour } n = \infty, \end{cases}$$

où $x = (x_0, x_1, x_2, \ldots) \in E^\infty$, et z est un élément fixé dans E. Pour $n \geq 0$, soit $\mathcal{S}_n$ la σ-algèbre $\subset \mathcal{B}^\infty$, définie par

$$\mathcal{S}_n = \theta_n^{-1}(\mathcal{B}^\infty)$$

et

$$\mathcal{S}_\infty = \bigcap_{n=0}^\infty \mathcal{S}_n.$$

Désignons par $\mu_{(n)}$, $\mu_{(\infty)}$ la restriction de μ à $\mathcal{S}_n$, $\mathcal{S}_\infty$.

Définition 2   *1. On dit qu'une mesure $\gamma \in \Gamma(\mu, \nu)$ est un couplage réussi si*

$$\lim_{n \to \infty} \gamma\{(x, y) \in E^\infty \times E^\infty : \theta_n x = \theta_n y\} = 1.$$

2. On dit qu'une mesure $\gamma \in \Gamma(\mu, \nu)$ est un couplage maximal *si*

$$\gamma\{(x, y) \in E^\infty \times E^\infty : \theta_n x = \theta_n y\} = 1 - \sigma(\mu_{(n)}, \nu_{(n)}), \textit{ pour tout } n \in \mathbb{N}.$$

Les notions de couplage réussi et maximal sont très utiles pour démontrer des théorèmes d'ergodicité des chaînes de Markov (cf. [3], [25], [26] et [41]).

Théorème 15 *Les propriétés suivantes sont équivalentes:*

1. Il existe un couplage $\gamma \in \Gamma(\mu, \nu)$ réussi.

2. $\mu = \nu$ sur $\mathcal{S}_\infty$.

3. $\lim_{n \to \infty} \sigma(\mu_{(n)}, \nu_{(n)}) = 0$.

DÉMONSTRATION DU THÉORÈME 15. La propriété *(3)* entraîne *(2)* provient de l'inégalité suivante

$$\sigma(\mu_{(\infty)}, \nu_{(\infty)}) \leq \sigma(\mu_{(n)}, \nu_{(n)}).$$

Montrons que *(2)* implique *(3)*. Supposons pour tout n, il existe un ensemble $A_n \in \mathcal{S}_n$ tel que

$$|\mu(A_n) - \nu(A_n)| \geq \alpha_{\infty},$$

avec

$$\alpha_n = \sigma(\mu_{(n)}, \nu_{(n)}), \text{ et } \alpha_{\infty} = \lim_{n \to \infty} \alpha_n.$$

Sans perdre de généralité, on peut supposer que

$$\mu(A_n) - \nu(A_n) \geq \alpha_{\infty} \quad \text{pour tout} \quad n.$$

Maintenant, introduisons l'espace de Hilbert

$$\mathcal{H} := L^2(E^{\infty}, \mathcal{E}^{\infty}, \mu + \nu),$$

et on note $< \cdot, \cdot >_{\mathcal{H}}$ son produit scalaire — rappelons que dans un espace de Hilbert toute suite bornée admet une sous-suite faiblement convergente —.
Soient $f_n = 1_{A_n}$, $g = \dfrac{d\mu}{d(\mu + \nu)}$ et $g' = \dfrac{d\nu}{d(\mu + \nu)}$. La suite $(f_n)_0^{\infty}$ est bornée, et alors il existe une sous-suite $(f_{k_n})_0^{\infty}$ qui converge faiblement vers $f_{\infty} \in [0, 1]$. D'autre part, les fonctionnelles linéaires

$$f \longrightarrow \int f \, d\mu = <f, g>_{\mathcal{H}}, \text{ et } f \longrightarrow \int f \, d\nu = <f, g'>_{\mathcal{H}}$$

sont bornées, et donc on obtient

$$
\begin{aligned}
\int f_{\infty} \, d\mu - \int f_{\infty} \, d\nu &= \lim_{n \to \infty} (<f_{k_n}, g>_{\mathcal{H}} - <f_{n_k}, g'>_{\mathcal{H}}) \\
&= \lim_{n \to \infty} \left(\int f_{k_n} \, d\mu - \int f_{k_n} \, d\nu \right) \\
&= \lim_{n \to \infty} (\mu(A_{k_n}) - \mu(A_{k_n})) \geq \alpha_{\infty};
\end{aligned}
$$

et comme l'espace $L^2(E^{\infty}, \mathcal{S}_n, \mu + \nu)$ est faiblement fermé, on a $f_{\infty} \in \mathcal{S}_n$, pour tout n, et donc $f_{\infty} \in \mathcal{S}_{\infty}$. Et par suite, on en déduit immédiatement le résultat.

Maintenant, prouvons *(3)* est équivalent à *(1)*. D'après le théorème 14 et l'égalité (21), il existe une mesure $\gamma_n \in \Gamma(\mu_{(n)}, \nu_{(n)})$ telle que

$$
\begin{aligned}
\sigma(\mu_{(n)}, \nu_{(n)}) &= \gamma_n \{(x, y) \in E^{\infty} \times E^{\infty} : \theta_n x \neq \theta_n y\} \\
&= \gamma \{(x, y) \in E^{\infty} \times E^{\infty} : \theta_n x \neq \theta_n y\};
\end{aligned}
$$

où $\gamma \in \Gamma(\mu, \nu)$ est le prolongement de la mesure γ_n sur $E^{\infty} \otimes E^{\infty}$. Et par conséquent, la mesure γ est un couplage maximal, si $\sigma(\mu_{(n)}, \nu_{(n)}) \longrightarrow 0$; ceci est équivalent à γ est un couplage réussi. Ce qui achève la démonstration.

$\square$

5 Appendice

Les théorèmes suivant sont parmi les versions géométriques du théorème de Hahn-Banach (cf. [10], [16], [29] et [34]).

Définition 3 *Soit E un espace vectoriel réel. Un point $x_0 \in A \subset E$ est appelé point intérieur de A si pour tout $y \in E$, il existe $\delta > 0$ tel que $x_0 + ty \in A$ pour tout $|t| < \delta$. Ou bien si $A - x_0$ est un ensemble absorbant.*

Théorème 16 (de séparation) *Soit E un espace vectoriel réel, M et N deux sous-ensembles non vides, disjoints et convexes de E. Supposons que M contient au moins un point intérieur, et N ne rencontre aucun point intérieur de M. Alors il existe une forme linéaire f sur E, non triviale, telle que*

$$\inf_{x \in N} f(x) \geq \sup_{y \in M} f(x).$$

Théorème 17 *Soit E un espace vectoriel réel et L un sous espace vectoriel de E. Soit M un sous-ensemble convexe de E. Supposons que M contient un point intérieur appartenant à L. Soit f une forme linéaire non nulle sur L, bornée supérieurement sur $L \cap M$. Alors f se prolonge en une forme linéaire g sur E, vérifiant*

$$\sup_{x \in M} g(x) = \sup_{y \in L \cap M} f(y).$$

DÉMONSTRATION DU THÉORÈME 17. Soit $\alpha := \sup_{x \in L \cap M} f(x)$ et $A := \{x \in L : f(x) > \alpha\}$. Alors L et M sont convexes, disjoints. Par conséquent, d'après le théorème 16 de séparation il existe une forme linéaire h sur E telle que

$$\inf_{x \in A} h(x) \geq \sup_{y \in M} h(y).$$

Soit $x_0 \in L \cap M$ le point intérieur de M. Alors $\sup_{x \in M} h(x) > h(x_0)$, d'où $h(y) > h(x_0)$ pour tout $y \in A$, et par suite h n'est pas une constante sur L. Soit $B := \{x \in L : h(x) = h(x_0)\}$. Remarquons que f est constante sur B. En effet, supposons que f prend deux valeurs différentes sur deux points quelconques de B. Alors f prend toutes les valeurs sur la droite joignant les deux points, et donc cette droite rencontre A. Mais $h = h(x_0)$ sur cette droite, contradiction. Soit $f = k$ sur B ($k \in \mathbb{R}$). Si $k = 0$, comme f n'est pas constante sur L, on a $0 \in B$ et $h(x_0) = 0$. Considérons $z \in L \setminus B$, $h(z) \neq f(z) \neq 0$ et $h(x) = h(z)f(x)/f(z)$ pour $x = z$ et pour tout $x \in B$, donc pour tout $x \in L$. D'autre part, si $k \neq 0$, alors 0 n'appartient pas à B, $h(x_0) \neq 0$ et $h(x) = h(x_0)f(x)/k$, pour tout $x \in B$ et pour $x = 0$, donc pour tout $x \in L$. Dans les deux cas, $h = \beta f$ sur L pour $\beta \neq 0$. Alors $g := h/\beta$ prolonge f sur E et vérifie

$$\sup_{x \in M} g(x) \leq \inf_{y \in A} h(y)/\beta = \inf_{y \in A} f(y) = \alpha.$$

$\square$

Références

[1] ABDELLAOUI, T. Détermination d'un couple optimal du problème de Monge Kantorovich. *C. R. Acad. Sci. Paris*, 319:981–984, 1994.

[2] ABDELLAOUI, T., ET HEINICH, H. Sur la distance de deux lois dans le cas vectoriel. *C. R. Acad. Sci. Paris*, 319:397–400, 1994.

[3] ALDOUS, D. J. Shift-coupling. *Stoch. Proc. Appl*, 44:1–14, 1993.

[4] CATTIAUX, P., ET F. GAMBOA. Large deviations and variational theorem for marginal problems. *Preprint*, 1996.

[5] CHOQUET, G. Forme abstraite du théorème de capacitabilité. *Ann. Inst. Fourier*, 9:83–89, 1959.

[6] CUESTA-ALBERTOS, J. A., AND MATRÁN, C. Notes on the Wasserstein metric in Hilbert spaces. *Ann. Probab.*, 17:1264–1276, 1989.

[7] CUESTA-ALBERTOS, J. A., MATRÁN, C, AND TUERO-DIAZ, A . On lower bounds for the l^2-Wasserstein metric in a Hilbert space. *J. of Theoretical Prob.*, 9:263–283, 1996.

[8] DALL'AGLIO. Fréchet classes and compatibility of distribution function. *Sym. Math.*, 9:131–150, 1972.

[9] DELLACHERIE, C., MEYER, P.A. *Probabilités et potentiel*. Herman, Paris, 1983.

[10] DIEUDONNÉ, J. Sur le théorème de Hahn-Banach. *Rev. Sci*, 79:642–643, 1941.

[11] DOEBLIN, W. Exposé de la théorie des chaînes simples constantes de Markov à un nombre fini d'états. *Rev. Math. Union Interbalkanique*, 2:77–105, 1938.

[12] DOWSON, D. C., LANDAU, B. V. The Fréchet distance between multivariate normal distribution. *J. Multivariate Anal.*, 12:450–455, 1982.

[13] DUDLEY, R. M. Distances of probability measures and random variables. *Ann. Math. Stat.*, 39:1563–1572, 1968.

[14] DUDLEY, R. M. *Probability and metrics*. Aarhus Univ., Aarhus, 1976.

[15] DUDLEY, R. M. *Real analysis and probability*. Chapman and Hall, New York London, 1989.

[16] DUNFORD, N., AND SCHWARTZ, J. T. *Linear Operators*. Interscience Publishers, a division of John Wiley and Sons, New York, t. I, 1958.

[17] EDWARDS, D. A. On the existence of probability measures with given marginals. *Ann. Inst. Fourier.*, 28:53–78, 1978.

[18] FERNIQUE, X. *Sur le théorème de Kantorovitch-Rubinstein dans les espaces polonais*. Lecture Notes in Mathematics 850., Springer, 1981.

[19] FRÉCHET, M . Sur les tableaux de corrélation dont les marges sont données. *Annales de l'université de Lyon, Sciences.*, 4:13–84, 1951.

[20] FRÉCHET, M . Sur la distance de deux lois de probabilité. *C. R. Acad. Sci. Paris.*, 244, 1957.

[21] GANGBO, W., AND MCCANN, R. J. The geometry of optimal transportation. *Acta. Math.*, 177:113–161, 1996.

[22] GELBRICH, M . On a formula for the l^2-Wasserstein metric between measures on Euclidean and Hilbert spaces. *Math. Nachr.*, 147:185–203, 1990.

[23] GIVENS, C. R., AND SHORTT, R. M . A class of Wasserstein metrics for probability distributions. *Michigan Math. J.*, 31:231–240, 1984.

[24] GOLDSTEIN, S. Maximal coupling. *Z. Wahrscheinlichkeitstheor. Verw. Geb.*, 46:193–204, 1979.

[25] GRIFFEATH, D. A maximal coupling for Markov chains. *Z. Wahrscheinlichkeitstheor. Verw. Geb.*, 31:95–106, 1975.

[26] GRIFFEATH, D. Uniform coupling of non-homogenous Markov chains. *J. Appl. Probability*, 12:753–762, 1975.

[27] HAMMERSLEY, I. M., AND HANDSCOMB, D. C. *Monte Carlo methods*. Meth, London, 1964.

[28] HANSEL, G, AND TROALLIC, J. P. Mesures marginales et théorème de Ford-Fulkerson. *Z. Wahrscheinlichkeitstheor. Verw. Geb.*, 43:245–251, 1978.

[29] HERMES, H., AND LASALLE, J.P. *Functional Analysis and Time optimal control*. Academic Press, New York and London, 1969.

[30] KAMAE, T., KRENGEL, U. AND O'BRIEN. Stochastic inequalities on partially ordered spaces. *Ann. Probab.*, 5:899–912, 1977.

[31] KANTOROVICH, L.V. On the translocation of masses. *C. R. (Doklady) Acad. Sci. URSS (N.S.)*, 37:199–201, 1942.

[32] KANTOROVICH, L.V. On a problem of Monge (in russian). *Uspekhi Math. Nauk*, 3:225–226, 1948.

[33] KELLERER, H. G. Duality theorems for marginal problems. *Z. Wahrscheinlichkeitstheor. Verw. Geb.*, 67:399–432, 1984.

[34] KELLEY, J. L., AND NAMIOKA, I. *Linear topological spaces*. D. Van Nostrand Company, Princeton, N. I, 1963.

[35] KNOTT, M., AND SMITH, C. S. On the optimal mapping of distributions. *J. Optim. Th. Appl.*, 43:39–49, 1984.

[36] LINDVALL, T. *Lectures on the coupling method*. Wiley, New York, 1993.

[37] MAJOR, P. On the invariance principle for sums of independent identically distributed random variables. *J. Multivariate Anal.*, 8:487–517, 1978.

[38] MARSHALL, A. W., OLKIN, I. *Inequalities theory of majorization and its applications*. Academic Press, New York, 1979.

[39] MONGE, G. Mémoire sur la théorie des déblais et des remblais. *Histoires de l'Académie Royale des Sciences de Paris, avec les mémoires de Mathématiques et de Physique pour la même année*, pages 257–263, 1781.

[40] OLKIN, I., AND PUKELSHEIM, F. The distance between two random vectors with given dispertion matrices. *Linear Algebra Appl.*, 48:257–263, 1982.

[41] PITMAN, J.W. On coupling of Markov chains. *Z. Wahrscheinlichkeitstheor. Verw. Geb.*, 35:315–322, 1976.

[42] RACHEV, S. T. The Monge Kantorovich mass transference problem and its stochastic applications. *Theory Prob. Appl.*, 29:647–676, 1984.

[43] RACHEV, S. T. On a problem of Dudley. *Soviet. Math. Dokl.*, 29:162–164, 1984.

[44] RACHEV, S. T. *Probability metrics and the stability of the stochastic models*. Wiley, New York, 1991.

[45] RACHEV, S. T., RÜSCHENDORF, L., AND SCHIEF, A. Uniformities for the convergence in law and in probability. *J. of Theoretical Prob.*, 5:33–44, 1992.

[46] RAMACHANDRAN, D., AND RÜSCHENDORF, L. A general duality theorem for marginal problems. *Probab. Theory Relat. Fields*, 101:311–319, 1995.

[47] RAMACHANDRAN, D., AND RÜSCHENDORF, L. Duality and perfect probability spaces. *Proceedings of the American mathematical society*, 124:2223–2228, 1996.

[48] ROCKAFELLAR, R. T. *Convex Analysis*. Princeton, Univ. Press, 1970.

[49] RÜSCHENDORF, L. Fréchet bounds and their applications. In Kotz S Dall'Aglio, G. and Salinetti G, editors, *Advances in probability distributions with given marginals: Beyond the Copulas*, pages 141–176. Dordrecht, Kluwer Academic Publishers, 1991.

[50] RÜSCHENDORF, L., AND RACHEV, S. A characterization of random variables with minimum l^2-distance. *J. of Multivariate Anal.*, 32:48–54, 1990.

[51] SKALA, H. G. The existence of probability measures with given marginals. *Ann. Probab.*, 21:136–142, 1993.

[52] SNIJDERS, T. A. B. Antithetic variates for Monte-Carlo estimation of probabilities. *Statistics Neerlandica*, 38:1–19, 1984.

[53] STRASSEN, V. The existence of measures with given marginals. *Ann. Math. Stat*, 36:423–439, 1965.

[54] SZUGLA, A. On minimal metrics in the space of random variables. *Theory Prob. Appl.*, 27:424–430, 1982.

[55] THORISSON, H. On maximal and distributional coupling. *Ann. Probab.*, 14:873–876, 1986.

[56] VALLENDER, S. S. Calculation of the Wasserstein distance between probability distributions on the line. *Theory. Prob. Appl.*, 18:784–786, 1973.

The distribution of local times of a Brownian bridge

Jim Pitman

1 Introduction

Let $(L_t^x, t \geq 0, x \in \mathbb{R})$ denote the jointly continuous process of local times of a standard one-dimensional Brownian motion $(B_t, t \geq 0)$ started at $B_0 = 0$, as determined by the occupation density formula [20]

$$\int_0^t f(B_s)\, ds = \int_{-\infty}^{\infty} f(x) L_t^x \, dx$$

for all non-negative Borel functions f. Borodin [7, p. 6] used the method of Feynman-Kac to obtain the following description of the joint distribution of L_1^x and B_1 for arbitrary fixed $x \in \mathbb{R}$: for $y > 0$ and $b \in \mathbb{R}$

$$P(L_1^x \in dy, B_1 \in db) = \frac{1}{\sqrt{2\pi}}(|x| + |b - x| + y)\, e^{-\frac{1}{2}(|x|+|b-x|+y)^2} \, dy \, db. \tag{1}$$

This formula, and features of the local time process of a Brownian bridge described in the rest of this introduction, are also implicit in Ray's description of the joint law of $(L_T^x, x \in \mathbb{R})$ and B_T for T an exponentially distributed random time independent of the Brownian motion [19, 23, 6]. See [14, 17] for various characterizations of the local time processes of Brownian bridge and Brownian excursion, and further references. These local time processes arise naturally both in combinatorial limit theorems involving the height profiles of trees and forests [1, 17] and in the study of level crossings of empirical processes [21, 4].

Section 2 of this note presents an elementary derivation of formula (1), based on Lévy's identity in distribution [15],[20, Ch. VI, Theorem (2.3)]

$$(L_t^0, |B_t|) \stackrel{d}{=} (M_t, M_t - B_t) \quad \text{where} \quad M_t := \sup_{0 \leq s \leq t} B_s, \tag{2}$$

and the well known description of this joint law implied by reflection principle [20, Ch. III, Ex. (3.14)], which yields (1) for $x = 0$. Formula (1) determines the one-dimensional distributions of the process of local times at time 1 derived from a Brownian bridge of length 1 from 0 to b, as follows: for $y \geq 0$

$$P(L_1^x > y \mid B_1 = b) = e^{-\frac{1}{2}((|x|+|b-x|+y)^2 - b^2)}. \tag{3}$$

Section 3 presents a number of identities in distribution as consequences of this formula. If x is between 0 and b then $|x| + |b - x| = |b|$, so the distribution of L_1^x for the bridge from 0 to b is the same for all x between 0 and b. Furthermore, assuming for simplicity that $b > 0$, the process $(L_1^x, 0 \le x \le b \mid B_1 = b)$ is both reversible and stationary. Reversibility follows immediately from the fact that if $(B_s^{0 \to b}, 0 \le s \le 1)$ denotes the Brownian bridge of length 1 from 0 to b then

$$(b - B_{1-s}^{0 \to b}, 0 \le s \le 1) \overset{d}{=} (B_s^{0 \to b}, 0 \le s \le 1). \tag{4}$$

To spell out the stationarity property, for each $x > 0$ and $\theta > 0$ with $0 < x + \theta \le b$, there is the invariance in distribution

$$(L_1^{x+a}, 0 \le a \le \theta \mid B_1 = b) \overset{d}{=} (L_1^a, 0 \le a \le \theta \mid B_1 = b). \tag{5}$$

Equivalently, for all such x and θ and every non-negative measurable function f which vanishes off the interval $[0, \theta]$

$$\int_0^1 f(B_s^{0 \to b} - x) ds \overset{d}{=} \int_0^1 f(B_s^{0 \to b}) ds. \tag{6}$$

Howard and Zumbrun [11] proved (6) for f the indicator of a Borel set, and (6) for general f can be established by their method. Alternatively, (6) can be deduced from the following invariance in distribution on the path space $C[0, 1]$: for each $0 \le x \le b$

$$({}^x B_s^{0 \to b}, 0 \le s \le 1) \overset{d}{=} (B_s^{0 \to b}, 0 \le s \le 1) \tag{7}$$

where $({}^x B_s^{0 \to b}, 0 \le s \le 1)$ is derived from the bridge $(B_s^{0 \to b}, 0 \le s \le 1)$ by the following path transformation:

$$
{}^x B_s^{0 \to b} := \begin{cases} B_{\sigma_x + s}^{0 \to b} - x & \text{if } 0 \le s \le 1 - \sigma_x \\ b - B_{1-s}^{0 \to b} & \text{if } 1 - \sigma_x < s \le 1 \end{cases}
$$

with σ_x the first hitting time of x by $(B_s^{0 \to b}, 0 \le s \le 1)$. The invariance (7) can be checked by a standard technique [5, 21]: the corresponding transformation on lattice paths is a bijection, which gives simple random walk analogs of (7), (6) and (5); the results for the Brownian bridge then follow by weak convergence. Formula (7) implies also that the process $(\sigma_x, 0 \le x \le b)$ derived from $(B_s^{0 \to b}, 0 \le s \le 1)$ has stationary increments. It is easily shown by similar arguments that the increments of this process are in fact exchangeable. According to the above discussion, for each $b > 0$ the process of bridge occupation times

$$\left(\int_0^1 1(0 \le B_s^{0 \to b} \le x) \, ds, 0 \le x \le b \right)$$

has increments which are stationary and reversible, but not exchangeable (due to continuity of the local time process).

See [9, 16] for other examples of stationary local time processes. In particular, it is known [16, Prop. 2] that if $({}^\delta L_t^x, t \ge 0, x \in \mathbb{R})$ is the process of local times of a Brownian motion with drift $\delta > 0$, say ${}^\delta B_t := B_t + \delta t, t \ge 0$, then the process $({}^\delta L_\infty^x, x \ge 0)$ is a stationary diffusion, with ${}^\delta L_\infty^x$ exponentially distributed with rate δ. The stationarity of this process is obvious by application of the strong Markov property of $^\delta B$ at its first hitting time of $x > 0$. It is also known [23, Theorem 4.5] that if T_δ is exponential with rate $\frac{1}{2}\delta^2$, independent of B, then for each $b > 0$

$$(B_t, 0 \le t \le T_\delta \,|\, B_{T_\delta} = b) \stackrel{d}{=} (^\delta B_t, 0 \le t \le {}^\delta\lambda_b) \tag{8}$$

where $^\delta\lambda_b$ is the time of the last hit of b by $^\delta B$, and hence

$$(L^x_{T_\delta}, 0 \le x \le b \,|\, B_{T_\delta} = b) \stackrel{d}{=} (^\delta L^x_\infty, 0 \le x \le b). \tag{9}$$

The stationarity of $(^\delta L^x_\infty, x \ge 0)$ then implies stationarity of $(L^x_{T_\delta}, 0 \le x \le b \,|\, B_{T_\delta} = b)$, which is implicit in Ray's description of this process [19, 23, 6]. This yields another proof of the stationarity of $(L^x_t, 0 \le x \le b \,|\, B_t = b)$ for arbitrary $t > 0$, by uniqueness of Laplace transforms.

2 Proof of formula (1).

As observed in the introduction, formula (1) for $x = 0$ is equivalent via (2) to Lévy's well known description of the joint distribution of M_1 and $M_1 - B_1$. The case of (1) with $x \ne 0$ will now be deduced from the case with $x = 0$. It clearly suffices to deal with $x > 0$, as will now be supposed. Let $\sigma_x := \inf\{t : B_t = x\}$, and set

$$^x B_t := B_{\sigma_x + t} - x \qquad (t \ge 0).$$

According to the strong Markov property of B, the process $^x B := (^x B_t, t \ge 0)$ is a standard Brownian motion independent of σ_x. Let $^x M_t$ and $^x L^0_t$ be the functionals of $^x B$ corresponding to the functionals M_t and L^0_t of B. Then for $y > 0$, $b \in \mathbb{R}$, and $a := |b - x|$, we can compute as follows:

$$
\begin{aligned}
P(L^x_1 > y, B_1 \in db)/db &= P(\sigma_x < 1)\tfrac{1}{2} P(^x L^0_{1-\sigma_x} > y, |^x B_{1-\sigma_x}| \in da)/da \\
&= P(\sigma_x < 1)\tfrac{1}{2} P(^x M_{1-\sigma_x} > y, {}^x M_{1-\sigma_x} - {}^x B_{1-\sigma_x} \in da)/da \\
&= \tfrac{1}{2} P(M_1 > x + y, M_1 - B_1 \in da)/da \\
&= P(L^0_1 > x + y, B_1 \in da)/da.
\end{aligned}
$$

The first and third equalities are justified by the strong Markov property, the second appeals to Lévy's identity (2) applied to $^x B$, and the fourth uses (2) applied to B. The formula (1) for $x > 0$ can now be read from (1) with $x = 0$.

3 Some identities in distribution.

Let R denote a random variable with the Rayleigh distribution

$$P(R > r) = e^{-\frac{1}{2}r^2} \qquad (r > 0). \tag{10}$$

According to formula (3),

$$(L^x_1 \,|\, B_1 = 0) \stackrel{d}{=} (R - 2|x|)^+ \tag{11}$$

where the left side denotes the distribution of L_1^x for a standard Brownian bridge. The corresponding result for the unconditioned Brownian motion, obtained by integrating out b in (1), is

$$L_1^x \overset{d}{=} (|B_1| - |x|)^+. \tag{12}$$

Lévy gave both these identities for $x = 0$. For the bridge from 0 to $b \in \mathbb{R}$ and $x > 0$ the events $(M_1 > x)$ and $(L_1^x > 0)$ are a.s. identical. So (3) for $y = 0$ reduces to Lévy's result that

$$P(M_1 > x \mid B_1 = b) = e^{-2x(x-b)} \qquad (0 \vee b < x).$$

Let

$$\varphi(z) := \frac{1}{\sqrt{2\pi}} e^{-\frac{1}{2}z^2}; \quad \bar{\Phi}(x) := \int_x^\infty \varphi(z) dz = P(B_1 > x).$$

Then the mean occupation density at $x \in \mathbb{R}$ of the Brownian bridge from 0 to $b \in \mathbb{R}$ is

$$E(L_1^x \mid B_1 = b) = \int_0^1 \frac{1}{\sqrt{s(1-s)}} \varphi\left(\frac{x - bs}{\sqrt{s(1-s)}}\right) ds = \frac{\bar{\Phi}(|x| + |b - x|)}{\varphi(b)}. \tag{13}$$

The first equality is read from the occupation density formula and the fact that $B_s^{0 \to b}$ has normal distribution with mean bs and variance $s(1-s)$. The second equality, which is not obvious directly, is obtained using the first equality by integration of (3). The case $b = 0$ of the second equality is attributed [21, p. 400, Exercise 3] to M. Gutjahr and E. Haeusler. See also [18] for another approach to this identity involving properties of the arc sine distribution. As a consequence of (13), for each $b > 0$ and each Borel subset A of $[0, b]$, the expected time spent in A by the Brownian bridge from 0 to b is

$$E\left(\int_0^1 1(B_s^{0 \to b} \in A) ds \mid B_1 = b\right) = |A| \frac{\bar{\Phi}(b)}{\varphi(b)} \tag{14}$$

where $|A|$ is the Lebesgue measure of A. Take $A = [0, b]$ to recover the standard estimate $\bar{\Phi}(b) < \varphi(b)/b$. For each $b \in \mathbb{R}$, the function of x appearing in (13) is the probability density function of $B_U^{0 \to b}$ for U a uniform$[0, 1]$ variable independent of the bridge. In particular, for $b = 0$, formula (13) yields

$$|B_U^{0 \to 0}| \overset{d}{=} \tfrac{1}{2} U R \tag{15}$$

where the Rayleigh variable R is independent of U. This and related identities were found in [1], where the reflecting bridge $(|B_s^{0 \to 0}|, 0 \leq s \leq 1)$ was used to describe the asymptotic distribution of a path derived from a random mapping.

Recall that σ_x is the first hitting time of x by the Brownian motion B. Let γ_1^x denote last time B is at x before time 1, with the convention $\gamma_1^x = 0$ if no such time, and set $\delta_1^x = (\gamma_1^x - \sigma_x)^+$. By well known first entrance and last exit decompositions [10], given B_1 and δ_1^x with $\delta_1^x > 0$, the segment of B between times σ_x and γ_1^x is a Brownian bridge of length δ_1^x from x to x. Therefore,

$$(L_1^x \mid B_1 = b) \overset{d}{=} (\sqrt{\delta_1^x} R \mid B_1 = b) \tag{16}$$

where the Rayleigh variable R is independent of δ_1^x, and R given $\delta_1^x > 0$ may be interpreted as the local time at 0 of the standard bridge derived by Brownian scaling of the segment of B between times σ_x and γ_1^x. By consideration of moments, formula (16) shows that the law of $(L_1^x \mid B_1 = b)$ displayed in formula (3) determines the law of $(\delta_1^x \mid B_1 = b)$, and vice versa. As indicated by Imhof [12], the distribution of δ_1^x given $B_1 = b$ can be derived by integration from the joint distribution of σ_x and γ_1^x given $B_1 = b$, which is easily written down. By comparison with the formula for the joint density of γ_1^0 and $|B_1|$, due to Chung [8, (2.5)], it turns out that

$$(\delta_1^x \mid B_1 = 0, \delta_1^x > 0) \stackrel{d}{=} (\gamma_1^0 \mid B_1 = 2x) \stackrel{d}{=} \frac{B_1^2}{B_1^2 + 4x^2} \tag{17}$$

where the second equality is obtained from Chung's formula by an elementary change of variable, as in [2, (6)-(8)], where the same family of distributions on $[0, 1]$ appears in another context. Set $a = 2x$ and combine (11), (16) and (17) to deduce the identity

$$\sqrt{\frac{B_1^2}{B_1^2 + a^2}} \; R \stackrel{d}{=} (R - a \mid R > a) \qquad (a \geq 0) \tag{18}$$

where R and B_1 are assumed independent. By consideration of moments, this identity amounts to the equality of two different integral representations for the Hermite function [13, (10.5.2) and Problem 10.8.1].

If ε is independent of B and exponentially distributed with rate 1, a variation of (16) gives

$$L_{2\varepsilon}^0 \stackrel{d}{=} \sqrt{\gamma_{2\varepsilon}^0} \, R \tag{19}$$

where $\gamma_{2\varepsilon}^0$ and R are independent, hence

$$\varepsilon \stackrel{d}{=} |B_1| R \tag{20}$$

where B_1 and R are independent. By consideration of moments, this classical identity is equivalent to the duplication formula for the gamma function [22]. Another Brownian representation of (20) is

$$|B_{2\varepsilon}^0| \stackrel{d}{=} \sqrt{2\varepsilon - \gamma_{2\varepsilon}^0} \, M_1 \tag{21}$$

where M_1 is the final value of a standard Brownian meander. Compare with [20, Ch. XII, Ex's (3.8) and (3.9)], and [5]. See also [3, p. 681] for an appearance of (20) in the study of random trees.

Acknowledgment

Thanks to Marc Yor for several stimulating conversations related to the subject of this note.

References

[1] D. Aldous and J. Pitman. Brownian bridge asymptotics for random mappings. *Random Structures and Algorithms*, 5:487–512, 1994.

[2] D.J. Aldous and J. Pitman. The standard additive coalescent. Technical Report 489, Dept. Statistics, U.C. Berkeley, 1997. To appear in *Ann. Probab.*. Available via http://www.stat.berkeley.edu/users/pitman.

[3] D.J. Aldous and J. Pitman. Tree-valued Markov chains derived from Galton-Watson processes. *Ann. Inst. Henri Poincaré*, 34:637–686, 1998.

[4] R. F. Bass and D. Khoshnevisan. Laws of the iterated logarithm for local times of the empirical process. *Ann. Probab.*, 23:388 – 399, 1995.

[5] J. Bertoin and J. Pitman. Path transformations connecting Brownian bridge, excursion and meander. *Bull. Sci. Math. (2)*, 118:147–166, 1994.

[6] Ph. Biane and M. Yor. Sur la loi des temps locaux Browniens pris en un temps exponentiel. In *Séminaire de Probabilités XXII*, pages 454–466. Springer, 1988. Lecture Notes in Math. 1321.

[7] A. N. Borodin. Brownian local time. *Russian Math. Surveys*, 44:2:1–51, 1989.

[8] K. L. Chung. Excursions in Brownian motion. *Arkiv fur Matematik*, 14:155–177, 1976.

[9] S.N. Evans and J. Pitman. Stopped Markov chains with stationary occupation times. *Probab. Th. Rel. Fields*, 109:425–433, 1997.

[10] P. Fitzsimmons, J. Pitman, and M. Yor. Markovian bridges: construction, Palm interpretation, and splicing. In E. Çinlar, K.L. Chung, and M.J. Sharpe, editors, *Seminar on Stochastic Processes, 1992*, pages 101–134. Birkhäuser, Boston, 1993.

[11] P. Howard and K. Zumbrun. Invariance of the occupation time of the Brownian bridge process. Preprint, Indiana Univerity. Available via http://www.math.indiana.edu/home/zumbrun, 1998.

[12] J. P. Imhof. On Brownian bridge and excursion. *Studia Sci. Math. Hungar.*, 20:1–10, 1985.

[13] N. N. Lebedev. *Special Functions and their Applications*. Prentice-Hall, Englewood Cliffs, N.J., 1965.

[14] C. Leuridan. Le théorème de Ray-Knight à temps fixe. In J. Azéma, M. Émery, M. Ledoux, and M. Yor, editors, *Séminaire de Probabilités XXXII*, pages 376–406. Springer, 1998. Lecture Notes in Math. 1686.

[15] P. Lévy. Sur certains processus stochastiques homogènes. *Compositio Math.*, 7:283–339, 1939.

[16] J. Pitman. Cyclically stationary Brownian local time processes. *Probab. Th. Rel. Fields*, 106:299–329, 1996.

[17] J. Pitman. The SDE solved by local times of a Brownian excursion or bridge derived from the height profile of a random tree or forest. Technical Report 503, Dept. Statistics, U.C. Berkeley, 1997. To appear in *Ann. Prob.*. Available via http://www.stat.berkeley.edu/users/pitman.

[18] J. Pitman and M. Yor. Some properties of the arc sine law related to its invariance under a family of rational maps. In preparation, 1998.

[19] D. B. Ray. Sojourn times of a diffusion process. *Ill. J. Math.*, 7:615–630. 1963.

[20] D. Revuz and M. Yor. *Continuous martingales and Brownian motion.* Springer. Berlin-Heidelberg, 1994. 2nd edition.

[21] G. R. Shorack and J. A. Wellner. *Empirical processes with applications to statistics.* John Wiley & Sons, New York, 1986.

[22] S. S. Wilks. Certain generalizations in the analysis of variance. *Biometrika.* 24:471–494. 1932.

[23] D. Williams. Path decomposition and continuity of local time for one dimensional diffusions I. *Proc. London Math. Soc. (3)*, 28:738–768, 1974.

PATHS OF **FINITELY ADDITIVE** BROWNIAN MOTION
NEED NOT BE BIZARRE

by

Lester E. Dubins

Abstract. Each stochastic process, in particular the Wiener process, has a finitely additive cousin whose paths are polynomials, and another cousin whose paths are step functions.

Notation. R is the real line; T is the half-ray of nonnegative moments of time; a path, w, is a mapping of T into R; W is the set of paths; I is the identity map of W onto itself.

Plainly, I is essentially the same as the one-parameter family of evaluation maps, $I(t)$ or $I(t, \,.\,)$, defined for t in T, by $I(t, w) = w(t)$.

Of course, once W, the space of paths, is endowed with a sufficiently rich probability measure, I becomes a stochastic process. Probabilities in this note are not required to be countably additive; those on W are assumed to be defined (at least) on F, the set of finite-dimensional (Borel) subsets of W. As always, to a stochastic process, X, is associated its family $J = J(X)$ of finite-dimensional joint distributions, one such distribution $J(t)$ for each n-tuple t of distinct moments of time. Of course, $J(X)$ is a consistent family, which has the usual meaning that, if t is a subsequence of t', then $J(t)$ is the t-marginal of $J(t')$.

Definition. Two stochastic processes are *cousins* if the J of one of the processes is the same as the J of the other process.

Of interest herein are those subsets H of W that satisfy:

Condition *. Each stochastic process X has a cousin almost all of whose paths are in H.

Throughout this note, J designates a consistent family of finite-dimensional joint distributions, and a stochastic process X is a J-process if $J(X) = J$.

Record here the following alternative formulation of Condition *.

Condition **. For each J, there is a J-process almost all of whose paths are in H.

That ** suffices for * is a triviality. That * suffices for ** becomes a triviality once one recalls that, for each J, there is a J-process. So the conditions are equivalent.

As a preliminary to characterizing the H that satisfy Condition *, introduce for each n-tuple t of distinct time-points, $t = (t_1, \ldots, t_n)$, and each n-tuple x of possible positions, $x = (x_1, \ldots, x_n)$, the set $S[t, x]$ of all paths w such that, for each i from 1 to n, $w(t_i)$ is x_i.

Condition ***. H has a nonempty intersection with each $S[t, x]$.

Proposition 1. *A set H of paths satisfies Condition * if and only if it satisfies Condition ***.*

Proof. Suppose H satisfies *. Then, for each probability P on F, these three equivalent conditions hold: [i] There is a probability Q that agrees with P on F for which $QH = 1$; [ii] H has outer P-probability 1; [iii] the inner P-probability of the complement of H is zero. As [iii] implies, for no finite-dimensional set S disjoint from H is $P(S)$ strictly positive. A fortiori, for no such S does $P(S) = 1$. In particular, no $S[t, x]$ disjoint from H has P-probability 1. This implies that there is no $S[t, x]$ disjoint from H. For, as is easily verified, for each $S[t, x]$ there is a P under which $S[t, x]$ has probability 1. Consequently, each $S[t, x]$ has nonempty intersection with H, or, what is the same thing, H satisfies ***.

For the converse, suppose that H satisfies ***, or equivalently, that no $S[t, x]$ is included in the complement, H' of H. Surely then, no nonempty union of the $S[t, x]$ is included in H'. Since, as is easily verified, each finite-dimensional set is such a union, no nonempty, finite-dimensional set is included in H'. Since the empty set is the only finite-dimensional set included in H', the only finite-dimensional set that includes H is the complement of the empty set, namely, W. Now fix a consistent family J, and let P be the corresponding probability on F. For this P, as for all P on F, the outer P-probability of H is necessarily 1. Therefore, P has an extension that assigns probability 1 to H. Equivalently, there is a J-process, almost all of whose paths are in H. So H satisfies *. ∎

A step function is one that, on each bounded time-interval, has only a finite number of values, each assumed on a finite union of intervals.

Theorem 1. *Each stochastic process, in particular the Wiener process, has a cousin almost all of whose paths are polynomials, another cousin almost all of whose paths are step functions that are continuous on the right (on the left), and a fourth cousin almost all of whose paths are continuous, piecewise-linear functions.*

Proof of Theorem 1. Plainly, each of the four sets of paths satisfies Condition ***. Therefore, Proposition 1 applies. ∎

A remark (informal). The (strong) Markov property need not be inherited by a cousin of a process, or, as is closely related, the existence of proper disintegrations (proper conditional distributions) of the future given the past need not transfer to the cousin. An example is provided by a cousin of Brownian Motion whose paths are polynomials. On the other hand, those properties are inheritable by those cousins of Brownian Motion whose paths are step functions, or piecewise-linear functions. Definition of proper, and of disintegration, may, amongst other places, be seen in the two references.

References

[1] David Blackwell and Lester Dubins. On existence and non-existence of proper, regular, conditional distributions. *Ann. Prob. 3* (1975), 741–752.

[2] Lester Dubins and Karel Prickry. On the existence of disintegrations. *Séminaire de Probabilités XXIX* (1995), 248–259.

A LIMIT THEOREM FOR THE PREDICTION PROCESS UNDER ABSOLUTE CONTINUITY

HIDEATSU TSUKAHARA

Abstract

Consider a stochastic process with two probability laws, one of which is absolutely continuous with respect to the other. Under each law, we look at a process consisting of the conditional distributions of the future given the past. Blackwell and Dubins showed in discrete case that those conditional distributions merge as we observe more and more; more precisely, the total variation distance between them converges to 0 a.s. In this paper we prove its extension to continuous time case using the prediction process of F. B. Knight.

1. Introduction. Let $(\mathbb{E}_n, \mathcal{E}_n)$, $n \in \mathbb{N}$ be measurable Lusin spaces and put $(\mathbb{E}, \mathcal{E}) = (\mathbb{E}_1 \times \mathbb{E}_2 \times \cdots, \mathcal{E}_1 \otimes \mathcal{E}_2 \otimes \cdots)$. Suppose that μ and ν are probability measures on $(\mathbb{E}, \mathcal{E})$ satisfying $\nu \ll \mu$. We then denote by $Z_n^\mu(x_1, \ldots, x_n)(\bullet)$ and $Z_n^\nu(x_1, \ldots, x_n)(\bullet)$ the regular conditional distributions for the future $(\mathbb{E}_{n+1} \times \cdots)$ given the past $\mathcal{E}_1 \otimes \cdots \otimes \mathcal{E}_n$ under μ and ν respectively. Blackwell and Dubins (1962) showed that those conditional distributions merges as n becomes large; more precisely, the total variation distance between them converges to 0 ν-a.s. as $n \to \infty$. In what follows, we prove its extension to continuous time case using the prediction process of F. B. Knight (1975, 1992).

We start with introducing the prediction process, which consists of suitable versions of conditional distributions of the future given the past in continuous time setting. For our purpose, it is unnecessary to make any topological assumption on the state space. Thus we need only the prediction process in a measure-theoretic setting as developed in Chapter 1 of Knight (1992). Let $(\mathbb{E}, \mathcal{E})$ be a *measurable* Lusin space and $\mathbb{M}_\mathbb{E}$ the space of $\mathcal{B}_+/\mathcal{E}$ measurable functions as before. Let us define the *pseudo-path* filtration $(\mathcal{F}_t')_{t \in \mathbb{R}_+}$ by

$$\mathcal{F}_t' \triangleq \sigma\left(\int_0^s f(w(u))\, du, \ s < t, \ f \in b\mathcal{E}\right),$$

and put $\mathcal{F}' = \mathcal{F}'_\infty = \bigvee_{t>0} \mathcal{F}_t'$. Note that each $\mathcal{F}_t'$ is countably generated and satisfies $\mathcal{F}_{t-}' = \mathcal{F}_t'$ for $t > 0$. We denote by Π the set of probability measures on $(\mathbb{M}_\mathbb{E}, \mathcal{F}')$, and set $\mathcal{G} \triangleq \sigma(z(A), \ A \in \mathcal{F}')$. The shift operator θ_t on $\mathbb{M}_\mathbb{E}$ is defined by $\theta_t w(s) = w(t+s)$ and is $\mathcal{F}_{t+s}'/\mathcal{F}_s'$ measurable for all $s, t \in \mathbb{R}_+$.

It is shown in Chapter 1 of Knight (1992) that the prediction process $(Z_t^z)_{t \in \mathbb{R}_+}$ on $(M_{\mathbb{E}}, \mathcal{F}')$ is P^z-a.s. uniquely defined by the requirements:

(i) The mapping $(z, s, w) \mapsto Z_s^z(\bullet, w)$ on $\Pi \times [0, t] \times M_{\mathbb{E}}$ is $\mathcal{G} \otimes \mathcal{B}[0, t] \otimes \mathcal{F}'_{t+\epsilon}/\mathcal{G}$ measurable for each $t \in \mathbb{R}_+$ and each $\epsilon > 0$.

(ii) For any $(\mathcal{F}'_{t+})$-optional T and $A \in \mathcal{F}'$, $Z_T^z(A) = P^z(\theta_T^{-1} A \mid \mathcal{F}'_{T+})$ on $\{T < \infty\}$.

Analogously, the left-limit prediction process $(Z_{t-}^z)_{t>0}$ on $(M_{\mathbb{E}}, \mathcal{F}')$ is P^z-a.s. uniquely defined by the requirements:

(i) The mapping $(z, s, w) \mapsto Z_{s-}^z(\bullet, w)$ on $\Pi \times [0, t] \times M_{\mathbb{E}}$ is $\mathcal{G} \otimes \mathcal{B}[0, t] \otimes \mathcal{F}'_t/\mathcal{G}$ measurable for each $t > 0$.

(ii) For any $(\mathcal{F}'_t)$-predictable $T > 0$ and $A \in \mathcal{F}'$, $Z_{T-}^z(A) = P^z(\theta_T^{-1} A \mid \mathcal{F}'_T)$ on $\{T < \infty\}$.

We note that even when the space $\mathbb{E}$ has been given the prescribed Lusin topology, the processes (Z_t^z) and (Z_{t-}^z) are not related to each other through that topology of $\mathbb{E}$ (see Knight (1992)).

Furthermore, employing the notation of Meyer (1976), we define the processes K_t^z and K_{t-}^z by

$$K_t^z(f \circ \theta_t) = Z_t^z(f), \quad K_{t-}^z(f \circ \theta_t) = Z_{t-}^z(f)$$

for $f \in b\mathcal{F}'$. Hence Π is the state space of the K^z's and they satisfy, besides measurability conditions,

$$K_T^z(A) = P^z(A \mid \mathcal{F}'_{T+}) \quad \text{on } \{T < \infty\},$$

for any $(\mathcal{F}'_{t+})$-optional T and $A \in \mathcal{F}'$, and

$$K_{T-}^z(A) = P^z(A \mid \mathcal{F}'_T) \quad \text{on } \{T < \infty\},$$

for any $(\mathcal{F}'_t)$-predictable $T > 0$ and $A \in \mathcal{F}'$,

Following Meyer (1976), we define the optional and predictable σ-fields as follows. The optional σ-field $\mathcal{O}$ is generated by the càdlàg processes adapted to $(\mathcal{F}'_{t+})$, and the predictable σ-field $\mathcal{P}$ is generated by the left continuous processes adapted to $(\mathcal{F}'_{t-})$. The utility of K_t^z and K_{t-}^z lies in the following result due to Meyer (1976).

Proposition 1.1 *For every bounded measurable process X on $(M_{\mathbb{E}}, \mathcal{F}')$ and for every $z \in \Pi$,*

$$(t, \omega) \mapsto \int K_t^z(dw, \omega) X_t(w)$$

defines an optional projection of X for z, and

$$(t,\omega) \mapsto \int K^z_{t-}(dw,\omega)\, X_t(w)$$

defines a predictable projection of X for z.

A simple monotone class argument proves the above theorem. We can actually improve on this result, using a similar monotone class argument. This is also due to Meyer (1976).

Proposition 1.2 Let $X(t,\omega,t',\omega')$ be a bounded function that is $\mathcal{O}\otimes\mathcal{B}(\mathbb{R}_+)\otimes$ $\mathcal{F}'$ measurable. Then an optional projection for z of the process $X(t,\omega,t,\omega)$ is given by

$$(t,\omega) \mapsto \int K^z_t(dw,\omega)\, X(t,\omega,t,w).$$

Similarly, if $X(t,\omega,t',\omega')$ is a bounded function that is $\mathcal{P}\otimes\mathcal{B}(\mathbb{R}_+)\otimes\mathcal{F}'$ measurable, then a predictable projection for z of the process $X(t,\omega,t,\omega)$ is given by

$$(t,\omega) \mapsto \int K^z_{t-}(dw,\omega)\, X(t,\omega,t,w).$$

Remark 1.3 In Dellacherie and Meyer (1980), VI.43, optional and predictable projections are defined under the "usual conditions". Here we are not assuming them, but in view of Lemma 7 of Dellacherie and Meyer (1980), Appendix I, we can choose a version of the optional projection which is optional relative to $(\mathcal{F}'_{t+})$, and a version of the predictable projection which is predictable relative to $(\mathcal{F}'_t)$. Thus according to our definition of the optional and predictable σ-fields , no complications on those projections arise.

2. Main result. For $z, z' \in \Pi$, the *total variation distance* $\rho_{TV}(z,z')$ on Π is defined by

$$\rho_{TV}(z,z') \triangleq \sup_{A\in\mathcal{F}'} |z(A) - z'(A)|.$$

Our main result is the following theorems.

Theorem 2.1 Let z and z' be two probabilities on $(\mathbb{M}_{\mathbb{E}},\mathcal{F}')$ satisfying $z' \ll z$ and let (Z^z_t) and $(Z^{z'}_t)$ be their prediction processes. Then the total variation distance between (Z^z_t) and $(Z^{z'}_t)$ converges to 0 as $t \to \infty$, $P^{z'}$-a.s.

Theorem 2.2 *Let z and z' be two probabilities on $(M_E, \mathcal{F}')$ satisfying $z' \ll z$ and let (Z_{t-}^z) and $(Z_{t-}^{z'})$ be their left-limit prediction processes. Then the total variation distance between (Z_{t-}^z) and $(Z_{t-}^{z'})$ converges to 0 as $t \to \infty$. $P^{z'}$-a.s.*

To prove the theorems, we need some preliminary results. Let $L(w) = \frac{dz'}{dz}(w)$. We would like to show first that $K_{t-}^{z'}$ [$K_t^{z'}$] is $P^{z'}$-a.s. absolutely continuous with respect to K_{t-}^z [K_t^z] and find a version of Radon-Nikodym derivative. The following general lemma is well known in the filtering theory.

Lemma 2.3 *Let $(\Omega, \mathcal{F}, P)$ be a probability space and $\mathcal{G}$ a sub-σ-field of $\mathcal{F}$. Suppose that we have another probability measure Q which is absolutely continuous with respect to P. Set $L = \frac{dQ}{dP}$. Then for any Q integrable $\mathcal{F}$ measurable V. we have*

$$E^Q(V \mid \mathcal{G}) = \frac{E^P(LV \mid \mathcal{G})}{E^P(L \mid \mathcal{G})}, \quad Q\text{-a.s.}$$

It follows from the above lemma that for any $f \in b\mathcal{F}'_\infty$,

$$E^{z'}(f \mid \mathcal{F}'_t) = \frac{E^z(fL \mid \mathcal{F}'_t)}{E^z(L \mid \mathcal{F}'_t)}, \quad P^{z'}\text{-a.s.}.$$

Let $L_t^z \triangleq K_t^z(L)$. Then L_t^z is a càdlàg version of the martingale $E^z(L \mid \mathcal{F}'_t)$ since it is an optional projection of the process L constant in t (L is not bounded, but it is positive). And we put $L_{t-}^z \triangleq \lim_{s \uparrow\uparrow t} L_s^z = K_{t-}^z(L)$, $t > 0$. By the same reasoning, L_{t-}^z is a predictable projection of L.

Proposition 2.4 *For $P^{z'}$-almost all ω, $K_t^{z'}(dw, \omega)$ is absolutely continuous with respect to $K_t^z(dw, \omega)$ with the Radon-Nikodym derivative $L(w)/L_t^z(\omega)$ for all t. Similarly, for $P^{z'}$-almost all ω, $K_{t-}^{z'}(dw, \omega)$ is absolutely continuous with respect to $K_{t-}^z(dw, \omega)$ with the Radon-Nikodym derivative $L(w)/L_t^z(\omega)$ for all $t > 0$.*

Proof. We give a proof for K_t^z case since K_{t-}^z case can be proved analogously. As in Chapter 1 of Knight (1992), we may assume that $E = [0, 1]$. There exists a metric on $M_{[0,1]}$ for which it is compact, and its Borel sets are $\mathcal{F}'$. Since $C(M_{[0,1]})$ is separable, we can find a sequence (f_j) in $C(M_{[0,1]})$ that is uniformly dense in $\{f \in C(M_{[0,1]}): 0 \le f \le 1\}$. By Lemma 2.3, for each $r \in \mathbb{Q}_+$ and $j \in \mathbb{N}$, there is a set $A_{r,j}$ with $P^z(A_{r,j}) = 1$ such that if $\omega \in A_{r,j}$,

(2.1)
$$K_r^{z'}(f_j, \omega) = \int f_j(w) \frac{L(w)}{L_r^z(\omega)} K_r^z(dw, \omega).$$

Set

$$A = \left(\bigcap_{r \in \mathbb{Q}_+} \bigcap_{j \in \mathbb{N}} A_{r,j} \right) \cap \left\{ K^{z'}(f_j), K^z(f_j L) \text{ and } K^z(L) \text{ are right continuous} \right\}$$

The processes $K^{z'}(f_j)$, $K^z(f_j L)$ and $K^z(L)$ are all right continuous $P^{z'}$-a.s. since they are optional projections of the processes which are constant in t. Thus we have $P^{z'}(A) = 1$. Since (f_j) is uniformly dense in $\{f \in C(\mathsf{M}_{[0,1]}) : 0 \leq f \leq 1\}$ and both sides of (2.1) are right continuous for $\omega \in A$, we conclude that if $\omega \in A$,

$$K_t^{z'}(f, \omega) = \int f(w) \frac{L(w)}{L_t^z(\omega)} K_t^z(dw, \omega),$$

for all $f \in C(\mathsf{M}_{[0,1]})$ and all $t \in \mathbb{R}_+$. In view of the fact that the continuous bounded functions separate the measures on a metric space, we see that the above is true for all $f \in b\mathcal{F}'$ and all $t \in \mathbb{R}_+$, which implies our assertion. ∎

To get a similar result for the Z^z, we need the splicing operator on $\mathsf{M}_\mathbb{E} \times \mathbb{R}_+ \times \mathsf{M}_\mathbb{E}$ onto $\mathsf{M}_\mathbb{E}$ defined by

$$(\omega/t/w)_s \triangleq \begin{cases} \omega(s), & s < t, \\ w(s-t), & s \geq t. \end{cases}$$

The mapping $(\omega, t, w) \mapsto \omega/t/w$ is continuous, so it is $\mathcal{F}' \otimes \mathcal{B}(\mathbb{R}_+) \otimes \mathcal{F}'/\mathcal{F}'$ measurable. It follows from Lemma 4 of Meyer (1976) that for P^z-almost every ω and all t, $K_t^z(dw, \omega)$ and $K_{t-}^z(dw, \omega)$ is concentrated on the atom of $\mathcal{F}_t'$ containing ω, which is the set of $w \in \mathsf{M}_\mathbb{E}$ such that $w(s) = \omega(s)$ for λ-a.e. $s \leq t$. Moreover, denoting the mapping $w \mapsto \omega/t/w$ by $\gamma_{\omega,t}(w)$, we have

$$K_t^z(f, \omega) = \int f(w/t/\theta_t w) K_t^z(dw, \omega) = \int f(\omega/t/\theta_t w) K_t^z(dw, \omega)$$

$$= \int f(\omega/t/u) Z_t^z(du, \omega) = Z_t^z(f \circ \gamma_{\omega,t}, \omega),$$

and similarly for K_{t-}^z and Z_{t-}^z. Thus $K_t^z(dw, \omega)$ and $K_{t-}^z(dw, \omega)$ are the image measures of $Z_t^z(dw, \omega)$ and $Z_{t-}^z(dw, \omega)$ respectively under $\gamma_{\omega,t}$. It then follows that $P^{z'}$-a.s.,

$$Z_t^{z'}(f, \omega) = \int f(w) \frac{L(\omega/t/w)}{L_t^z(\omega)} Z_t^z(dw, \omega), \quad f \in b\mathcal{F}',$$

for all $t \in \mathbb{R}_+$, and

$$Z_{t-}^{z'}(f,\omega) = \int f(w) \frac{L(\omega/t/w)}{L_t^z(\omega)} Z_{t-}^z(dw,\omega), \quad f \in b\mathcal{F}',$$

for all $t > 0$.

Proof of Theorem 2.1. It is easy to see that if $z' \ll z$, then, with $\ell = dz'/dz$,

$$\rho_{TV}(z, z') = \int_{\{\ell > 1\}} (\ell - 1) \, dz.$$

Using this, we can evaluate the total variation distance between Z_t^z and $Z_t^{z'}$. We have, for any $\epsilon > 0$,

$$\begin{aligned}
\rho_{TV}\left(Z_t^z, Z_t^{z'}\right) &= \int_{\left\{w: \frac{L(\omega/t/w)}{L_t^z(\omega)} - 1 > 0\right\}} \left(\frac{L(\omega/t/w)}{L_t^z(\omega)} - 1\right) Z_t^z(dw.\omega) \\
&\leq \epsilon + \int_{\left\{w: \frac{L(\omega/t/w)}{L_t^z(\omega)} - 1 > \epsilon\right\}} \frac{L(\omega/t/w)}{L_t^z(\omega)} Z_t^z(dw,\omega) \\
&= \epsilon + Z_t^{z'}\left(\left\{w: \frac{L(\omega/t/w)}{L_t^z(\omega)} - 1 > \epsilon\right\}, \omega\right) \\
&= \epsilon + K_t^{z'}\left(\left\{w: \frac{L(\omega/t/\theta_t w)}{L_t^z(\omega)} - 1 > \epsilon\right\}, \omega\right).
\end{aligned}$$

Note that for $P^{z'}$-almost all ω and all t, we have $\omega/t/\theta_t w = w/t/\theta_t w = w$, $K_t^{z'}(\bullet.\omega)$-a.e. Thus we get

$$(2.2) \qquad \rho_{TV}\left(Z_t^z, Z_t^{z'}\right) \leq \epsilon + K_t^{z'}\left(\left\{w: \frac{L(w)}{L_t^z(\omega)} - 1 > \epsilon\right\}, \omega\right).$$

Now let $f(x,y) = 1_{\{(x,y): y/x - 1 > \epsilon\}}(x,y)$ for $(x,y) \in \mathbb{R}^2$. Clearly this is $\mathcal{B}(\mathbb{R}^2)$ measurable. Put $X(t,\omega,w) = f(L_t^z(\omega), L(w))$, so that we have

$$X(t,\omega,w) = 1_{\left\{(t,\omega,w): \frac{L(w)}{L_t^z(\omega)} - 1 > \epsilon\right\}}(t,\omega,w).$$

$(t,\omega) \mapsto L_t^z(\omega)$ is $\mathcal{O}$ measurable since $L_t^z(\omega) = K_t^z(L,\omega)$ is the optional projection of L for P^z. Also L is $\mathcal{F}'$ measurable. Thus $(t,\omega,w) \mapsto (L_t^z(\omega), L(w))$ is $\mathcal{O} \otimes \mathcal{F}'/\mathcal{B}(\mathbb{R}^2)$ measurable, and hence $X(t,\omega,w)$ is $\mathcal{O} \otimes \mathcal{F}'$ measurable. It follows from Proposition 1.2 that

$$(2.3) \qquad \int K_t^{z'}(\omega, dw) X(t,\omega,w) = K_t^{z'}\left(\left\{w: \frac{L(w)}{L_t^z(\omega)} - 1 > \epsilon\right\}.\omega\right)$$

is the optional projection of the process $X(t, \omega, \omega)$.

By the martingale convergence theorem, as $t \to \infty$, L_t^z converges to L, P^z-a.s. and hence $P^{z'}$-a.s. Since $L > 0$, $P^{z'}$-a.s., L/L_t^z converges to 1, $P^{z'}$-a.s. as $t \to \infty$. This implies that $X(t, \omega, \omega) \to 0$, $P^{z'}$-a.s. Finally, by Dellacherie and Meyer (1980), VI.50 c), the optional projection of $X(t, \omega, \omega)$, i.e. (2.3) converges to 0, $P^{z'}$-a.s. Therefore the right-hand side of (2.3) converges to 0 as $t \to \infty$, $P^{z'}$-a.s., and the theorem follows from (2.2). ∎

By an entirely analogous argument using the predictable counterparts, one can prove Theorem 2.2. When $z' \ll z$ does not necessarily hold, we still have the following.

Corollary 2.5 *Let* $T = \inf\{t \geq 0 : Z_t^{z'} \ll Z_t^z\}$, *where* z *and* z' *are any two probabilities on* $(\mathbf{M}_{\mathbf{E}}, \mathcal{F}')$. *Then the conclusions of Theorem 2.1 and Theorem 2.2 holds on* $\{T < \infty\}$.

Proof. Since Π is the set of probability measures, the set $\Lambda = \{(z_1, z_2) \in \Pi^2 : z_1 \ll z_2\}$ belongs to $\mathcal{G}^2$. T is then the first entrance time into Λ by the process $(Z_t^{z'}, Z_t^z)$, so it is $\mathcal{F}'_{t+}$-optional.

Next we quote the following result due to Yor and Meyer (1976): For any $\mathcal{F}'_{t+}$-optional $S < \infty$ and any $z \in \Pi$, $Z_{S+t}^z(\bullet, w) = Z_t^{Z_S^z(w)}(\bullet, \theta_S w)$ for all $t \geq 0$, P^z-a.s. The corollary then easily follows from this, together with Theorems 2.1 and 2.2, and the section theorem, ∎

Acknowledgements. The author would like to thank Frank Knight, Catherine Doléans-Dade and Ditlev Monrad for their helpful comments.

REFERENCES

Blackwell, D. and Dubins, L. (1962). Merging of opinions with increasing information, *Ann. Math. Statist.* **33**, 882-886.

Dellacherie, C and Meyer, P.-A. (1980). *Probabilités et Potentiel: Théorie des Martingales*, Chapitres V à VIII, Hermann, Paris.

Knight, F.B. (1975). A predictive view of continuous time processes, *Ann. Probab.* **3**, 573-596.

Knight, F.B. (1992). *Foundations of the Prediction Process*, Oxford University Press, Oxford.

Meyer, P.-A. (1976). La théorie de la prédiction de F. Knight. *Séminaire de Probabilités X. Lect. Notes in Math.* **511**. Springer-Verlag, Berlin Heidelberg New York, 86-103.

Yor, M. and Meyer, P.-A. (1976). Sur la théorie de la prédiction, et le problème de décomposition des tribus $\mathcal{F}^0_{t+}$, *Séminaire de Probabilités X. Lect. Notes in Math.* **511**, Springer-Verlag, Berlin Heidelberg New York, 104-117.

Processus Gouvernés par des Noyaux

Karl CHRETIEN   David KURTZ   Bernard MAISONNEUVE

Résumé. Ce travail, intimement lié au théorème de Ionescu Tulcea, est consacré aux processus gouvernés par des noyaux, d'abord en temps discret et relativement à une filtration, puis pour un ensemble d'indices quelconque. Il est extrait d'un Travail d'Etude et de Recherche effectué en maîtrise par les deux premiers auteurs.

1 Processus gouvernés par des noyaux (temps discret)

Soit $(E, \mathcal{E})$ un espace mesurable arbitraire et pour tout $n \in \mathbb{N}^*$ soit ν_n un noyau de E^n dans E. On note $\mathcal{M}_+(E)$ l'ensemble des fonctions mesurables positives sur E. Nous introduisons la définition suivante

1.1 Définition

Un processus $(X_n)_{n \in \mathbb{N}}$ à valeurs dans $(E, \mathcal{E})$, défini sur un espace $(\Omega, \mathcal{F}, P)$, est gouverné par les noyaux ν_n, relativement à une filtration $(\mathcal{F}_n)$, s'il est adapté à $(\mathcal{F}_n)$ et si pour tout $n \in \mathbb{N}$ on a

$$P^{\mathcal{F}_n}_{X_{n+1}} = \nu^{\overline{X}_n}_{n+1}, \ o\grave{u} \ \overline{X}_n = (X_0, \cdots, X_n) \ , \tag{1}$$

ce qui signifie : $E(f(X_{n+1})|\mathcal{F}_n) \underset{p.s.}{=} \nu^{\overline{X}_n}_{n+1} f \ (f \in \mathcal{M}_+(E)) \ .$

Par exemple, si $E = \mathbb{R}$ et si $(\mathcal{F}_n)$ est la filtration naturelle associée à une suite (U_n) de v.a.r. indépendantes de loi uniforme sur $]0, 1[$, un processus gouverné par les ν_n relativement à $(\mathcal{F}_n)$, de loi initiale donnée μ_0 peut être construit de la manière suivante. On considère les inverses continues à gauche G_0, $G^{x_0, \cdots, x_n}_{n+1}$ des fonctions de répartition de μ_0, $\nu^{x_0, \cdots, x_n}_{n+1}$ respectivement et l'on définit par récurrence la suite (X_n) telle que

$$X_0 = G_0(U_0), X_n = G^{\overline{X}_{n-1}}_n(U_n), (n \geq 1) \ .$$

Ceci fournit un procédé de simulation des processus réels gouvernés par des noyaux.

Dans la situation générale la mention de la filtration est naturellement omise lorsqu'il s'agit de $\mathcal{F}_n = \sigma(X_0, \cdots, X_n)$. Dans ce cas (1) équivaut à : $P_{\overline{X}_{n+1}} = P_{\overline{X}_n} \otimes \nu_{n+1} \ (n \in \mathbb{N})$.

Ainsi tout processus (X_n) à valeurs dans un espace lusinien (borélien de compact métrisable), en particulier polonais (cf. [1], [2]) peut être considéré comme étant gouverné par des noyaux. Inversement, si les ν_n sont donnés (E est maintenant arbitraire), il existe une probabilité unique sur $(W, \mathcal{G}) = (E, \mathcal{E})^{\mathbb{N}\otimes}$ pour laquelle le processus (ξ_n) des coordonnées est gouverné par les ν_n, $n \geq 1$ avec pour mesure initiale μ_0 donnée. Cette probabilité fournie par le théorème de Ionescu Tulcea (voir par exemple [5]) est notée $\mu_0 \otimes \nu_1 \otimes \nu_2 \otimes \cdots$

Dans la suite (X_n) est un processus gouverné par des noyaux ν_n relativement à une filtration $(\mathcal{F}_n)$.

1.2 Prédiction

Pour $n \in \mathbb{N}$ et $x \in E^{n+1}$ on note Q_n^x la mesure $\delta_{x_0} \otimes \cdots \otimes \delta_{x_n} \otimes \nu_{n+1} \otimes \nu_{n+2} \otimes \cdots$ sur $(W, \mathcal{G})$.

THÉORÈME — *La famille $Q_n = (Q_n^x)_{x \in E^{n+1}}$ est un noyau de E^{n+1} dans W et l'on a*

$$P_X^{\mathcal{F}_n} = Q_n^{\overline{X}_n}, \ n \in \mathbb{N}, \ \text{où } X = (X_k)_{k \in \mathbb{N}} \ . \tag{2}$$

DÉMONSTRATION — Il suffit de montrer que

(a) $Q_n h$ est mesurable,

(b) $E(h(X)|\mathcal{F}_n) \underset{p.s.}{=} Q_n^{\overline{X}_n} h$,

pour toute fonction $h \in \mathcal{M}_+(W)$ ne dépendant que de $\xi_0, \cdots, \xi_p$ ($p \geq n$). Les conditions (a) et (b) sont évidentes pour $p = n$. Supposons les vérifiées pour un $p \geq n$; elles le sont encore à l'ordre $p + 1$ grâce aux calculs suivants, où l'on choisit h de la forme $g(\overline{\xi}_p)f(\xi_{p+1})$ avec $f \in \mathcal{M}_+(E)$ et $g \in \mathcal{M}_+(E^{p+1})$:

$$(*) \qquad Q_n(h) = Q_n(g(\overline{\xi}_p)\nu_{p+1}^{\overline{\xi}_p}f) \ , \ \text{par définition des } Q_n^x,$$

$$E(h(X)|\mathcal{F}_p) = g(\overline{X}_p)\nu_{p+1}^{\overline{X}_p}f \ , \ \text{en vertu de (1)},$$

$$E(h(X)|\mathcal{F}_n) = Q_n^{\overline{X}_n}(g(\overline{\xi}_p)\nu_{p+1}^{\overline{\xi}_p}f) \ , \ \text{par hypothèse de récurrence},$$

$$= Q_n^{\overline{X}_n}(h) \ , \ \text{d'après } (*).$$

COROLLAIRE — *Soit $\theta_n = (\xi_{n+k})_{k \in \mathbb{N}}$. On a les formules suivantes de prédiction du futur (large ou strict) :*

$$P_{\theta_n(X)}^{\mathcal{F}_n} = P_n^{\overline{X}_n}, \ P_{\theta_{n+1}(X)}^{\mathcal{F}_n} = \Pi_{n+1}^{\overline{X}_n}, \ (n \in \mathbb{N}) \ ,$$

$$\text{où } P_n = \theta_n(Q_n) \text{ et } \Pi_{n+1} = \theta_{n+1}(Q_n) \ . \tag{3}$$

Les formules (2) et (3) s'étendent à un *temps d'arrêt* T de $(\mathcal{F}_n)$. Par exemple on a

$$P_{\theta_{T+1}(X)}^{\mathcal{F}_T} = \Pi_{T+1}^{\overline{X}_T}, \ \text{sur } \{T < \infty\} \ , \tag{4}$$

où $\theta_\infty = \theta_0$ (ou tout autre choix permettant de donner un sens à la v.a. $\theta_{T+1}(X)$).

REMARQUE — En posant $\xi^n = (\xi_{k \wedge n})_{k \in \mathbb{N}}$, $X^n = \xi^n(X)$, $\alpha'_{n+1} = \xi^{n+1}(Q_n^{\bar{\xi}_n(\cdot)})$ (c'est un noyau sur $(W, \mathcal{G})$), on voit d'après (2) que (X^n) est une chaîne de Markov relativement aux noyaux $\alpha_n, n \geq 1$ et à la filtration $(\mathcal{F}_n)$.

1.3 Caractérisation des temps d'arrêt de (X_n)

Notre objectif est d'adapter aux processus gouvernés par des noyaux les résultats de [3] sur la caractérisation des temps d'arrêt. Les choses sont naturellement plus simples en temps discret et ne nécessitent aucune hypothèse sur $(E, \mathcal{E})$. *On suppose ici que $\mathcal{F}_n = \sigma(X_0, \cdots, X_n)$ et $\mathcal{F} = \sigma(X_n, n \in \mathbb{N})$.* On considère aussi la filtration $(\widetilde{\mathcal{F}}_n)$ complétée ($\widetilde{\mathcal{F}}_n$ est engendrée par $\mathcal{F}_n$ et la famille des ensembles $\mathcal{F}$-mesurables et de mesure nulle).

THÉORÈME— *Soit T un temps aléatoire à valeurs dans $\overline{\mathbb{N}}$. Pour que T soit un temps d'arrêt de $(\widetilde{\mathcal{F}}_n)$, il faut et il suffit qu'il vérifie*

$$P^{\mathcal{F}_T}_{\theta_{T+1}(X)} = \Pi^{\overline{X}_T}_{T+1}, \text{ sur } \{T < \infty\} \ .$$

Rappelons que $\mathcal{F}_T$ est engendrée par les Z_T, où $(Z_n)_{n \in \overline{\mathbb{N}}}$ est un processus réel adapté.

DÉMONSTRATION — La condition est évidemment nécessaire. Inversement, si T vérifie (4) et si $f = u(\bar{\xi}_n)v(\theta_{n+1})$ pour un entier n avec $u \in \mathcal{M}_+(E^{n+1})$ et $v \in \mathcal{M}_+(W)$, on a

$$E[f(X), T = n] = E[u(\overline{X}_n)\Pi^{\overline{X}_n}_{n+1}v, T = n], \text{ d'après (4)},$$

$$= E[Q_n^{\overline{X}_n} f \, Z_n], \text{ où } Z_n = P(T = n | \mathcal{F}_n) \ ,$$

$$= E[f(X)Z_n], \text{ d'après (2)},$$

égalité qui s'étend à $f \in \mathcal{M}_+(W)$, donc $\mathbb{1}_{\{T=n\}} \underset{p.s.}{=} Z_n$. Ainsi, T est un temps d'arrêt de $(\widetilde{\mathcal{F}}_n)$.

REMARQUE — Pour une filtration générale $(\mathcal{F}_n)$ on montre de manière analogue qu'un temps aléatoire T est un temps d'arrêt de $(\widetilde{\mathcal{F}}_n)$ si et seulement si pour toute v.a.n. ≥ 0 H et $H_n = E(H|\mathcal{F}_n)$ on a $E(H|\mathcal{F}_T) \underset{p.s.}{=} H_T$ sur $\{T < \infty\}$.

1.4 Exemples

On suppose toujours que $(\mathcal{F}_n)$ est la filtration naturelle de (X_n) et que $\mathcal{F} = \bigvee_n \mathcal{F}_n$. Soit T un temps aléatoire.

1) Si (X_n) est une chaîne de Markov relativement à un noyau ν sur $(E, \mathcal{E})$, elle est aussi gouvernée par les noyaux ν_n tels que $\nu_n^{x_0, \cdots, x_{n-1}} = \nu^{x_{n-1}}$. Dans ce cas $\Pi_{n+1}^{x_0, \cdots, x_n}$ ne dépend que de x_n et est notée Π^{x_n}. Par suite T est un temps d'arrêt de $(\widetilde{\mathcal{F}}_n)$ si et seulement si

$$P^{\mathcal{F}_T}_{\theta_{T+1}(X)} = \Pi^{X_T} \text{ sur } \{T < \infty\} \ .$$

2) En particulier supposons que les X_n soient indépendantes, de même loi μ et que $P(T < \infty) > 0$. Alors T est un temps d'arrêt de $(\widetilde{\mathcal{F}}_n)$ si et seulement si, sous $Q = P(\,.\,|T < \infty)$, $\theta_{T+1}(X)$ est indépendante de $\mathcal{F}_T$ et a pour loi $\pi = \mu^{\mathbb{N}\otimes}$. Il faut prendre garde que la condition d'indépendance seule ne suffit pas. Par exemple si $0 < \mu(C) < 1$, alors $T = \inf\{n \geq 0 : X_{n+1} \in C\}$ est p.s. fini, $\mathcal{F}_T$ et $\theta_{T+1}(X)$ sont P-indépendantes, mais T n'est pas un temps d'arrêt de $(\widetilde{\mathcal{F}}_n)$: la loi de $\theta_{T+1}(X)$ est $\mu(\,.\,|C) \otimes \mu \otimes \mu \otimes \cdots$ au lieu de π.

2 Construction de probabilités sur un espace produit infini

Le théorème qui suit contient à la fois le théorème de Kolmogorov et le théorème d'existence d'un produit quelconque de probabilités (*sans conditions* sur les espaces facteurs, cf. [5] prop. IV-1-2).

Soit I un ensemble infini d'indices et pour tout $i \in I$ soit $(E_i, \mathcal{E}_i)$ un espace mesurable *arbitraire*. Pour $J \subset I$ non vide on note $(E_J, \mathcal{E}_J) = \prod_{j \in J}(E_j, \mathcal{E}_j)$ et X_J la projection canonique de E_I sur E_J; $X_{\{i\}}$ est noté X_i pour tout $i \in I$. Soit ϕ l'ensemble des parties *finies* de I.

2.1 Le théorème de construction

Soit $(\mu_J)_{J \in \phi}$ un système de probabilités tel que pour tout $J \in \phi$ et $i \in I \setminus J$:

- *μ_J est une probabilité sur E_J*
- *$\mu_{J \cup \{i\}}$ s'identifie à $\mu_J \otimes \nu_{J,i}$ où $\nu_{J,i}$ est un noyau de E_J dans E_i.*

Alors il existe une probabilité P unique sur $(\Omega, \mathcal{F}) = (E_I, \mathcal{E}_I)$ telle que $P_{X_J} = \mu_J$ $(J \in \phi)$.

DÉMONSTRATION — Soit Δ la famille des parties infinies dénombrables de I. Pour $D \in \Delta$, numérotée $d_0, d_1, \cdots$ on pose $\mathcal{F}_D = \sigma(X_D)$ et on note μ_D la probabilité sur $(\Omega, \mathcal{F}_D)$ telle que (X_{d_n}) ait pour loi initiale μ_{d_0} et soit gouverné par les noyaux $\nu_n = \nu_{\{d_0, \cdots, d_{n-1}\}, d_n}$ [4]. On a $X_J(\mu_D) = \mu_J$ $(J \in \phi, J \subset D)$, donc μ_D ne dépend pas de la numérotation choisie et de plus $\mu_{D'} = \mu_D$ sur $\mathcal{F}_D$ pour $D' \in \Delta$, $D \subset D'$. Comme $\mathcal{F} = \bigcup_{D \in \Delta} \mathcal{F}_D$, il reste à poser $P = \mu_D$ sur $\mathcal{F}_D$ pour obtenir la probabilité cherchée.

2.2 Applications

1) Soit $(\mu_i)_{i \in I}$ une famille de probabilités sur les $(E_i, \mathcal{E}_i)$. En appliquant le théorème précédent aux $\mu_J = \bigotimes_{i \in J} \mu_i$, $J \in \phi$ on trouve la probabilité P (unique) sur $(\Omega, \mathcal{F})$ telle que les X_i soient *indépendantes*, de lois respectives μ_i, $i \in I$.

---

[4] noter que les considérations de la première partie s'étendent au cas où X_n prend ses valeurs dans un espace dépendant de n.

2) Si les E_i sont lusiniens et si $(\mu_J)_{J \in \phi}$ est un système projectif de mesures de probabilités sur les E_J, l'existence des noyaux $\nu_{J,i}$ est satisfaite, donc le *théorème de Kolmogorov* apparaît comme un deuxième cas particulier.

3) Le théorème s'applique également si les tribus $\mathcal{E}_i$ sont engendrées par des partitions dénombrables, car l'existence des noyaux $\nu_{J,i}$ est alors automatique.

Références

1. N. BOURBAKI, *Eléments de Mathématiques, Topologie Générale, Chapitre IX*, Hermann, 1958.

2. C. DELLACHERIE, P.A. MEYER, *Probabilités et Potentiel, Chapitres I à IV*, Hermann, 1975.

3. F. KNIGHT, B. MAISONNEUVE, *A characterization of stopping times*, The Annals of Probability, Vol. 22 N°3, 1994, 1600-1606.

4. B. MAISONNEUVE, *Construction de probabilités sur un espace produit*, Exposé au séminaire de statistiques de Grenoble, 1984 (non publié).

5. J. NEVEU, *Bases Mathématiques du Calcul des Probabilités*, Masson, 1964.

SUR L'HYPERCONTRACTIVITE DES SEMI-GROUPES ULTRASPHERIQUES.

A. BENTALEB

En s'inspirant d'une démarche développée par M. LEDOUX dans le cadre gaussien [5], nous retrouvons le résultat d'hypercontractivité de D. BAKRY et M. EMERY pour les semi-groupes ultrasphériques [1].

Définitions et notations.

Pour $n > 0$, on considère, sur l'intervalle $[-1, +1]$, la famille des mesures ultrasphériques $\mu_n(dx) = K_n(1 - x^2)^{\frac{n}{2}-1}dx$, où $K_n = \Gamma((n-1)/2)/\sqrt{\pi}\Gamma\left(\frac{n}{2}\right)$ est la constante de normalisation qui fait de μ_n une mesure de probabilité. On associe à ces mesures la famille des opérateurs L_n définis par :

$$L_n f(x) = (1 - x^2)f''(x) - nxf'(x).$$

Nous commençons par introduire des notations que nous utiliserons par la suite. L'intégrale d'une fonction par rapport à μ_n sera désignée par $< f >$, et, nous noterons $L^2(\mu_n)$ l'espace des fonctions f telles que $< |f|^2 > < \infty$. La norme dans cet espace sera notée $\|f\|_2$ et le produit scalaire de deux fonctions f et g sera noté $< f, g >$. Les notations $<>$ et $<,>$ seront remplacées par $<>_{(n)}$ et $<,>_{(n)}$ s'il y a des ambiguïtés sur la valeur de n.

Propriété de symétrie et de dissipativité.

Pour deux fonctions f et g de classe C^2 sur $[-1, +1]$, on vérifie sans peine, à l'aide d'une intégration par parties, les formules de symétrie et de dissipativité :

$$< L_n f, g > = < f, L_n g > = - < \Gamma(f, g) >,$$

où Γ est la forme bilinéaire symétrique positive qui vaut

$$\Gamma(f, g)(x) = \frac{1}{2}\{L_n(fg) - fL_ng - gL_nf\} = (1 - x^2)f'(x)g'(x).$$

En appliquant la formule précédente avec $g = 1$, on obtient

$$< L_n f > = 0, \ \forall f \in C^2[-1, +1].$$

D'un autre côté, pour toute fonction $f \in C^2[-1, +1]$, cette identité implique

$$< L_n f, f > \leq 0.$$

Décomposition chaotique.

Sur la définition de l'opérateur L_n, nous voyons que, pour tout entier k, l'image par L_n d'un polynôme de degré k est à nouveau un polynôme de degré inférieur ou égal à k. L_n apparaît alors comme un opérateur symétrique sur l'espace vectoriel (de dimension finie) des polynômes de degré inférieur ou égal à k et donc il est diagonalisable dans une base orthonormée avec des valeurs propres réelles négatives. On obtient ainsi une suite $(G_k)_{k \geq 0}$ de polynômes de degré k orthogonaux dans $L^2(\mu_n)$ et par suite les polynômes $(G_k / \|G_k\|_2)_{k \geq 0}$ en forment une base orthonormale.

Appelons $-\lambda_k$ la $k^{ième}$ valeur propre de L_n asociée au $k^{ième}$ vecteur propre G_k : en écrivant

$$L_n G_k = -\lambda_k G_k$$

et en identifiant les coefficients des termes du plus haut degré. il vient

$$\lambda_k = k(k + n - 1).$$

Remarquons que pour tout $n > 0$, le monôme x est le premier vecteur propre associé à la première valeur propre $-\lambda_1^{(n)}$, de L_n.

Si une fonction f de $L^2(\mu_n)$ se décompose sous la forme $f = \sum_{k \geq 0} a_k G_k$ avec $\sum_{k \geq 0} a_k^2 \|G_k\|_2 < \infty$, alors :

$$L_n f = -\sum_{k \geq 0} \lambda_k^{(n)} a_k G_k.$$

Version intégrale du critère Γ_2.

Le procédé qui définit Γ à partir du produit permet de la même manière de définir la forme bilinéaire symétrique Γ_2 à partir de Γ :

$$\Gamma_2(f, g) = \frac{1}{2} \left\{ L_n \Gamma(f, g) - \Gamma(f, Lg) - \Gamma(g, Lf) \right\}.$$

Cet opérateur a été introduit par D. BAKRY et M. EMERY dans leur étude de diffusions hypercontractives dans un cadre général, et nous renvoyons les lecteurs intéresssés au cours présenté par le premier auteur à l'École d'Été de Probabilité de Saint-Flour [2] pour mieux en comprendre l'intérêt. Il contient en outre une abondante bibliographie.

L'un des traits frappants des travaux de ces deux auteurs est une approche directe de la propriété d'hypercontractivité par l'introduction d'une hypothèse maniable (critère Γ_2). Ce point de vue a permis de retrouver dans le cas des semi-groupes ultrasphériques (de dimension n) la propriété d'hypercontractivité pour $n \geq 1/4$ (leur méthode bute sur le cas $n < 1/4$). Nous voudrions montrer ici que l'on peut obtenir un résultat analogue en utilisant seulement la version intégrale du critère Γ_2 (voir corollaire 1 de [1]) : il se présente sous la forme suivante

$$\forall f \in C^2[-1,1], \quad < e^f, \Gamma_2(f,f) > -\lambda_1^{(n)} < e^f, \Gamma(f,f) > \; \geq 0. \quad (1)$$

Dans ces conditions, nous obtenons une inégalité de Sobolev logarithmique avec constante $2/\lambda_1^{(n)}$ ([2], p. 101), dont on sait par ailleurs qu'elle est optimale.

Pour établir ce critère intégral, nous suivons la même démarche que celle de M. LEDOUX [5] pour le semigroupe d'Hermite, en écrivant explicitement une relation de commutation du générateur L_n avec la dérivation. En effet, il est aisé de voir que, pour tout réel positif n,

$$\frac{d}{dx} L_n = [L_{n+2} - \lambda_1^{(n)} I] \frac{d}{dx}.$$

Cette formule de commutation est l'ingrédient crucial qui, jointe aux propriétés de symétrie et de dissipativité, va nous permettre d'obtenir ce critère intégral lorsque $n \geq 1/4$.

Preuve.

En utilisant la définition de Γ_2, le membre de gauche de (1) peut se réécrire,

$$\frac{1}{2} < e^f, L_n \Gamma(f,f) > - < e^f, \Gamma(f, L_n f) > .$$

À nouveau, par la formule de dissipativité, le premier terme de cette différence est

$$-\frac{1}{2} < (1-x^2)f'e^f, (\Gamma(f,f))' >$$

$$= \; < (1-x^2)xe^f, f'^3 > - < (1-x^2)^2 f'^2 e^f f'' >$$

$$= \; \frac{K_n}{K_{n+2}} \frac{1}{\lambda_1^{(n+2)}} < -L_{n+2}(x), e^f f'^3 >_{(n+2)} - < (1-x^2)^2 f'^2 e^f f'' >$$

$$= \; \frac{1}{\lambda_1^{(n+2)}} < (1-x^2)^2 e^f [3f'^2 f'' + f'^4] > - < (1-x^2)^2 e^f f'^2 f'' >,$$

la dernière égalité provenant cette fois-ci de la symétrie de l'opérateur L_{n+2}. Le deuxième terme s'écrit quant à lui

$$< e^f, (1-x^2)f'(L_n f)' > = \frac{K_n}{K_{n+2}} < e^f f', L_{n+2}(f') - \lambda_1^{(n)} f' >_{(n+2)}$$

$$= \; \frac{K_n}{K_{n+2}} < L_{n+2}(f'), f'e^f >_{(n+2)} - \lambda_1^{(n)} < e^f, \Gamma(f,f) >$$

$$= \; - < (1-x^2)^2 e^f (f'^2 f'' + f''^2) > - \lambda_1^{(n)} < e^f, \Gamma(f,f) >,$$

où nous avons fait usage à la fois de la propriété de dissipativité et de la formule de commutation.

En définitive, il nous reste, après avoir arrangé les termes,

$$< e^f, \Gamma_2(f,f) - \lambda_1^{(n)} \Gamma(f,f) >$$

$$= \; < (1-x^2)^2 e^f [f''^2 + \frac{1}{\lambda_1^{(n+2)}}(3f'^2 f'' + f'^4)] >$$

$$= \; \frac{1}{n+2} < (1-x^2)^2 e^f [(\sqrt{n+2}f'' + \frac{3}{\sqrt{n+2}}f'^2)^2 + \frac{n-1/4}{n+2}f'^4] >$$

qui est positive lorsque $n \geq 1/4$. La condition $n \geq 1/4$ n'est pas nécessaire pour obtenir une inégalité de Sobolev Logarithmique qui, rappelons-le, est équivalente à la propriété d'hypercontractivité (voir [7] et [3]). Mais, par cette méthode, nous n'avons pas su la montrer pour $0 < n < 1/4$, et on retrouve ainsi l'obstruction rencontrée dans [1] dans leur approche directe à l'aide du semigroupe associé à L_n. Peut-être un lecteur sera-t-il plus habile que nous ?

L'auteur tient à remercier les Professeurs D. BAKRY et M. LEDOUX pour leur accueil au sein du laboratoire de Statistique et Probabilités de l'Université Paul Sabatier de Toulouse durant le mois de mars 1998.

414

Références bibliographiques

[1] D. BAKRY et M. EMERY, Diffusions hypercontractives, Lecture Notes in Maths 1123 (1985), 177-206, Springer-Verlag.

[2] D. BAKRY, L'hypercontractivité et son utilisation en théorie des semigroupes, Ecole d'Eté de Probabilités de Saint-Flour 1992, Lecture Notes in Maths 1581 (1994), 1-114, Springer-Verlag.

[3] A. BENTALEB, Inégalités de Sobolev pour l'opérateur ultrasphérique, C. R. Acad. Sci. Paris, Série I, t.317 (1993), 187-190.

[4] A. BENTALEB, Développement de la moyenne d'une fonction pour la mesure ultrasphérique, C. R. Acad. Sci. Paris, Série I, t.317 (1993), 781-784.

[5] M. LEDOUX, On an Integral Criterion for Hypercontractivity of Diffusion Semi-groups and Extremal Functions, J. Func. Anal. 105 (1992), 444-465.

[6] M. LEDOUX, L'algèbre de Lie des gradients itérés d'un générateur markovien. Développements de moyennes et entropies, Ann. Scient. Ec. Norm. Sup. 28 (1995), 435-460.

[7] C. MUELLER et F. WEISSLER, Hypercontractivity for the Heat Semigroup for Ultraspherical Polynomials and on the n-sphere, J. Funct. Anal. 48 (1982), 252-283.

AN ADDENDUM TO A REMARK ON SLUTSKY'S THEOREM

FREDDY DELBAEN

In [D], I gave a counter-example to the following statement. If X_n is a sequence of measurable functions taking values in a Polish space E, and converging almost surely to a measurable function X, then for every Borel function h defined on E, $h(X_n)$ converges a.s. to $h(X)$. In the case of convergence in probability, the statement holds provided the image measures (or distributions) form a relatively weakly compact sequence (Slutsky's theorem). After the paper was printed, I discovered in the paper by Dellacherie, Feyel and Mokobodzki, [DFM], that the counterexample was already known. In fact there the authors show:

Theorem. *If $(X_n)_{n \geq 1}$ is a sequence of measurable mappings and if X is a measurable function, all defined on a probability space $(\Omega, \mathcal{F}, \mathbb{P})$ and taking values in the Polish Space $(E, \mathcal{T})$, then the following are equivalent*

1. *for each real valued Borel function $h: E \to \mathbb{R}$ we have that $h(X_n)$ tends to $h(X)$ almost surely*
2. *X_n tends to X almost surely in a stationary way, i.e. for almost every $\omega \in \Omega$ there is n_0 (depending on ω) such that for all $n \geq n_0$ we have $X_n(\omega) = X(\omega)$*

The aim of this addendum is not only to give credit to Dellacherie, Feyel and Mokobodzki but also to give some extra background information. As an example we will see that the Slutsky result can be stated in a different way. To fix notation, let $(E, \mathcal{T}, \mathcal{E})$ be a Polish space equipped with its topology $\mathcal{T}$ and its Borel structure $\mathcal{E}$. Let μ_n be a sequence of probability measures on $(E, \mathcal{E})$. We recall that the sequence μ_n tends weak* to the probability μ if for every $\mathcal{T}$-continuous bounded function f on E we have $\int f \, d\mu_n$ tends to $\int f \, d\mu$. We say that μ_n tends to μ weakly if for every Borel function f on E we have that $\int f \, d\mu_n$ tends to $\int f \, d\mu$. In this case we have that μ_n tends to μ weakly in the sense of the topology $\sigma(\mathcal{M}, \mathcal{M}^*)$. By a result of Grothendieck, to have weak convergence, it is sufficient to ask that for every open set $O \in \mathcal{T}$ we have that

1991 *Mathematics Subject Classification.* 28A20, 28A33 60B10, 60B12.
Key words and phrases. Slutsky's Theorem, weak compactness.

$\mu_n(O)$ tends to $\mu(O)$. But for Polish spaces there are many topologies that give the same Borel structure, in fact every finer topology $\mathcal{T}'$, which is still Polish, gives, by Blackwell's theorem, the same Borel sets $\mathcal{E}$ on E. The relation between the weak* and the weak convergence becomes clearer thanks to the following result, see [S] page 91–93.

Theorem. *If A_n is a sequence of Borel sets in E, then there exists a finer topology $\mathcal{T}'$ on E, still Polish and such that each A_n is an open-closed set in $\mathcal{T}'$.*

Corollary. *If f_n is a sequence of Borel functions $f_n : E \to \mathbb{R}$, then there is a finer, still Polish, topology $\mathcal{T}'$ on E such that each f_n is continuous.*

This result gives us the following theorem.

Theorem. *For a sequence of probability measures on $(E, \mathcal{E})$, the following two properties are equivalent*

(1) *the sequence μ_n converges weakly to μ,*
(2) *for every Polish topology $\mathcal{T}'$, finer than $\mathcal{T}$, the sequence μ_n converges weak* to μ.*

The result of Dellacherie, Feyel and Mokobodzki can now be rephrased as

Theorem. *For a sequence of measurable mappings defined on a probability space $(\Omega, \mathcal{F}, \mathbb{P})$ and taking values in a Polish space $(E, \mathcal{T})$, the following are equivalent*

(1) *the sequence X_n tends to X is a stationary way (as above)*
(2) *for every Polish topology $\mathcal{T}'$ finer than $\mathcal{T}$, we have that X_n tends to X almost surely.*

References

[D]   F. Delbaen, *A remark on Slutsky's Theorem*, Séminaire de Probabilités XXXII, Lecture Notes in Mathematics **1686**, (1998), Spinger, 313–315.

[DFM]   C. Dellacherie, D. Feyel, G. Mokobodzki, *Intégrales de capacités fortement sous-additives*, Séminaire de Probabilités XVI, Lecture Notes in Mathematics **920** (1982), Spinger, 8–28.

[S]   S. M. Srivastava, *A course on Borel sets* (1998), Springer.

THEOREMES LIMITES POUR LES TEMPS LOCAUX
D'UN PROCESSUS STABLE SYMETRIQUE

Nathalie Eisenbaum

Ce rectificatif concerne le paragraphe 2) de la Section II, intitulé :
Passage du temps exponentiel indépendant à un temps déterministe.
En effet, la preuve de ce passage s'appuie sur la Remarque (i) (Section
II, 1)). Cette Remarque (i) nécessite assurément des justifications plus
étendues. Aussi nous proposons l'argument suivant qui présente le double
avantage de : - prouver le passage du temps exponentiel indépendant à un
temps déterministe sans utiliser la Remarque (i)
- fournir une preuve de la Remarque (i).

En premier lieu, on remarque qu'à partir du Théorème 2, on obtient
facilement le corollaire suivant par un argument de convergence dominée :

Corollaire 2.1 : *Pour $y_1, y_2, \ldots, y_n$ n réels distincts , on a :*

$$\left(X_T , L_T , \left(\frac{1}{\varepsilon^{(\beta-1)/2}}\left(L_T^{\varepsilon x+y_k} - L_T^{y_k}\right) \ ; \ x\in\mathbb{R} , 1\leq k\leq n\right) \right)$$

$$\xrightarrow[\varepsilon\to 0]{(d)} \left(X_T , L_T , \left(\sqrt{c_\beta} \ B_{\frac{y_k}{2L_T^{y_k}}}^{[y_k]}(x) \ ; \ x\in\mathbb{R}, \ 1\leq k\leq n \right) \right)$$

où $\{ B_u^{[y_k]}(x), x\in\mathbb{R}, 1\leq k\leq n, u\geq 0 \}$ *est un système gaussien indépendant de*
X et de T, composé de n draps browniens fractionnaires d'indice $(\beta-1)$,
tous indépendants.

On rappelle alors que la famille des lois de

$$\left(\left(X_t , L_t, \left(\frac{1}{\varepsilon^{(\beta-1)/2}}\left(L_t^{\varepsilon x+y_k} - L_t^{y_k}\right); x\in\mathbb{R}, 1\leq k\leq n\right)\right) \ t\geq 0 \right) \text{ est faiblement}$$

relativement compacte. Donc pour toute suite $(\varepsilon_m)_{m>0}$ de réels strictement
positifs tendant vers 0, quitte à extraire une sous-suite, on sait que :

$$\left(\left(X_t , L_t, \frac{1}{\varepsilon_m^{(\beta-1)/2}}\left(L_t^{\varepsilon_m x+y_k} - L_t^{y_k}\right) \ ; \ x\in\mathbb{R}, 1\leq k\leq n\right) \ t\geq 0 \right) \text{ converge en loi}$$

quand m tend vers l'infini vers une loi de probabilité ν. On note de façon évidente : $\nu = (\nu(t) , t \geq 0)$.

Grâce au Corollaire 2.1, on obtient pour toute fonctionnelle continue bornée F de $(\mathbb{R}, \mathscr{C}(\mathbb{R}, \mathbb{R}), \mathscr{C}(\mathbb{R}, \mathbb{R}^n))$ dans $\mathbb{R}$, et tout $\lambda > 0$:

$$\int_0^{+\infty} dt e^{-\lambda t} \nu(t)(F) = \int_0^{+\infty} dt e^{-\lambda t} E[F(X_t, L_t, (\sqrt{c_\beta} \; B^{[y_k]}_{\frac{y_k}{2L_t}}(x) \; ; x \in \mathbb{R}, 1 \leq k \leq n))].$$

On en deduit que dt-p.s. $\nu(t)$ est la loi de

$(X_t, L_t, (\sqrt{c_\beta} \; B^{[y_k]}_{\frac{y_k}{2L_t}}(x) \; ; x \in \mathbb{R}, 1 \leq k \leq n))$. Par scaling , c'est donc vrai pour

tout $t > 0$. D'où pour tout $t > 0$:

$$\left(X_t , L_t , \left(\frac{1}{\varepsilon^{(\beta-1)/2}}(L_t^{\varepsilon x + y_k} - L_t^{y_k})\right) ; x \in \mathbb{R}, 1 \leq k \leq n)\right)$$

$$\xrightarrow[\varepsilon \to 0]{(d)} \left(X_t , L_t , (\sqrt{c_\beta} \; B^{[y_k]}_{\frac{y_k}{2L_t}}(x) \; ; x \in \mathbb{R}, 1 \leq k \leq n)\right)$$

La Remarque (i) s'obtient alors simplement par convergence dominée.

NOTE TO CONTRIBUTORS

Contributors to the Séminaire are reminded that their articles should be formatted for the Springer Lecture Notes series.

The dimensions of the printed part of a page without running heads should be:

15.3 cm × 24.2 cm if the font size is 12 pt (or 10 pt magnified 120%),

12.2 cm × 19.3 cm if the font size is 10 pt.

Page numbers and running heads are not needed. Author(s)' address(es) should be indicated, either below the title or at the end of the paper.

Packages of TEX macros are available from the Springer-Verlag site

http://www.springer.de/math/authors/b-tex.html

Druck: Strauss Offsetdruck, Mörlenbach
Verarbeitung: Schäffer, Grünstadt

Lecture Notes in Mathematics

For information about Vols. 1–1525
please contact your bookseller or Springer-Verlag

Vol. 1526: J. Azéma, P. A. Meyer, M. Yor (Eds.), Séminaire de Probabilités XXVI. X, 633 pages. 1992.

Vol. 1527: M. I. Freidlin, J.-F. Le Gall, Ecole d'Eté de Probabilités de Saint-Flour XX – 1990. Editor: P. L. Hennequin. VIII, 244 pages. 1992.

Vol. 1528: G. Isac, Complementarity Problems. VI, 297 pages. 1992.

Vol. 1529: J. van Neerven, The Adjoint of a Semigroup of Linear Operators. X, 195 pages. 1992.

Vol. 1530: J. G. Heywood, K. Masuda, R. Rautmann, S. A. Solonnikov (Eds.), The Navier-Stokes Equations II – Theory and Numerical Methods. IX, 322 pages. 1992.

Vol. 1531: M. Stoer, Design of Survivable Networks. IV, 206 pages. 1992.

Vol. 1532: J. F. Colombeau, Multiplication of Distributions. X, 184 pages. 1992.

Vol. 1533: P. Jipsen, H. Rose, Varieties of Lattices. X, 162 pages. 1992.

Vol. 1534: C. Greither, Cyclic Galois Extensions of Commutative Rings. X, 145 pages. 1992.

Vol. 1535: A. B. Evans, Orthomorphism Graphs of Groups. VIII, 114 pages. 1992.

Vol. 1536: M. K. Kwong, A. Zettl, Norm Inequalities for Derivatives and Differences. VII, 150 pages. 1992.

Vol. 1537: P. Fitzpatrick, M. Martelli, J. Mawhin, R. Nussbaum, Topological Methods for Ordinary Differential Equations. Montecatini Terme, 1991. Editors: M. Furi, P. Zecca. VII, 218 pages. 1993.

Vol. 1538: P.-A. Meyer, Quantum Probability for Probabilists. X, 287 pages. 1993.

Vol. 1539: M. Coornaert, A. Papadopoulos, Symbolic Dynamics and Hyperbolic Groups. VIII, 138 pages. 1993.

Vol. 1540: H. Komatsu (Ed.), Functional Analysis and Related Topics, 1991. Proceedings. XXI, 413 pages. 1993.

Vol. 1541: D. A. Dawson, B. Maisonneuve, J. Spencer, Ecole d´ Eté de Probabilités de Saint-Flour XXI - 1991. Editor: P. L. Hennequin. VIII, 356 pages. 1993.

Vol. 1542: J.Fröhlich, Th.Kerler, Quantum Groups, Quantum Categories and Quantum Field Theory. VII, 431 pages. 1993.

Vol. 1543: A. L. Dontchev, T. Zolezzi, Well-Posed Optimization Problems. XII, 421 pages. 1993.

Vol. 1544: M.Schürmann, White Noise on Bialgebras. VII, 146 pages. 1993.

Vol. 1545: J. Morgan, K. O'Grady, Differential Topology of Complex Surfaces. VIII, 224 pages. 1993.

Vol. 1546: V. V. Kalashnikov, V. M. Zolotarev (Eds.), Stability Problems for Stochastic Models. Proceedings, 1991. VIII, 229 pages. 1993.

Vol. 1547: P. Harmand, D. Werner, W. Werner, M-ideals in Banach Spaces and Banach Algebras. VIII, 387 pages. 1993.

Vol. 1548: T. Urabe, Dynkin Graphs and Quadrilateral Singularities. VI, 233 pages. 1993.

Vol. 1549: G. Vainikko, Multidimensional Weakly Singular Integral Equations. XI, 159 pages. 1993.

Vol. 1550: A. A. Gonchar, E. B. Saff (Eds.), Methods of Approximation Theory in Complex Analysis and Mathematical Physics IV, 222 pages, 1993.

Vol. 1551: L. Arkeryd, P. L. Lions, P.A. Markowich, S.R. S. Varadhan. Nonequilibrium Problems in Many-Particle Systems. Montecatini, 1992. Editors: C. Cercignani, M. Pulvirenti. VII, 158 pages 1993.

Vol. 1552: J. Hilgert, K.-H. Neeb, Lie Semigroups and their Applications. XII, 315 pages. 1993.

Vol. 1553: J.-L- Colliot-Thélène, J. Kato, P. Vojta. Arithmetic Algebraic Geometry. Trento, 1991. Editor: E. Ballico. VII, 223 pages. 1993.

Vol. 1554: A. K. Lenstra, H. W. Lenstra, Jr. (Eds.), The Development of the Number Field Sieve. VIII, 131 pages. 1993.

Vol. 1555: O. Liess, Conical Refraction and Higher Microlocalization. X, 389 pages. 1993.

Vol. 1556: S. B. Kuksin, Nearly Integrable Infinite-Dimensional Hamiltonian Systems. XXVII, 101 pages. 1993.

Vol. 1557: J. Azéma, P. A. Meyer, M. Yor (Eds.), Séminaire de Probabilités XXVII. VI, 327 pages. 1993.

Vol. 1558: T. J. Bridges, J. E. Furter, Singularity Theory and Equivariant Symplectic Maps. VI, 226 pages. 1993.

Vol. 1559: V. G. Sprindžuk, Classical Diophantine Equations. XII, 228 pages. 1993.

Vol. 1560: T. Bartsch, Topological Methods for Variational Problems with Symmetries. X, 152 pages. 1993.

Vol. 1561: I. S. Molchanov, Limit Theorems for Unions of Random Closed Sets. X, 157 pages. 1993.

Vol. 1562: G. Harder, Eisensteinkohomologie und die Konstruktion gemischter Motive. XX, 184 pages. 1993.

Vol. 1563: E. Fabes, M. Fukushima, L. Gross, C. Kenig, M. Röckner, D. W. Stroock, Dirichlet Forms. Varenna, 1992. Editors: G. Dell'Antonio, U. Mosco. VII, 245 pages. 1993.

Vol. 1564: J. Jorgenson, S. Lang, Basic Analysis of Regularized Series and Products. IX, 122 pages. 1993.

Vol. 1565: L. Boutet de Monvel, C. De Concini, C. Procesi, P. Schapira, M. Vergne. D-modules, Representation Theory, and Quantum Groups. Venezia, 1992. Editors: G. Zampieri, A. D'Agnolo. VII, 217 pages. 1993.

Vol. 1566: B. Edixhoven, J.-H. Evertse (Eds.), Diophantine Approximation and Abelian Varieties. XIII, 127 pages. 1993.

Vol. 1567: R. L. Dobrushin, S. Kusuoka, Statistical Mechanics and Fractals. VII, 98 pages. 1993.

Vol. 1568: F. Weisz, Martingale Hardy Spaces and their Application in Fourier Analysis. VIII, 217 pages. 1994.

Vol. 1569: V. Totik, Weighted Approximation with Varying Weight. VI, 117 pages. 1994.

Vol. 1570: R. deLaubenfels, Existence Families, Functional Calculi and Evolution Equations. XV, 234 pages. 1994.

Vol. 1571: S. Yu. Pilyugin, The Space of Dynamical Systems with the C⁰-Topology. X, 188 pages. 1994.

Vol. 1572: L. Göttsche, Hilbert Schemes of Zero-Dimensional Subschemes of Smooth Varieties. IX, 196 pages. 1994.

Vol. 1573: V. P. Havin, N. K. Nikolski (Eds.), Linear and Complex Analysis – Problem Book 3 – Part I. XXII, 489 pages. 1994.

Vol. 1574: V. P. Havin, N. K. Nikolski (Eds.), Linear and Complex Analysis – Problem Book 3 – Part II. XXII, 507 pages. 1994.

Vol. 1575: M. Mitrea, Clifford Wavelets, Singular Integrals, and Hardy Spaces. XI, 116 pages. 1994.

Vol. 1576: K. Kitahara, Spaces of Approximating Functions with Haar-Like Conditions. X, 110 pages. 1994.

Vol. 1577: N. Obata, White Noise Calculus and Fock Space. X, 183 pages. 1994.

Vol. 1578: J. Bernstein, V. Lunts, Equivariant Sheaves and Functors. V, 139 pages. 1994.

Vol. 1579: N. Kazamaki, Continuous Exponential Martingales and *BMO*. VII, 91 pages. 1994.

Vol. 1580: M. Milman, Extrapolation and Optimal Decompositions with Applications to Analysis. XI, 161 pages. 1994.

Vol. 1581: D. Bakry, R. D. Gill, S. A. Molchanov, Lectures on Probability Theory. Editor: P. Bernard. VIII, 420 pages. 1994.

Vol. 1582: W. Balser, From Divergent Power Series to Analytic Functions. X, 108 pages. 1994.

Vol. 1583: J. Azéma, P. A. Meyer, M. Yor (Eds.), Séminaire de Probabilités XXVIII. VI, 334 pages. 1994.

Vol. 1584: M. Brokate, N. Kenmochi, I. Müller, J. F. Rodriguez, C. Verdi, Phase Transitions and Hysteresis. Montecatini Terme, 1993. Editor: A. Visintin. VII. 291 pages. 1994.

Vol. 1585: G. Frey (Ed.), On Artin's Conjecture for Odd 2-dimensional Representations. VIII, 148 pages. 1994.

Vol. 1586: R. Nillsen, Difference Spaces and Invariant Linear Forms. XII, 186 pages. 1994.

Vol. 1587: N. Xi, Representations of Affine Hecke Algebras. VIII, 137 pages. 1994.

Vol. 1588: C. Scheiderer, Real and Étale Cohomology. XXIV, 273 pages. 1994.

Vol. 1589: J. Bellissard, M. Degli Esposti, G. Forni, S. Graffi, S. Isola, J. N. Mather, Transition to Chaos in Classical and Quantum Mechanics. Montecatini Terme, 1991. Editor: 2S. Graffi. VII, 192 pages. 1994.

Vol. 1590: P. M. Soardi, Potential Theory on Infinite Networks. VIII, 187 pages. 1994.

Vol. 1591: M. Abate, G. Patrizio, Finsler Metrics – A Global Approach. IX, 180 pages. 1994.

Vol. 1592: K. W. Breitung, Asymptotic Approximations for Probability Integrals. IX, 146 pages. 1994.

Vol. 1593: J. Jorgenson & S. Lang, D. Goldfeld, Explicit Formulas for Regularized Products and Series. VIII, 154 pages. 1994.

Vol. 1594: M. Green, J. Murre, C. Voisin, Algebraic Cycles and Hodge Theory. Torino, 1993. Editors: A. Albano, F. Bardelli. VII, 275 pages. 1994.

Vol. 1595: R.D.M. Accola, Topics in the Theory of Riemann Surfaces. IX, 105 pages. 1994.

Vol. 1596: L. Heindorf, L. B. Shapiro, Nearly Projective Boolean Algebras. X, 202 pages. 1994.

Vol. 1597: B. Herzog, Kodaira-Spencer Maps in Local Algebra. XVII, 176 pages. 1994.

Vol. 1598: J. Berndt, F. Tricerri, L. Vanhecke, Generalized Heisenberg Groups and Damek-Ricci Harmonic Spaces. VIII, 125 pages. 1995.

Vol. 1599: K. Johannson, Topology and Combinatorics of 3-Manifolds. XVIII, 446 pages. 1995.

Vol. 1600: W. Narkiewicz, Polynomial Mappings. VII, 130 pages. 1995.

Vol. 1601: A. Pott, Finite Geometry and Character Theory. VII, 181 pages. 1995.

Vol. 1602: J. Winkelmann, The Classification of Three-dimensional Homogeneous Complex Manifolds. XI, 230 pages. 1995.

Vol. 1603: V. Ene, Real Functions – Current Topics. XIII, 310 pages. 1995.

Vol. 1604: A. Huber, Mixed Motives and their Realization in Derived Categories. XV, 207 pages. 1995.

Vol. 1605: L. B. Wahlbin, Superconvergence in Galerkin Finite Element Methods. XI, 166 pages. 1995.

Vol. 1606: P.-D. Liu, M. Qian, Smooth Ergodic Theory of Random Dynamical Systems. XI, 221 pages. 1995.

Vol. 1607: G. Schwarz, Hodge Decomposition – A Method for Solving Boundary Value Problems. VII, 155 pages. 1995.

Vol. 1608: P. Biane, R. Durrett, Lectures on Probability Theory. Editor: P. Bernard. VII, 210 pages. 1995.

Vol. 1609: L. Arnold, C. Jones, K. Mischaikow, G. Raugel, Dynamical Systems. Montecatini Terme, 1994. Editor: R. Johnson. VIII, 329 pages. 1995.

Vol. 1610: A. S. Üstünel, An Introduction to Analysis on Wiener Space. X, 95 pages. 1995.

Vol. 1611: N. Knarr, Translation Planes. VI, 112 pages. 1995.

Vol. 1612: W. Kühnel, Tight Polyhedral Submanifolds and Tight Triangulations. VII, 122 pages. 1995.

Vol. 1613: J. Azéma, M. Emery, P. A. Meyer, M. Yor (Eds.), Séminaire de Probabilités XXIX. VI, 326 pages. 1995.

Vol. 1614: A. Koshelev, Regularity Problem for Quasilinear Elliptic and Parabolic Systems. XXI, 255 pages. 1995.

Vol. 1615: D. B. Massey, Le Cycles and Hypersurface Singularities. XI, 131 pages. 1995.

Vol. 1616: I. Moerdijk, Classifying Spaces and Classifying Topoi. VII, 94 pages. 1995.

Vol. 1617: V. Yurinsky, Sums and Gaussian Vectors. XI, 305 pages. 1995.

Vol. 1618: G. Pisier, Similarity Problems and Completely Bounded Maps. VII, 156 pages. 1996.

Vol. 1619: E. Landvogt, A Compactification of the Bruhat-Tits Building. VII, 152 pages. 1996.

Vol. 1620: R. Donagi, B. Dubrovin, E. Frenkel, E. Previato, Integrable Systems and Quantum Groups. Montecatini Terme, 1993. Editors:M. Francaviglia, S. Greco. VIII, 488 pages. 1996.

Vol. 1621: H. Bass, M. V. Otero-Espinar, D. N. Rockmore, C. P. L. Tresser, Cyclic Renormalization and Auto-morphism Groups of Rooted Trees. XXI, 136 pages. 1996.

Vol. 1622: E. D. Farjoun, Cellular Spaces, Null Spaces and Homotopy Localization. XIV, 199 pages. 1996.

Vol. 1623: H.P. Yap, Total Colourings of Graphs. VIII, 131 pages. 1996.

Vol. 1624: V. Brinzanescu, Holomorphic Vector Bundles over Compact Complex Surfaces. X, 170 pages. 1996.

Vol.1625: S. Lang, Topics in Cohomology of Groups. VII, 226 pages. 1996.

Vol. 1626: J. Azéma, M. Emery, M. Yor (Eds.), Séminaire de Probabilités XXX. VIII, 382 pages. 1996.

Vol. 1627: C. Graham, Th. G. Kurtz, S. Méléard, Ph. E. Protter, M. Pulvirenti, D. Talay, Probabilistic Models for Nonlinear Partial Differential Equations. Montecatini Terme, 1995. Editors: D. Talay, L. Tubaro. X, 301 pages. 1996.

Vol. 1628: P.-H. Zieschang, An Algebraic Approach to Association Schemes. XII, 189 pages. 1996.

Vol. 1629: J. D. Moore, Lectures on Seiberg-Witten Invariants. VII, 105 pages. 1996.

Vol. 1630: D. Neuenschwander, Probabilities on the Heisenberg Group: Limit Theorems and Brownian Motion. VIII, 139 pages. 1996.

Vol. 1631: K. Nishioka, Mahler Functions and Transcendence.VIII, 185 pages.1996.

Vol. 1632: A. Kushkuley, Z. Balanov, Geometric Methods in Degree Theory for Equivariant Maps. VII, 136 pages. 1996.

Vol.1633: H. Aikawa, M. Essén, Potential Theory – Selected Topics. IX, 200 pages.1996.

Vol. 1634: J. Xu, Flat Covers of Modules. IX, 161 pages. 1996.

Vol. 1635: E. Hebey, Sobolev Spaces on Riemannian Manifolds. X, 116 pages. 1996.

Vol. 1636: M. A. Marshall, Spaces of Orderings and Abstract Real Spectra. VI, 190 pages. 1996.

Vol. 1637: B. Hunt, The Geometry of some special Arithmetic Quotients. XIII, 332 pages. 1996.

Vol. 1638: P. Vanhaecke, Integrable Systems in the realm of Algebraic Geometry. VIII, 218 pages. 1996.

Vol. 1639: K. Dekimpe, Almost-Bieberbach Groups: Affine and Polynomial Structures. X, 259 pages. 1996.

Vol. 1640: G. Boillat, C. M. Dafermos, P. D. Lax, T. P. Liu, Recent Mathematical Methods in Nonlinear Wave Propagation. Montecatini Terme, 1994. Editor: T. Ruggeri. VII, 142 pages. 1996.

Vol. 1641: P. Abramenko, Twin Buildings and Applications to S-Arithmetic Groups. IX, 123 pages. 1996.

Vol. 1642: M. Puschnigg, Asymptotic Cyclic Cohomology. XXII, 138 pages. 1996.

Vol. 1643: J. Richter-Gebert, Realization Spaces of Polytopes. XI, 187 pages. 1996.

Vol. 1644: A. Adler, S. Ramanan, Moduli of Abelian Varieties. VI, 196 pages. 1996.

Vol. 1645: H. W. Broer, G. B. Huitema, M. B. Sevryuk, Quasi-Periodic Motions in Families of Dynamical Systems. XI, 195 pages. 1996.

Vol. 1646: J.-P. Demailly, T. Peternell, G. Tian, A. N. Tyurin, Transcendental Methods in Algebraic Geometry. Cetraro, 1994. Editors: F. Catanese, C. Ciliberto. VII, 257 pages. 1996.

Vol. 1647: D. Dias, P. Le Barz, Configuration Spaces over Hilbert Schemes and Applications. VII. 143 pages. 1996.

Vol. 1648: R. Dobrushin, P. Groeneboom, M. Ledoux, Lectures on Probability Theory and Statistics. Editor: P. Bernard. VIII, 300 pages. 1996.

Vol. 1649: S. Kumar, G. Laumon, U. Stuhler, Vector Bundles on Curves – New Directions. Cetraro, 1995. Editor: M. S. Narasimhan. VII, 193 pages. 1997.

Vol. 1650: J. Wildeshaus, Realizations of Polylogarithms. XI, 343 pages. 1997.

Vol. 1651: M. Drmota, R. F. Tichy, Sequences, Discrepancies and Applications. XIII, 503 pages. 1997.

Vol. 1652: S. Todorcevic, Topics in Topology. VIII, 153 pages. 1997.

Vol. 1653: R. Benedetti, C. Petronio, Branched Standard Spines of 3-manifolds. VIII, 132 pages. 1997.

Vol. 1654: R. W. Ghrist, P. J. Holmes, M. C. Sullivan, Knots and Links in Three-Dimensional Flows. X, 208 pages. 1997.

Vol. 1655: J. Azéma, M. Emery, M. Yor (Eds.), Séminaire de Probabilités XXXI. VIII, 329 pages. 1997.

Vol. 1656: B. Biais, T. Björk, J. Cvitanic, N. El Karoui, E. Jouini, J. C. Rochet, Financial Mathematics. Bressanone, 1996. Editor: W. J. Runggaldier. VII, 316 pages. 1997.

Vol. 1657: H. Reimann, The semi-simple zeta function of quaternionic Shimura varieties. IX, 143 pages. 1997.

Vol. 1658: A. Pumarino, J. A. Rodríguez, Coexistence and Persistence of Strange Attractors. VIII, 195 pages. 1997.

Vol. 1659: V, Kozlov, V. Maz'ya, Theory of a Higher-Order Sturm-Liouville Equation. XI, 140 pages. 1997.

Vol. 1660: M. Bardi, M. G. Crandall, L. C. Evans, H. M. Soner, P. E. Souganidis, Viscosity Solutions and Applications. Montecatini Terme, 1995. Editors: I. Capuzzo Dolcetta, P. L. Lions. IX, 259 pages. 1997.

Vol. 1661: A. Tralle, J. Oprea, Symplectic Manifolds with no Kähler Structure. VIII, 207 pages. 1997.

Vol. 1662: J. W. Rutter, Spaces of Homotopy Self-Equivalences – A Survey. IX, 170 pages. 1997.

Vol. 1663: Y. E. Karpeshina: Perturbation Theory for the Schrödinger Operator with a Periodic Potential. VII, 352 pages. 1997.

Vol. 1664: M. Väth, Ideal Spaces. V, 146 pages. 1997.

Vol. 1665: E. Giné, G. R. Grimmett, L. Saloff-Coste, Lectures on Probability Theory and Statistics 1996. Editor: P. Bernard. X, 424 pages, 1997.

Vol. 1666: M. van der Put, M. F. Singer, Galois Theory of Difference Equations. VII, 179 pages. 1997.

Vol. 1667: J. M. F. Castillo, M. González, Three-space Problems in Banach Space Theory. XII, 267 pages. 1997.

Vol. 1668: D. B. Dix, Large-Time Behavior of Solutions of Linear Dispersive Equations. XIV, 203 pages. 1997.

Vol. 1669: U. Kaiser, Link Theory in Manifolds. XIV, 167 pages. 1997.

Vol. 1670: J. W. Neuberger, Sobolev Gradients and Differential Equations. VIII, 150 pages. 1997.

Vol. 1671: S. Bouc, Green Functors and *G*-sets. VII, 342 pages. 1997.

Vol. 1672: S. Mandal, Projective Modules and Complete Intersections. VIII, 114 pages. 1997.

Vol. 1673: F. D. Grosshans, Algebraic Homogeneous Spaces and Invariant Theory. VI, 148 pages. 1997.

Vol. 1674: G. Klaas, C. R. Leedham-Green, W. Plesken, Linear Pro-*p*-Groups of Finite Width. VIII, 115 pages. 1997.

Vol. 1675: J. E. Yukich, Probability Theory of Classical Euclidean Optimization Problems. X, 152 pages. 1998.

Vol. 1676: P. Cembranos, J. Mendoza, Banach Spaces of Vector-Valued Functions. VIII, 118 pages. 1997.

Vol. 1677: N. Proskurin, Cubic Metaplectic Forms and Theta Functions. VIII, 196 pages. 1998.

Vol. 1678: O. Krupková, The Geometry of Ordinary Variational Equations. X, 251 pages. 1997.

Vol. 1679: K.-G. Grosse-Erdmann, The Blocking Technique. Weighted Mean Operators and Hardy's Inequality. IX, 114 pages. 1998.

Vol. 1680: K.-Z. Li, F. Oort, Moduli of Supersingular Abelian Varieties. V, 116 pages. 1998.

Vol. 1681: G. J. Wirsching, The Dynamical System Generated by the 3n+1 Function. VII, 158 pages. 1998.

Vol. 1682: H.-D. Alber, Materials with Memory. X, 166 pages. 1998.

Vol. 1683: A. Pomp, The Boundary-Domain Integral Method for Elliptic Systems. XVI, 163 pages. 1998.

Vol. 1684: C. A. Berenstein, P. F. Ebenfelt, S. G. Gindikin, S. Helgason, A. E. Tumanov, Integral Geometry, Radon Transforms and Complex Analysis. Firenze, 1996. Editors: E. Casadio Tarabusi, M. A. Picardello, G. Zampieri. VII, 160 pages. 1998.

Vol. 1685: S. König, A. Zimmermann, Derived Equivalences for Group Rings. X, 146 pages. 1998.

Vol. 1686: J. Azéma, M. Émery, M. Ledoux, M. Yor (Eds.), Séminaire de Probabilités XXXII. VI, 440 pages. 1998.

Vol. 1687: F. Bornemann, Homogenization in Time of Singularly Perturbed Mechanical Systems. XII, 156 pages. 1998.

Vol. 1688: S. Assing, W. Schmidt, Continuous Strong Markov Processes in Dimension One. XII, 137 page. 1998.

Vol. 1689: W. Fulton, P. Pragacz, Schubert Varieties and Degeneracy Loci. XI, 148 pages. 1998.

Vol. 1690: M. T. Barlow, D. Nualart, Lectures on Probability Theory and Statistics. Editor: P. Bernard. VIII, 237 pages. 1998.

Vol. 1691: R. Bezrukavnikov, M. Finkelberg, V. Schechtman, Factorizable Sheaves and Quantum Groups. X, 282 pages. 1998.

Vol. 1692: T. M. W. Eyre, Quantum Stochastic Calculus and Representations of Lie Superalgebras. IX, 138 pages. 1998.

Vol. 1694: A. Braides, Approximation of Free-Discontinuity Problems. XI, 149 pages. 1998.

Vol. 1695: D. J. Hartfiel, Markov Set-Chains. VIII, 131 pages. 1998.

Vol. 1696: E. Bouscaren (Ed.): Model Theory and Algebraic Geometry. XV, 211 pages. 1998.

Vol. 1697: B. Cockburn, C. Johnson, C.-W. Shu, E. Tadmor, Advanced Numerical Approximation of Nonlinear Hyperbolic Equations. Cetraro, Italy, 1997. Editor: A. Quarteroni. VII, 390 pages. 1998.

Vol. 1698: M. Bhattacharjee, D. Macpherson, R. G. Möller, P. Neumann, Notes on Infinite Permutation Groups. XI, 202 pages. 1998.

Vol. 1699: A. Inoue,Tomita-Takesaki Theory in Algebras of Unbounded Operators. VIII, 241 pages. 1998.

Vol. 1700: W. A. Woyczyński, Burgers-KPZ Turbulence,XI, 318 pages. 1998.

Vol. 1701: Ti-Jun Xiao, J. Liang, The Cauchy Problem of Higher Order Abstract Differential Equations, XII, 302 pages. 1998.

Vol. 1702: J. Ma, J. Yong, Forward-Backward Stochastic Differential Equations and Their Applications. XIII, 270 pages. 1999.

Vol. 1703: R. M. Dudley, R. Norvaiša, Differentiability of Six Operators on Nonsmooth Functions and p-Variation. VIII, 272 pages. 1999.

Vol. 1704: H. Tamanoi, Elliptic Genera and Vertex Operator Super-Algebras. VI, 390 pages. 1999.

Vol. 1705: I. Nikolaev, E. Zhuzhoma, Flows in 2-dimensional Manifolds. XIX, 294 pages. 1999.

Vol. 1706: S. Yu. Pilyugin, Shadowing in Dynamical Systems. XVII, 271 pages. 1999.

Vol. 1707: R. Pytlak, Numerical Methods for Optical Control Problems with State Constraints. XV, 215 pages. 1999.

Vol. 1708: K. Zuo, Representations of Fundamental Groups of Algebraic Varieties. VII, 139 pages. 1999.

Vol. 1709: J. Azéma, M. Émery, M. Ledoux, M. Yor (Eds), Séminaire de Probabilités XXXIII. VIII, 418 pages. 1999.

Vol. 1710: M. Koecher, The Minnesota Notes on Jordan Algebras and Their Applications. IX, 173 pages. 1999.

Vol. 1711: W. Ricker, Operator Algebras Generated by Commuting Projections: A Vector Measure Approach. XVII, 159 pages. 1999.

Vol. 1712: N. Schwartz, J. J. Madden, Semi-algebraic Function Rings and Reflectors of Partially Ordered Rings. XI, 279 pages. 1999.

Vol. 1713: F. Bethuel, G. Huiksen, S. Müller, K. Steffen, Calculus of Variations and Geometric Evolution Problems. Cetraro, 1996. Editors: S. Hildebrandt, M. Struwe. VII, 293 pages. 1999.

Vol. 1714: O. Diekmann, R. Durrett, K. P. Hadeler, P. Maini, H. L. Smith, Mathematics Inspired by Biology. Martina Franca, 1997. Editors: V. Capasso, O. Diekmann. VII, 268 pages. 1999.

Vol. 1715: N. V. Krylov, M. Röckner, J. Zabczyk, Stochastic PDE's and Kolmogorov Equations in Infinite Dimensions. Cetraro, 1998. Editor: G. Da Prato. VIII, 239 pages. 1999.

Vol. 1716: J. Coates, R. Greenberg, K. A. Ribet, K. Rubin, Arithmetic Theory of Elliptic Curves. Cetraro, 1997. Editor: C. Viola. VIII, 260 pages. 1999.

4. Lecture Notes are printed by photo-offset from the master-copy delivered in camera-ready form by the authors. Springer-Verlag provides technical instructions for the preparation of manuscripts. Macro packages in T_EX, L^AT_EX2e, $L^AT_EX2.09$ are available from Springer's web-pages at

http://www.springer.de/math/authors/b-tex.html.

Careful preparation of the manuscripts will help keep production time short and ensure satisfactory appearance of the finished book.

The actual production of a Lecture Notes volume takes approximately 12 weeks.

5. Authors receive a total of 50 free copies of their volume, but no royalties. They are entitled to a discount of 33.3 % on the price of Springer books purchase for their personal use, if ordering directly from Springer-Verlag.

Commitment to publish is made by letter of intent rather than by signing a formal contract. Springer-Verlag secures the copyright for each volume. Authors are free to reuse material contained in their LNM volumes in later publications: A brief written (or e-mail) request for formal permission is sufficient.

Addresses:

Professor F. Takens, Mathematisch Instituut,
Rijksuniversiteit Groningen, Postbus 800,
9700 AV Groningen, The Netherlands
E-mail: F.Takens@math.rug.nl

Professor B. Teissier, DMI, École Normale Supérieure
45, rue d'Ulm,
F-7500 Paris, France
E-mail: Teissier@ens.fr

Springer-Verlag, Mathematics Editorial, Tiergartenstr. 17,
D-69121 Heidelberg, Germany,
Tel.: *49 (6221) 487-701
Fax: *49 (6221) 487-355
E-mail: lnm@Springer.de